Contents

KT-160-284

The Changing Global Environment

Edited by
Neil Roberts

BLACKWELL
Oxford UK & Cambridge USA

Copyright © Basil Blackwell Ltd 1994

First published 1994

Blackwell Publishers
238 Main Street, Suite 501
Cambridge, Massachusetts 02142
USA

108 Cowley Road
Oxford OX4 1JF
UK

Library of Congress Cataloging-in-Publication Data

The Changing global environment/edited by Neil Roberts.
     p.   cm.
   Includes bibliographical references and index.
   ISBN 1-55786-271-0 (alk. paper). — ISBN 1-55786-272-9 (alk. paper
pbk.)
   1. Man—Influence on nature. 2. Climatic changes. 3. Physical
geography. I. Roberts, Neil, 1953– .
GF75. C45 1993
304.2'8—dc20 92-35490 CIP

British Library Cataloguing in Publication Data

A CIP catalogue record for this book is available from the British Library.

Typeset in 10 on 11½ pt Sabon by TecSet Ltd, Wallington, Surrey
Printed in Great Britain by T.J. Press Ltd, Padstow, Cornwall

This book is printed on acid-free paper

Contributors

Professor Richard Battarbee
Environmental Change Research Centre, Department of
   Geography, University College London, 26 Bedford Way,
   London, England

Dr Grace S. Brush
Department of Geography and Environmental Engineering, The
   Johns Hopkins University, Baltimore, Maryland 21218, USA

Dr John Dearing
Centre for Environmental Science Research and Consultancy,
   Coventry University, Priory Street, Coventry, England

Professor Edward Derbyshire
Department of Geography, Royal Holloway and Bedford New
   College, Egham Hill, Surrey, England

Professor Ian Douglas
Department of Geography, University of Manchester,
   Manchester, England

Dr Peter A. Furley
Department of Geography, University of Edinburgh, High School
   Yards, Edinburgh, Scotland

Professor Andrew Goudie
School of Geography, Mansfield Road, Oxford, England

Dr Martin Haigh
Geography Section, Oxford Brookes University, Headington,
   Oxford, England

Dr Roy Haines-Young
Department of Geography, Nottingham University, University
   Park, Nottingham, England

Professor Ann Henderson-Sellers
School of Earth Sciences, Macquarie University, North Ryde,
   New South Wales 2109, Australia

Dr Nick Hulton
Department of Geography, University of Edinburgh, High School
  Yards, Edinburgh, Scotland

Dr Mike Hulme
Climatic Research Unit, School of Environmental Sciences,
  University of East Anglia, Norwich, England

Professor Eduard Koster
University of Utrecht, Geographical Institute, Heidelberglaan 2,
  PO Box 80115, NL-3508, TC Utrecht, Netherlands

Professor Geoff Petts
Freshwater Environments Group, Department of Geography,
  Loughborough University, Leicestershire, England

Dr Neil Roberts
Department of Geography, Loughborough University,
  Leicestershire, England

Dr Erhard Schulz
Geographisches Institut, University of Würzburg, Am Hubland,
  D-8700 Würzburg, Germany

Dr Tom Spencer
Department of Geography, University of Cambridge, Downing
  Place, Cambridge, England

Dr Philip Stott
Geography Department, School of Oriental and African Studies,
  Malet Street, London, England

Dr F. Alayne Street-Perrott
Environmental Change Unit, Oxford University, Mansfield Road,
  Oxford, England

Professor David Sugden
Department of Geography, University of Edinburgh, High School
  Yards, Edinburgh, Scotland

Dr Michael J. Tooley
Environmental Research Centre, Department of Geography,
  University of Durham, Science Laboratories, South Road,
  Durham, England

Professor J. T. Wang
Geological Hazards Research Institute, Gansu Academy of
  Sciences, Lanzhou, P. R. China

Abbreviations and Acronyms

AGCM	atmospheric general circulation models
AVHRR	advanced very high resolution radiometer
CCC	Canadian Climate Center
CEC	cation exchange capacity
CEGB	Central Electricity Generating Board (U.K.)
CENTURY	ecosystem production model
CET	Central England temperature
CFCs	chlorofluorocarbons
CLIMAP	Climate, Long-range Investigation, Mapping and Prediction Project
COADS	Comprehensive Ocean – Atmosphere Data Set
COHMAP	Cooperative Holocene Mapping Project
DSS	decision support models
ENSO	El Niño-Southern Oscillation
GCM	general circulation model
GFDL	Geophysical Fluid Dynamics Laboratory
GIS	Geographical Information Systems
GISS	Goddard Institute of Space Studies
GPCP	Global Precipitation Climatology Project
HIRIS	high resolution imaging spectrometer
ICSU	International Council of Scientific Unions
IGBP	International Geosphere-Biosphere Programme
INPA	Brazilian Amazon Research Institute
IPCC	Intergovernmental Panel on Climate Change
IUCN	International Union for the Conservation of Nature
LRTAP	long range transboundary air pollution
MATT	mean annual air temperature
MAGIC	Model of Acidification of Groundwaters in Catchments
MAPST	mean annual permafrost surface temperature
MODIS	moderate resolution imaging spectrometer
MSST	mean (annual) snow surface temperature
MSUs	microwave sounding units
NCAR	National Center for Atmospheric Research

NDVI	Normalized Difference Vegetation Index
NOAA	National Oceanic and Atmospheric Administration (U.S.)
OGCM	Oceanic General Circulation Models
PAM	plant available moisture
PAN	plant availabel nutrients
PAR	photosynthetically active radiation
PYRO	process response model
RAIN	Reversing Acidification in Norway
RSPM	resource system planning models
SAR	synthetic aperture radar
SBC	sum of base cations
SEASAT	Sea Satellite
SeaWIFS	Sea Wide Field Sensor
SHRUBKILL	Microcomputer based advisory programme with three key modules; BURNECON BURNTIME BURNWAYS
SLAR	side looking airborne radar
SSA	sum of strong acids
SST	sea surface temperatures
TM	thematic mapper
TREE	Tropical Rainforest Ecology Experiment
TRMM	Tropical Rainfall Monitoring Mission
TVA	Tennessee Valley Authority
UKMO	United Kingdom Meteorological Office
UN	United Nation Organisation
UNCOD	United Nations Conference on Desertification
UNEP	United Nations Environment Program
USLE	universal soil loss equation
WCRP	World Climate Research Programme
WMO	World Meteorological Organisation
WWF	World Wide Fund for Nature

Preface and Acknowledgements

If books have lives, the moment of conception for this one was during the night of October 2, 1989. The setting, however, was a decidedly unromantic hospital bed in Leicester Royal Infirmary where I lay awake recovering from a minor operation. Perhaps it was the after-effects of the general anaesthetic; perhaps it was a feeling of guilt at abandoning my colleagues at the start of a new academic year. But whatever the reason, my thoughts drifted to the deficiencies of existing student texts in physical geography, which seemed to have become stuck in the rut of a rather abstract and mechanistic "systems" approach to analyzing environmental processes. Their discussion of environmental systems often seemed sterile and some way removed from the pressing and exciting environmental problems facing the contemporary world (cf. Newson, 1992). Systems analysis in physical geography had been pioneered by Richard Chorley and Barbara Kennedy, but while most texts continued to draw slavishly on their 1971 book *Physical Geography: a Systems Approach*, research had moved on from this starting-point to apply systems ideas to real-world problems about our natural environment. The systems approach has underpinned important progress which has been made in the last twenty years in understanding issues like soil erosion and sediment transport, sea-surface fluctuations, forest dynamics and climate change. Could a selected group of physical geographers at the forefront of research in these and other areas be persuaded to distil and synthesize developments in their own subject to create an edited text intelligible to a student and non-specialist readership?

This germ of an idea might have been stillborn had John Davey not made one of his periodic transatlantic phone calls as I convalesced at home a few days later, but with that call and John's encouragement the book started its long gestation toward becoming reality. From the outset, two features of the book were to have priority; the first being the need for authors to expound good science. There is now a plethora of semi-popular books around the theme of global environmental change, but few, if any, take the trouble to explain the scientific issues (and uncertainties) behind the prognoses about the Earth's poor state of health. In particular, we cannot hope to comprehend what we, as human beings, are doing to modify our planetary home unless and until we achieve an understanding of how environmental systems function under "natural" conditions. For this reason, many of the chapters start with a discussion of underlying scientific principles – the physics of glacier mass balances, the chemistry of acid precipitation, or the biology of savanna grasses – before moving on to assess human impact upon these processes. The second priority in the book's design was coherence; twenty chapters and twenty-two authors do not a textbook make. To achieve chapters of a similar "level" and length, and to avoid gaps in coverage or unnecessary duplication, have needed rather strict editorial guidance. I am grateful to contributing authors for agreeing to this discipline and for accepting with such good grace my liberal use of a red pen in modifying chapter drafts.

My thanks are due to colleagues in the Department of Geography at Loughborough University for their support, in particular to Keith Boucher and Phil Barker for their comments, to Gwynneth Barnwell, Val Pheby and Ruth Austin for help with creating a single, unified bibliography file, to Helen Ruffell for assistance with the index, and to Erica Millwain and Peter Robinson for drafting or re-drafting the formidable total of 145 diagrams. Thanks too, to John Davey and Jan Chamier at Blackwell, and to picture researcher Debbie Azurdia who showed admirable bargaining as well as investigative skills with photographic agencies. Finally this list of thanks would be incomplete without Sylvie and Sara, who have helped in ways too numerous to mention.

For permission to reproduce diagrams and tables, the publisher and authors are grateful to the following:

Professor N. J. Shackleton (figure 3.1); Ecological Bulletins (figure 11.8); Dr Brian Huntley (figure 3.2); *Geographical Magazine* (figure 12.6); Professor W. L. Graf (figure 12.6); A. A. Balkema (figure 17.2); Royal Society of Edinburgh (table 3.2, figure 3.1); Blackwell Scientific (figure 13.2); Pergamon Press (figure 19.2); University of Colorado, *Arctic and Alpine Research* (figure 6.1); IGBP Secretariat, Royal Swedish Academy of Sciences (table 1.1); Blackwell-Heinemann Publishers (figures 8.4 and 8.7); Butterworth Publishers (table 14.1); CAB International (figure 14.3); Professor C. Lorius (figure 3.8); Professor Richard Fairbanks (figure 8.9); Edward Arnold, *Progress in Physical Geography* (figure 13.8); IUCN (table 14.5); Springer-Verlag (figures 9.8 and 14.3, table 9.3); University of Hawaii Press (figure 9.8); Catena Verlag (figure 19.6); WASWC (table 15.3); Dr Kerry Emanuel (figure 9.6); Professor U. Hellden (figure 2.4); Geological Society of America (figure 7.2); American Geophysical Union (figures 5.3, 5.4, 5.6,

6.10 and 8.2); Routledge (table 4.1); Cambridge University Press (figures 4.11, 4.13, 5.5, 5.7, 5.8, 5.9, 6.5, 7.3, 8.5, 8.6, 9.1 and 14.4, tables 1.2, 7.1 and 7.2); Royal Meteorological Society (figure 5.1); Academic Press (figure 4.3); Dr Keith Briffa (figures 4.1 and 3.9); Professor John Kutzbach (figures 3.3, 3.4, 3.5 and 3.7); Macmillan Magazines, *Nature* (figures 3.3, 3.7, 3.8, 3.9, 4.1, 8.9 and 9.6); John Wiley (figures 11.1, 11.4, 11.5 and 11.7); Royal Society of London (figure 10.6); National Research Council of Canada, *Canadian Geotechnical Journal* (figure 6.9, table 6.3); *Journal of Biogeography* (figures 13.2 and 13.3); Kluwer Academic Publishers (figures 10.2 and 10.8); American Society of Civil Engineers (figure 6.8); American Association for the Advancement of Science (figures 3.2, 3.4 and 3.5); HMSO (figure 10.3); ENSIS (figures 10.4 and 10.15); IAHS (figure 11.9); Water Resources Publications (table 16.2); Reed Business Publishing, *Water Power and Dam Construction* (tables 12.1 and 12.2); Quaternary Research Center AK-60, University of Washington, *Quaternary Research* (figures 7.1, 11.2 and 11.4); Professor Nicholas Polunin, Foundation for Environmental Conservation (figure 11.3a); Taylor and Francis *International Journal of Remote Sensing* (figure 2.2); American Society of Limnology and Oceanography (figure 18.6); Ann Arbor Science (table 12.3); US Geological Survey (figure 7.4 and table 12.3); Methuen (figure 4.2); Friends of the Earth (figures 14.1 and 14.2); IUBS *Biology International* (figures 13.3, 13.5, 13.6, 13.7 and 13.9).

The publisher and authors are also grateful to the following for permission to use, and for their help in supplying, photographs:

Dr Tim Allott (10.4); Ambio (14.2); Steve Archer (13.6); Dr M. G. L. Baillie (4.1); Professor Richard Batterbee (10.1, 10.3); Joy Belsky, Cornell University (13.2, 13.3); Bodleian Library (4.4); Dr John Boardman (11.1) British Antarctic Survey and EOSAT (3.2 Dr D. A. Peel, 7.1); Dr John Dearing (11.2); David Couling (4.3); Michael Dent (8.1); Professor E. Derbyshire (19.1, 19.2, 19.3, 19.5, 19.6, 19.7); Professor Ian Douglas (15.3); Professor Andrew Goudie (16.1, 16.3, 16.4); Dr Martin Haigh (20.1, 20.2, 20.5, 20.6); Dr Roy Haines-Young (2.2, 2.3); Sonia Halliday Photographs (3.1 James Wellard); Steve Haynes (20.4); The Image Bank (12.2 Francois Dardelet, 12.3 Sobel/Klonsky); Klute with kind permission (17.6); Professor E. Koster (6.1, 6.2, 6.3, 6.4, 6.5, 6.6); Life File Photographic Library (9.1, 9.2 Jeremy Hoare); The Mansell Collection, London (4.2); Maryland Historical Society, Baltimore (18.1); NASA (2.1); Natural History Photo Agency (9.4 A. Krummins); Tero Niemi (10.5); Novosti Photo Library (12.4); H. Preisig, Grande Dixence (12.1); R. J. Scholes, University of the Witwatersrand (13.5); Dr Erhard Schulz (17.1, 17.2, 17.3, 17.4, 17.5, 17.6, 17.7, 17.8, 17.9); Soil Conservation Service (16.2); Dr Philip Stott (13.1, 13.5); Charles Swithinbank (7.3); US Navy for

US Geological Survey (7.2 Charles Swithinbank); E. Webster (11.4); David Williamson, London (11.3, 12.5); The World Wide Fund for Nature (9.3 Eva Cropp, 14.1, 14.4 Mark Edwards, 14.3 Mauri Rautkari, 15.1, 15.2 Michele Depraz, 20.3 Nikom Putta); BBL (10.2).

PART I

The Nature of Environmental Change

CHAPTER ONE

The Global Environmental Future

Neil Roberts

Look abroad through nature's range,
Nature's mighty law is change.
                              Robert Burns

Introduction

There has never been greater concern about the global environmental future than there is at present. Problems such as deforestation and land degradation, pollution of marine and fresh waters, depletion of the ozone layer, and – above all – greenhouse-gas induced climatic warming, are rightly viewed with grave anxiety by both the scientific community and the public at large. Global warming, which may already be underway, would have wide-ranging consequences for the global environmental system, affecting storm and flood frequencies, plant and animal distributions, glacier and ice sheet dynamics, and sea levels. Individual issues related to climate change are addressed in detail in the first half of this book (chapters 3 to 9, and chapter 13).

However, there is considerable scientific uncertainty in any predictions about the global environmental future. One important reason for this uncertainty is that the Earth system is not, and never has been, free from change. Natural variability is an intrinsic part of all environmental systems, and only the short time span of direct human observation deceives us into thinking of the Earth as being in a "steady-state." The atmospheric concentration of greenhouse gases, for example, has fluctuated through geological time, most recently during glacial–interglacial cycles of the Quaternary ice age (see Street-Perrott and Roberts, chapter 3). Furthermore, ice-core data show that they oscillated in close harmony with global temperature changes during the past 150 000 years, indicating that the two are almost certainly related (Lorius et al., 1990). In order to understand the natural variability of the Earth's climate system, it is therefore also necessary to investigate its environmental history.

A second complicating factor in environmental prediction is the fact that human action has already seriously disturbed many Earth-surface systems. Any future greenhouse-gas induced changes will consequently be superimposed on other anthropogenic changes to the planet, so compounding them. Although the global environment has an intrinsic resilience to disturbance, the nature and rate of human-induced change are such that its effects may be outstripping the Earth's potential capacity for recovery. To use Andrew Goudie's phrase, the Earth is becoming "groggy."

In some environmental systems, it is now difficult to be sure what their natural, pre-disturbance condition was like. For instance, acid deposition as a result of industrial emissions has meant that there may be no pristine lake ecosystems remaining in the boreal forest zone of the Northern Hemisphere. As Richard Battarbee shows in chapter 10, even remote mountain lakes in Canada and northwest Europe have had their pH levels reduced since the time of the Industrial Revolution, often crossing a critical biological threshold at around pH 5.5 in the process. As well as acid deposition, other major environmental issues addressed in the second half of the book include large-scale manipulation of rivers and other freshwaters, dryland degradation and the clearance of tropical forests. Human impacts near to the ground are critical in their own right, and cumulatively may be as significant as those caused by disturbance to the global climate system.

Global environmental change is thus a complex problem. In order to understand how Earth-surface systems function and respond to forcing, a number of major international scientific research programs have been initiated. One example of a program which is attempting to understand the nature and dynamics of the global environmental system and the effect of human impact upon it, is the International Geosphere–Biosphere Programme (IGBP) of the International Council of Scientific Unions (ICSU). In their "Global Change" report (IGBP, 1990) a number of key questions about environmental change were proposed (table 1.1), forming the core of the research program. The "Mission to Planet Earth" (Space Science Board, 1988; NASA, undated; Asrar, 1990) is another program examining environmental change that is occurring both naturally and in response to human activity. This is bringing together the space effort of a number of countries in order to pursue the new "Earth system science" using remote sensing technology.

Greenhouse gases and global warming

The issue which has, more than any other, galvanized opinion about changes to the natural environment is the prospect of global climatic warming, which is predicted to occur because of the increased emission of so-called "greenhouse gases." The most

Table 1.1   Key questions about environmental change posed by the
International Geosphere–Biosphere Global Change Program

1	How is the chemistry of the global atmosphere regulated and what is the role of biological processes in producing and consuming trace gases?
2	How do ocean biogeochemical processes influence and respond to climatic change?
3	How do changes in land use, sea level and climate alter coastal ecosystems, and what are the wider consequences?
4	How does vegetation interact with the physical processes of the hydrological cycle?
5	How will global changes affect terrestrial ecosystems?
6	What significant climatic and environmental changes have occurred in the past and what were their causes?

Source: modified after IGBP, 1990

important of the greenhouse gases – water vapour, carbon dioxide (CO_2) and methane (CH_4) – are present naturally in the atmosphere (table 1.2), and indeed without them the mean global temperature would be lower by around 33°C (Houghton et al., 1990). Notwithstanding this, recent centuries have witnessed an exponential increase in the concentration of some of the greenhouse gases as a result of human activities. Twentieth-century observational records and measurements on air bubbles in polar ice cores show a 25–30 percent increase of CO_2 on pre-industrial levels, while CH_4 has more than doubled (figure 1.1).

Greenhouse gases other than CO_2 are usually converted to their CO_2-equivalent for the purposes of climatic prediction, such as the doubled carbon dioxide ($2 \times CO_2$) scenario. On the other hand, it needs to be borne in mind that the increase in different gases has not been caused by the same factors. This is particularly relevant when it comes to the formulation of policies to control their emission. Carbon dioxide levels have risen principally because of fossil-fuel burning since the nineteenth century, with reduced photosynthesis due to deforestation as an important secondary cause. By contrast, the main anthropogenic sources of methane are intensive rice cultivation and cattle-raising. Methane is a particularly effective greenhouse gas and is increasing in the atmosphere more rapidly than carbon dioxide (table 1.2). Water vapour, the most important of all atmsopheric greenhouse gases, has not been increased directly by human activities as have other gases, but it may play an important role none the less, by amplifying any future climate change (see below).

Greenhouse gases affect the Earth's heat balance by permitting incoming short-wave radiation to pass largely unimpeded through the atmosphere, but absorbing most of the outgoing long-wave terrestrial radiation emitted from the Earth's surface. All other

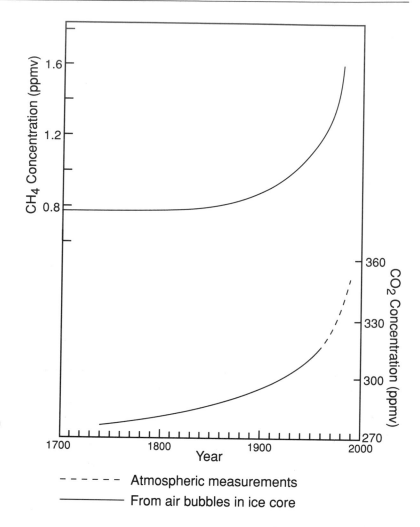

Figure 1.1 Historical increases in the atmospheric concentration of carbon dioxide and methane (data after Pearman and Fraser, 1988; Neftel et al., 1985; Friedli et al., 1986; Keeling et al., 1989).

– – – – – – Atmospheric measurements

———————— From air bubbles in ice core

things being equal, the increasing concentration of these gases ought to lead to a global temperature rise, but by how much they will rise and how quickly is far from certain. The Intergovernmental Panel on Climate Change (IPCC) has estimated that under a "business-as-usual" situation for greenhouse gas emissions, the global mean temperature during the next century will rise by about 0.3°C per decade (Houghton et al., 1990). In other IPCC scenarios involving increasing controls on pollution, the rate of temperature rise is reduced. However, because the atmospheric concentrations of carbon dioxide, CFCs and nitrous oxide adjust only slowly to changes in emissions (table 1.2), we are already committed to a rise in these greenhouse gases during the next century. This will have environmental consequences even if there were to be no effect on climate (a most unlikely case), particularly on vegetation whose growth will be stimulated by the fertilization effect of CO_2.

Table 1.2   Key greenhouse gases affected by human activities

	Carbon dioxide	Methane	CFC-11	CFC-12	Nitrous oxide
Atmospheric concentration	p.p.m.v.	p.p.m.v.	p.p.t.v.	p.p.t.v.	p.p.b.v.
Pre-industrial (1750–1800)	280	0.8	0	0	288
Present day (1990)	353	1.72	280	484	310
Current rate of increase per year (percent)	0.5	0.9	4	4	0.25
Atmospheric lifetime (years)	50–200	10	65	130	150
Greenhouse forcing, W m^{-2} (1765–1990)	1.50	0.56	0.06	0.14	0.10
	(63%)	(24%)	(3%)	(6%)	(4%)

p.p.m.v. = parts per million by volume. p.p.b.v. = parts per billion (thousand million) by volume. p.p.t.v. = parts per trillion (million million) by volume.
Source: from Houghton et al., 1990

It is certain that any global warming will also change the global water balance. This is because as temperatures rise the atmosphere can hold more water vapour, itself enhancing the greenhouse effect. Increased ocean to atmosphere moisture flux would lead to higher precipitation, but there might not be any overall increase in effective (e.g. soil) moisture, because of the greater evaporative losses as temperature rises (Rind et al., 1990). Even so, the rate of many Earth-surface processes, such as chemical weathering, would be likely to speed up under such a warmer, wetter climate.

Modelling environmental systems

In order to understand the interactions between the various components which make up the global environmental system, it is necessary to model them. Models provide the essential theoretical framework within which environmental change can be measured and analyzed. Models vary in type from the conceptual to the numerical, and function over a hierarchy of spatial scales from local to global (see table 1.3). At local and regional scales, the watershed ecosystem and island ecosystem have provided valuable conceptual and empirical contexts for measuring environmental processes, in particular because they permit system boundaries to be delimited "on the ground" (Bormann and Likens, 1970). In the former case, the analysis of lake sediments adds a time dimension to the spatially aggregated record of catchment changes, creating a lake-watershed ecosystem framework within which Earth-surface processes can be studied over a range of timescales (Oldfield, 1977).

Table 1.3 A hierarchy of environmental systems

Scale	System type	Example	Chapter authors
Local–regional	Lake-watershed ecosystem	L. Busjösjö, Sweden	Dearing, ch. 11
	Estuary and catchment	Chesapeake Bay	Brush, ch. 18
Sub-continental	Large river basin	Yellow River	Derbyshire and Wang, ch. 19
	Biome	Savanna	Stott, ch. 13
Global	Earth system	GCM	Henderson-Sellers, ch. 5
	Gaia model		

At a global scale, the potential interaction between different components of the Earth system makes any comprehensive model become enormously complex, as figure 1.2 illustrates. A conceptually simpler and more elegant global model is that based on the Gaia hypothesis proposed by James Lovelock (1979, 1988). It has long been recognized that life on Earth has been maintained through its ability to adapt to ever-changing conditions. According to the Gaia model, however, life has not simply responded passively to external stimulus but has played an active part in shaping its and the planet's destiny. Through its links to the atmosphere, oceans and other components of the Earth system, Lovelock argues, the biosphere has played a crucial role in maintaining a global steady-state equilibrium. If Lovelock is correct, then Gaia's own self-regulatory mechanisms will ensure that the environmental status quo is maintained in the future.

On the other hand, the dangers of over-optimistic trust in the power of mother nature are highlighted by Lovelock's prediction in the 1970s that the ozone layer would be unaffected by CFC emissions. In this case subsequent events have not proved him right. And while the Gaia model has undoubtedly stimulated new perceptions of global environmental change, the Earth's own climatic history does not support a dynamic steady-state model for the Earth system. In recent geological time, many environmental systems have exhibited a bi-modal behavior, associated with oscillations between glacial and interglacial states (Broecker et al., 1985). Furthermore, the switch between these meta-stable equilibrium states seems to have been controlled primarily by geophysical agencies, such as ice cover and eustatic sea level, rather than by the biological ones involved in the Gaia model (e.g. Ives et al., 1975; Ruddiman and McIntyre, 1981).

A substantially different type of models involves those which attempt to simulate numerically a segment of the global

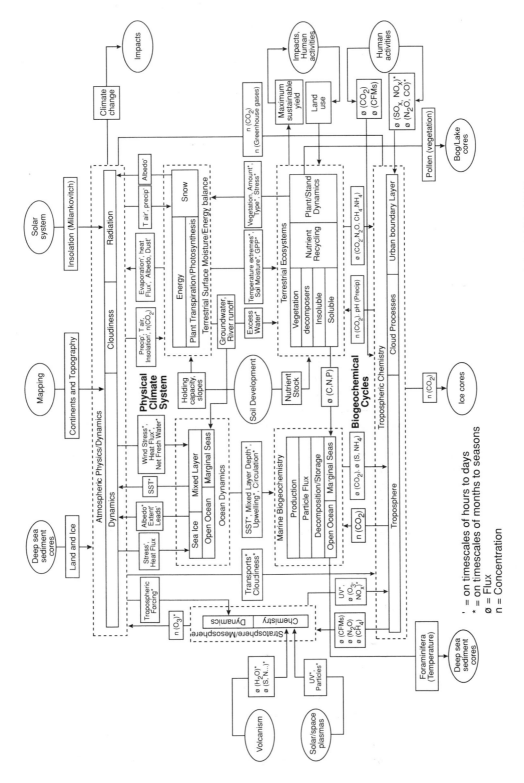

Figure 1.2 Conceptual model of Earth system operating over timescales from decades to centuries (after NASA, 1988).

environmental system. Computer-driven numerical models, like conceptual ones, function over a wide range of scales, from the modelling of erosion on an individual hillslope (e.g. Thornes, 1990) to simulations of the entire global climate system. Pre-eminent among the global models are general circulation models (or GCMs), discussed in chapter 5 by Ann Henderson-Sellers. These provided the basis for the IPCC predictions about the response of climate to greenhouse-gas forcing.

All GCMs predict some warming, but they differ in its magnitude, with estimates between 1.9 and 5.2°C for a doubling of CO_2-equivalent (recent UK Meteorological Office simulations suggest a somewhat smaller temperature rise than these). They also exhibit little coherence in the simulated geographical patterns of climate change over the Earth's surface under a $2 \times CO_2$ scenario. The models agree that land areas would warm more rapidly than the oceans, and that northern high latitudes would experience a temperature rise greater than the global mean, at least in winter. For other factors, such as effective soil moisture, there is no such agreement, with individual regions predicted to become wetter in some GCM experiments but drier in others. This divergence is caused by the fact that GCMs, like all numerical simulation models, are simplifications of the real world. GCMs simulate the atmosphere in a three-dimensional grid with a horizontal spacing of between 250 and 800 km. In reality, many atmospheric processes take place at much smaller scales than can be resolved by the model grid, and have to be represented in other, simplified ways.

Non-climatic factors that critically affect the climate system, such as oceanic heat flux, also need to be included in any model, and ideally require modelling in their own right. Thus while some GCMs run with prescribed sea surface temperatures (SSTs), others make allowance for oceanic heat transport and some are now coupled to fully dynamic ocean models (e.g. Stouffer et al., 1989). With some factors, it is not even clear how they will influence any future global warming. Cloud cover, for example, would be expected to increase with the greater atmospheric moisture content at higher surface temperatures, but whether this will amplify or damp down warming is unclear. On the one hand clouds reflect incoming solar radiation, hence reducing warming on the ground, but on the other they can also transfer heat from their upper to their lower surfaces, hence acting as heat pumps (Maddox, 1990).

Given that all the main North American and UK GCMs are able to replicate the Earth's present climate with considerable success (Mitchell and Zeng Qingcun, 1991), how are we to choose between their often very different projections for the future? What we need to know is how well they simulate climate *change*; that is, conditions significantly different from those of today. One approach which has been adopted has been to test different GCMs against past climates, in order to calibrate them (see Street-Perrott and Roberts, chapter 3). Thus, COHMAP (1988) compared the simulated global circulation for time periods since the last glacial maximum around 18 000 years ago, with paleoclimatic reconstructions based on pollen, lake levels and ocean cores. The combination of numerical modelling and empirical test is a powerful one, for it allows calibration of the model, and it directs empirical field data collection to where it is most needed.

A good example of a coupled empirical/modelling approach to studying environmental change is David Sugden and Nick Hulton's study of ice sheet behavior in chapter 7. In this case, field data have been used to constrain a numerical model of the Greenland ice sheet, and to test its response to past climatic forcing. Their ice sheet model is a dynamic one, unlike most GCM experiments which are equilibrium simulations of a stable, although different, climate. Dynamic models are important because we know that the way most environmental systems respond to forcing is time dependent. Figure 1.3 shows a range of different possible system responses to gradual, linear forcing. These could be showing the rise in temperature induced by greenhouse-gas forcing, or the changing mass balance of an ice sheet in response to climatic warming. The linear responses in figure 1.3a and b are perhaps the least commonly encountered ones in real environmental systems, although it should be noted that even in these cases there will be random variations, or "noise," around any underlying trend.

Because most environmental systems are complex, their response to forcing is more likely to be of the type illustrated in figure 1.3c

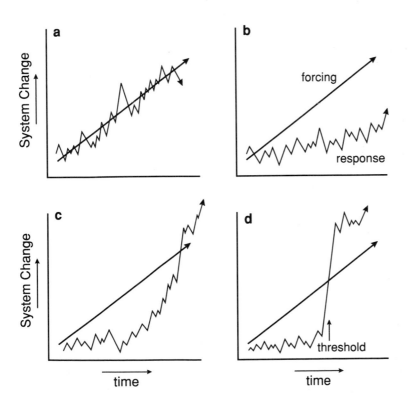

Figure 1.3 Schematic alternative responses of an environmental system to forcing (see text for explanation).

and d. For example, a non-linear ice sheet response to warming would result from the interplay of only two factors, namely accumulation and ablation. Starting from a temperature of $-20°C$, warming would in fact gradually *increase* ice volume until about $-12°C$, at which point ablation would begin to account for significant melting, a process which would then accelerate exponentially (see figure 7.3 of Sugden and Hulton, chapter 7). Bedrock topography can introduce a further significant non-linearity in the response of ice sheets to climate change (Sugden and Payne, 1990).

In the case of figure 1.3d, the system contains an internal threshold which, once crossed, leads to positive rather than negative feedback and a consequent rapid shift to a new equilibrium state. As we have already noted, there is evidence from recent Earth history of rapid phases of climatic transition, for instance during the last deglaciation (14 000–7500 years ago), emphasizing the importance of such thresholds within the global system.

A non-linear response can also result from the interaction of fast (e.g. atmospheric) and slow (e.g. deep ocean) components of the Earth system, or from geographically separate, but interconnected, sub-systems. The cryospheric response to warming once again provides us with an example. An important distinction exists between relatively "warm" non-polar, mountain glaciers which lie close to the melting threshold and which are small enough to respond to climate change within a few decades, and polar ice sheets whose low temperatures and massive bulk buttress them against significant ice loss through ablation. On this basis the IPCC report concluded that the Greenland and Antarctic ice sheets would make almost no contribution to sea-level rise during the next century if global warming were to occur (Warrick and Oerlemanns, 1990; see also Tooley, chapter 8).

However, the West Antarctic ice sheet, in particular, is vulnerable to ice calving because it is grounded below sea level. So, while it may be meta-stable in itself, any sea-level rise brought about by melting of mountain glaciers elsewhere in the world and by thermal expansion of ocean water (the "steric" effect), might render unstable the floating ice shelves which protect the West Antarctic ice sheet. If this were to happen, ice sheet gradients would steepen and ice velocities increase, leading to significant loss of ice volume, which in turn would further raise sea levels, reinforcing the process. Mercer (1978), among others, has proposed that a rapid deglaciation and catastrophic rise in sea levels of 5–7 m may be imminent as a result of the increasing atmospheric concentration of greenhouse gases. Satellite imagery confirms that some of the West Antarctic ice shelves have indeed been breaking up in recent decades (Doake and Vaughan, 1991), although this could reflect natural cyclical changes rather than a response to anthropogenic impact.

Timescales of analysis

The changes occurring in West Antarctica bring into sharp focus the different timescales over which environmental processes operate and over which they consequently need to be studied. Over time periods of a few decades or less, the methods of observation and monitoring can be employed, either through direct measurement or via remote sensing. Satellite imagery is proving to be an especially vital tool in monitoring global and regional changes in the extent of ice and snow, vegetation cover and oceanographic conditions (see Haines-Young, chapter 2). However, many environmental systems change over timescales longer than those which can be observed directly, and it is therefore necessary to turn to human and natural archives to study these (figure 1.4).

Over intermediate timescales (i.e. decades to centuries), documentary sources of data can be used, although these data were not always collected in a form ideal for the analysis to which we wish to subject them. The length of the "historical" time period also varies greatly in different parts of the world. Whereas we have written records from ancient Sumer, including data on early problems of land degradation, back to 3000 BC, prehistory in the Papua New Guinea highlands only ended in the 1930s. In regions with short archival records the importance of the third group of techniques becomes especially important. These, operating over Holocene or

Figure 1.4 Changing sources of environmental data over different timescales.

Data Sources

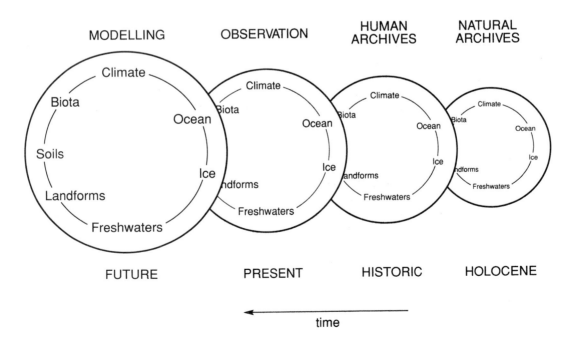

even longer timescales, are proxy methods of paleoenvironmental reconstruction like tree-ring analysis and ^{14}C dating (Berglund, 1986; Roberts, 1989). A Holocene timescale is valuable if we want to establish how typical present-day rates of environmental change are compared to those of the past.

The record of past changes provides us with clues about the geomorphological and biological responses to future climate change, notably by examining how these Earth surface systems have behaved during previous warming phases such as that at the end of the last glaciation. Paleoecological data on early postglacial migrations indicate that most biota (e.g. trees) take centuries or even millennia to adjust to a substantial temperature rise, although some groups (e.g. insects, weeds) have been effectively able to keep pace with any climate change (Adams and Woodward, 1992). These proxy records indicate that the response to warming was strongly individualistic by species, both because of ecological disequilibrium and because of the changing combinations of climatic factors. It is therefore probable that familiar communities of animals and plants would break up in a greenhouse world and that new species mixtures would emerge (Huntley, 1990b; Peters, 1991).

In chapter 9, Tom Spencer uses the same approach to investigate the possible effect of sea-level rise on coral reef islands, which may be among the most vulnerable of all environmental systems to global warming. By studying coral reefs that were growing during the early Holocene marine transgression, it is possible to show the maximum rates of sea-level rise with which corals could keep pace, before being forced to give up. These in fact lie well above the most rapid rate of rise postulated by the IPCC (Warrick and Oerlemanns, 1990), suggesting that coral growth would be able to match sea-level rise. On the other hand, coral ecosystems would be put under another threat under a warmer climate, because of the increasing frequency and intensity of tropical storms. This example serves to show that it is the *rate* of environmental change as much as its magnitude which will determine how well biotic and geomorphic processes adapt to greenhouse-gas induced climatic forcing.

Not surprisingly, the difficulty of recognizing equilibrium conditions in a world modified by human action has led to the search for recent departures from former baseline states. For climate, meteorological records show a warming of the globe of around 0.5°C over the past century (Hulme, chapter 4). This temperature rise has not been steady, however, with most of the warming occurring between 1910 and 1940, and since 1975. Global precipitation over land areas has also fluctuated since the mid-nineteenth century, but without such a pronounced underlying trend (Bradley et al., 1987). The rise in temperatures over the past century is broadly consistent with that calculated by GCM experiments for greenhouse-gas forcing so far, although it lies towards the lower end of the estimated range (Houghton et al., 1990). On the other hand, this warming is also of the same magnitude as natural climate variability during the past thousand years. Unequivocal detection of a greenhouse-gas induced warming is consequently not likely for some decades to come. Indeed, examination of tree-ring and other proxy climate records for the past millennium shows that the nineteenth and early twentieth century warming formed part of the climatic amelioration at the end of the so-called "Little Ice Age" (Grove, 1988; Bradley and Jones, 1992). Separating natural climatic oscillations from any underlying trend is thus not easy, and is heavily dependent on the timescale of observation.

The measurement and understanding of other types of environmental change, for example soil erosion, are also heavily scale-dependent (Clayton, 1991). One of the most important potential consequences of impending climatic change is its effect upon the landscape, and the intensity of soil erosion provides perhaps the clearest and simplest indication of landscape stability or instability (Roberts and Barker, 1993). Land degradation through accelerated soil erosion is in any case already a pressing environmental problem, especially in many of the Third World tropics (Douglas, chapter 15). In any given situation, it is important to know how far rates of soil loss are typical of the long-term geological norm, to what degree they are climatically controlled, how far they have been anthropogenically accelerated, and if so over what time period they have been changed.

The measurements of the rate of soil erosion which are needed to answer those questions are critically affected by the temporal and spatial scale of analysis (figure 1.5). Erosion can be measured by observation and experiment over timescales of less than about ten years, as well as by a range of proxy methods over longer periods.

Figure 1.5 Different approaches to the measurement of soil erosion over a range of temporal and spatial scales.

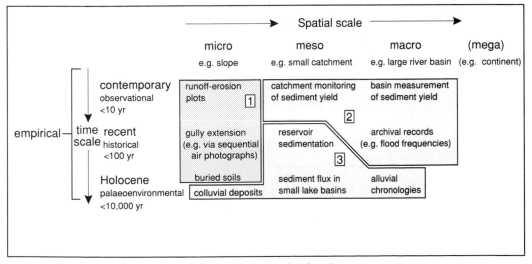

1. measurement at point of erosion
2. measurement of flux during transport
3. measurement of flux after deposition

Many of the latter are stratigraphical in approach, using the changing rate and character of sedimentation in colluvial, alluvial and lacustrine sequences to determine erosion histories (Dearing, chapter 11). These different approaches to measurement involve different levels of data resolution, with the trend from long to short timescales generally bringing greater temporal precision. For example, old air photographs can be used to establish rates of gully retreat over periods longer than is possible by direct observation. On the other hand, because they provide a discontinuous time series, it is not possible to know whether the mean rate of gully headwall retreat was a result of a few high magnitude runoff events, or more regular soil removal.

Similar problems affect erosion rates measured over different spatial scales. They range from erosion plot studies on individual slopes (micro), through the small catchment monitoring studies (meso), to the measurement of sediment yield from entire drainage basins (macro), and even beyond to mega-scale variations in denudation rate between different tectonic provinces. This spectrum of spatial scales incorporates two opposite tendencies, namely increasing problems of sediment storage moving from a small to large scale (i.e. declining downstream sediment delivery ratios), and decreasing spatial representativeness as one moves from macro to micro. An example of the latter would be donga-type gully erosion, where an area of a few square metres experiences very rapid rates of soil loss, while the rest of the landscape is relatively stable. Erosion values also depend on where measurement is made; namely, at the point of erosion, during transport as sediment, and after deposition in a lake, alluvial plain or delta (figure 1.5).

This book includes macro-scale case studies of erosion in two river basins with some of the world's highest sediment yields (table 1.4). Geomorphological and climatic factors have created high natural erosion rates in the catchments of the Yellow River (Huang He) (Derbyshire and Wang, chapter 19) and the Ganges–Brahmaputra (Haigh, chapter 20). However, in both cases soil losses have undoubtedly been accelerated by human activities in the upper parts of the watersheds, related to farming in China's loess

Table 1.4 Sediment yields of major world rivers (t $\times$ 10^6 year^{-1})

River	Suspended sediment	Solute load
Ganges–Brahmaputra	1670	152
Yellow River (Huang He)	1080	34
Amazon	900	292
Yangtze	478	166
Hsi Chiang (China)	469	132
Irrawaddy	265	91
Magdalena (Colombia)	220	28
Mississippi	210	131
Orinoco (Venezuela)	210	49

Source: Degens et al., 1991

plateau region and to Himalayan deforestation, respectively. The spatial linkages *within* catchment ecosystems mean that the human and environmental effects of land degradation are not confined to the headwater zone, but extend downstream to lowland river floodplains. In the case of these river basins, their floodplain lowlands are home to some of the largest rural populations anywhere on earth, which are then threatened by river aggradation and flooding.

Global change: regional response?

It may be useful to draw a distinction between global change, on the one hand, and global environmental problems, on the other. Global changes are those which are inherently worldwide in character and which have to be studied primarily at a global scale. The most obvious and important example of these is climatic change. Global environmental problems, in contrast, also include large-scale regional phenomena, usually common to several different geographical contexts. An example of this would be atmospheric acid deposition, which, as a result of industrial pollution, now affects large areas of North America and northern Europe.

This book deals with *both* sets of issues, and it does so in large measure because there is an interaction between the two which is essential to our understanding of how the Earth system responds to forcing. Because of the need to examine environmental changes over a hierarchy of scales, some chapters adopt a global and others a regional scale of analysis (figure 1.6). For example, while Andrew Goudie (chapter 16) offers a global review of dryland degradation, Erhard Schulz (chapter 17) provides a regional case study of recent environmental change in the central Sahara. If, as Schulz argues, we cannot objectively define the spatial limits to the desert as an environmental system, how can we know whether these limits are changing through climate change or through local resource depletion? Both authors agree, however, that "desertification" (an increasingly criticized umbrella term for a range of different processes) is less a process of shifting sands than it is of impoverishment, both cultural and environmental. Significantly this impoverishment includes not just plant and animal species but also indigenous human knowledge about their uses.

A key feature of global change is that its effects over the Earth are geographically uneven, and our ability to specify the regional consequences of future climatic change represents a major limitation to predicting its impact on society (Warwick and Farmer, 1990). Among the most responsive areas are those on or close to natural margins, such as the ecotones that separate different vegetation formations. The transitional zone from woodland to grassland within savanna ecosystems would be one example of an ecotone likely to respond sensitively to global warming (Stott, chapter 13).

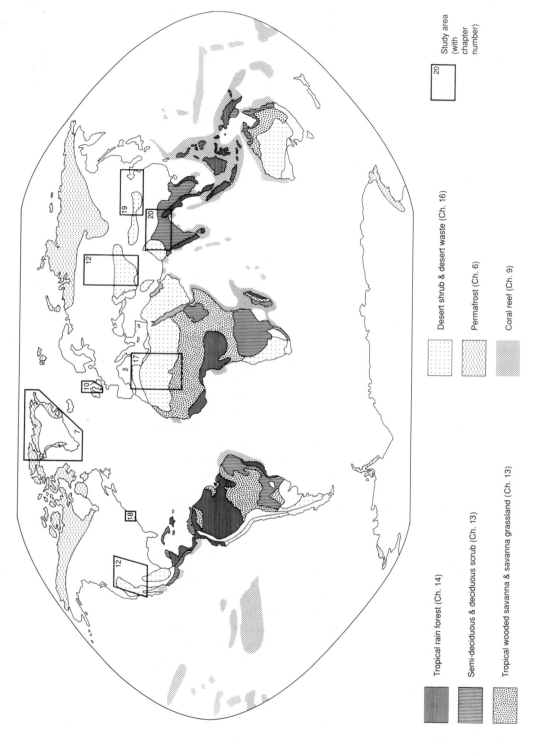

Figure 1.6   World map showing biomes and regional case study areas discussed in this book.

Tropical rain forest (Ch. 14)

Semi-deciduous & deciduous scrub (Ch. 13)

Tropical wooded savanna & savanna grassland (Ch. 13)

Desert shrub & desert waste (Ch. 16)

Permafrost (Ch. 6)

Coral reef (Ch. 9)

Study area (with chapter number)

Another vital margin is the coastline, although some coastal areas would be much more vulnerable than others to any future sea-level rise, notably coral atolls (Spencer, chapter 9), mangrove swamps and low-lying deltas. Many deltas are intensively cultivated and support dense rural populations, which may be put at risk by sea-level rise during the coming century. Milliman et al. (1989) estimate that by AD 2100, *local* sea levels in the Nile and Ganges–Brahmaputra deltas could be 3.3–4.5 m higher than today, representing 26 and 34 percent of the habitable land in Egypt and Bangladesh, respectively. This is not only a function of projected global sea-level rise, but also of natural and accelerated subsidence in the deltas (the latter related to groundwater abstraction), and a decrease in fluvial sediment influx because of interception by dams like that at Aswan. Dam construction on major tropical rivers looks set to increase significantly in the next few decades (Petts, chapter 12).

As noted earlier, the transient response of land ice to global warming, which would contribute to sea-level rise, is likely to involve the melting of smaller valley glaciers rather than of major ice sheets. These glaciers are located in mountainous terrain such as the Himalaya, Rockies and Alps, and often lie adjacent to important agricultural lowlands such as the Ganges plains and the Po valley. The rivers that flow across these plains are largely regulated by spring snowmelt and summer glacier ablation. In cases like the Grand Dixence HEP scheme in Switzerland, elaborate hydro-technology has been put in place to exploit more systematically glacial meltwater (Petts, chapter 14). With global warming causing reduced winter snowfall and enhanced summer ablation, these water resources would be put at risk. In California, the snowpack on the Sierras shrank during the late 1980s and early 1990s, so that in the Sierra Nevada it was only 13 percent of "normal" by February 1991. As spring runoff provides the water which feeds the irrigation systems of California's fruit and vegetable farmers, the threat of glacier retreat has economic implications which are all too obvious.

Is environmental change simply a matter of global forcing and regional response? In reality, global environmental changes are themselves caused by an agglomeration of smaller-scale influences, such as changes in surface albedo or carbon dioxide emissions. The destruction of tropical moist forests, for example, makes a significant contribution to global climate change via the release of carbon into the atmosphere (Myers, 1988b; Detwiler and Hall, 1988), as well as being a critical environmental issue in its own right through the loss of biodiversity. On both counts, a sustainable *and* realistic program for conserving tropical forest resources needs to be put into practice as a matter of urgency (Furley, chapter 14).

Additionally, study of global changes in the past suggests that restricted geographical areas may be critical in triggering much wider modifications to the environment. In the case of Quaternary glaciations, certain sensitive areas, like the Labrador–Ungava plateau of northern Canada, seem to have been capable of initiating ice sheet growth (Flohn, 1974; Ives et al., 1975). Climatic conditions here are always near to those which produce glaciation, because a small anomaly in the snow cover can lead to the powerful ice-albedo positive feedback being put into operation, thereby leading to further atmospheric cooling and snowfall.

A similar situation will probably apply to greenhouse-gas forcing, with "sensitive" areas holding the key as to whether positive or negative feedback will prevail, and global warming be amplified or damped down. Apart from the tropical forests, one of the other major terrestrial carbon stores which could potentially be released to the atmosphere is that in high latitude (e.g. tundra) soils, peats and wetlands (Adams et al., 1990; Jenkinson et al., 1991). Much of this carbon store is trapped in seasonally or permanently frozen ground, but its decomposition would be greatly accelerated if global warming were to continue. In addition to the local effects of permafrost melting on buildings, roads and other installations, there is a real danger that periglacial landscapes will significantly enhance the greenhouse effect through the release of CO_2 and CH_4 (Koster, chapter 6).

Conclusion

Understanding the changing global environment means being able to shift up or down from local, through regional and subcontinental to planetary scales of study, and to move over a range of timescales; that is, through a nested hierarchy of environmental systems. Analysis over these different spatial and temporal scales reveals that environmental change is a natural occurrence, although the amplitude of change increases the longer timespan of observation. Separating natural variations from those induced by human activities is not always straightforward.

Although global environmental change is not a new phenomenon, greenhouse-gas forcing is pushing the climate to a condition warmer than any experienced in recent the Earth history. Consequently, we are moving into unknown territory as far as the response of environmental systems is concerned. There is now a scientific consensus that the limited warming of the climate which has already occurred during this century will continue and probably accelerate during the next (Houghton et al., 1990). But a rise in temperature will not be the sole, or even the most important, change to take place; water-balance and sea-level changes may be equally significant, for example. It should also be borne in mind that not all the consequences of global environmental change will be negative ones. There will be winners as well as losers from the new distribution of natural resources. In the case of farming, areas where crop growth is temperature-dependent, like parts of Canada, will find conditions more favorable (Parry, 1990). By contrast, many agricultural lands close to the limit of available rainfall may experience declining soil moisture levels and falling yields. There will, of course, be a key role to be played by human ingenuity in adapting to the new environmental conditions. It is likely, however, that those economically best able to adapt will benefit most (or lose least), while those unable to adapt will be at greatest risk. The challenge of global enviromental change facing humanity is not only a scientific one.

Further reading

Goudie, A.S. (1990) The global geomorphological future, *Zeitschrift für Geomorphologie, Suppl-Bd.*, 79, 51–62.

Jäger, J. and Ferguson, H.L. (eds) (1991) *Climate Change: Science, Impacts and Policy.* Cambridge: Cambridge University Press (Proceedings of the second world climate conference).

Leggett, J. (ed.) (1990) *Global Warming. The Greenpeace Report.* Oxford: Oxford University Press.

Parry, M. (1990) *Climate Change and World Agriculture.* London: Earthscan.

Roberts, N. (1989) *The Holocene. An Environmental History.* Oxford: Blackwell.

Schneider, S.H. (1990) *Global Warming.* London: Lutterworth.

Wyman, R.L. (ed.) (1991) *Global Climate and Life on Earth.* London and New York: Routledge.

CHAPTER TWO

Remote Sensing of Environmental Change

Roy Haines-Young

> Once a photograph of the Earth from outside is available –
> once the sheer isolation of the Earth becomes known – a new
> idea as powerful as any in history will be let loose.
>
> *Fred Hoyle, 1948*

Monitoring environmental change

Our understanding of the world around us is, to a large extent, determined by the technology available to us. In the seventeenth century the microscope began to revolutionize perceptions of the "micro-world" and whole new areas of study were opened up. The invention of the telescope had similar revolutionary effects. In the twentieth century the ability to view the Earth from space has also changed the way we look at the world.

The first images showing the isolation of Earth in space were not widely available until the Apollo missions of the late 1960s, but as Hoyle foretold, their effect on scientific and public perceptions has been profound (plate 2.1). In the public imagination images of the Earth from space have emphasized the scale of the human impact and the fragility of the planet we inhabit. In the scientific arena space missions have stimulated study of the Earth as a single integrated system, composed of the core, mantle, lithosphere, atmosphere, biosphere and oceans. Much of the focus of this research is on the changes which are taking place within the so-called "Earth system."

The purpose of this chapter is to explore how remote sensing, and the associated technologies for handling geographical data, can help us study environmental change. It will describe some of the essential features of the technology and opportunities it offers for the study of the Earth. It will also describe some of the problems we face in monitoring environmental change using measurements from space.

Plate 2.1   Apollo 17 mission view of the full Earth, taken by astronauts on the way to the Moon, showing Africa and the Indian Ocean.

What do we need to know about environmental change?

The question of what we need to know about environmental change seems simple, but it is as difficult to answer as any of the major issues which have confronted science. Difficulties arise because of the complexity of environmental processes themselves and the speed and scale of the changes which may be in prospect. Although environments have rarely been constant during the course of Earth history, there is growing concern that as the result of human impact the pace of environmental change will mean that biological, social, and economic systems may find it difficult to adjust. The implications of such change are so far reaching that answers are needed urgently if remedial actions are to be taken.

One way of beginning to look at a complex issue such as environmental change is to take a pragmatic approach and ask how its detrimental effects manifest themselves. The *World Conservation Strategy* (WCS) (IUCN, 1980, 1991), for example, has addressed the issue in this way and has identified three "conservation criteria" by which the impact of human activity can be judged. These criteria focus on the need to preserve the genetic resources of the biosphere, the conservation of the essential life support systems of the planet, and the importance of achieving sustainable development. Although the purpose of these criteria was to provide a framework for development, they also highlight the sorts of information we need about environmental change more generally.

The need to preserve the genetic resource base of the biosphere arises from the vital role which living things play in the world economy. The major reservoirs of genetic resources are natural and semi-natural ecosystems on which human activity is having a major impact. Thus, information is required on the resources they contain and the continuing integrity of these areas. Such data form an essential component in the list of things we need to know about environmental change. Although living things contribute significantly to the systems which support life, the need to ensure their integrity goes much wider than the preservation of living resources themselves. The loss of tropical rain forests, for example, not only leads to the extinction of species but also results in the release of significant quantities of carbon dioxide, which may affect a whole range of climatic parameters. A second important category in the list of things we need to know about environmental change thus concerns the operation, dynamics, and condition of systems essential for life as we know it.

The final issue identified in the World Conservation Strategy is that of "sustainable development." The concept of sustainable development suggests that, while the use of resources and environmental change are inevitable as a result of the pressures of population and human aspirations, it is the "quality" of that development which is significant and which must be changed. The WCS asserts, therefore, that development must be sustainable in the sense that, whatever its character, it does not undermine the future use of the biosphere through the loss of genetic resources and the collapse of essential life support systems. The idea of sustainable development thus defines the third important category of information we require about environmental change. The environmental impact of current developments needs to be monitored. Future plans need to be modelled, and once implemented, their effects must be audited so as to detect any unpredicted and unwanted environmental spin-offs.

The problem of environmental change which seems to confront us is therefore a complex one. The three areas identified, concerning the genetic resource base, the maintenance of essential life support systems and sustainable development, help us begin to focus on the important elements of the task in hand. In this chapter I will use them as a framework to illustrate how the associated technologies of satellite remote sensing and Geographical Information Systems (GIS) can help provide the sorts of data we need to plan for the future.

What sorts of data do remote sensing systems provide?

Remote sensing is the acquisition of information about the land, sea and atmosphere by sensors located at some distance from the target of study. The sensors generally depend on the spectral properties of the target as the basis of the measurements they make. Many such sensors have been designed and deployed, and I cannot possibly describe them all in detail here. An extensive treatment can, for example, be found in Colwell (1985). In this chapter I will ignore conventional aerial photography, although some reference must be made to photographic images brought back from manned flights such as Space Shuttle. The concern will be mainly be with space-borne sensors which provide data in digital rather than photographic form. With the advent of computer processing these systems have done much to revolutionize the way in which we study the Earth. Some of their important characteristics can be described by reference to a selection of the more important optical systems, listed in table 2.1.

Table 2.1 Characteristics of some space- and air-borne optical sensors

Satellite	No. of channels	Spectral range (IFOV)	Spatial resolution (IFOV)	Swath width
Geostationary satellites				
Meteosat	3	0.40–12.5 μm	2.4–5 km	full visible face of Earth
NOAA-AVHRR sensor	5	0.58–12.5 μm	1.1 km	3000 km
Sun synchronous, polar-orbiting satellites				
Landsat MSS	4	0.50–1.1 μm	80 m	185 km
Landsat TM	7	0.45–12.25 μm	30 m (visible IR) 120 m (thermal IR)	185 km
SPOT HRV	4	0.5–0.9 μm (multispectral)	20 m	60 km
		0.5–0.73 μm (panchromatic)	20 m	60 km
Aircraft				
ATM (Dadelus)	11	0.50–2.2 μm	variable depending on altitude	

General platform characteristics

One way of grouping the various types of system listed in table 2.1 is according to the general character of the craft, or "platform," used to carry them. The airborne systems are carried by aircraft and the nature of these platforms should be familiar to most people. The space platforms are perhaps less so. These can be grouped according to whether they have a "geostationary" or "polar" orbit.

A number of the geostationary satellites were launched to provide information for weather analysis and forecasting. They orbit the Earth at distances of about 36 500 km in such a way that they keep pace with the rotation of the Earth below. They can therefore be used to image the same portion of the Earth continuously. Meteosat, for example, sits above the point where the Greenwich meridian crosses the equator, and makes observations every 30 minutes.

Complete coverage of the Earth using geostationary platforms can only be achieved by the launch of a whole suite of satellites. The polar orbiting group, by contrast, are designed to give this more complete coverage by a single sensor, but this time only on a repeat cycle of some days. Landsat-5 is an example of one such platform. Landsat-5 carries a sensor known as Thematic Mapper (TM). The satellite orbits the Earth at an altitude of about 705 km, in a near-polar, circular pattern which has been designed so that the satellite is synchronous with the sun. This means that the satellite passes over the area being imaged at the same sun time on each repeat cycle. At the equator this is approximately 09.30 hours. The orbit gives repetitive coverage of the Earth's surface every 16 days, with as constant an illumination geometry as possible. The TM sensor records data for a swath along its track about 185 km wide. For the purposes of distribution these data are divided into scenes depicting an area 185 km × 185 km on the ground.

Polar orbiting satellites have been launched for the purposes of making weather observations, such as the NOAA series, or with the aim of making more general Earth observations. Like the Landsat system described above, the more recently launched SPOT sensor was designed for Earth observation purposes. This satellite has a repeat cycle of 26 days but, since the sensor can be pointed at targets, repeat coverage can be achieved much more frequently, depending on the latitude of the target.

Spectral resolution

The sensors listed in table 2.1 can all be described as "passive" systems since they record the intensity of electromagnetic radiation reflected back into space by objects on the Earth's surface. The sensors generally record information selectively across the electromagnetic spectrum, returning the information as data for a series of "spectral bands" or channels. The position of these bands, together

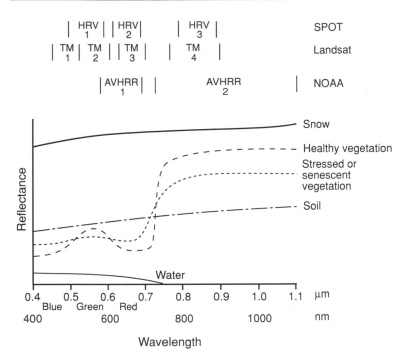

Figure 2.1   The position of the spectral bands for Landsat TM, SPOT and the AVHRR sensors.

with their width and number, define what is known as the "spectral resolution" of the sensor.

Figure 2.1 shows the position of the spectral bands for Landsat TM, SPOT and the AVHRR (Advanced Very High Resolution Radiometer) sensor carried by the NOAA weather satellites. TM records data in seven spectral bands ranging from the visible through the near infra-red to the thermal. SPOT records three bands in its multispectral mode. The design of the AVHRR sensors carried by the NOAA series has varied. The most recently launched systems record data on five bands.

The position and width of the bands recorded by a sensor are largely determined by the nature of the intended application and the spectral characteristics of the intended targets. The selection of those bands intended to record properties of vegetation has been made, for example, on the basis of information about how radiation in different parts of the electromagnetic spectrum is absorbed and reflected by leaf canopies.

Spatial resolution

The spatial resolution of a sensor is a measure of the size of the smallest object which can be detected. Although a considerable amount of detail is often evident there is a size threshold below which objects cannot be seen. This threshold gives a measure of the spatial resolution of the sensor.

To see how the issue of spatial resolution is important we need to consider the way in which remotely sensed images produced by optical scanners are constructed. The images are built up of a number of picture elements, or "pixels." The sensor scans the scene, recording for each pixel the reflectance levels across the various bands to which it is sensitive. Landsat TM imagery, for example, has a pixel size which corresponds to a cell measuring 30 m × 30 m on the ground. SPOT imagery has a pixel size of 20 m × 20 m, while AVHRR data are much coarser with a pixel size of 1.1 km × 1.1 km in "local" mode and 5 km × 3 km in "global" mode. In each case, the features on the ground in the area scanned by the sensor "stand out" by virtue of the contrast in reflectance recorded from one pixel to another across the scene. Clearly, the larger the pixel size, the more general or cruder the representation of features.

Unfortunately there is no simple relationship between pixel size and the spatial resolution of the sensor since the detectability of objects depends not only on the characteristics of the sensor themselves, but also on the nature of the surface being imaged (Townshend, 1981). Features smaller than the nominal resolution of a system may be detected on an image if they show a great contrast with their surroundings. Conversely, a large object may not be discernible even though it is larger than the theoretical resolving power of the system if it shows little contrast with its surroundings. It is this effect which is responsible for the fact that high-contrast features, such as rivers, canals and roads, may be detectable on Landsat multi-spectral scanner (MSS) images even though they are narrower than 79 m, the resolution of this particular sensor (Mather, 1987). The effect can be very significant even

Figure 2.2 Variation in the minimum classifiable area for a range of sensors (after Fuller et al., 1989).

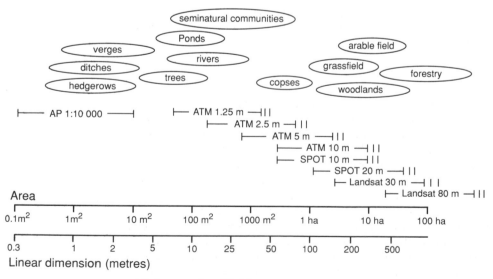

Minimum classifiable area

on imagery with apparently very good spatial resolution. Stolz (1988), for example, reports that in the case of space photographs provided by the Metric and Large Format Cameras objects up to 5 m wide (e.g. footbridges, railways and roads) are detectable in areas of the image which show good contrast. In areas of low contrast objects up to 30 m in size (e.g. roads, river channels, fields) may not be observable.

Figure 2.2 shows the variations in the minimum classifiable area for a range of sensors suggested by Fuller et al. (1989) for different types of landscape feature in the UK. Although the overall detectability of features is a function of the ability to discriminate spectral differences between the objects, the spatial resolution of the sensor clearly must set a general limit to the usefulness of these image data in environmental work.

How can we find out about environmental change?

On the face of things the detection of change in any environmental system is straightforward. It requires the comparison of the state of the system at two or more points in time. We then have to form a judgement about whether any differences detected are large enough for us to say some change has occurred. For example, we might try to overlay the remotely sensed image onto another and by a process of "subtraction" try to identify where change has taken place. Unfortunately, when we try to assess change in environmental systems, especially when we rely on sophisticated technologies like remote sensing to make our measurements, the judgements we have to make are often very difficult ones. In order to highlight the sorts of issue involved we will look at the process of measurement and analysis using remote sensing and then consider the complications introduced by our understanding of the behavior of ecological systems themselves.

Problems of measurement and analysis

When we use remote-sensing data to measure some properties of the Earth's surface we tend to assume that the sensor represents faithfully what is actually on the ground. Equally, when we compare images from two points in time and detect change, we tend to assume that differences are real ones rather then artifacts introduced by variations in measurement technique. Our confidence comes about, perhaps, as a result of our familiarity with conventional aerial photography. As Duggin and Robinove (1990) have shown, however, the extent to which satellite sensors accurately represent the real world depends on a number of factors. They show that when we use imagery to describe conditions on the Earth's surface, or by comparing imagery try to detect change, we make a number of assumptions about the measurement process.

The soundness of our judgements will depend on whether these assumptions are correct.

Important issues identified by Duggin and Robinove (1990) include the fact that while we assume there is a strong relationship between the radiance properties recorded on the image and the optical properties of the features on the ground, this may not always be the case. For example, in the case of Landsat MSS imagery, although pixel size is reported as 79 m × 79 m, the center-to-center distance along the scan lines is only 56 m. Thus there is a 23 m overlap between pixels. This produces an averaging effect, and the radiance measured may not correspond in any simple way to the spectral properties of the features on the ground.

The way in which the satellite sensor records information about the spatial variations in surface properties is further affected by such things as platform drift and jitter. These factors, coupled with fluctuations in atmospheric transmission, back-scatter, and self-radiance, may degrade the spatial resolution and also lead to an averaging effect between pixels.

An understanding of atmospheric conditions at the time of the satellite overpass is especially important if images taken at different times are to be used to assess changes on the ground. We may assume that the atmosphere is "transparent" so far as the sensor is concerned but this is not always the case. Kaufman (1989) has described how atmospheric effects may be significant. Not only may the atmosphere affect wavelengths differentially, but it may also alter the spatial distribution of the reflected radiation and hence the spatial resolution of the sensor. It may also change the apparent brightness of an object, affecting measurements of albedo and brightness. Finally, atmospheric effects may introduce spatial variations in surface brightness due to the presence of subpixel-sized clouds. Since the strength and significance of all these factors may change under different atmospheric conditions, the comparison of images over time to detect change may be difficult. Kaufman (1989) goes on to review how the problem of atmospheric correction might be approached.

In making comparisons of images at different times or from different sensors, the issue of radiometric calibration is also highly important. As the sensor records conditions on the ground the output voltages from the instruments are converted to a "digital value" which is sent back to the ground. For example, in the case of Landsat MSS, the strength of the reflectance recorded on each of the spectral channels is represented on a 64-point scale, with 0 indicating no recorded reflectance and 63 the maximum. If images are to be compared then these digital values have to be converted back to measures of physical radiance values, and this assumes that we can calibrate the sensor. A particularly graphic illustration of how differences in calibration may be important is given by Slater (1979). He showed that there were slight differences in the array of six detectors of Landsat MSS. This resulted in significant differences between the sensors in the recorded radiance for the same target. The effect can sometimes be seen as so-called "six-line banding" across an image. In their review of the assumptions implicit in using remotely sensed data, Duggin and Robinove (1990) document a number of studies which show how radiometric calibration might be achieved.

In addition to radiometric calibration, comparison between images from different sensors or for different times also requires corrections for any variation in viewing illumination or viewing geometry. As noted above, sun-synchronous satellites such as Landsat or SPOT pass overhead at the same time each day. However, with the passage of the seasons, sun angle will vary and the illumination conditions of the scene will differ. Thus apparent differences between images may be due to no more than illumination effects. Such problems are compounded when data from different sensors are to be compared. Not only may they be recorded under different illumination conditions but their different optical characteristics, such as field of view, will mean that the measurements of surface properties are not being made under the same conditions.

A particularly important "geometric" factor which must be considered when images collected at different times or from different sensors are to be compared concerns the fact that we need to be sure that they are properly registered, one with another. As Mather (1992) has noted, remotely sensed images are not maps. Their coordinate systems, consisting of scan lines and pixel number, have only a weak correspondence with real geographical coordinate systems such as latitude and longitude. From information about the orbital geometry of the platform the calculation of geographical coordinates may be possible. Other techniques of geometric correction involve correlating the location of features shown on the image with features on the map whose real coordinates are known. The precision of such calculations will depend on the fullness and accuracy of information at our disposal.

The technical issues described so far are all largely to do with the properties of the sensors themselves and the factors which affect their performance. In their review Duggin and Robinove (1990) go on to describe a range of assumptions which relate to issues of how the user intends to use the data, rather than properties of the data themselves. For example, it is a widely accepted (but rarely tested) assumption that the image acquisition conditions are optimal, or at least adequate, for measurements of surface conditions to be made. Other assumptions include the belief that the scale of the image is appropriate for the features of interest to be detected, and the idea that the correlation between recorded radiance and the optical properties of the objects themselves does not vary spatially. This last factor is particularly significant. As we have seen in the case of linear features, detectability depends on their contrast with neighboring features. Since their geographical context may differ, detectability may also vary across a scene.

Duggin and Robinove (1990) stress that in using remotely sensed data to measure surface properties a final key assumption concerns the appropriateness of the analytical techniques used for the task in hand. A review of the full range of image processing techniques available cannot be given here. Mather (1987) provides an extensive treatment of the topic, and Eastman and McKendry (1991) and Singh (1989) focus on those techniques useful for change detection. The review by Singh is particularly important because he shows that the various procedures of change detection can produce different results even in the same environment. Examples of the application of some of these digital change techniques to the problem of mapping land-use change in savanna areas is provided by Randell (1990).

Analytical uncertainties, coupled with all the other factors affecting the accuracy of the measurement process, mean that the use of remotely sensed data to record surface conditions and detect change must be approached critically. The scientist must demand an independent test of the accuracy of measurements or the output from analytical routines before statements about the environment can be accepted with any degree of confidence.

Ecosystem dynamics

The problems of measurement and analysis using remote sensing systems may seem formidable, but generally they can be overcome so that we can be reasonably certain that the data do represent the real world, if only in an approximate way. It is when we turn to the problems which arise from our understanding of the structure and dynamics of environmental systems themselves that the most substantial issues arise.

The difficulty of assessing change in complex systems like the environment is that often data on the current state of the system are difficult to acquire or ambiguous when they are to hand. If we are interested in change then we may also be frustrated because data on past conditions are lacking. More fundamentally, problems can also arise because our understanding of the system is partial and we simply cannot agree about what we ought to measure, how measurements of specific variables are related to the condition of the system as a whole, or even what interpretation we can put on particular results. Figure 2.3a clarifies some aspects of the problem.

Suppose, for the moment, that the condition of an environmental system could be represented by just one variable. For an aquatic system it might be a water quality parameter. For a terrestrial system it might be a variable describing biomass or productivity. If there were some long- or medium-term change induced by human agency it would probably be superimposed upon the natural variability of the system. Thus to say that change had occurred, we would have to have some notion of a threshold beyond which a difference is regarded as significant. This assumes, of course, that the system was at least fluctuating about a stable equilibrium before human intervention – which is not always the case. In these situations we may need to turn to other techniques (see chapter 1) in order to judge the significance of contemporary measurements.

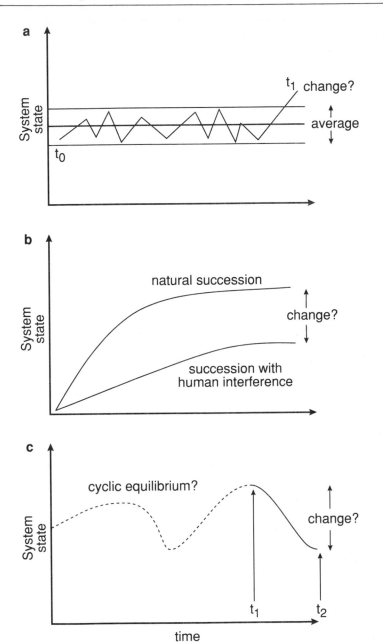

Figure 2.3 Some problems of detecting change in environmental systems (see text for discussion).

Ecosystems, particularly successional ones, may not be at equilibrium, and human activity is not the only factor responsible for environmental change, although it may be modifying the sort of change that is occurring (figure 2.3b). In these kinds of situation, when human influence may be superimposed on some longer, natural change the assessment of the importance of human impact may be difficult to make. Even if the system were in some form of equilibrium before human intervention, however, the situation may not be as simple at that shown in in figures 2.3a or b. The form of that equilibrium might, in the long-term, for example, be a cyclic one (figure 2.3c). Thus a difference in average condition between two sets of measurement at different time periods might not be as significant as it first seems.

So how do we measure environmental change? The purpose of figure 2.3a–c is to show that measurement is not simply a matter of collecting data, even if we are using the latest, most expensive technology. Measurements clearly have to be based on an understanding of the structure and dynamics of the system that we are measuring. In other words we have to possess some model of the system that we are interested in.

The point that all measurement of reality is made in the context of some conceptual model cannot be emphasized too strongly. The model reduces the complexity of the real world down to a basic set of parameters, or key variables, by which the behavior of the system as a whole can be understood. These variables are often things which are easy to measure and which represent some more fundamental property of the system. The model specifies just how we think these "surrogate" variables link into the system as a whole. They also play an essential role in the interpretation of measurements, such as differences in measured state over time. Models are based on our theoretical understanding of the system, and it is through such abstraction that the significance of change is judged. In figure 2.3c, for example, the difference between the state of the system at times t_1 and t_2 would not be judged "significant" if we held that the system exhibited some form of cyclic equilibrium. If, on the contrary, we *thought* that a static equilibrium applied then the interpretation of the observations would be quite different.

The systems illustrated in figure 2.3a–c, are, of course, extremely simple and a more realistic model would be far more complex. Significantly, many of the parameters required to operationalize complex global models, such as those dealing with aspects of moisture and energy budgets and variations in ground cover, can only be investigated by space-borne sensors. Over the past two decades or so, the scientific community has been learning how to exploit the visible, infra-red and microwave portions of the electro-magnetic spectrum to view the atmosphere, land surface and oceans. The development of Earth system science has, in fact, depended fundamentally on access to remote sensing and the associated data handling technologies. Without the data which such sensors can provide the development of models describing the operation of the Earth system could be neither constructed nor tested.

An understanding of the role of models in the detection and monitoring of environmental change is essential. Although the arguments developed above are also relevant to measurements made *in situ* on the ground by conventional methods, they are just as important when dealing with the more novel technologies of remote sensing. The large volumes of data which these systems can provide are no substitute for a critical analysis of our ideas about the ways in which we think environmental systems work.

What can we find out using remote sensing data?

I cannot describe all the ways in which remote sensing has been or could be used to study environmental change. Instead just three areas of application will be described. These correspond to the areas of potential concern highlighted by the *World Conservation Strategy* and described above.

The genetic resource base

The progressive loss of typical forests constitutes one of the major threats to the Earth's genetic resource base. These areas support some of the most species-rich ecosystems on this planet. The genetic resources found there have, for example, been important in efforts to breed new and better agricultural crops or to develop new pharmaceutical products. Most of the organisms which are found in the forests are still to be described, however. It has been argued widely that if deforestation continues at present rates then much of the forest will be cleared within the next few decades, leading to species extinctions on a massive scale. Estimates of the rates of forest loss are, however, difficult to make and Sayer and Whitmore (1991) have suggested that earlier estimates of the rate of loss need to be revised downwards. In the future access to data from remote-sensing systems may help to establish a more accurate picture.

The pressures on the tropical forest areas come from agriculture, commercial logging, industry, urbanization, and the development of road networks which allow people to penetrate the forest areas (see Furley, chapter 14; Hecht and Cockburn, 1990; Repetto, 1990). As Nelson and Holben (1986) have shown, Landsat MSS imagery provides a particularly dramatic view of the process in the state of Rondonia in the Brazilian Amazon. In a study area of about 10 000 km^2 clearings can be seen spreading out from a grid-iron road net. A somewhat similar effect can be seen in the tropical cloud forests of the Peruvian Andes. Plate 2.2 shows the effect of forest clearance on the ground. The Landsat MSS image shown in plate 2.3 gives an overview of part of central Peru. Intact forests are indicated by the darker tones. The loss of the forests shows up the lighter tones along the major valleys.

Although remote-sensed imagery cannot allow us to track the fate of individual forest species in detail, these systems provide an important set of tools for monitoring the rates of loss. These data can be vital in developing better management programs. As Nelson and Holben (1986) point out, the collection of reliable data is difficult by traditional means, and many developing countries may

Plate 2.2   Cloud forests showing the effects of clearance on the eastern slopes of the Peruvian Andes.

not have an accurate idea of the distribution and rates of forest loss within their own borders.

Nelson and Holbenn (1986) show that the local area coverage (LAC) provided by AVHRR data can be used to delimit forest and non-forest areas in their test areas within Rondonia. Malingreau et al. (1989) have subsequently used such data to monitor forest disturbance over a much larger area covering the Territory of Acre and the states of Rondonia, Mato Grosso, Para, and Goias. The thermal channel 3 (3.5–3.9 µm) proved to be the most effective for the discrimination of deforestation patterns. Atmospheric moisture, haze, and smoke meant that atmospheric transmission in the wavelengths recorded by channels 4 and 5 was poor. Deforestation was measured by the identification of "disturbance areas," that is areas where the activities associated with deforestation are occurring. The techniques allowed an assessment of the state of disturbance and deforestation in the whole southern fringe of the Amazon to be made. Using a time series of AVHRR imagery since 1982, these workers suggested that deforestation in the state of Rondonia had increased exponentially since 1985 with the opening of the main highway to central and southern Brazil. Such data will be vital in the future for monitoring the effectiveness of recent

Plate 2.3 Landsat MSS image of the cloud forests areas of the central Peru. The effect of forest clearance on the cloud forests which reach up from the Amazon Basin into the Andes of central Peru can be seen. Intact forests are indicated by the darker tones. The loss of the forests shows up the lighter tones along the major valleys. The area shown centers on the city of Huanuco.

attempts by the Brazilian government to limit burning of the forests in these areas and thus take steps to conserve something of their living resources.

Studies such as those dealing with deforestation illustrate how remotely sensed data can be used to monitor the threat to the genetic resource base by the analysis of general ecosystem characteristics. Recent studies have shown that these types of data can also be used to monitor performance at the species scale. Thus, for example, Sader et al. (1991) have used classified Landsat TM imagery to map habitats important for migratory bird species in Costa Rica, where deforestation rates of about 4 percent per year threaten over-wintering sites for a number of important neotropical land birds.

In contrast to these probems caused by deforestation, Avery and Haines-Young (1990) have recently described how Landsat MSS data can be used to assess the impact of afforestation on wading bird populations in the Flow Country of Caithness and Sutherland,

Scotland. The area is of international significance for wading birds and has been affected by extensive commercial forestry between 1978 and 1988. In the study, bird distributions were predicted from information about spatial variations in ground wetness obtained from Landsat MSS data. Such studies illustrate how archive (i.e. old) remotely sensed data can be a valuable tool, not only for ecological modelling, but also in environmental impact and environmental audit studies.

Essential life support systems

The loss of tropical rain forests has significant environmental consequences at the local and regional levels, but may also have a global significance. Woodwell (1984) has argued that increasing concentrations of CO_2 resulting from forest burning may contribute to increasing global temperatures via the greenhouse effect. Changes in surface albedo brought about by deforestation may, by contrast, tend to reduce temperatures. Given the potential consequences of land use changes on such massive scale, the scientific community is now devoting considerable resources to monitoring the effects of these transformations. Remote sensing has a vital role to play in this.

Justice et al. (1985) have shown how coarse-resolution AVHRR data can be used to monitor seasonal vegetation dynamics. The technique they use is the so-called "normalized difference vegetation index" (NDVI), calculated as follows:

$$NDVI = (ch2 - ch1)/(ch2 + ch1)$$

where ch1 and ch2 are the spectral reflectance values for a given pixel on two of the spectral bands recorded by the AVHRR sensor. By calculating the NDVI for each pixel, the index allows us to produce a new image on which certain patterns in the vegetation are shown more clearly. The ecological basis of the index is in the fact that channel 1 lies in the range of wavelengths where chlorophyll causes considerable absorption. Channel 2 is in that spectral region where the spongy mesophyll tissue in leaves causes reflectance. The data are combined in a ratio to reduce the effects of surface topography and to compensate for differences in sun elevation for different parts of the scene. The NDVI has been found to be highly correlated with vegetation parameters such as green-leaf biomass and green-leaf area, and thus to be of value in mapping vegetation patterns. As we have seen, an understanding of global distribution of the major vegetation types, their biomass, seasonal dynamics, and vulnerability to human disturbance form important input models which seek to predict long-term global climatic change.

Such studies as those of Justice et al. (1985) illustrate the potential importance of remote sensing technology in monitoring the operation of the systems essential to the support of life on this planet. But, as the study of Townshend and Justice (1986) has shown, their use is not without its problems. The results of their study of NDVI data for Africa suggest that interpretation is assisted by stratifying imagery by cover types. Similar values of NDVI can be obtained for quite different vegetation communities. The challenge for the future, it would seem, is to discover how remotely sensed data can best be integrated with a whole range of other environmental information so that more robust ecological models can be developed.

Sustainable development

For sustainable development to be achieved human activities must neither undermine the genetic resource base of the biosphere nor disrupt essential life support systems. Although remote sensing can provide much of the information required in the pursuit of such a goal, it is important to emphasize that success requires more than the provision of good data. The goal of sustainable development can only be achieved if information can be used on the ground in appropriate decision-making and planning procedures. Thus as we look to the development of remote sensing as a tool for environmental management, we are increasingly forced to look at institutional and organizational issues.

The work reported by Hellden (1987) provides a case study which illustrates the way in which data collection and analysis can be integrated into the planning process in a developing country. The study concerned the assessment of woody biomass, community forests, land use, and soil erosion in Ethiopia. Hellden concluded that there are no reliable data available for these variables, and that the only feasible way to obtain such information was to use remotely sensed imagery. They would be used as one input to a GIS, where they could be integrated with a whole range of other environmental and socio-economic data for planning purposes (figure 2.4).

The GIS described by Hellden allows, for example, the supply and demand for fuelwood to be modelled. Fuelwood is important in developing countries such as Ethiopia where it forms a major domestic source of energy in rural areas. If hazards such as soil erosion are to be avoided, the availability of the resource needs to be assessed and appropriate planning measures taken to minimize and manage the human impact on natural ecosystems. Landsat MSS, TM and SPOT were used to assess the distribution and abundance of woody biomass in two test areas. When these data were integrated with population statistics and likely demand information, areas where the fuelwood resource was in excess or deficit could be identified.

On the basis of his study Hellden went on to recommend a configuration for a multi-purpose information and planning system to operate at the national scale. The costs of such a system for a developing country such as Ethiopia, in terms of both capital outlay

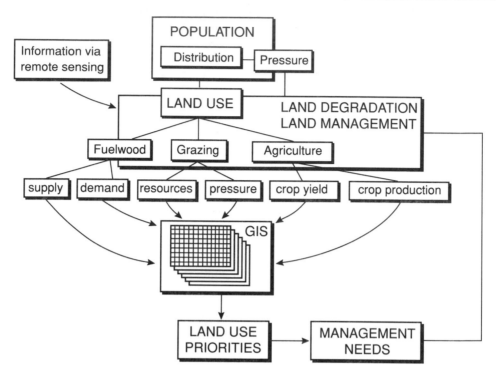

Figure 2.4   The integration of remotely sensed data in a GIS (after Hellden, 1987).

and trained staff, are considerable. As Elkington (1987) has pointed out, the extent to which remote sensing and the associated techniques for handling spatial information are themselves "sustainable" is an issue which needs to be considered carefully.

The future

The analysis of environmental change requires the observation of a range of processes on different temporal and spatial scales. As we have seen, global scale coverage for some important variables is available through the operational, geostationary, weather satellites such as NOAA. Earth observing satellites have been useful for more detailed local and regional coverage. Over the past two decades or so, such sensors have done much to advance our understanding of the problems of environmental change but much remains to be done. Many of the measurements are, for example, only loosely coordinated in time and space, and some are poor in quality. What of the future?

In the next ten to fifteen years we are likely to see a massive investment in new remote sensing systems. With a major focus of the effort directed towards the problem of change at global scales, these programs will be more highly integrated initiatives than we have been accustomed to. Any global observing system has, by its very nature, to be conceived as a coordinated program which provides for data from both space platforms and *in situ* ground measurements.

In his review of the potential of remote sensing for the study of global change, Rasool (1987) suggested that measurements are required in three major areas: on the physical and chemical state of the Earth's surface, atmosphere and oceans; on the chemical changes in the atmosphere, ocean and land; and on the biological state of the Earth's surface and oceans. The first components, he suggested, are largely taken into account by the World Climate Research Programme (WCRP). The last two fall under the remit of the Global Change Programme of ICSU. In looking at the space components of these programs Rasool found it convenient to distinguish between the near term (up to 1995) and the long term (beyond 1995).

In the medium term, data for the WCRP will be collected by the existing operational and planned weather satellites and specialized space research missions such as TOPEX/POSEIDON. The latter will use precise laser altimeters to measure height variations in the sea surface. These data can reveal patterns important in the study of the oceanic circulation and sea floor geological structures. In the medium term, the needs of the Global Change Programme will be met from the continuing operation of the meterological satellites such as GOES, METEOSAT, IRS, GMS, NOAA, METOR, and planned research missions such as ERS-1, UARS, TOPEX, N-ROSS and JERS.

ERS-1 is one of the more important near-term research initiatives. It is intended to be the forerunner of a series of European remote sensing satellites expected to become operational in the mid-1990s. Launched in 1991, ERS-1 combines three missions: first, to collect data on short- to medium-term weather and sea state conditions; second, to collect continuous information on a long-term basis on a range of climatological variables; third, to provide an all-weather imaging capability using a synthetic aperture radar instrument for monitoring land and sea ice characteristics, coastal zone processes, and land characteristics (UK EODC, 1991).

For the Global Change Programme one of the major challenges will be to combine data from a range of satellites in order to provide a global context for process studies. For example ocean color (ocean chlorophyll) data provided by SeaWIFS (Sea Wide Field Sensor), when combined with surface winds from instruments such as those aboard ERS-1, will be essential for the study of large-scale marine biosphere–atmosphere exchange. The SeaWIFS will be deployed on missions planned for the mid-1990s.

The requirements of simultaneity, accuracy, and consistency of data in global change studies are only likely to be met in the long term by the launch of the polar orbiting platforms and advanced geostationary platforms, and specialized research missions envisaged under the EOS program. The low-orbit polar platforms, the first of which is planned for 1997, will provide coverage of the entire Earth with constant illumination conditions at two fixed times each day. The important feature of such platforms is that they will carry a much wider range of sensors than is possible with the current generation of satellites. The instrumentation will, for example, constitute the follow-up to the present day operational satellites of NOAA and will allow measurements of the Earth radiation budget, vegetation indices, ocean color, ozone and temperature profiles in the stratosphere, and ice and snow cover.

In addition to these low resolution, global scale measurements made by the polar platforms, other sensors will be capable of giving high resolution data which will extend the capabilities of the current generation of Earth observing satellites. The High Resolution Imaging Spectrometer (HIRIS), for example, will have a spatial resolution similar to SPOT and TM but a much higher spectral resolution. It will acquire data for 192 continuous 10 nm wide spectral bands in the 0.4–2.25 µm region. The sensor will be pointable so that frequent coverage of selected sites can be achieved. Data from HIRIS will complement information from the Moderate Resolution Imaging Spectrometer (MODIS) which will be available at resolutions of 1 km on a repeat cycle of once or twice each month. As Wickland (1989) notes, the additional spectral information provided by imaging spectrometers will allow a level of detail in vegetation classification to be achieved that is not possible with the current generation of sensors, particularly when they are combined with microwave data.

Conclusions

In an assessment of the priorities for global change research, it has been suggested that current work is "data limited" (IGBP, 1990, pp. 9–3). The IGBP report suggests that it is now important to ensure that global change research can be supported by the necessary measurement and data sets. This chapter has emphasized the contribution which remote sensing systems can make in such an undertaking. As we have seen, these systems can provide global, regional, and local coverage for a range of key environmental processes.

Access to good data is not the only factor that will allow us to understand environmental change and the impact of human activities on the process. We also need to become involved in the development of new and more complex theories and models, for these provide the conceptual framework in which measurements are taken and results interpreted. The development and testing of such models will require, in turn, the development of more sophisticated approaches to handling data. Thus the associated technology of Geographical Information Systems will come to play an ever increasing role in our thinking.

The problem of environmental change is as difficult and complex as any which has challenged science. Hoyle showed great insight in his prediction that space observation would have a profound effect on the way in which we viewed our world. When future generations look back they may see that not only did it transform our scientific thinking, but that it also changed the way we managed the Earth's fragile resources.

Further reading

Colwell, R.N. (1985) *Manual of Remote Sensing*, 2nd edn. Falls Church, VA: American Society of Photogrammetry.

Duggin, M.J. and Robinove, C.J. (1991) Assumptions implicit in remote sensing data acquisition and analysis. *International Journal of Remote Sensing*, 11, 1669–94.

Eastman, J.R. and McKendry, J.E. (1991) *Change and Time Series Analysis*. UNITAR, Explorations in Geographic Information Systems Technology, Volume 1. Geneva: United Nations Institute for Training and Research.

IUCN (1980) *World Conservation Strategy*. Gland, Switzerland: International Union for the Conservation of Nature, United Nations Development Programme, and World Wildlife Fund.

IUCN (1991) *Caring for the Earth: a Strategy for Sustainable Living*. Gland, Switzerland: International Union for the Conservation of Nature, United Nations Development Programme, and World Wide Fund for Nature.

IGBP (1990) *The International Geosphere–Biosphere Programme: a Study of Global Change*. Report No. 12: The Initial Core Projects. Stockholm: Royal Swedish Academy of Sciences.

Mather, P.M. (1987) *Computer Processing of Remotely-sensed Images*. Chichester: Wiley.

Singh, A. (1989) Digital change detection techniques using remotely-sensed data. *International Journal of Remote Sensing*, 10, 989–1005.

PART II

Global Climate Change

CHAPTER THREE

Past Climates and Future Greenhouse Warming

F. Alayne Street-Perrott and Neil Roberts

Introduction

The Earth's climate system looks set for change. Numerical models predict that this will take place at a rate faster, and will achieve an equilibrium warmer, than anything encountered during the entire timespan of human life on the planet (Hansen et al., 1988). Because projected greenhouse-gas induced change takes us beyond the range of our observed experience of the climate system, there is considerable uncertainty – and even skepticism – about the reliability of the scientific predictions being made (e.g. Idso, 1989a). Thus, while there has undoubtedly been a rise in global mean temperature of about 0.5°C during the twentieth century, this is still within the "natural" range of climatic variability over recent centuries (Briffa et al., 1990; Houghton et al., 1990; Hulme, chapter 4, this volume). Furthermore, the recent 0.5°C warming is only about half that calculated for the same period by atmospheric models, primarily because of sequestering of heat by the oceans (Schlesinger, 1989).

Given these uncertainties and complications, there is a need to examine the longer-term behavior of climate. In the more distant past, climates were sufficiently different from that of the present day to help in the difficult art of predicting what the world's climate may be like in the future as a result of increased emissions of greenhouse gases. Just as Ed Deevey (1969) "coaxed history to conduct experiments" with the sedimentary record of lakes, so past (or paleo-) climates may be used to provide insights into processes not operative or not visible today. The ways in which paleoclimates have been used – as analogues for the future, as a means of testing numerical climate models, and as a means of probing the behavior of atmospheric, oceanic and Earth-surface systems during periods of major climatic transition – form the basis of this chapter.

Quaternary paleoclimates

Instrumental or documentary records of climatic change are available for about the past 1000 years in Europe, and somewhat longer in parts of Asia such as China. Because this is too short to sample the full range of natural variation in climate, the history of climate for older time periods has to be reconstructed from other, "proxy" indicators (see Bradley, 1984; Roberts, 1989). Land-based evidence such as moraines and related glacial landforms provided the initial basis for recognizing the existence of a former "ice age" (Agassiz, 1840). The later discovery of buried soils and deposits containing a fauna indicative of a warm climate beneath or between glacial tills and outwash gravels, showed that Pleistocene climate had alternated between cold glacials and warmer interglacials (Chamberlin, 1895; Penck and Brückner, 1909). In most mid- and high-latitude regions, conditions during interglacial periods were similar to those of today, although at their peak, temperatures may have been up to 2°C above what they are now.

The study of Quaternary climatic history was revolutionized and made more rigorous by the advent around 1950, first, of deep-sea coring and second, of oxygen-isotope analysis (Emiliani, 1955). The latter primarily reflects past global ice volumes, and as such provides a "paleo-glaciation" index. It can be applied to a range of materials, including the group of marine microfossils known as foraminifera, whose calcareous shells are preserved in deep ocean sediments. Deep-sea cores represent effectively unbroken stratigraphic records which now cover the entire Quaternary (Prell, 1982; Shackleton et al., 1988). Oxygen-isotope analysis of foraminifera in these cores has revealed a continually changing pattern of climate over the past several million years of Earth history, but which includes a significant break-point at 0.8–0.9 million years, after which climatic cycles became longer and of larger amplitude (figure 3.1). Eight major glacial–interglacial cycles have occurred since this time, each cycle lasting about 100 000 years.

The deep-sea record of climate is of sufficient precision to allow it to be tested against different theories of the causes of climatic change. Specifically, the CLIMAP (Climate, Long-range Investigation, Mapping and Prediction) project, established in 1970, was able to use deep-sea core data to test the so-called astronomical or "Milankovitch" theory (Hays et al., 1976). It had been suggested as early as the mid-nineteenth century that relatively minor perturbations in the Earth's orbit caused by the gravitational interaction of the bodies in the solar system were responsible for ice-age climatic fluctuations. Adhémar (1842) proposed that climate would follow

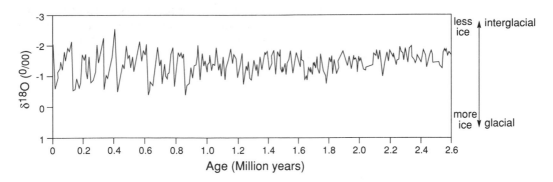

Figure 3.1   Oxygen- isotope record of Quaternary fluctuations in temperature and global ice volume, from ocean core ODP 677 (based on Shackleton et al., 1990). For most of the past one million years the dominant glacial–interglacial cycle has had a periodicity of about 100 000 years.

the 21 000 year long precessional cycle, with glaciations occurring alternately in the Northern and Southern Hemispheres – although he weakened his case somewhat by proposing that a huge tidal wave would accompany each transition! To the precessional cycle were later added the 41 000 year axial-tilt and the approximately 100 000 year orbital-eccentricity cycles (table 3.1).

It was the Serbian mathematician Milutin Milankovitch who, in the early years of the twentieth century, combined these three cycles of different periodicities to provide a composite sequence of variations in the Earth's receipt of solar radiation over the past 650 000 years (Milankovitch, 1930). Unfortunately, in the absence of adequate Quaternary climatic data, testing of Milankovitch's theory had to wait for the CLIMAP project 50 years later. Then, in a spectral analysis of paleoclimatic indicators in Indian Ocean deep cores, it was revealed that the cores included dominant periodicities almost identical to those predicted by Milankovitch (table 3.1) (Hays et al., 1976). It was announced (Imbrie and Imbrie, 1979) that the mystery of ice age changes in climate had been solved. In reality, however, many questions remain unresolved. Why, for example, did the 100 000 year cycle, which is intrinsically weak in

Table 3.1   Astronomically induced cycles of Quaternary climatic change

	Milankovitch predictions of cycle length (years)	Cycles identified by Hays et al. (1976) from deep sea cores (years)
Orbital eccentricity	100 000	100 000
Axial tilt (or obliquity)	41 000	43 000
Precession of the equinoxes	23 000 19 000	24 000 19 000

orbital forcing, suddenly become amplified about 0.8 million years ago, to replace the 41 000 year cycle which had been dominant during the early Pleistocene?

None the less, the Milankovitch theory has proved invaluable in providing critical boundary conditions for past climates (i.e. variations in incoming radiation as a function of latitude and season), notably over timespans between 10^3 and 10^6 years (Berger et al., 1984; Berger, 1992). The theory also allows us to make specific forecasts about the future natural course of climate. It is clear that we are nearing the end of an interglacial period (the Holocene), and that if allowed to run its course the Earth's climate would enter the next major glacial phase some 25 000 years from now (Berger et al., 1991). This, of course, makes no allowance for anthropogenically induced climatic warming, which looks likely to create a "super interglacial" well before temperatures fall of their own accord (Imbrie and Imbrie, 1979).

Past climates as analogues for a warmer Earth

One possible justification for studying past climates has been because of their potential similarity to a future world warmed by anthropogenic increases in greenhouse gases (Webb and Wigley, 1985). Among the potential "analogue" candidates have been the Early Medieval Warm Epoch (ca.800–1200 AD), the early to middle Holocene (9000–5000 years BP), the last ("Eemian") interglacial (ca.125 000 years BP), the late Pliocene (ca.3.8 million years) and various earlier climates as far back as the Cretaceous (ca.100 million years) (Flohn, 1977; Crowley, 1988).

The mid-Holocene (6000–5000 years BP) has been a particularly attractive candidate. At this time, coastlines, ice sheets, and other physical boundary conditions were close to modern ones, there are a wealth of well-dated paleoclimatic data available, and most of these indicators (e.g. pollen) appear to have been in equilibrium with climate (Huntley, 1990b). This time period was believed to incorporate an interval known as the Hypsithermal (or Altithermal or climatic optimum), when temperatures may have been 1.0–2.0°C higher than today. The evidence for such a Hypsithermal initially came from biotic indicators, such as the pond tortoise (*Emys orbicularis*), whose former limits in Scandinavia lay well to the north of those of today. Similarly, a poleward extension of the treeline in Canada and an upward displacement of ecological zones in the mountains of the Alps and western North America was used to imply a longer and/or warmer growing season (LaMarche, 1973; Ritchie, 1987).

Attempts at using the climate of the mid-Holocene as an analogue for a greenhouse-gas warmed Earth have been made by Kellogg (1978), Butzer (1980), Borzenkova and Zubakov (1984), and Budyko and Izrael (1987) (see Folland et al., 1990, pp. 203–6, for summary). Budyko has estimated mean annual Northern

Hemisphere temperature anomalies of approximately +1°C for the mid-Holocene, +2°C for the last interglacial and +3 to +4°C for the late Pliocene, and believes that they provide analogues for the climate in the years AD 2000, 2025 and 2050, respectively. He has also argued that the warming increased with latitude in the Northern Hemisphere, so that northern Siberia was 4–8°C warmer in summer during the last interglacial, for example, whereas the temperature in most of Russia rose by only 1–3°C. Latitudinal differences in summer temperature change were also apparent in Huntley and Prentice's (1988) study of mid-Holocene Europe, in which the Mediterranean region actually recorded a temperature decrease (figure 3.2). Some tropical areas may also have been cooler around this time on account of increased cloud cover and soil moisture.

Given that not all areas of the world experienced a significant temperature rise during the mid-Holocene, and that research is increasingly showing any warming to have been restricted to the summer months, the concept of a global Hypsithermal interval requires some revision. The enhanced seasonality and continentality

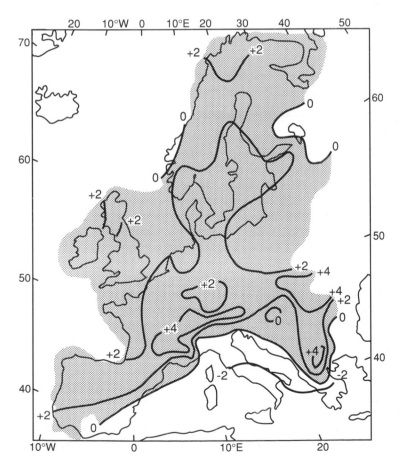

Figure 3.2 Difference in summer temperatures (Δ°C) across Europe, 6000 years BP versus present day, as estimated from pollen data (after Huntley and Prentice, 1988, copyright AAAS). Stipple shows the area which experienced higher temperatures than today during the mid-Holocene.

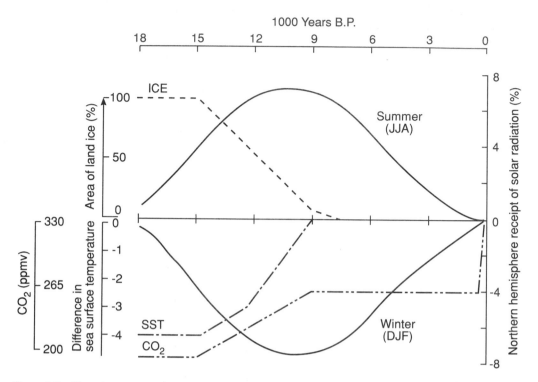

Figure 3.3 Changing receipt of solar radiation in the Northern Hemisphere due to astronomical variations, and other changes in climatic controls, since the last glacial maximum (after Kutzbach and Street-Perrott, 1985).

of the early-to-mid-Holocene also provides a clue as to the mechanism underlying these differences in climate. At the beginning of the Holocene, the Earth lay closest to the sun in July rather than January because of the 21 000 year precessional cycle. As a result, the Northern Hemisphere received almost 8 percent more solar radiation during the summer months at the beginning of the Holocene than it does today, although correspondingly less in winter (figure 3.3). These large seasonal variations stand in contrast to the radiative forcing caused by greenhouse gases in which the annual mean value rises, but with little change in seasonality. In order to compare these two types of forcing, Mitchell (1990) has run two climate model experiments with the same general circulation model (GCM), one involving a doubling of CO_2 and the other with the Earth's orbital parameters set to 9000 years BP. He found the resulting global patterns of temperature and moisture change to be quite different, and that such similarity as occurred, for example over the northern continents, was largely coincidental.

The effect of Milankovitch astronomical forcing also severely limits the usefulness of the last interglacial as an analogue for the future, while the more distant pre-Pleistocene climates were influenced by the very different continental configurations and topographies of those times, which in turn modified atmospheric and oceanic circulations. These early climates also lacked some or all of the polar ice cover which, through inertia, will persist through at least the first few centuries (and probably through many millennia) of any future global warming (Crowley, 1988). In the case of the Medieval Warm Epoch, there is considerable debate about whether a truly global warm phase occurred at all (Grove, 1988), and even if it did, its magnitude was very much less than that projected for future greenhouse-gas induced warming. In short, neither the early-to-mid-Holocene nor any of the other paleo-climates cited provides a rigorous analogue for a greenhouse-gas warmed climate, because either forcing factors (e.g. orbital varia-tions) or boundary conditions (e.g. ice cover) or both were dis-similar to those prevailing at the present day (Mitchell et al., 1990, p. 159).

Climate sensitivity and GCM validation

Paleodata compilation

General circulation models are without doubt the most important tool for predicting the future course of climate. All of the principal GCMs are able to simulate the gross features of the Earth's contemporary atmospheric circulation, but they are at present unable to reproduce many of the derived or secondary features of the climate system. For example, there is almost no agreement between the five main GCMs about soil moisture values in the Northern Hemisphere mid- and high-latitudes (Gates et al., 1990, p. 112). Parameters such as soil moisture are especially important as it is these, rather than, for example, precipitation *per se*, which determine the species composition of terrestrial ecosystems, water-resource availability and agricultural potential (Parry, 1990). If we are to plan for the consequences of climate change "on the ground," it is vital that modelled estimates of derived parameters become more reliable. Similarly, the various equilibrium GCM experiments for $2 \times CO_2$ reveal disturbing discrepancies between the behavior of different models, even though their simulations of the present climate appear equally realistic. For example, the amount of warming predicted under a $2 \times CO_2$ atmosphere diver-ges significantly according to different GCMs, with estimated increases in mean global temperature ranging from 1.9 to 5.2°C (Henderson-Sellers, chapter 5, this volume).

A promising source of information for testing the validity of GCMs under climatic conditions substantially different from to-day's is provided by the Earth's climate in the past. Paleoclimates can be used, not as direct analogues for the future, but as means of testing the sensitivity of climate to altered boundary conditions and forcing functions, and for model validation. In this exercise, past climates colder than at present may be just as useful as those from warmer epochs. To be of maximum utility paleoclimatic recon-structions should therefore be presented in a form that can be used to test GCM experiments numerically. The data on which these reconstructions are based ideally ought to have a worldwide distribution with no major gaps in coverage, and should be amenable to climatic calibration by statistical methods such as transfer functions (CLIMAP Project Members, 1984) or pollen–climate response surfaces (Bartlein et al., 1986).

However, paleoclimatic reconstruction involves a diverse array of techniques yielding results that cannot easily be integrated into a single data set. For instance, paleotemperature data are derived from glacier fluctuations, isotopic analyses, fossil pollen and Coleoptera (beetles), and former tree lines. In comparing the estimates provided by these various sources, it is difficult to know how much the resulting variations are due to climate and how much to other factors, such as different response times or an incomplete knowledge of the physics or ecology of the indicator in question. Because of this, a number of strategies have been developed for testing the results of paleoclimatic simulations, depending on the nature, coverage, and quality of the relevant data. Simple comparisons of pattern can be quite effective when the data do not lend themselves to quantitative assessments. Thus, Manabe and Hahn (1977) and Manabe and Broccoli (1985) compared the former global distribution of sand dunes and loess, which are good indicators of aridity, with the simulated anomalies of $P - E$ (precipitation − evaporation) and soil moisture, respectively, for the last glacial maximum.

A more systematic attempt to establish a proper methodology for handling paleoclimatic data began in the late 1970s with the Cooperative Holocene Mapping Project (COHMAP). COHMAP members established a global paleoclimatic database from three main sources, namely pollen and lake-level fluctuations on the continents, and marine microfossils (notably foraminifera) on the oceans. For the later part of the Quaternary, land-based paleoclimatic data often achieve a temporal precision higher than that which can be obtained from the oceanic record (e.g. pollen-based sequences of vegetation and climate from sites such as Grande Pile and Les Echets in France (Guiot et al., 1989)). The data types selected by COHMAP are widely distributed and lend themselves to the quantitative reconstruction of specific climatic parameters. Pollen diagrams or lake-level curves from individual sites and dated by radiocarbon have been used to produce data at particular time intervals from the present day back to the last glacial maximum, 18 000 years BP (Wright et al., 1993). These data can then be climatically calibrated, mapped and compared with numerical simulations of climates with boundary conditions different from those operating today.

Numerical simulations have been carried out for every 3000 year interval between the last glacial maximum and the present day using the boundary condition changes shown in figure 3.3 (COHMAP Members, 1988). At 18 000 years BP, the Earth's orbital configuration was similar to today; the differences in climate between then and now are mainly attributable to the presence of

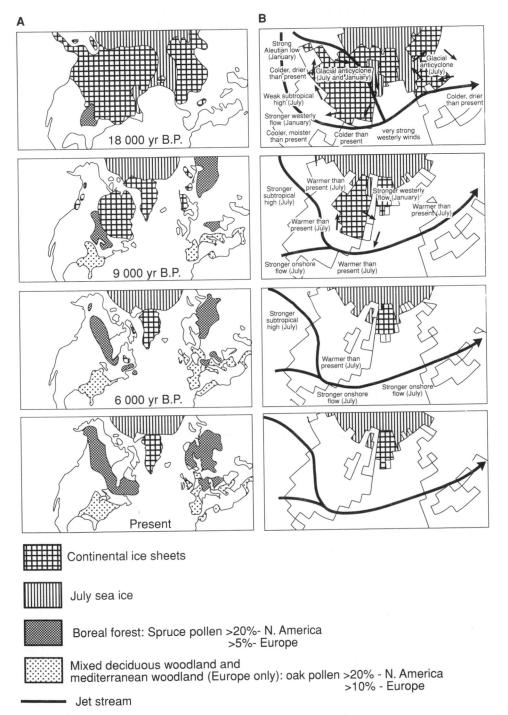

Figure 3.4 Climatic changes in the North Atlantic sector since the last glacial maximum (modified after COHMAP members, 1988, copyright AAAS). (a) Environmental conditions as observed at present or as reconstructed from geological, pollen and ocean core records; (b) numerical simulations of present and past atmospheric circulation. Boundary conditions for the GCM simulations are as shown in figure 3.3.

large ice sheets over North America and Eurasia. Glaciation resulted in a lowering of about 120 m in eustatic sea level, widespread decreases in sea-surface temperatures (SSTs), and additional increases in surface albedo owing to expanded oceanic pack ice, increased seasonal snow cover and reduced vegetation cover on land (Denton and Hughes, 1981; Fairbanks, 1989). These simulations show that the simulated bifurcation of the upper airflow around the North American ice sheet at the last glacial maximum is generally compatible with the paleoclimatic evidence (figure 3.4), giving some validation to the NCAR (National Center for Atmospheric Research) GCM which was used to produce them (Kutzbach and Guetter, 1986).

Tropical and subtropical paleoclimates

It had been recognized since the early years of the twentieth century that the Saharan, Arabian, and Thar deserts were not always the harsh, arid wildernesses they are today. Rock paintings from the Tibesti mountains of the Sahara which depict herds of cattle, gazelle and giraffe (plate 3.1), and bone remains of these and other large fauna, suggested a well-watered land which, like the modern savannas, was rich in animal life (Lhote, 1959; Petit-Maire and Riser, 1983). More recently, pollen diagrams like that from Oyo in the heart of the eastern Sahara record plants that are typical of the savanna zone 500 km to the south (Ritchie et al., 1985; Ritchie and

Plate 3.1 Neolithic rock painting from Tassili in the central Sahara, depicting rich animal life (in this case mouflon) and reflecting a landscape and climate much less arid than at present.

Haynes, 1987). Tropical plants such as *Hibiscus*, whose pollen is rather poorly dispersed, provide reliable evidence that these plants were growing there locally at the time. Pollen diagrams from West Africa and from salt lakes in Rajasthan, northwest India, indicate that the same northward shift of tropical vegetation belts took place over a broad sector of the Old World tropics and subtropics (Singh et al., 1974; Lézine; 1989; Street-Perrott et al., 1991).

It was originally believed that the "pluvial" (or wet) conditions responsible for these climatic changes in lower latitudes matched up with the glacial (or cold) periods of high latitudes – and in a few subtropical areas, notably the American Southwest, this is indeed the case (Smith and Street-Perrott, 1983). But for many more regions, particularly those within the tropics, ^{14}C dating has now shown that the last wet period occurred during the terminal Pleistocene and the first half of the Holocene (12 500–5000 years BP). At this time the Sahara could hardly be called a desert at all; rather it was a land of lakes. And lakes – or at least their remains in the form of lacustrine deposits – provide the best evidence we have for the changing climate and hydrology of the world's dry zones over the past 20 000 years. In particular, lakes lacking an outlet adjust their surface area and water depth dynamically in response to changing inputs in the form of rainfall and runoff, and to changing losses in the form of evaporation (Street-Perrott and Harrison, 1985). Lake sediments and shorelines left high and dry above the modern water surface thus testify to times when the water balance was more favorable than it is today.

A second, equally valuable, source of information about the paleohydrology of non-outlet lakes is water chemistry, for as lake levels fall the remaining water is concentrated chemically, initially becoming brackish, later fully saline, and eventually hypersaline like the Dead Sea. For this reason most arid-zone lakes are salt lakes. This spectrum from fresh to saline water not only influences the character of the sediments deposited in the lake, but also largely controls the organisms living within it, many of whose remains are preserved in the sediment of the lake bed. Among the most useful of these indicator organisms are diatoms, ostracods, and molluscs (Gasse et al., 1983; de Deckker et al., 1988; Fritz et al., 1991). Paleosalinity indicators, such as diatoms, show that many salt lakes contained fresh water during the early Holocene, and were able to support rich and varied aquatic faunas.

A range of different methods has been used to calculate early Holocene precipitation levels for individual sites or regions. For lakes, the two principal methods are the simple water balance and combined water- and energy-balance approaches. In the former, a given paleolake area is used to calculate total evaporation losses from the water surface, which can then be balanced against the input of precipitation plus runoff from the catchment (Street, 1980). In the second approach, the evaporation term is eliminated between the water- and energy-balance equations, but it is necessary to estimate certain surface properties such as albedo and

Bowen ratio, based on the former vegetation (Kutzbach, 1980). Rainfall estimates can also be obtained directly from pollen data, either through comparison with modern vegetation or, more rigorously, by means of statistical techniques such as transfer functions (e.g. Swain et al., 1983; Bonnefille et al., 1990). Finally, Street-Perrott et al. (1991) have used an area-based approach to estimate early Holocene changes in surface albedo and precipitation for a 520 000 km^2 block of land in Mali based on a paleogeographical map constructed by Petit-Maire et al. (1988).

The paleoprecipitation estimates resulting from these different techniques are shown in table 3.2. They indicate an average annual increase in rainfall of at least 250 $\pm$ 130 mm for the Sahara, East Africa, and South Asia at 9000 years BP. The northwest Indian pollen data of Singh et al. (1974) have been calibrated climatically by season using statistical techniques, and show clearly that it was summer monsoonal rainfall which was responsible for bringing the increased wetness and plant life; winter rains changed little compared with those at present (Swain et al., 1983). The suggestion that the wetter early Holocene climate was brought about by an intensification and northward displacement of the monsoonal

Table 3.2 Paleoprecipitation estimates for the early Holocene based on pollen and lake-level data

Site	Latitude	Increase in precipitation (mm year^{-1})	Remarks
Lunkaransar, India	28° 30'N	230	P
Sambhar, India	27° 00'N	>200	WE
Chemchane, Mauritania	20° 56'N	440	P
Oyo, Sudan	19° 16'N	395	P
Chad, Chad/Nigeria/Niger	13° 00'N	>300	WE
Ziway-Shala, Ethopia	7° 45'N	>450 >268	W (0) W (−2)
Bosumtwi, Ghana	6° 30'N	0	W(0)
Turkana, Kenya	5° 00'N	>110	WE
Nakuru/Elmenteita, Kenya	0° 25'S	>625 >505 >280	W(2–3) W(−2) WE
Naivasha, Kenya	0° 41'S	⩾225 ⩾130	W(0) WE
Victoria, Tanzania/Uganda/Kenya	1° 00'S	220	P
Manyara, Tanzania	3° 37'S	237	W(0)
Mean		253 ± 132	

P refers to a pollen-based estimate, W to the simple lake water-balance approach (assumed temperature change in °C given in parentheses) and WE to the combined water- and energy-balance approach. The inequality (>) indicates that the estimate is a minimum value because overflow was not taken into account. Mean is based on P, WE and more conservative W estimates.
Source: after Street-Perrott et al., 1990.

circulations is supported by the pattern of moisture change over the Old World tropics. The results of the COHMAP reconstructions for 18 000 years BP, 9000 years BP, and the present day in the Indian Ocean (i.e. monsoonal) sector are shown in figure 3.5. As

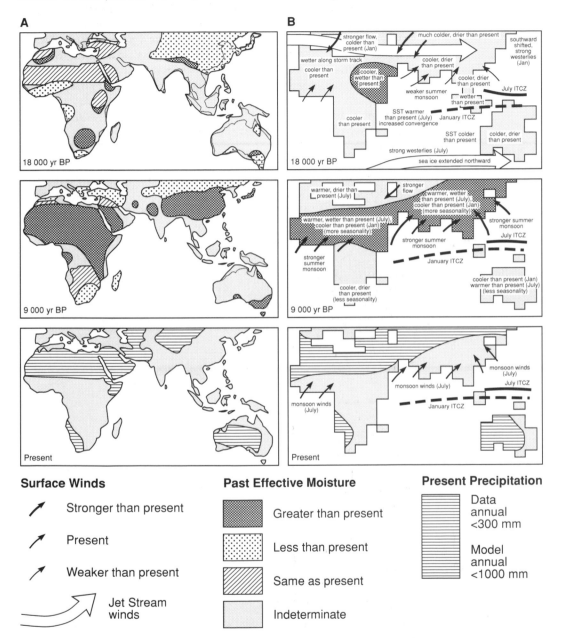

Figure 3.5 Climatic changes in the Indian Ocean sector since the last glacial maximum (modified after COHMAP members, 1988, copyright AAAS). (a) Environmental conditions as observed at present or as reconstructed from lake-level, pollen and ocean core records; (b) numerical simulations of present and past atmospheric circulation. Boundary conditions for the GCM simulations are as shown in figure 3.3. Note the weakening of the monsoon system at the last glacial maximum and its strengthening during the early Holocene.

this shows, the band of high lake levels expanded and spread northwards at 9000 years BP to cover the area between 5°S and 30°N, but was not recorded significantly to the south and west of the East African Rift.

Nor do the early Holocene monsoon rains appear to have reached the Mediterranean region with any regularity; in areas such as Turkey and western Iran pollen and lake levels suggest a drier rather than a wetter climate than at present (Roberts and Wright, 1993). The same is true of some other extratropical regions like the American Midwest, where the prairie/forest boundary has shifted backwards and forwards during the Holocene in response to changing moisture regimes (Webb et al., 1984). Figure 3.6 indicates

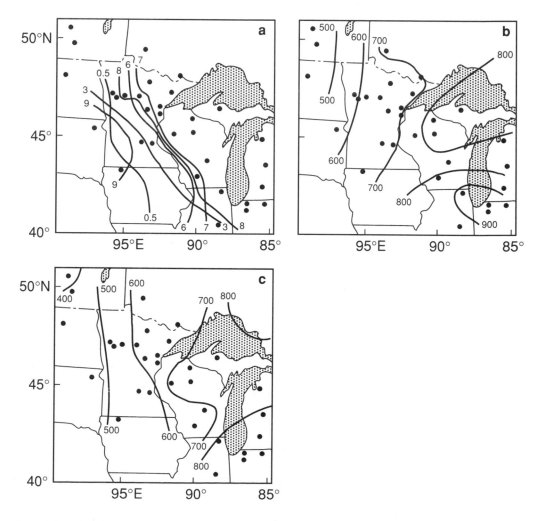

Figure 3.6 Holocene changes of vegetation and climate in the American Midwest reconstructed from pollen data (adapted from Bartlein et al., 1984). (a) Isochron map showing the former position of the prairie–forest boundary (contours are in thousands of years BP); (b) mean annual precipitation (mm) at present day; (c) estimates of annual precipitation at 6000 years BP. Dots show locations of pollen sites used in the study.

how this boundary migrated eastwards up until 7000 years BP, reflecting increasing moisture stress at the forest limit, but later retreated back westwards as the climate became more humid.

Pollen and lake-level data for the early-to-mid-Holocene serve to remind us that some of the most critical changes in future climate will relate to moisture stress and drought, rather than purely to warming (Rind et al., 1990). These paleoclimatic data are impressive, for they show that large changes in the extent of the world's drylands were associated with relatively small changes in global mean temperature. In fact, it was not the mean temperature but its seasonal distribution that modified rainfall regimes so dramatically. For example, the strength of the South Asian monsoon circulation is a function of heating of air over the Eurasian land mass; with higher summer temperatures, the South Asian low pressure area deepened during the early Holocene. This sucked in more moist air from the tropical oceans causing more intense and widespread monsoon rains. The linkage between enhanced rainfall in the northern tropics and increased summer heating of the Northern Hemisphere continents points to ultimate causation by the seasonal receipt of solar radiation.

The NCAR GCM simulations clearly show that the enhanced summer insolation at the start of the Holocene greatly intensified the northern monsoons (figure 3.5) (Kutzbach and Guetter, 1986; COHMAP Members, 1988). The simulated values for effective moisture (precipitation minus evaporation) also show good correspondence with the paleolake-level record during the past 18 000 years (figure 3.7) (Kutzbach and Street-Perrott, 1985). On the other hand, the NCAR model underestimated the severity of the continental aridity so evident from the widespread occurrence of sand dunes and low lake levels at 15 000 years BP. This suggests that other factors, such as changes in oceanic circulation, were able to

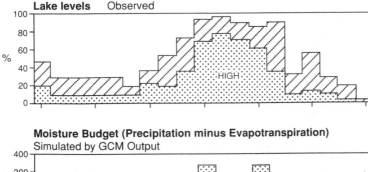

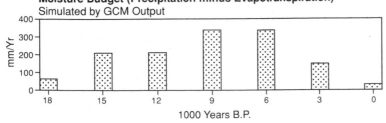

Figure 3.7 Changes in world water balance since the last glacial maximum as observed in the frequency of high water levels in non-outlet lakes (above), and as simulated by GCM output (below) using the boundary conditions shown in figure 3.3. Both show a marked increase in moisture availability during the early Holocene associated with greater seasonality in incoming Northern Hemisphere solar radiation (after Kutzbach and Street-Perrott, 1985).

override the influence of astronomical forcing during this transitional stage of deglaciation. Similarly, a lag is observed at 6000 years BP, when seasonality was much weaker than at 9000 or 12 000 years BP, but when tropical climates remained relatively wet. In this case the persistence of well-vegetated land surfaces in regions such as the Sahara may have maintained albedo values at sufficiently low values to extend the early Holocene wet period beyond its allotted time span, as determined by astronomical forcing (Street-Perrott et al., 1991).

Paleoclimate dynamics

As well as helping to test the validity of different GCM "snapshot" experiments for equilibrium climates, the study of Late Quaternary paleoclimates can also offer insights on the dynamics of climate change.

Past changes in greenhouse gases

Plate 3.2 Cores, such as this, drilled through accumulated polar ice contain a record of climate, atmospheric deposition (e.g. of dust and pollutants), and greenhouse gas concentrations.

For the last full glacial–interglacial cycle, one of the most valuable paleoclimatic records is contained in the ice sheets themselves (plate 3.2). In particular, the Vostok core, taken through more than two vertical kilometers of ice in Antarctica, provides detailed information spanning the past 160 000 years (Lorius et al., 1990). Not only has this ice core generated a paleotemperature curve (from oxygen-

isotope and deuterium–hydrogen ratios), but the changing atmospheric concentration of greenhouse gases has been reconstructed
from ancient air bubbles trapped within the ice. The resulting
sequence of fluctuations of carbon dioxide and methane matches
those in temperature rather well (figure 3.8). Thus during interglacials CO_2 levels were around 270–280 p.p.m.v. while under full
glacial conditions they fell to 190–200 p.p.m.v. While these data
strongly support a link between greenhouse gases and global
climate over long time spans, they should not be taken to imply that
the variations in CO_2 and other gases were the main cause of
glacial–interglacial shifts in climate. Rather, it is likely that the key
role played by greenhouse gases has been to amplify the Milankovitch astronomical signal and to make climate transitions sharp
rather than gradual.

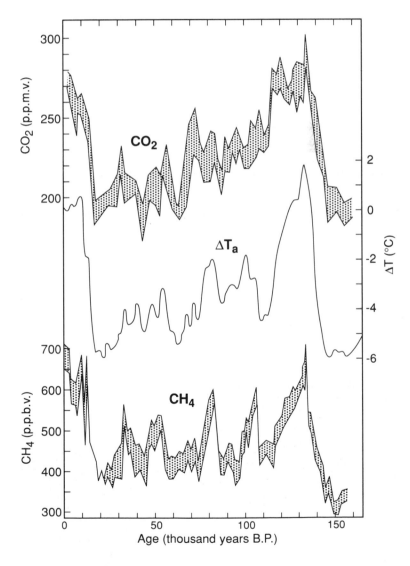

Figure 3.8 Variations in carbon dioxide, methane, and temperature in the Vostok ice core from the East Antarctic ice sheet, during the last full glacial–interglacial cycle (after Lorius et al., 1990).

While astronomical forcing appears to have acted as a "pace-maker" for climate change during the recent geological past (Hays et al., 1976; Imbrie et al., 1989), there is doubt about whether the variations in solar radiation input on their own would have been sufficient to account for the very considerable temperature fluctuations which took place during the Quaternary; (the 6–8°C range between glacials and interglacials observed at Vostok is typical of many extratropical land areas). Climate-modelling experiments confirm that that the observed glacial cooling of sea-surface temperatures, especially in the Southern Hemisphere, cannot be explained without reference to feedback effects such as variations in greenhouse-gas and aerosol concentrations (Manabe and Broccoli, 1985). If astronomical variations acted primarily as the trigger for larger-scale changes, then shifts in atmospheric composition could have provided a positive feedback mechanism amplifying the Milankovitch forcing, particularly during glacial–interglacial transitions.

The Vostok ice-core record shows that the atmospheric content of CO_2 at the last glacial maximum was 80 p.p.m.v. lower than its pre-industrial level of 270–280 p.p.m.v. This cannot be accounted for by increased carbon storage on land areas; indeed, owing to colder and generally drier conditions, storage of carbon in terrestrial vegetation and soils may have *decreased* at this time to about 0.96×10^{12} t compared with 2.3×10^{12} t today (Adams et al., 1990). The glacial lowering in atmospheric CO_2 must therefore be explained by an enhanced carbon flux from the atmosphere to the oceans, most plausibly related to increased marine productivity. Various mechanisms, such as an influx of iron in wind-borne dust (Martin and Fitzwater, 1988), have been proposed to explain greater marine plankton productivity, but empirical evidence of the assumed increase in productivity has so far been inconclusive (Mix, 1989; Mortlock et al., 1991). Nevertheless, this does clearly highlight the critical role that the oceans have played in modulating past global climate change.

Atmospheric concentrations of methane may also have been affected by oceanic processes, specifically by the burial and release of methane hydrates in deep-ocean sediments. However, in this case, terrestrial sources and sinks also appear to have been of major significance. These include the Arctic, where there exist enormous stores of methane hydrates trapped by permafrost (Sebacher et al., 1986; Nisbet, 1989) and herbaceous wetlands of the Northern Hemisphere tropics. In the latter, papyrus swamps, in particular, have very high primary productivities and rates of CH_4 emission (Street-Perrott, 1992). While both the Arctic and the semi-arid tropics experienced enormous changes as a result of glacial–interglacial climate fluctuations, the timing of hydroclimatic shifts in the latter appears to correlate the more closely with the record of CH_4 variations from Vostok (Petit-Maire et al., 1991). Fluctuations in the intensity of monsoonal circulations and in atmospheric CH_4 both show a strong precessional signal, that is, a periodicity of around 21 000 years (table 3.1). Release of methane would thus have been associated with periods of wet climate in the tropics, most notably in areas like the Sahel, where extensive swamps developed in the Sudd marshlands of the middle reaches of the Nile and around Lake Chad.

The last glacial–interglacial transition

Of all the periods of past climatic change, the last glacial–interglacial transition (18 000 to ca.6000 years BP) seems most relevant to understanding the speed and nature of global warming. Paleoclimatic data indicate that neither temperature nor rainfall increased steadily during deglaciation in line with the gradual shift in the Earth's radiation balance after 18 000 years BP. For example, Coleoptera, which respond swiftly and sensitively to climatic changes, show that temperatures in Northwest Europe rose abruptly soon after 13 000 years BP (figure 3.9) (Atkinson et al., 1987). The climate cooled briefly between 11 000 and 10 000 years BP, falling 7°C in only 50 years according to the ice-core record (Dansgaard et al., 1989), before warming again more permanently at the start of the Holocene. A similar two-step glacial-to-interglacial transition is evident in the eustatic sea-level record (Fairbanks, 1989), in North Atlantic deep-sea cores (Ruddiman and McIntyre, 1981), and in the North African lake-level record of precipitation (Gasse et al., 1990).

Feedback mechanisms may thus have not only amplified the climatic response to astronomical forcing, but also caused that response to be non-linear in character. One positive feedback

Figure 3.9 Summer and winter temperatures for Britain during the last glacial-to-interglacial transition, based on Coleoptera (after Atkinson et al., 1987). Note the existence of two periods of rapid warming separated by a return to cold conditions during the Younger Dryas stadial.

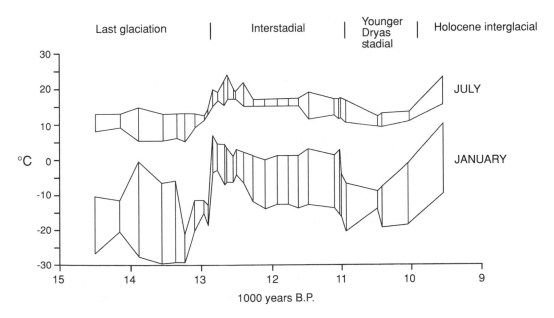

mechanism which helped to make deglaciation non-linear was the rapid release into the atmosphere of CO_2 and CH_4 (see above), while another feedback mechanism involved ice-sheet melting. Once initiated, the latter led to sea-level rise which itself caused further ice-sheet melting as a result of iceberg calving (Ruddiman and McIntyre, 1981). In the Northern Hemisphere, the northern margins of the Eurasian and Laurentide ice sheets were most prone to attack because they were marine- rather than land-based, and surrounded by floating ice shelves which were rendered unstable by rapid sea-level rise (Hughes et al., 1977). The Ross and Weddell Sea ice shelves fulfill such a protective function around the Antarctic ice sheets at the present day and would be vulnerable to similar attack if global sea levels were to rise significantly in the near future. Under certain circumstances ice sheet disintegration can be catastrophic; this was the case, for instance, with the Laurentide ice sheet when the sea broke into Hudson's Bay soon after 8000 years BP (Denton and Hughes, 1981).

The influx of cold, fresh meltwater into the North Atlantic following the collapse of the ice sheet over Hudson's Bay may have been instrumental in causing an African drought ca.8000–7200 years BP, as indicated by an abrupt fall in many lake levels at this time (Street-Perrott and Roberts, 1983; Roberts 1990). A similar lake regressional event is associated with the cold "Younger Dryas stadial," 11 000–10 000 years BP. The link between these apparently unconnected events may be found in North Atlantic deep-water formation, which has a fundamental effect on world oceanic circulation and tropical moisture flux, and which may be vulnerable to being suddenly switched off by pulses of cold freshwater from the continents (Strett-Perrott and Perrott, 1990; Broecker and Denton, 1990). In the case of the Younger Dryas event, the freshwater pulse was derived from the rapid change of meltwater from proglacial Lake Agassiz when its outlet was diverted from the Mississippi to the St Lawrence, and hence from the Gulf of Mexico to the northern North Atlantic (Broecker, 1987; Teller, 1990). Abrupt perturbations in the hydrological cycle such as those occurring around 11 000–10 000 and 8000–7200 years BP were essentially the product of contingency; that is, a particular combination of historical circumstances which may or may not be repeated (Gould, 1990, chapter 4). They interrupted the otherwise more gradual and predictable changes driven by Milankovitch forcing, and could make any future projections about climate subject to a considerable degree of error (Broecker, 1987).

Conclusions

Paleoclimates, such as those of the early-to-mid-Holocene, cannot match the boundary conditions and forcing functions associated with a modern world experiencing a rise in greenhouse gases; that is, higher temperatures but with no significant change in seasonality, major ice continental sheets such as Antarctica and Greenland still extant, and temperatures at least 1–2°C higher than those typical of the past 100 years. In consequence, they cannot provide realistic analogues for a future, greenhouse-gas warmed earth. On the other hand, past climates provide a way for modellers to validate their computer simulations of the global atmospheric circulation and to test the sensitivity of the climate system. Indeed, testing paleoclimate experiments against the geological record has come to be seen as one of the few means of verifying the ability of GCMs to simulate climates very different from today. The richness of forcing and response and the abundance of paleoclimatic data have made the past 18 000 years particularly suitable for climatic experiments.

The paleoclimatic record also informs us about the dynamical response to forcing through time of the atmosphere, biosphere, ice, and ocean, the numerical modelling of which is effectively beyond the range of existing GCMs (see Henderson-Sellers, chapter 5, this volume). It is clear that the pace of global adjustment depends critically on feedback mechanisms which can damp down or amplify minor shifts in climatic forcing. Among these are changes in atmospheric composition, which ice core records show were about 30 percent lower than their pre-industrial levels for CO_2 and about 45 percent lower for CH_4 at the last glacial maximum. Paleoclimatic evidence indicates that the last major global warming, namely the transition from glacial to interglacial between 18 000 and 6000 years BP, was characterized by abrupt shifts in both temperature and precipitation. This in turn created temporary disequilibrium between climatic, biotic and geomorphic processes (Street-Perrott et al., 1985), which, if it were to be repeated, might lead to accelerated erosion or to species extinction (Huntley, 1990b; Adams and Woodward, 1992; Roberts and Barker, 1993). If past climate change is any guide, the future response of climate to forcing by greenhouse gases will be neither simple nor easily predicted.

Further reading

Broecker, W.S. and Denton, G.H. (1990) The role of ocean-atmosphere reorganisations in glacial cycles. *Quaternary Science Reviews*, 9, 305–42.

COHMAP Members (1988) Climatic changes of the last 18,000 years: observations and model simulations. *Science*, 241, 1043–52.

Huntley, B. (1990) Lessons from climates of the past. In J. Leggett (ed.), *Global Warming. The Greenpeace Report.* Oxford: Oxford University Press, 133–48.

Kutzbach, J.E. and Street-Perrott, F.A. (1985) Milankovitch forcing of fluctuations in the level of tropical lakes from 18 to 0 kyr BP. *Nature*, 317, 130–4.

Lorius, C., Jouzel, J., Raynaud, D., Hansen, J. and Le Treut, H. (1990) The ice-core record: climate sensitivity and future greenhouse warming. *Nature*, 347, 139–45.

Mitchell, J.F.B. (1990) Greenhouse warming: is the Holocene a good analogue? *Journal of Climate*, 3, 1177–92.

CHAPTER FOUR

Historic Records and Recent Climatic Change

Mike Hulme

Introduction

One of the most important challenges confronting the science of climatology is to be able to explain changes through time in global-mean temperature and precipitation. Are they due to known natural forcing mechanisms of global climate (such as solar variability or the effects of major volcanic eruptions) or to the inherent inter-annual and inter-decadal variability of the global climate system, or are they the initial indications of the enhancement of the Earth's natural greenhouse effect due to anthropogenic emissions of greenhouse gases? In attempting to answer this question it is essential to work with time series of meteorological data covering large parts of the Earth's surface. To obtain these time series requires the study of climate history.

The human history of direct weather observations is a long one and reveals our close dependency on climate for a wide range of activities, whether they be for wealth creation, procreation, or recreation. This history can conveniently be divided into two eras: the pre-instrumental and the instrumental, the distinction being whether direct observations of weather were made and recorded by the human senses (e.g. recording a gale in a ship's log or a snowfall in a personal diary), or whether a purposely designed instrument was used to record the weather event in detail. The dividing line between these two eras is poorly defined (for example, the Greeks used wind vanes to measure wind direction, and cloud cover is still widely observed by eye), but we may take the late seventeenth and early eighteenth centuries as a convenient boundary. More recently, the instrumental era has spawned a new period in which direct satellite-based observations of weather are made. Since the early 1960s, Earth-orbiting and (slightly later) geostationary satellites have been measuring both long-established (e.g. temperature) and new (e.g. outgoing long-wave radiation) characteristics of the Earth's climate (see chapter 2, Haines-Young).

These three methods of obtaining *direct* information about weather (human senses, ground-based instruments, and satellite-based instruments) and hence about climate, are complemented by *indirect* climate information; what are called "proxy climate variables." In these instances, information is obtained either about some physical entity which is affected by climate (e.g. tree-ring width, glacier extent, or pollen composition; plate 4.1), or about some artifact or cultural practice which is influenced largely by climate (e.g. a walkway over a swamp, an oral legend of great famine, or frost fairs on a frozen river; plate 4.2). From such secondary information we may deduce knowledge about our ultimate goal, namely the prevailing climatic conditions.

Plate 4.1   These two overlapping tree ring sequences derive from different places in Ireland, but show the same "bar-code" signature related to regional climate history. The thick ring marked with arrows, for example, formed during AD 1580, which was a good year for tree growth, just before the climatic deterioration which led into the "Little Ice Age."

Plate 4.2 Frost fair on the Thames during the Great Frost of AD 1683 (detail from painting by Jan Griffier the Elder).

The further back through time we wish to reconstruct climate for a given region, the more we have to rely on non-instrumental (and therefore subjective) direct observations of weather and/or indirect proxy measures of climate. To improve further these pre-instrumental estimates of climate, a period of overlap between an instrumental series and a non-instrumental or proxy series is highly desirable to enable more accurate calibration of the subjective or proxy information. Chapter 3 has investigated the role of paleoclimates in understanding the Earth's climate system and in identifying potential future climate change. Such paleoclimate reconstructions rely very heavily on proxy methods.

In this chapter I will outline the more recent climatic history of the Earth, delving briefly into the pre-instrumental period back to AD 1000, but focusing largely on the instrumental era since the late seventeenth century. The concern is twofold. First, by reconstructing climate in the pre-instrumental era, we are able to demonstrate the extent of climate variability which occurred independently of any anthropogenic forcing; the human population prior to the seventeenth century was less than 500 million and its technological and cultural development such as to generate little interference in the natural climate system. Second, by using the more accurate instrumental observations of climate over the past 200–300 years, a quantified assessment of recent climate change can be made. This latter goal is important for at least two reasons. First, it establishes the variability of climate on the timescales of decades and centuries, timescales upon which it is reasonable for humans to plan their economic and socio-political activities. Second, and more specifically, it enables us to quantify the magnitude of any global-scale change in climate which may have occurred over this period of increasing human imprint upon the natural world. Such detailed diagnostic climate information is essential if global-scale warming which may have occurred because of the enhanced greenhouse effect is to be detected. Such detection is in turn necessary to validate climate models (see chapter 5). These currently predict that owing to the 30 percent rise in atmospheric greenhouse-gas concentration since the latter half of the nineteenth century, global-mean temperature should have risen by between 0.5 and 1.1°C (Houghton et al., 1990).

The pre-instrumental era

Although the major emphasis of this chapter is on climate change in the instrumental era, a brief overview of climate variability over the past 1000 years deduced from non-instrumental and proxy records is provided. In the pre-instrumental era it is important to distinguish clearly between "regional" and "global" climate reconstructions. One of the inherent problems of climate reconstruction using non-instrumental and/or proxy methods is that the evidence is generally restricted to specific locations or, at best, regions. Inferring global-scale climatic behavior from such spatially intermittent records is not generally possible. This is a problem shared with the earlier decades of the instrumental era in the eighteenth and nineteenth centuries, when meteorological recording stations were still restricted to Europe, North America, and perhaps some coastal regions of Africa, southeast Asia, and South America.

Climate variability over the past millennium

Two major prolonged climate fluctuations are frequently identified in the climatic history of the Earth over the past 1000 years. These may be described as the Medieval Warm Epoch centered on ca.1200, and the Little Ice Age, usually described as affecting the sixteenth to nineteenth centuries (Williams and Wigley, 1983). To these two fluctuations may perhaps now be added a third: twentieth century warming. This latter will be discussed in some detail later in the chapter, and we will limit our attention for the moment to the two major historical fluctuations. Reducing the history of climate variability over the past 1000 years to just these two climatic events may justly be criticized as over-simplifying the behavior of the Earth's climate system. The designation of these events is often based on highly selected regional data and the emphasis is very often on the behavior of temperature without considering fluctuations in other climate variables. Nevertheless, these two climate episodes provide a convenient backdrop against which to view recent natural variability in global climate.

One of the longest continuous proxy records with annual resolution of a regional climate has been constructed by Briffa et al.

Figure 4.1   Reconstructed northern Fennoscandian temperature anomalies (°C) (April–August mean) relative to 1951–70. (a) Year-by-year values, with measured temperatures smoothed with 10-year low-pass filter shown after 1876; (b) filtered with 10-year low-pass filter; (c) filtered with 30-year low-pass filter; (d) filtered with 100-year low-pass filter. (Source: Briffa et al., 1990).

(1990). This consists of mean summer (April–August) surface air temperature for northern Scandinavia for each year from AD 500 to the present, reconstructed from living and remnant Scots pine (*Pinus sylvestris*). This record is shown in figure 4.1 This

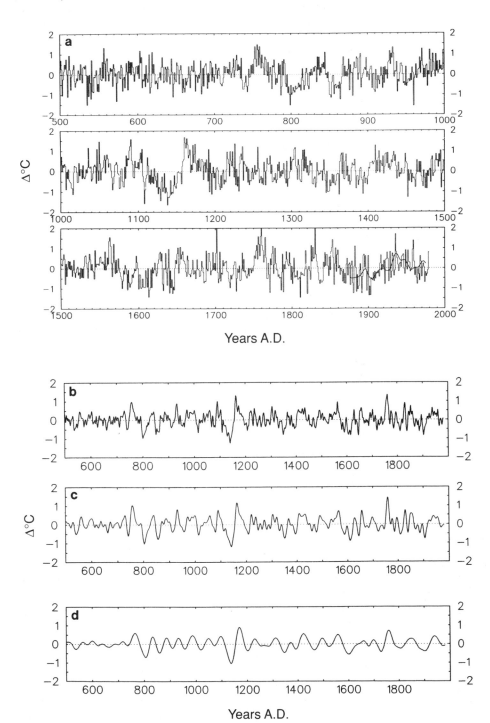

reconstruction provides a useful reference against which the climate variability of the past 1000 years can be discussed.

The Medieval Warm Epoch

The Medieval Warm Epoch is believed to have lasted, in Europe at least, from about AD 1000 to 1300 (Lamb, 1982). Substantial local and regional documentary and proxy evidence can be compiled in support of such a view. For example, the expansionary activities in the North Atlantic of the Vikings in the eleventh and twelfth centuries, the northerly expansion of viticulture in England at the time of the Domesday Book, and the altitudinal extension of tillage in some of England's moorlands, all point toward sustained conditions of warmth. The evidence is not just restricted to Europe. Tree growth of Bristlecone pines in California, cave temperatures in New Zealand derived from oxygen isotope studies, and reconstructed glacier movements over both hemispheres (Rothlisberger, 1986) also support the contention of a warmer planet during the twelfth and thirteenth centuries. Placing magnitudes and specific temporal and spatial limits on this warming, however, is much harder, although Lamb (1966) suggests that between 1200 and 1300 mean annual temperatures over central England were between 0.4 and 0.8°C warmer than during the surrounding centuries. The Scots pine tree-ring record for northern Scandinavia (figure 4.1) suggests that the peak summer anomaly in that region occurred between 1150 and 1200 and was perhaps 1°C above the 1951–70 average. This proxy time series for high latitudes shows a much more complex climatic history for the Medieval Warm Epoch than is commonly supposed, with no sustained century timescale warming and a marked cool period between 1100 and 1150. This is a good illustration of how hard it is to reconstruct global or even hemispheric temperatures from pre-instrumental or proxy sources.

The Little Ice Age

The term "Little Ice Age" seems first to have appeared in the scientific literature with Matthes in 1939, who described it as an "epoch of renewed but moderate glaciation which followed the warmest part of the Holocene" (Matthes, 1939). Again, however, it remains a poorly resolved climate fluctuation both temporally and spatially. Figure 4.1 shows that in ca.1570 summer temperatures over northern Scandinavia fell more than 0.5°C below the 1951–70 average, before recovering after 1650. Conditions were near the average from 1660 to 1750 with the following two decades being extremely warm. In contrast, the century from 1650 to 1750 is the period most frequently quoted as representing the peak of the Little Ice Age, certainly for Europe (e.g. figure 4.2), and perhaps also in the tropics (Thompson et al., 1986).

Such contrasts again highlight the non-uniform spatial manifestation of recent global climate fluctuations. This should lead us to argue that such differences in regional climatic history are a characteristic of periods of global climate fluctuations rather than

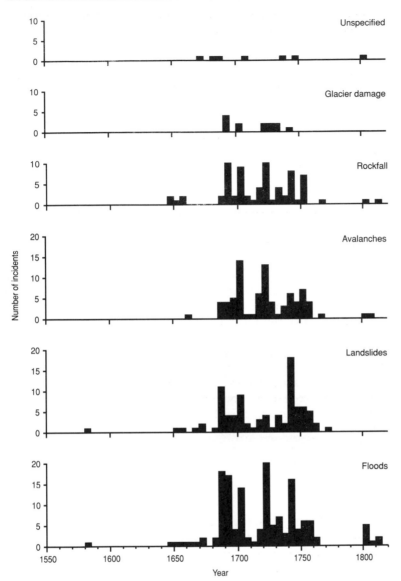

Figure 4.2 The incidence of mass movements and floods, over five-year periods, in the Norwegian parishes of Oppstryn, Nedstryn, Loen, and Olden revealed in *avtak* records. (Source: Grove, 1988).

to doubt that global-mean temperature fluctuations of up to ± 0.5°C have occurred within the past millennium. Indeed, the twentieth century instrumental record shows that spatially complex and contrasting temperature trends clearly *do* occur within periods of sustained global-scale temperature anomalies.

In view of these regional contrasts, how might the global temperature anomaly associated with events like the Little Ice Age be quantified? One approach addresses the possibility that such fluctuations might be attributable to variations in solar output. Solar output varies on many timescales, the most widely known one being associated with the 11-year sunspot cycle. Recent satellite measurements have demonstrated that received solar radiation at the top of the atmosphere in fact varies by only $\pm\,0.2$ percent during a full sunspot cycle. There are also proposed longer-term variations in solar output associated with changes in the amplitude of the sunspot cycle itself. The best documented of these is the so-called Maunder minimum; between 1645 and 1715 sunspot activity was so low that the 11-year sunspot cycle was virtually undetectable (Eddy, 1976).

To attempt to quantify the relationship between solar variability and global-mean temperature requires a two-stage analysis. First, the link between global-mean temperature and glacier advance/retreat is established, and, second, the link between solar output and isotope ^{14}C production in the upper atmosphere is determined. The hemispheric and global record of glacier movements over the past 10 000 years compiled by Rothlisberger (1986) and the instrumental global-mean temperature record of the past 140 years of Jones et al. (1986) enable an estimate to be made of the magnitude of Little Ice Age global cooling (Wigley, 1988). If this 0.4°C cooling was associated with the Maunder minimum between 1645 and 1715, then using ^{14}C as a proxy for solar output (Stuiver and Quay, 1980), the Little Ice Age can be shown to represent a decline in solar irradiance of about 6 W m^{-2}, or a reduction in net radiative forcing at the top of the troposphere of about 1 W m^{-2}. The absolute magnitude of this net tropospheric forcing is substantially less than the estimated net forcing due to greenhouse-gas emissions between 1765 and 1990 of +2.2 W m^{-2}, and the projected additional greenhouse forcing between 1990 and 2030 of +3 W m^{-2} (Wigley and Kelly, 1990). If this analysis of Little Ice Age global climate is close to being correct, then the enhanced greenhouse effect is likely to swamp any future climate fluctuation which may occur as a result of natural solar variability.

Tropical climates

Most reconstructions of the Earth's climate over the past millennia focus on surface temperature. When considering recent climate variability in the tropics, precipitation is a far more appropriate variable to consider. The relatively low seasonal and inter-annual variability of temperature there, together with the greater constraint on human activity imposed in these latitudes by moisture availability, means that attempts to reconstruct historical tropical climates should address changes in precipitation and water balance.

The discharge of the River Nile is one of the best proxy measures of tropical climate over recent centuries. The records of annual high and low water levels at the Roda Nilometer at Cairo were collected together by Tousson (1925) (plate 4.3). The series is practically complete from AD 641 to 1451, but patchy thereafter until the nineteenth century. Various problems exist with this record, although these have been addressed by numerous authors (e.g. Popper, 1951). It is generally accepted that the early summer, or low, flood level reflects precipitation falling over the Equatorial

Plate 4.3 The Roda Nilometer at Cairo was used to record the flood height in the Nile, and provides a time series, unequalled in length, of river discharge and rainfall over the Ethiopian Highlands.

Lakes Plateau (i.e. around Lake Victoria), whereas the late summer, or high, flood level responds to precipitation supply over the Ethiopia Highlands. Hassan (1981) has performed a detailed anlaysis on this time series and demonstrates generally high late summer floods in 1351–1470, 1737–70, 1850–1900, and 1950–64, which corresponds roughly with periods of glacial advance in northwest Europe. Unfortunately, the Roda records are sparse for the main periods of Little Ice Age glacier advance in the final decades of the sixteenth and seventeenth centuries.

Elsewhere in Africa, Maley (1977) identified high and low stages of Lake Chad using radiocarbon dating. He distinguishes a high level of the lake from 1550 to 1650 which he attributes mainly to an increase in the discharge of rivers entering the south of the Sahel zone from equatorial Africa. Using documentary, oral and some physical evidence, Nicholson (1978) has thrown more light on the precipitation variations of recent centuries in the West African Sahel. Relatively wet conditions prevailed from the ninth through fourteenth centuries followed by drier conditions in the fifteenth century. The sixteenth and seventeenth centuries were generally wet, in agreement with the Lake Chad record, but the period 1680–1830 was again drier with particularly dry decades in the 1730s and 1740s (figure 4.3). We will see later how the instrumental record for this region since 1900 continues this sequence of wet and dry phases.

Figure 4.3 Chronology of famine and drought in Chad, Senegambia, the Niger Bend and northern Algeria, 1500–1900. The relative level of Lake Chad is shown at the top (source: Nicholson, 1978).

The precipitation record for the West African Sahel differs considerably, therefore, from the record for Northeast and East Africa derived from the Roda Nilometer (T. Evans, 1990). The historical climatic sequences of these two tropical regions seem quite out of phase, with the late Pleistocene synchronism of cold mid-latitudes and dry tropics (Roberts, 1989) seemingly reflected more in the Sahel than in East Africa.

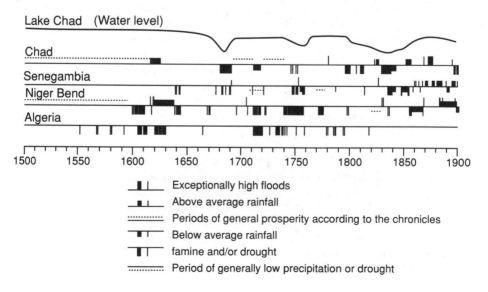

The instrumental era

Most of the basic ground-based meteorological instruments had been invented by 1700, and during the eighteenth century these came into regular use for daily weather observations in Europe and eastern North America (Lamb, 1977). The longest continuous station records commence in the early 1700s, but the extension of routine weather observations to sufficient regions of the Earth to enable hemispheric and global estimates of surface temperature and precipitation to be made did not occur before the late 1800s. The Vienna Meteorological Congress of 1873 acted as a catalyst for many countries to establish their own meteorological agencies.

The advent of a new generation of meteorological instruments placed on satellite platforms in the 1960s, and their continued development into the 1990s, has now given the potential for "true" global estimates to be made of a much wider variety of climatic variables than previously. For example, monthly estimates of Northern Hemisphere snow cover and Antarctic sea ice extent are now made routinely by satellite and exist back to 1973 (NOAA, 1990), and considerable effort is currently being made to integrate satellite and ground-based precipitation measurements to generate "true" global estimates of monthly precipitation (Rudolf et al., 1991). One difficulty with using these satellite-derived variables in climate change studies is the restricted length of the resulting time series. Much effort will therefore be required for the foreseeable future to increase the length of instrumental climatic time series for climate change detection purposes.

The longest instrumental records

One of the essential requirements of instrumental records if they are to be used for climate change detection purposes is that they remain *homogeneous* through time. That is, the records must reflect *only* day-to-day changes in weather and longer-term changes in climate. Factors such as the relocation of the station site, changed observing practices, and changes in instrumentation or local environmental conditions, can all cause inhomogeneities in station records which must be eliminated or corrected for the record to be useful in climate change detection. This question of homogeneity will surface several times in the following sections.

Station time series

Table 4.1 lists some of the stations with the longest continuous records of monthly mean temperature and precipitation. Nearly all these stations are located between 40°N and 64°N. The longest Southern Hemisphere temperature record is from Rio de Janeiro, which commenced recording in 1832. Ensuring the homogeneity of these records is usually accomplished by comparing their values with those of neighboring stations. For the very early years of these

Table 4.1 List of some early instrumental temperature and precipitation records

First year	Site	Country	Latitude	Longitude
Temperature				
1701	Berlin	Germany	52.5	13.4E
1706	de Bilt	Netherlands	52.1	5.2E
1743	Boston	USA	42.2	71.0W
1743	St Petersburg	Russia	60.0	30.3E
1753	Geneva	Switzerland	46.2	6.2E
1756	Stockholm	Sweden	59.4	18.1E
1757	Paris	France	48.4	2.5E
1761	Trondheim	Norway	63.4	10.5E
1763	Milan	Italy	45.4	9.2E
1768	Copenhagen	Denmark	55.7	12.6E
Precipitation				
1725	Padua	Italy	45.4	12.0E
1735	Hoofdorp	Netherlands	52.3	4.7E
1738	Charleston	USA	32.8	79.9W
1740	St Petersburg	Russia	60.0	30.3E
1748	Lund	Sweden	55.7	13.2E
1748	Marseille	France	43.3	5.4E
1764	Milan	Italy	45.4	9.2E
1770	Paris	France	48.8	2.5E
1770	Seoul	South Korea	37.3	127.0E
1774	Uppsala	Sweden	59.9	17.6E

Source: Jones and Bradley, 1992

time series this obviously is not feasible and careful assessment of the siting, instrumentation, and recording procedures can only be a partial alternative (e.g. Schaake (1982) for the homogenization of the long Berlin temperature record).

Regional time series
While individual station records are useful in their own right, it is often of greater significance to investigate climatic changes over larger space-scales than can be represented by single stations. Some examples are provided of long regional climate time series which use different methods to aggregate individual station time series.

The longest, and perhaps most well known, such regional climatic record is the temperature series constructed by Gordon Manley (1953, 1974). Known as the Central England Temperature (CET) record, this consists of a time series starting in 1659 of monthly mean surface air temperatures broadly representative of the English Midlands. The temperatures during the early decades of this record are accurate only to the nearest 1°C (to 1670) and 0.5°C (to 1720) and are based on a variety of short temperature records and journal entries for locations in the English Midlands. One of the earliest daily temperature sequences, for example, is from March 1666 to May 1667, recorded at Oxford by John Locke. Since 1721 the CET is regarded as accurate to ± 0.1°C, and from

1815 Manley used the average of the Radcliffe Observatory, Oxford (plate 4.4), and a value representative of several stations on the Lancashire Plain. This latter derivation is still broadly followed today.

A parallel CET series of mean daily temperatures from 1772 has also been constructed (Storey et al., 1986). Although Manley died in 1981, the daily and monthly series are now regularly updated by the UK Met Office, currently using temperatures from Rothamsted, Malvern, and the average of Squires Gate (Blackpool) and Ringway (Manchester). Recent work on assessing the homogeneity of the monthly and daily CET record has been completed by Legg (1989), who recommended an adjustment of $-0.1°C$ to allow for urban warming effects since the 1940s, especially around Malvern.

The CET record has been analyzed by numerous authors (e.g. Probert-Jones, 1984). Average annual mean temperature over the past 30-year period is about 0.5°C above the full 330-year mean, and almost 1.0°C warmer than the coldest decades at the end of the 1600s (figure 4.4). It should be noted from the earlier discussion about the Medieval Warm Epoch that the period covered by the CET is, however, probably cooler than the mean of the past 1000 years. The recent warming in the CET record has been more evident in winter than in summer. The Little Ice Age seems largely restricted to the seventeenth century (the coldest decade was the 1690s), although summer temperatures show little evidence of any secular trend. It is interesting to note that the single warmest year in the CET record occurred in 1990 (mean annual temperature 10.65°C) and the warmest pair of years was 1989 and 1990.

An equivalent monthly time series representing England and Wales precipitation was constructed by Nicholas and Glasspole

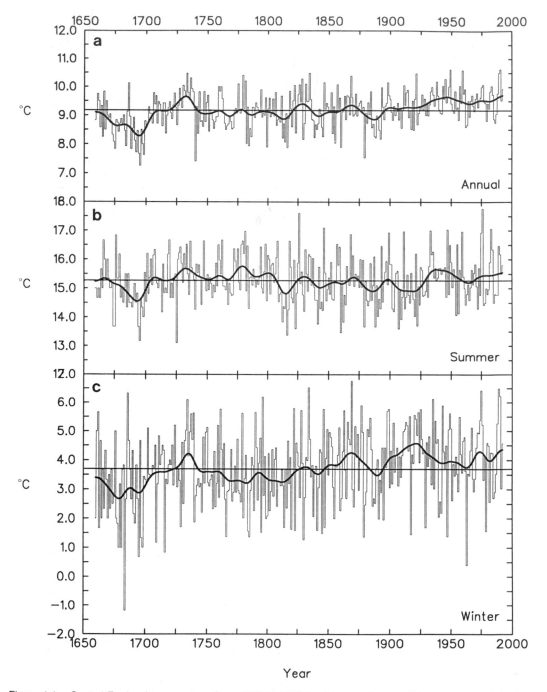

Figure 4.4 Central England temperature from 1659 to 1992 for (a) annual mean, (b) summer and (c) winter seasons. Smooth curves are Gaussian filters which suppress variations of less than 30 years. Horizontal lines represent full record means (adapted from Manley, 1974, updated by the UK Met Office).

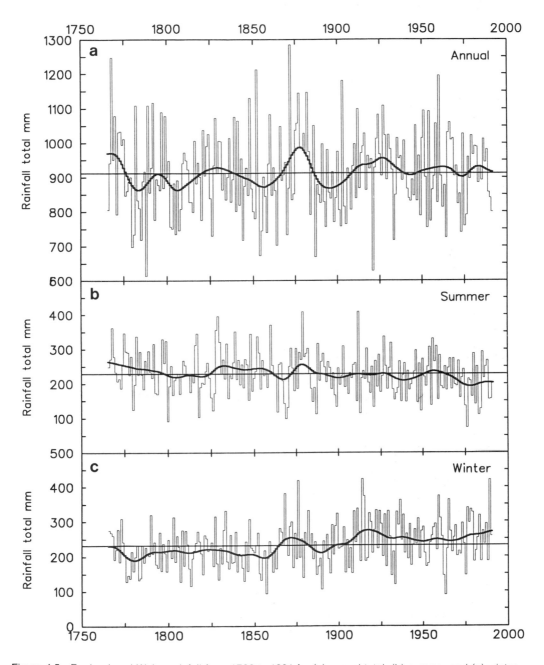

Figure 4.5 England and Wales rainfall from 1766 to 1991 for (a) annual total, (b) summer and (c) winter seasons. Smooth curves are Gaussian filters which suppress variations of less than 30 years. Horizontal lines represent the full record means (adapted from Wigley et al., 1984, and updated).

(1931). This record commences in 1727, although in its early decades estimates are based on fewer than ten gauges. This is a rather small number of gauges for determining regional precipitation with its high spatial variability, and other problems of homogeneity in the Nicholas and Glasspole record have been revealed by Wigley et al. (1984). These latter authors calculated instead their own homogeneous England and Wales monthly precipitation record commencing in 1766 and based on up to 35 gauges spread evenly over the region. This record is updated routinely and probably represents the longest homogeneous regional precipitation record in existence. In contrast to Manley's CET, there is little evidence of a long-term trend in annual precipitation since 1766, although the seasonal totals suggest some tendency toward wetter winters and drier summers (figure 4.5). The record reveals a 30-year variability of about ± 10 percent (annual) or ± 20 percent (seasonal).

As shown by table 4.1, few station records in the tropics commence before the late 1800s. Regional climate series for tropical latitudes based on instrumental data therefore commence later than those described above for mid-latitudes. Two of the longest regional precipitation series for the subtropics are those representative of the Indian sub-continent (from 1871) and of South Africa (from the late 1880s). Work in the latter region has been performed largely by Tyson (e.g. 1986), and the authoritative precipitation record described there is based on over 40 gauges in the late 1880s rising to over 150 gauges through the twentieth century. The record shows a remarkable regular 18-year oscillation in annual precipitation with dry phases around 1913, 1932, 1950, 1970, and 1986. This oscillation also affects other countries in southern Africa, although to a lesser extent.

Global data sets and time series

Numerous global climatologies of temperature, precipitation, radiation, and other surface variables have been compiled in recent years from instrumental records. The majority of these climatologies use monthly station means based either on a specified 30-year period (e.g. 1931–60) or on all available data irrespective of length or period. For the purposes of climate monitoring and climate change detection, such mean climatologies are of little value; rather, time series of these variables are required for as long a period as possible. One of the largest global data sets of station time series has been compiled by the Climatic Research Unit at the University of East Anglia in collaboration with the University of Massachusetts (Bradley and Jones, 1985). Monthly mean temperature and monthly precipitation time series for, respectively, just under 4000 and over 8000 stations worldwide have been accumulated, with the longest time series commencing in the early 1700s.

Figures 4.6 and 4.7 give an indication of the spatial and temporal extent of this data set. For determining global-scale climate change in the instrumental era such global data sets are essential, yet the

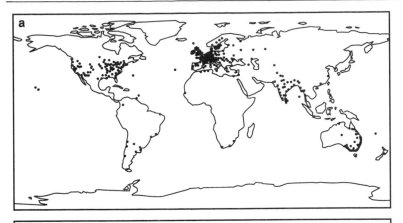

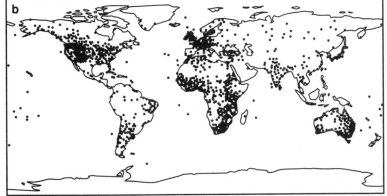

Figure 4.6 Distribution of stations containing monthly precipitation values for 90 percent of months between (a) 1871 and 1900, and (b) 1931 and 1960, in the Jones et al. (1985, and updated) data set.

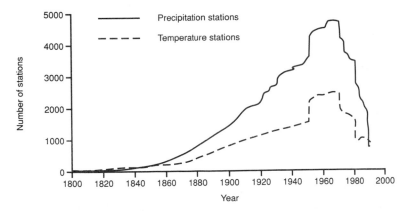

Figure 4.7 Annual frequency of stations with temperature and precipitation data present 1800–1990 in the Jones et al. (1985, and updated) data set.

problems of uneven station coverage (e.g. figure 4.6) and ensuring regular updates of these time series from around the world (e.g. figure 4.7) are clearly evident. Although some of the station time series commence in the eighteenth century, by 1800 only 40 temperature and 18 precipitation stations are included, these totals

Table 4.2 The number of stations with temperature and precipitation records in the global station data set of Jones et al. (1985, and updated) at 50-year intervals for the world, and for Northern and Southern Hemispheres

	Precipitation stations			Temperature stations, global
	NH	SH	Global	
1700	1	—	1	–
1750	8	—	8	4
1800	18	—	18	40
1850	169	5	174	154
1900	1063	363	1426	774
1950	2253	1380	3633	1543
1975	2289	1445	3734	1789

rising to 774 and 1426 by 1900 (table 4.2). These earliest records are heavily biased to Europe and North America (figure 4.6a), creating particular problems when constructing historical global estimates of global temperature and precipitation.

The global temperature record

The longest and most comprehensive assessment of global-mean surface air temperature based on instrumental records has been that undertaken by Jones et al. (1986; see later updates, Jones, 1988a; Jones and Wigley, 1991). The resulting time series of monthly mean hemispheric and global temperature commences in 1854 and combines data from nearly 2000 land-based stations (Jones et al., 1985) with millions of individual ship observations of marine air temperature and sea-surface temperature (Bottomley et al., 1990). The only comparable large-scale analyses have been performed by Hansen and Lebedeff (1987) and Vinnikov et al. (1990), although both of these temperature assessments were based on land records only, and in the latter case only for the Northern Hemisphere.

Constructing a reliable and accurate global temperature record
In the global station data set described above, homogeneity was assessed by comparing each station's records with those of stations from a few tens to a few hundred kilometers distant. Jumps or trends in the temperature recorded by one station that were not reflected at the others were regarded as signs of inhomogeneity. This homogeneity screening reduced the number of usable station time series to just under 2000. These records, together with the screened marine data set compiled at the UK Met Office (Bottomley et al., 1990), were combined to generate temperature time series on a regular latitude/longitude grid of resolution 5° by 10°. Gridding the data in this way eliminated any bias in hemispheric and global means which might arise due to the concentration of station records in certain regions. Before gridding, all temperatures were converted to anomalies (from the 1951–80 mean) which prevented spurious

warming or cooling trends arising owing to changing station networks.

Two remaining major potential sources of inhomogeneity in the gridded global temperature time series were a warming bias in the record due to urban warming effects (Karl and Jones, 1989), and substantial changes in the measurement procedure for temperature observations on board ships (Farmer et al., 1989). These two issues are discussed briefly in turn.

It is well known that urban areas generate their own heat, which results in urban temperatures frequently being detectably higher than those of surrounding rural areas (Oke, 1978). Some studies have suggested that such "heat islands" are detectable for urban centers with populations as low as 5000 (Landsberg, 1981). As a substantial number of the land-based temperature records are located in or around urban areas, many of which have generally expanded over the last century, the possibility exists that part of any global-mean warming trend may be due to this non-climatic cause. In response to criticism of the global temperature record (e.g. Balling and Idso, 1989), Jones et al. (1990) have recently estimated the likely upper limit of the magnitude of this urban warming bias in the global record. The temperature time series for four major regions (USA, western USSR, eastern China, and eastern Australia) based only on rural stations were compared with the temperature trend derived from the full station network for these regions. It was concluded that the upper limit of the urban warming bias in the global-mean temperature record over the past 100 years is 0.05°C, and it may be as low as 0.01°C.

The second potential source of inhomogeneity results from changes in the observational practice of marine temperature measurements. Although an international agreement on standardizing marine meteorological observations was signed by the major maritime nations in Brussels in 1853, differences in marine temperature observing practice continued well into the present century (Folland and Parker, 1990). The most serious difference was in the type of buckets which ships used to obtain the sample of surface ocean water in which thermometers were stood (sea-surface temperature (SST) is a good approximation for surface air temperature because of the uniformity of the boundary layer over large oceans). Although the 1853 Brussels agreement specified that wooden buckets should be used (these are good insulators and would prevent significant evaporation occurring from the water sample), many ships continued to use tin, other metals, and especially canvas buckets for many years. The last in particular result in lower temperatures being obtained owing to the evaporative cooling which occurs when the bucket is left standing on deck for several minutes. This cooling effect has been shown experimentally to be as much as 0.7°C under certain meteorological conditions (Folland et al., 1984), although it is usually much less than that. A substantial effort has therefore been made to correct marine temperatures for this effect (e.g. Farmer et al., 1989).

The consequence of unknown bucket designs in the late 1800s leads to an uncertainty of about 0.2°C in the global-mean ocean temperature up to the turn of the century (Folland et al., 1990). Nevertheless, comparisons of the land and marine temperature time series demonstrate a remarkable parallelism. Even on an annual timescale the hemispheric averages over land and ocean are strongly correlated; for longer period fluctuations the two almost completely agree. The excellent correlation between the independent Hansen and Lebedeff (1987) land-based global temperature record and the combined land and marine record of Jones et al. (1986) corroborates such a conclusion (Folland et al., 1990).

Describing and explaining recent global temperature changes
The most widely accepted global-mean temperature curve is that shown in figure 4.8. The curves are shown from 1854, although owing to rather sparse coverage the estimates before 1881 should be treated with caution (Jones et al., 1986). There are a number of conclusions to be drawn from this important indicator of the Earth's climate over the past 13 or 14 decades. Overall the planet has warmed at the surface by 0.45 ± 0.15°C since the middle of the nineteenth century. This warming, however, has been continuous neither through time nor over space. There have been two periods

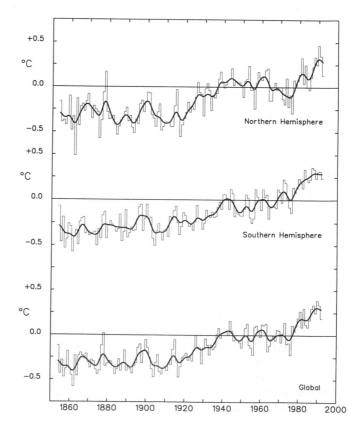

Figure 4.8 Global and hemispheric time series of mean annual surface air temperature from 1854 to 1992. Smooth curves are Gaussian filters which suppress variations of less than 10 years. All values are anomalies from the 1951–80 average. Both land and marine data are included (source: P. D. Jones, Climatic Research Unit).

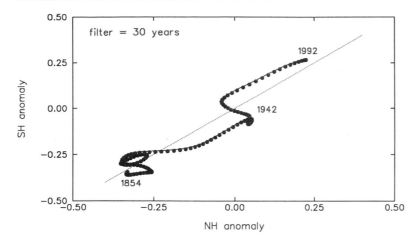

Figure 4.9 Northern Hemisphere (NH) versus Southern Hemisphere (SH) mean annual surface air temperature for 1854 to 1992. Data are smoothed with a 30-year padded Gaussian filter. Dots indicate the mid-year of the filter period. Where the line is near-vertical (horizontal) the Southern (Northern) Hemisphere warms while the Northern (Southern) Hemisphere temperature remains constant. If the two hemispheres had warmed in unison the line would be a diagonal from lower left to upper right as shown by the dotted line (data from figure 4.8).

of relatively rapid global warming (from the 1910s to the 1940s, and again since the mid-1970s to date), preceded respectively by two or three decades of fairly constant (1860s to 1900s) or slightly declining (1940s to 1970s) temperature.

The spatial discontinuity of this overall global-scale warming is evident even when one disaggregates to the hemispheric scale: the two hemispheres of the Earth have not warmed or cooled in unison. This is illustrated in figure 4.9, which plots the two hemispheric temperature time series on opposing axes. What one sees are periods where only the Northern Hemisphere (1911–30) or Southern Hemisphere (1950–70) is warming. This inter-hemispheric asymmetry of global temperature is an important diagnostic feature of the temperature record of the past 140 years and has led to consideration of the role of sulphate aerosols, which have different hemispheric loadings, in global climate regulation (Wigley, 1989). If the temperature data are examined on smaller regional scales, the spatial and temporal mosaic of warming and cooling becomes even more intricate. For example, between 1931–60 and 1961–90 much of the North Atlantic and northwest Europe experienced a cooling of about 0.3°C, whereas a large area of continental Russia warmed by over 0.6°C (figure 4.10a).

How well does the above temperature record compare with other assessments of global-mean air temperature? The other main compilation of global-mean temperature from ground-based instruments has been made by Hansen and Lebedeff (1987), although their domain is land only. Over the period 1881–1980 the correlation in the annual means between these two data sets is +0.93; an excellent agreement which suggests that although the land area represents only about 30 percent of the Earth's surface, air temperature variations over the land are a good indicator of the planetary air temperature variations. Additional corroborative evidence for this global warming comes from the proxy record of European land glacier movements over the past 300 years (figure

Figure 4.10 Global
change between 1931–60
and 1961–90 in (a) mean
annual surface air
temperature (°C), and (b)
annual precipitation
(percent). Areas with
insufficient data remain
unshaded. (Source:
Hulme and Marsh, 1990).

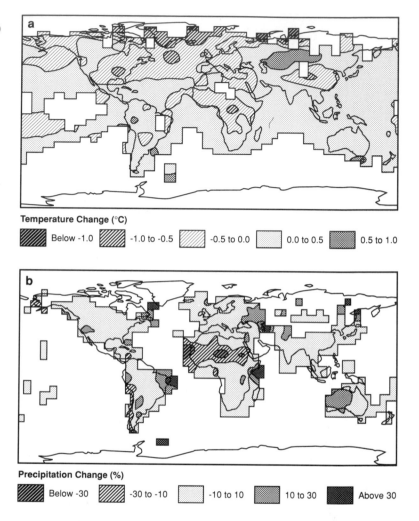

4.11). Of the glaciers which have been monitored over that time, there has been a clear majority which have retreated since the eighteenth century (Grove, 1988).

More recently, a record of global-mean temperature has been constructed using data derived from satellite-based radiometers rather than from surface thermometers (Spencer and Christy, 1990). This record is obtained from NOAA-6 and NOAA-7 satellites and is therefore available only for the period since 1979. Furthermore, because of the sensitivity of the microwave sounding units (MSUs) installed on these satellites, these satellite-derived estimates are measurements of mid-tropospheric temperature between 850 and 500 hPa rather than of surface air temperature. Nevertheless, the correlation between the monthly values of the surface and satellite-derived time series over a 10-year period is still good considering their different domains (table 4.3). The correla-

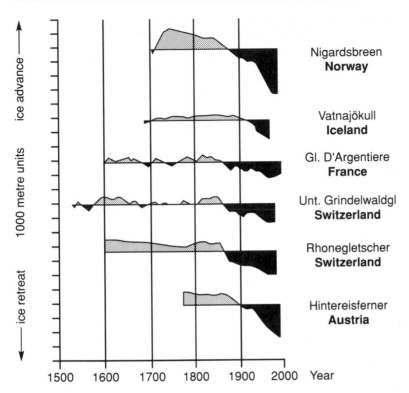

Figure 4.11 Variations of some selected glaciers as measured by their length. (Source: Houghton et al., 1990).

tions rise from only +0.33 when land-only monthly temperatures for the Southern Hemisphere are correlated, to +0.93 when annual land and marine temperatures for the whole world are compared.

Although the question of what has caused global temperature changes in recent centuries cannot as yet be answered with complete certainty (Wigley and Barnett, 1990), the effects of some of the main forcing mechanisms can be quantified and thereby

Table 4.3 Monthly, seasonal, and annual correlations between the mid-troposphere temperature derived from MSU satellite and the surface air temperature estimated from ground stations, for the two hemispheres and for the globe

Surface air temperature data set		NH	SH	Global
Monthly	Land	0.48	0.33	0.56
	Land and marine	0.50	0.45	0.62
Seasonal	Land	0.62	0.56	0.69
	Land and marine	0.66	0.64	0.76
Annual	Land	0.82	0.85	0.86
	Land and marine	0.84	0.77	0.93

All correlations calculated over 1979–88.
Source: Jones and Wigley, 1990

Figure 4.12 Schematic
representation of some of
the main forcing
mechanisms on global
temperature in terms of
their timescale (horizontal
axis) and their likely
magnitude of global-mean
temperature change
(vertical axis). Scales are
logarithmic.

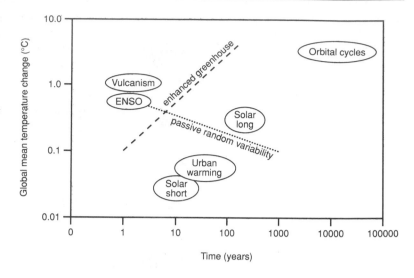

Figure 4.12 Schematic representation of some of the main forcing mechanisms on global temperature in terms of their timescale (horizontal axis) and their likely magnitude of global-mean temperature change (vertical axis). Scales are logarithmic.

eliminated as likely causes of the observed warming. Some of these forcing mechanisms are represented schematically in figure 4.12 in terms of both their timescale of operation and their likely magnitude of resultant global-mean temperature change. Changes in the orbital parameters of the Earth have not been discussed in this chapter (see Street-Perrott and Roberts, chapter 3) since they operate on very long timescales. The El Niño–Southern Oscillation (ENSO) phenomenon is included as a separate forcing mechanism since it represents the most dramatic component of the natural variability of the coupled ocean–atmosphere system and leads to marked fluctuations in tropical, and to a lesser extent global, air temperature and precipitation.

If the variability in the global-mean temperature record is divided into low-frequency (timescale >10 years) and high-frequency variability (<10 years), the respective standard deviations over the period 1861–1988 are 0.15°C and 0.08°C (Wigley and Raper, 1990). About a third of this high-frequency variation can be identified with the ENSO phenomenon (Jones, 1988b). A small fraction of the residual high-frequency variability is due to the effects of volcanic eruptions, which have an identifiable cooling signal in the global temperature record, but lasting only a few years (Kelly and Sear, 1984; Hansen and Lacis, 1990). Also causing high-frequency variability in global-mean temperature are high-frequency oscillations in solar output. As we have seen, the output of solar radiation is not constant over time. The past 12 years have seen accurate measurement of received solar radiation at the top of the atmosphere from satellite platforms (NIMBUS 7, the Solar Maximum Mission). These measurements have now monitored a complete 11-year sunspot cycle and they show that on this time-scale solar variability is only of the order of ± 0.2 percent, inadequate to alter global-mean temperature by more than 0.1°C

(Hansen and Lacis, 1990). The remaining high-frequency variation is probably due largely to the internal variability of the atmosphere, manifest at least partly as global-scale changes in the cloud radiation budget (Ramanathan et al., 1989).

Because of the thermal inertia of the oceans, these high-frequency variations will enhance the low-frequency variability of the climate system and, without any human intervention, generate detectable decadal and centennial timescale variations. Such low-frequency variations may be called the internal "noise" of the ocean–atmosphere system and they make attributing any given temperature change to an external cause problematic. Nevertheless, attempts to model the ocean thermal inertia effect have resulted in an upper limit of $\pm 0.3°C$ being placed on trends generated by this "passive internal variability" on a century timescale (Wigley and Raper, 1990).

From the above discussion we may deduce that the observed global warming over the past 140 years of $0.45 \pm 0.15°C$ is unlikely to be explicable by ENSO events, vulcanism, decadal-scale solar variability, or internal variability within the climate system. Furthermore, it was earlier suggested that century timescale variability in solar output, although a possible cause of the Little Ice Age temperature fluctuation, is unlikely to exceed the climate forcing anticipated in future decades owing to the rising concentration of atmospheric greenhouse gases. The revelations of the instrumental global temperature record over the past 140 years are therefore broadly consistent with the theoretical predictions of the enhanced greenhouse effect.

The instrumental global precipitation record

Surface precipitation is the other meteorological variable of greatest concern for the changing global environment. As with temperature, global-scale assessments of precipitation change based on instrumental data are restricted to the past 100–150 years. However, two differences from the global temperature record need to be recognized. First, there is no marine precipitation data equivalent to the marine temperature data set. Although data sets such as the Comprehensive Ocean–Atmosphere Data Set (COADS; Woodruff et al., 1987) contain historical estimates of monthly rainday frequencies along shipping routes, the conversion of these to actual precipitation totals is problematic. Consequently, global precipitation change assessments are restricted to terrestrial regions of the Earth. The second difference is that, currently, there is no published global precipitation time series estimated from satellite. Various methods of estimating precipitation from space have been developed (Arkin and Ardanuy, 1989) and these will undoubtedly be refined over the next decade (e.g. the Tropical Rainfall Monitoring Mission, TRMM), but a genuinely global precipitation time series is awaited. This is most likely to be produced through the Global Precipitation Climatology Project (GPCP) of the World

Meteorological Organization (WMO), which runs through to 1995. This project has the specific goal of integrating land-based and satellite-derived precipitation estimates on a global scale (Rudolf et al., 1991).

The most complete assessment of global precipitation change using instrumental data has been undertaken by Bradley et al. (1987), Diaz et al. (1989), and Hulme (1992), using the station time series data set discussed previously. A similar assessment, but restricted to extra-tropical land areas of the Northern Hemisphere, has also been completed by Vinnikov et al. (1990), who used much additional precipitation data for the former Soviet Union.

Precipitation data homogeneity

As with the temperature station records, a number of homogeneity problems exist with observed monthly precipitation totals. Using nearby stations to eliminate inhomogeneous precipitation series is more problematic than with temperature, however. Precipitation is more variable than surface temperature both in time and in space and therefore a considerably higher spatial density of precipitation records is needed to detect station inhomogeneities using the station-comparison approach.

A further problem with detecting precipitation change which is unique to ground-based precipitation measurements arises from the fact that the efficiency of the collection of precipitation by rain gauges varies with gauge siting, construction and climate (Sevruk, 1982; Folland, 1988). Major influences on collection efficiency are wind speed during rain events, the size distribution of precipitation particles, and the exposure of the rain gauge site. Of particular concern is the measurement of snowfall by conventional gauges, where undercatch of up to 40 percent can occur. Various corrections for wetting loss, snow undercatch, and gauge design have been performed on some regional precipitation records (Vinnikov et al., 1990; Legates and Willmott, 1990). Although the magnitude of some of the required corrections is large, when corrected and uncorrected records have been used to determine long-term precipitation trends over what was the USSR important variations are apparent in both data sets (Bradley and Groisman, 1989). Nevertheless, the lack of bias corrections in most precipitation records outside the former USSR is a severe impediment to the quantitative assessment of global precipitation trends, and makes the task of calibrating satellite-derived precipitation estimates that much harder.

Temporal trends in precipitation

The large-scale analyses of precipitation referred to above have demonstrated little change in overall global precipitation. For example, Hulme (1992) found no trend in global-mean precipitation between 1951 and 1980, nor any evidence that precipitation variability had changed significantly over that time period. There have, however, been substantial precipitation changes in certain latitudes and in certain regions, most notably an increase in Northern Hemisphere mid-latitude precipitation and a decrease in Northern Hemisphere subtropical precipitation. Figure 4.13 shows standardized anomaly time series of annual precipitation over the past 100 years for four regions. Over the former USSR annual precipitation has shown a fairly steady increase throughout this period, even after correcting for gauge undercatch. There is also some evidence for increases over northern Europe. In the tropics both the East African and Indian monsoon regions show little long-term trend, annual precipitation being dominated by high inter-annual (e.g. dry 1987 over India) and inter-decadal (wet 1960s over East Africa) variability.

The region of the world which has demonstrated the single most substantial change in precipitation in the instrumental period is the African Sahel. The decreasing supply of precipitation to the Sahel in recent years is shown clearly in figure 4.13b, a trend which has continued through 1990. The relative measurement scale in figure 4.13 is widely used for large-area precipitation monitoring. It is robust to the problems of station network changes, missing data, and high spatial precipitation variability previously mentioned. To express these changes in absolute terms, however, requires a different approach. Hulme et al. (1992) generated two independent

Figure 4.13 Standardized regional annual precipitation anomalies for (a) the former USSR, (b) the African Sahel, (c) East Africa, and (d) the Indian sub-continent (Source: Folland et al., 1990). Standardized anomalies are precipitation values expressed as standard deviations away from the mean.

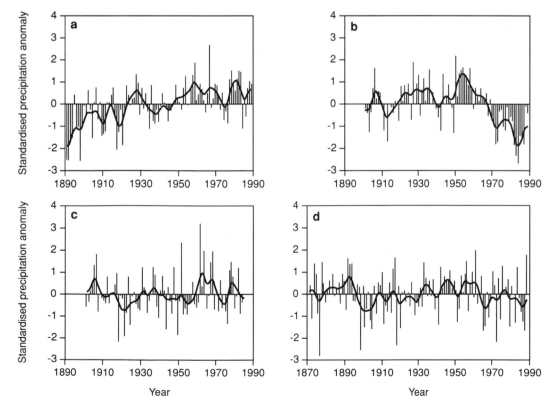

30-year climatologies of global precipitation for the periods 1931–60 and 1961–90. These show that over most of the Sahel the supply of annual precipitation has fallen by 30 percent or more between these two periods, and that this region dominates the global precipitation change field (figure 4.10b). This prolonged drought has contributed in recent years to the desertification of the drylands immediately south of the Sahara (see chapter 16, Goudie, and chapter 17, Schulz).

Although the full explanation for this decline in Sahel precipitation is not known, Folland et al. (1990) have shown that at least 60 percent of the decadal timescale precipitation variation of figure 4.13b can be explained by worldwide SST variations. Street-Perrott and Perrott (1990) have taken this partial explanation one stage further and demonstrated that north–south temperature gradients in Atlantic SSTs may well have been associated with African precipitation variations on timescales ranging from the decadal (twentieth century) to millennial (the last glaciation).

The monitoring and analysis of global precipitation in the instrumental era is harder than for global temperature. This is due to the paucity of precipitation observations over the oceans, the instrumental biases of ground-based gauges, and the high spatial and temporal variability of precipitation which necessitates a higher density of stations to capture adequately large-scale variations in precipitation. Nevertheless, no long-term global-scale trend in precipitation has yet been detected, although some significant regions have demonstrated changes in annual precipitation of more than ± 10 percent in recent decades. Climate models (see chapter 5, Henderson-Sellers) suggest that global precipitation should increase by between 2 and 3 percent for each degree of global-mean warming. In view of the difficulties of accurately measuring global precipitation, and its high natural variability, there remains little possibility of confirming or denying this relationship from the observational record for many years to come.

Recent climatic change

Natural variations in global climate have occurred since the end of the last glaciation (see chapter 3, Street-Perrott and Roberts). The historic record of climate change over the past 1000 years shows that the primary indicator of global climate, surface air temperature, has varied on a decadal to centennial scale by up to ± 0.6°C. Over the past 140 years, worldwide instrumental data reveal a global-mean warming of 0.45 ± 0.15°C and little associated change in global-mean precipitation. Significant regional precipitation trends have occurred in some parts of the globe. The Little Ice Age of the sixteenth to eighteenth centuries witnessed a global cooling of probably comparable magnitude to the warming experienced since about 1850. It is possible that some of the recent warming reflects the cessation of Little Ice Age conditions (Idso,

1989b), and that the rather rapid increase in global-mean temperature between 1920 and 1940 therefore may have had a largely natural origin. Nevertheless, the overall warming of about 0.45°C since the middle of the nineteenth century is suggestive of a long-term underlying cause of warming, for which the enhanced greenhouse effect is the primary candidate. The observed warming to date is at the lower end of the range predicted by climate models (0.5–1.1°C), which may suggest that the enhanced greenhouse effect is smaller than predicted by current models. It is also possible, however, that the enhanced greenhouse effect is larger than the models suggest and has merely been partially offset by natural variations in climate or other anthropogenic influences.

Unambiguous detection of the enhanced greenhouse effect from the observational record is premature at the present time. Three things are needed: first, continued careful monitoring of surface climate using conventional ground-based gauges; second, continued and expanded monitoring of surface and free atmosphere climate using current and new satellite instrumentation; and, third, using these improved observational data more sophisticated multivariate statistical techniques should enable unequivocal detection of the unique three-dimensional "fingerprint" of the enhanced greenhouse effect within 10–25 years (Wigley and Barnett, 1990). The world is currently at a temperature representing the upper limit of natural variability over the past 1000 years. Waiting 10–25 years before being sure that human pollution is the cause of the warming, and hence acting to reduce emissions, will, of course, have committed us to an even larger temparature rise in the second half of the twenty-first century.

Further reading

1 A recent book covering global climate during the last five centuries with contributions from numerous authors in Europe and the United States is: Bradley, R.S. and Jones, P.D. (eds) (1992) *Climate since AD 1500*. London: Routledge.

2 For more detailed information on the period covered by the Little Ice Age there is nothing better than: Grove, J.M. (1988) *The Little Ice Age*. London: Methuen.

3 The Intergovernmental Panel on Climate Change (IPCC) reported in 1990 and the full scientific version has been published. The chapter on recent climate variations provides good coverage of twentieth century climate variability: Folland, C.K., Karl, T.R. and Vinnikov, K. Ya. (1990) Observed climate variations and change. In J.T. Houghton, G.J. Jenkins and J.J. Ephraums (eds) *Climate Change: the IPCC Scientific Assessment*. Cambridge: Cambridge University Press, 195–238.

4 For a good concise summary of how global-mean temperature is monitored and what the record reveals see: Jones, P.D. and Wigley, T.M.L. (1990) Global warming trends. *Scientific American*, August, 84–91.

5 Many scientific papers have addressed the question of causal mechanisms of recent climate change. Perhaps one of the best overview papers is: Hansen, J. and Lacis, A.A. (1990) Sun and dust versus greenhouse gases: an assessment of their relative roles in global climate change. *Nature*, 346, 713–19.

CHAPTER FIVE

Numerical Modelling of Global Climates

Ann Henderson-Sellers

Introduction

It is generally accepted that there are three methods of attempting to improve understanding of specific phenomena: laboratory investigation, examination of analogues, and theoretical modelling. Investigation of the global climate precludes the laboratory method and, while – as concluded in chapter 3 – the search for historical or geological analogues of future climates is worthwhile, it has yet to be demonstrated that adequately complete data sets can be derived for full climatic reconstruction. An obvious, but crucially important, missing parameter is cloud amount and type. Thus, it is commonly argued, the investigation of future climates *must* rely heavily on the theoretical model.

One can, however, doubt whether this is a true picture, for in reality the art of numerical modelling of the global climate incorporates many aspects of laboratory and analogue techniques as well as the theoretical derivations upon which it is supposed to be solely founded. Numerical models of the global climate are based upon fundamental physical laws which are well understood and readily demonstrated but the computers that execute the calculations from which the simulations flow are, in essence, numerical "laboratories" in which experiments of considerable complexity are conducted. The complementary laboratory of the global atmosphere and oceans, often sensed by satellites, provides the majority of the data used to initiate and verify these computational experiments. Analogues are drawn upon extensively during the construction of the numerical models; for example, the exchanges of heat and mass (water vapor) between the surface and the atmosphere are formulated by analogy with the exchange of momentum. Indeed most of the parameterizations of physical processes in numerical climate models depend upon empirical studies of analogue processes.

The study of global climates by means of numerical climate models unites investigations from the separate kingdoms of the laboratory, analogy, and theory. The means by which this unification is achieved and the level of success, as measured by the realism of the simulated climates, is the subject of this chapter.

Numerical climate models

For several reasons a model must be a simplification of the real world. The processes of the climate system are not fully understood, but they are known to be complex. Furthermore, they interact with one another, producing feedbacks, so that any solution of the governing equations must involve a great deal of computation. Climate models can be slow to run and costly to use, even on the fastest computer, and the results can only be approximations. The solutions that are produced start from some initialized state and investigate the effects of changes in a particular component of the climate system. The boundary conditions, for example the solar radiation or sea-surface temperatures, are set from observational data but these data are rarely complete or of adequate accuracy to specify completely the environmental conditions, so that there is inherent uncertainty in the results.

Large-scale global climate models, designed to simulate the climate of the planet, must take into account the whole climate system. All of the interactions between components must be integrated in order to develop a climate model. This presents great problems, because the various interactions operate on different timescales. For example, the effects of deep ocean overturning may be very important when considering climate averaged over decades to centuries, while local changes in wind direction may be unimportant on this timescale. If, on the other hand, monthly timescales are of concern, the relative importance would be reversed.

The simplifications that must be made to the laws governing climatic processes can be approached in several ways. Consequently there are numerous different global-scale climatic models available. In general, two sets of simplifications need to be made. The first involves the processes themselves: it is usually possible to treat in detail some of the processes, specifying their governing equations fully, while other processes must be incorporated in an approximate way, either because of our lack of understanding or because there are inadequate computer resources to deal with them. For example, it might be decided to calculate the radiation fluxes in great detail, but only approximate the horizontal energy flows associated with regional-scale winds. The approximation may be approached either by using available observational data – the analogue approach – or through specification of the physical laws involved – the theoretical approach.

The second set of simplifications involves the resolution of the model in both time and space. While it is generally the case that the finer the spatial resolution, the more reliable the results, constraints of both data availability and computational time may dictate that a model may have to have, for example, latitudinally averaged values as the basic input. In addition, too fine a resolution may be inappropriate because processes acting on a smaller scale than the model is designed to resolve may be inadvertently incorporated. Similar considerations are involved in the choice of temporal resolution. Most computational procedures require a "timestep" approach to calculations. The processes are allowed to act for a certain length of time after which the new conditions are calculated. The process is then repeated using these new values. This continues until the conditions at the pre-specified "end time" have been established. Again, although accuracy potentially increases as the timestep decreases, there are constraints imposed by data, computational ability, and the design of the model.

Although models are designed to aid in predicting future climates, their performance can only be tested against the past or present climate. Usually when a model is developed an initial objective is to ascertain how well its simulations compare with the present climate. Thereafter it may be used to simulate past climates, not only to see how well it performs, but also to gain insight into the causes of these climates. Although such past climates are by no means well known, this comparison provides a very useful step in establishing the validity of the modelling approach (e.g. Guetter and Kutzbach, 1990). After such tests, the model may be used to derive projections of possible future climates.

The important components to be considered in constructing or understanding a model of the climate system are: (a) *radiation*, the way in which the input and absorption of solar radiation and the emission of infrared radiation are handled; (b) *dynamics*, the movement of energy and mass around the globe (specifically from low to higher latitudes) and vertical movements (e.g. convection); (c) *surface processes*, inclusion of land/ocean/ice and the resultant change in albedo, emissivity, and surface–atmosphere energy interchanges; and (d) *resolution in both time and space*, the timestep of the model and the horizontal and vertical scales resolved.

The relative importance of these processes and the physical and/or the analogue (empirical) basis for parameterizations employed in their incorporation can be discussed using the conceptual framework of a "climate modelling pyramid" (figure 5.1). The edges represent three of the basic elements of models (radiation, dynamics, surface processes), and resolution in time and space is shown increasing upwards. Around the base of the pyramid are the simpler climate models which incorporate only one primary process.

There are four basic types of numerical climate model. Energy balance models (EBMs) are one-dimensional models (close to the base of the modelling pyramid) predicting the variation of the surface (strictly the sea-level) temperature with latitude. Simplified relationships are used to calculate the terms contributing to the

Figure 5.1 The climate
modelling pyramid
showing the relative
positions of the basic
model types. The position
of a model on the pyramid
indicates the complexity
with which the three
primary processes
interact. The base of the
pyramid can be
considered hollow since
there is essentially no
interaction between the
primary processes.
Progression up the
pyramid leads to greater
interaction between each
primary process. The
vertical axis is not
intended to be
quantitative (modified
from Shine and
Henderson-Sellers, 1983).

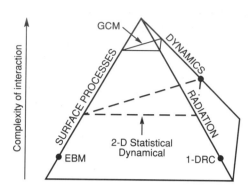

Figure 5.1 The climate modelling pyramid showing the relative positions of the basic model types. The position of a model on the pyramid indicates the complexity with which the three primary processes interact. The base of the pyramid can be considered hollow since there is essentially no interaction between the primary processes. Progression up the pyramid leads to greater interaction between each primary process. The vertical axis is not intended to be quantitative (modified from Shine and Henderson-Sellers, 1983).

energy balance in that latitude zone (e.g. Budyko, 1969; Sellers, 1969). Also close to the base of the pyramid are one-dimensional radiative–convective models, which compute the vertical (usually globally averaged) temperature profile by explicit modelling of the radiative processes and a "convective adjustment" which re-establishes a predetermined lapse rate (e.g. Manabe and Wetherald, 1967; Hansen et al., 1981). Two-dimensional statistical dynamical models lie about half-way up the climate modelling pyramid. They deal explicitly with surface processes and dynamics in a zonally averaged framework and have a number of atmospheric layers (e.g. Potter et al., 1979; MacCracken and Ghan, 1988).

The fourth type of numerical climate model is the closest to the apex of the climate modelling pyramid, incorporating all three fundamental processes with relatively high spatial and temporal resolution. This, the most complex type of climate model, is the general circulation model, or GCM, in which the three-dimensional nature of the atmosphere and/or ocean is incorporated. In these models an attempt is made to represent most physical processes believed to be important for the climate. The spatial resolution is, however, too coarse for synoptic and mesoscale processes to be fully captured (e.g. Hansen et al., 1983; Washington and Meehl, 1984). A distinction is often drawn between oceanic general circulation models and atmospheric general circulation models by terming them OGCMs and AGCMs. Since both the oceans and the atmosphere are crucial for the long-period evolution of the climate, considerable effort is currently being invested in coupling OGCMs and AGCMs into a more complete numerical model of the global climate (e.g. Schlesinger et al., 1985; Manabe and Bryan, 1985; Washington and Meehl, 1989).

Today's numerical climate models offer an incomplete representation of the real climate system. In particular, the incorporation of ocean processes and sea ice, important components of the long-term climate, is still very simplistic, and clouds and surface features are modelled only very sketchily at present. Moreover, the spatial resolution is poor (grid elements are approximately 500 km × 500 km) so that details of relatively small atmospheric phenomena, such as thunderstorms, cannot be described explicitly in these models, although modellers go to great pains to capture the ensemble effects of this "sub-gridscale weather" over a few years and over broad regions. In one sense, this coarse resolution is unimportant: since global climate models are designed to be useful for long periods of time, they do not have to forecast the exact timing and shape of each weather system precisely; rather they only have to capture the statistics of a large number of these events.

Although it is not important for numerical climate models to capture the evolution of specific, individual meteorological phenomena precisely, it is vitally important that they capture the overall sensitivity of the climate system correctly. The concept of the sensitivity of a system is illustrated in figure 5.2. The likelihood of and type of response to an imposed forcing, such as the addition of greenhouse gases to the atmosphere or a reduction in solar luminosity, depends upon both the stability of the system and its sensitivity to change. If the system is completely stable (lowest line in figure 5.2) its sensitivity is unimportant; at the other extreme, if the system is unstable (top line in figure 5.2) the sensitivity controls the initial response but not the system's ultimate destiny.

We know that the climate system can acquire at least two different, stable states (similar to the line second from the top in figure 5.2): glacial and interglacial. Thus the sensitivity of the system to an external perturbation becomes very important. At present, we do not understand the sensitivity of the real-world climate system. Moreover, numerical climate models exhibit a wide range of sensitivities to, for example, doubling of atmospheric carbon dioxide. Establishing the correct sensitivity of the climate system seems likely to prove to be a very difficult task.

The poor spatial resolution of current AGCMs is the direct result of the very large number of calculations required for each model

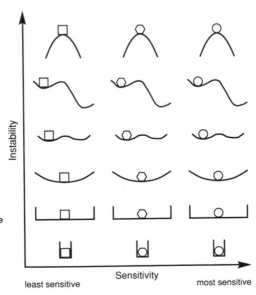

UNSTABLE
Small perturbation
(function of sensitivity)
causes radical change

DIFFERENT STABLE STATES
If a large enough perturbation
(function of sensitivity) is applied
a new stable state is achieved

IDENTICAL STABLE STATES
Perturbation can cause a shift
to a new identical stable state

EQUILIBRIUM IS STABLE
Perturbation changes state
but equilibrium in original state
is always achieved

EFFECTIVELY STABLE
Perturbation can alter state but
no effective change is discernable

STABLE
Nothing except system
destruction can cause
perturbation

Instability

Sensitivity

least sensitive   most sensitive

Figure 5.2 The relationship between sensitivity and stability of a system (or a model). Primary goals of climatic science are to establish the sensitivity and stability of the Earth's climate by careful observation and to simulate these responses in models.

timestep. At present the spatial resolution is as fine as is possible using the fastest supercomputers in the world. Computer power places other constraints on global model simulations too. The real-world experiment which humanity is currently conducting with the Earth involves a gradual increase in greenhouse gases. All but the most recent computer simulations have, in contrast, been "instantaneous doubling" experiments, that is, the usual method of simulating the climate in a doubled CO_2 world is to double the amount of CO_2 between two timesteps. Thus the model suddenly has twice as much radiative heating due to CO_2 as it had the day before. This sudden switch-on of extra heat causes immediate disequilibrium so that the model has to be run on for a number of years until the atmosphere, surface, and upper ocean catch up with the instantaneous doubling. Although this time for re-equilibration is a few years, it is still *very* much shorter than a simulation that allows a gradual increase in CO_2 from the middle of the nineteenth century to the middle of the twenty-first. "Transient" experiments (Hansen et al., 1988) which follow the real-world gradual increases are very much more expensive in computer time and hence costs than the much more common "switch-on doubling" experiments.

The constraint of computer power is such that, to date, only a handful of numerical models have been used to perform a transient greenhouse warming experiment (Hansen et al., 1988; Washington and Meehl, 1989; Stouffer et al., 1989) and no modelling group can possibly contemplate the much longer integrations required to simulate non-anthropogenic climatic shifts such as the Milankovitch-induced glacial cycles.

Greenhouse warming

The greenhouse effect is a theory which is well understood by atmospheric and climatic scientists and which successfully predicts temperatures on the Earth and on other planets. The greenhouse theory is based upon the fact (readily demonstrated by experiment) that while gases in the Earth's atmosphere are transparent to incoming solar radiation, some of them absorb outgoing thermal (or heat) radiation emitted from the Earth's surface. This radiative interaction between selected gases in the atmosphere (termed the greenhouse gases) and the outgoing heat radiation causes those gases to warm and, consequently, they themselves re-radiate heat in all directions. Some of this re-radiated energy travels back down through the atmosphere to the surface and it is this additional heating of the surface, over and above the heating due to the absorption of solar radiation, which is termed the greenhouse effect. "Natural" greenhouse heating turns the Earth from a planet unable to support life with a global mean temperature of $-18°C$ into its present habitable state with a global mean temperature of $+15°C$.

In the Earth's atmosphere, unlike those of Mars and Venus, only trace gases (less than a few percent of the total atmosphere) contribute to this greenhouse warming. These trace gases are water vapor, carbon dioxide (CO_2), methane (CH_4), nitrous oxide (N_2O), tropospheric ozone (O_3) and the chlorofluorocarbons (CFCs), although other gases such as carbon monoxide and sulphur dioxide also make very small radiative contributions to the surface warming.

The most significant greenhouse agent in the Earth's atmosphere is water vapor, which currently contributes about 100 of the 148 watts of additional, radiatively induced, heating of each square meter of the surface termed the greenhouse effect. Unfortunately, the amount of water vapor in the atmosphere is essentially uncontrollable as the major source is evaporation from the oceans. Human activities do, however, modify the atmospheric water vapor loading, since as temperatures rise in response to increases in CO_2 and other greenhouse gases, more water evaporates into the atmosphere, prompting greater temperature increases and thereby inducing further evaporation. This strong positive feedback coupling between temperature and water vapor produces much larger temperature changes than would CO_2 and other gas increases alone. Ramanathan (1988a) provides an excellent review of the physical and chemical processes underlying our current understanding of greenhouse theory and of humanity's impact on future greenhouse gas amounts.

The effects of radiative transfer of energy have been observed and described empirically for well over a hundred years. Indeed heat was recognized as similar, and later as identical, to light (i.e. a part, albeit invisible, of the electromagnetic spectrum) by Herschel in 1800 (Scott Barr, 1961). Although theoretical understanding of the interaction between gases and radiation was impossible before the development of quantum theory in the early years of this century, the observation that certain gases absorbed heat radiation prompted the recognition over 150 years ago that there is considerable potential for mankind's activities to modify the natural radiative budget of the planet. Carbon dioxide was the first obvious "culprit" and is, indeed, still used by most climate modelling groups as a surrogate for all radiatively active gases; hence experiments are still described as "doubled CO_2" simulations.

The initial research associated with the greenhouse theory concentrated on the understanding of the role of carbon dioxide in relation to atmospheric processes and radiative transfer. The emphasis throughout the nineteenth century and early in the twentieth century focused strongly on long-term geological implications of changes in the carbon dioxide content; in other words, as a means to understanding cyclical glacial theory (Chamberlin, 1899). Toward the end of the nineteenth century an increasing interest focused upon the atmospheric role of carbon dioxide. Langley (1884) appreciated the absorptive properties of the atmospheric gases and their beneficial effects on maintaining Earth surface temperatures at their present levels.

In 1895 Svante Arrhenius presented a paper to the Royal Swedish Academy of Sciences on "The influence of carbonic acid [carbon dioxide] in the air upon the temperature of the ground." This paper was later communicated (Arrhenius, 1896) to the Philosophical Magazine and published. On the basis of his calculations, Arrhenius concluded that past glacial epochs may have occurred largely because of a reduction in atmospheric carbon dioxide. Very recent research, notably on ice cores from Antarctica (e.g. Lorius et al., 1990), has confirmed that atmospheric CO_2 and CH_4 levels were indeed significantly lower during glacial periods of the Pleistocene. Arrhenius (1903) further suggested that most of the excess carbon dioxide from fossil fuel combustion may have been transferred to the oceans, displaying a remarkable awareness of cycling processes.

Between 1897 and 1899, T. C. Chamberlin presented a series of three papers expounding the geological implications of carbon dioxide theory. In 1897 he reviewed the current hypothesis of climatic change and in the following two years (Chamberlin, 1898, 1899) postulated the effects limestone-forming periods (e.g. Carboniferous, Jurassic and Cretaceous) may have had in contributing to subsequent glacial epochs.

By the late 1930s discussion of the heating role of atmospheric carbon dioxide had re-emerged in the literature. Callendar (1938) estimated that between 1890 and 1938 around 150 million tons of carbon dioxide had been pumped into the atmosphere from the combustion of fossil fuels, of which 75 percent had remained in the atmosphere. Furthermore, if all other factors remained in equilibrium, then anthropogenic activities would increase the mean global temperature by 1.1°C per century, another remarkable example of foresight given that current estimates from climate models indicate that global temperatures should have risen by about 1°C since about 1880.

It was around the beginning of the 1960s that the misnomer "greenhouse effect" was used to support an educational analogy of the Earth's atmosphere (see Hare, 1961, p. 15). Unfortunately the term soon became popularized and by 1963 there were already vain attempts (Fleagle and Businger, 1963) to clarify the confusion between a "greenhouse" and "atmospheric" effect.

Numerical climate modelling of human-induced greenhouse warming

Möller (1963) provided the first attempt at a one-dimensional atmospheric model using fixed relative humidity and cloudiness. An increase in temperature of 1.5°C was estimated for a doubling of carbon dioxide from 300 to 600 p.p.m.v. He concluded that cloudiness diminished the radiation effects but not the temperature changes and that the effect of a 10 percent increase in carbon dioxide could be compensated for completely by a change in atmospheric water content of 3 percent or in cloudiness of 1 percent (of its value). In 1967, Manabe and Wetherald provided quantitative results for carbon dioxide induced warming on the basis of a one-dimensional radiative–connective model with fixed relative humidity and cloudiness. They estimated a mean surface temperature increase of 2.4°C for an instantaneous doubling of atmospheric carbon dioxide. In 1971, Manabe was able to refine the Manabe and Wetherald (1967) model with a more elaborate treatment of infra-red radiation transfer. The early 1970s also saw sufficient concern about climate modification to initiate the SMIC (1971) report. In the same year, Rasool and Schneider (1971) observed that over the past few decades atmospheric carbon dioxide had increased by 7 percent and the aerosol content of the lower atmosphere by as much as 100 percent. They also calculated the change in tropospheric temperature through a doubling of carbon dioxide as being 0.8°C using a one-dimensional planetary radiation balance model with a fixed lapse rate, relative humidity, stratospheric temperature, and cloudiness.

Perhaps one of the most significant advances in the estimation of temperature change and a doubling of carbon dioxide was the application to this problem in 1975 of one of the earliest three-dimensional global climate models reported by Manabe and Wetherald (1975). In comparison with modern GCMs, this limited area model with idealized topography, fixed cloudiness and no heat transport by the oceans represented a very simplistic view. The study was, however, able to provide some indication of how an increase in carbon dioxide concentration might affect the distribution of temperatures in the atmosphere, as opposed to simply Earth surface temperatures. Model results suggested that while the air temperatures in the troposphere would be likely to increase, the stratospheric temperatures would be likely to decrease. Furthermore, for the first time a model suggested that the hydrological cycle would be likely to increase in intensity. The 1975 model of Manabe and Wetherald now forms the basis of the current Princeton University Geophysical Fluid Dynamics Laboratory (GFDL) model.

In 1981, Hansen et al., at the Goddard Institute of Space Studies (GISS), examined the main processes known to influence climate model sensitivity by inserting fixed absolute humidity, relative humidity, cloud formation (from fixed temperatures), and a moist adiabatic lapse rate individually into a one-dimensional radiative–convective (1-D RC) model. By doing so, they were able to examine the effects of various climatic feedback mechanisms, particularly clouds and water vapor. They concluded that by the end of the century, the anthropogenic carbon dioxide warming factor will have risen above the "noise level" of natural climatic variability. Natural climatic variations are caused by a range of agencies, including volcanic aerosols.

The importance of the atmospheric water vapor feedback mechanism was further underlined by the findings of Ramanathan et al. (1983) using the National Center for Atmospheric Research (NCAR) GCM. For the first time they were able to examine important mechanisms such as the way in which the vertical distribution of water vapor minimizes the lower stratospheric long-wave cooling and the dependence of the emissivity of cirrus clouds on liquid water content. The latter suggested that not only are clouds important in the modelling of atmospheric processes but also the type of cloud and the radiative properties are crucial. In particular, high level cirrus clouds appeared to enhance the radiative cooling of the polar troposphere significantly. Mitchell (1989) provides an excellent overview of equilibrium greenhouse simulations.

In the first of the more realistic, transient (gradually increasing CO_2) experiments, Hansen et al. (1988) provided three different trace gas growth scenarios to the GISS GCM. Scenario A assumes that the annual growth rates in the emissions of CO_2, CFCs 11 and 12, CH_4 and N_2O essentially remain at their present values of 1.5, 3, 1.5, and 0.4–0.9 percent, respectively. Projected future B assumes that the annual growth rates in the emissions decrease in time to zero in 2010 for CO_2 and CFCs 11 and 12, 0.5 percent per year in 2000 for CH_4, and 0.5 percent per year in 2010 for N_2O. Projected future C assumes that the annual emissions for all these gases decrease to zero in 2000. The global-mean surface air temperature predictions generated with the three assumed greenhouse scenarios are shown in figure 5.3. Hansen et al. (1988) concluded that scenario B, doubling carbon dioxide equivalent by AD 2060 (from 1958 levels), appeared the most plausible.

This selection between future greenhouse-gas emission scenarios is probably the most difficult part of any simulation of future climates. It is particularly important to realize that selection of emission scenarios depends upon social, economic, demographic, industrial, and technological changes over the next 50–100 years and for the globe as a whole. Clearly selection of such emission scenarios is not within the realm of expertise of numerical climate modellers.

In addition to generating globally averaged temperature curves, the GISS model (Hansen et al., 1988) was also able to distinguish some macro-regional scale variations in climate and hence identify regions where an unambiguous warming is likely to appear earliest (figure 5.4). In the GISS model, these "early warming" regions included the low latitude oceans, ocean areas near Antarctica, China and the interior areas of Asia.

It is worthwhile comparing some results from the more common "instantaneous doubling" greenhouse experiments before looking in detail at the GISS transient experiment. Figure 5.5 shows the changes in global-mean equilibrium surface air temperature simulated by 22 experiments with 17 numerical models in which atmospheric CO_2 has been instantaneously doubled. All the models

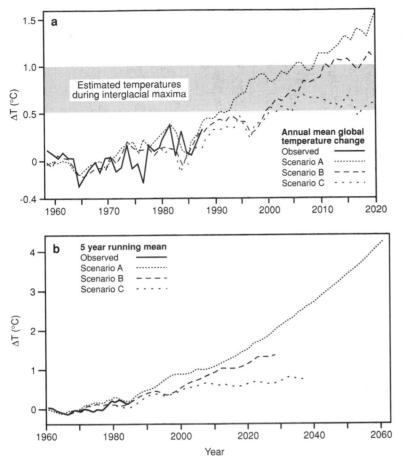

Figure 5.3   Annual-mean global surface air temperature computed for GISS scenarios A, B, and C. (a) Annual mean global temperature change, 1958–2019; (b) five-year running mean, 1960–2060 (after Hansen et al., 1988). Observational data are from Hansen and Lebedeff (1987, 1988). The shaded range in (a) is an estimate of global temperature during the peak of the current and previous interglacial periods, about 6000 and 120 000 years BP, respectively. The zero point for observations is the 1951–80 mean (Hansen and Lebedeff, 1987); the zero point for the model is the control run mean.

are AGCMs incorporating a range of parameterization schemes. Some are linked to simple mixed-layer ocean models but none includes the deep ocean circulation. Figure 5.5 also shows the magnitude of the intensification in the global hydrological cycle in terms of the percentage change in precipitation as compared with the $1 \times CO_2$ value. As might be anticipated as temperatures increase, evaporation increases and so does precipitation globally. The size of the predicted precipitation increase (in figure 5.5) increases with the size of the predicted temperature increase as a result of the Clausius–Clapeyron relation between the saturation vapor pressure of water vapor and temperature. Model numbers 20, 21, and 22 are among the most recent simulations and have the highest spatial resolution (approximately 300 km × 400 km). They might, therefore, offer the most up-to-date view of the impact of an instantaneous doubling of atmospheric CO_2. The range in predictions of temperature and precipitation increases is considerable but a "consensus" result can be identified as ~3.5–4.0°C and about an 8 percent increase in precipitation.

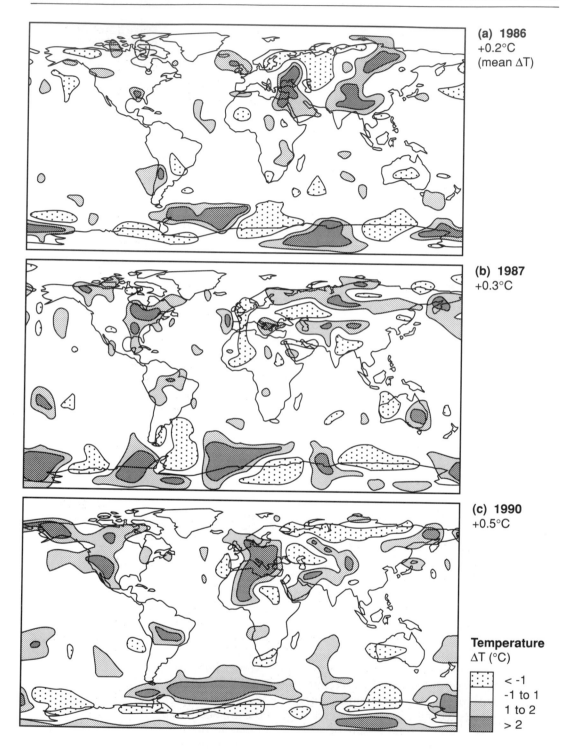

(a) 1986
+0.2°C
(mean ΔT)

(b) 1987
+0.3°C

(c) 1990
+0.5°C

Temperature
ΔT (°C)

⋯	< -1
	-1 to 1
▒	1 to 2
▓	> 2

Figure 5.4 The geographical distributions of the July surface air temperature changes relative to July 1958 simulated by Hansen et al. (1988) for scenario B for the years 1986, 1987, 1990, 2000, 2015, and 2029. Note

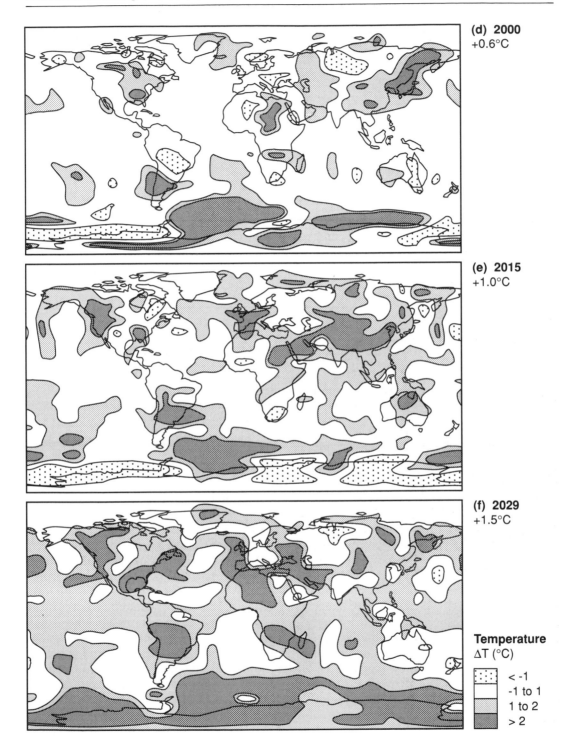

(d) 2000
+0.6°C

(e) 2015
+1.0°C

(f) 2029
+1.5°C

Temperature
ΔT (°C)

< -1
-1 to 1
1 to 2
> 2

that in the early transitional phases of this simulated greenhouse warming many regions of the world are colder than the present due to natural climatic variability. (After Hansen et al., 1988.)

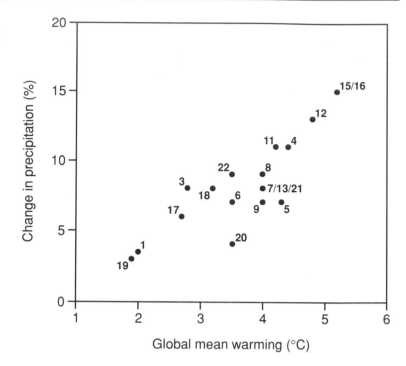

Figure 5.5 Scatterplot of global mean warming and percentage change in globally and annually averaged precipitation as simulated in 22 instantaneous doubling of CO_2 experiments conducted using 17 different AGCMs. Numbers refer to the different experiments/ models used. The three highest resolution experiments are numbers 20–22. (After Mitchell et al., 1990.)

The range in the predictions of temperature and precipitation is, in part, explained by figure 5.6, which shows the globally averaged warming for a collection of doubled CO_2 experiments plotted as a function of the mean surface air temperature in the $1 \times CO_2$ experiments. The regression line, which has been derived for the models without cloud phase and/or optical depth feedbacks (i.e. excluding UKMO (b), (c), (d)), indicates that warmer "control" ($1 \times CO_2$) models produce smaller predictions of temperature increase for doubled CO_2.

Figure 5.6 suggests that CO_2-induced changes in temperature and precipitation simulated by these models would be in better agreement if their $1 \times CO_2$ global-mean surface air temperatures were in better agreement. Furthermore, it is tempting to conclude that the resulting common model temperature and precipitation changes of about +4.5°C and +12 percent would be correct if the common $1 \times CO_2$ surface air temperature were in agreement with the observed temperature. Such an agreement, between observations and model simulations of the present-day climate, is one of the current aims of the numerical climate modelling community. However, it will not, by itself, guarantee accurate predictions for at least two reasons: (a) the observations are not precisely known and (b) agreement among models about the sensitivity to a prescribed forcing does not necessarily imply that the correct sensitivity has been achieved (cf. figure 5.2).

The topic of feedbacks within the real climate system and in numerical climate models is complex (see Lorius et al., 1990). The

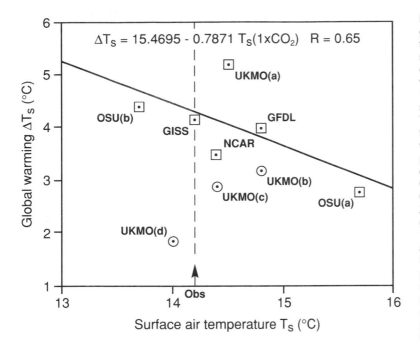

$$\Delta T_S = 15.4695 - 0.7871 \, T_S(1 \times CO_2) \quad R = 0.65$$

Global warming ΔT_S (°C)

Surface air temperature T_S (°C)

Figure 5.6 The global-mean surface air temperature warming simulated for a CO_2 doubling by six GCMs versus their simulated $1 \times CO_2$ global-mean surface air temperature. The regression line is for the model simulations without cloud phase-change and optical-depth feedbacks GISS, GFDL, NCAR, OSU(a) (Schlesinger and Zhao, 1989). OSU(b) (Schlesinger and Xu, 1989), UKMO(a) (Wilson and Mitchell, 1987). The UKMO model results with phase-change feedback (Mitchell et al., 1989) are shown by UKMO (b) (Heymsfield-type parameterization) and UKMO(c), and with additional optical-depth feedback by UKMO(d). "Obs" indicates the observed global-mean surface air temperature based on the data of Jenne (1975). (Adapted from Cess and Potter, 1988.)

degree of complexity can be illustrated by considering cloud feedback effects as currently captured in numerical climate models. The direct radiative forcing due to a doubling of CO_2 would increase the Earth's surface air temperature by about 1.2°C. That the model-predicted temperature increase is larger than this value is the result of feedbacks in the climate system, positive feedbacks enhancing the warming and negative feedbacks dampening it. Current numerical climate models suggest that positive feedback effects dominate the overall response. The two largest feedbacks are due to addition of water vapor to the atmosphere by evaporation (e.g. Ramanathan, 1988a) and due to the melting of snow and ice, which decreases the Earth's albedo. There are also important positive and negative feedbacks due to changes in cloud amount and micro-physical structure (e.g. changes in the cloud drop size distribution or changes from liquid water drops to frozen ice crystals).

When the troposphere warms as a result of increased CO_2, clouds located where temperatures lie between 0°C and about −15°C change from ice clouds to water clouds. Because of this and the assumed higher cloud water-to-precipitation conversion rate for ice clouds than for water clouds, the cloud amount increases. The solar effect of water clouds dominates their infra-red effect and so there is a reduction in the net radiation at the top of the atmosphere. This reduces the CO_2-induced warming: a negative feedback. There are other cloud feedback effects too. As the atmosphere warms, there is more liquid water and ice in clouds, thus increasing the optical thickness of the clouds. For water clouds, the solar effect dominates the infra-red effect, thereby leading to a reduction in the warming, that is, a negative feedback. For ice clouds, the feedback can be either negative or positive depending upon the emissivity of the clouds. Capturing the correct set of cloud feedback effects in numerical climate models depends upon improved observational data and improved parameterization schemes. The current range of uncertainty in the final sensitivity (see figure 5.2) of one AGCM (i.e. UKMO) is illustrated in figure 5.6.

The considerable residual uncertainty in the climate sensitivity due to cloud feedbacks is evidenced in figure 5.6 by the simulation results from the UKMO model. With a treatment of clouds similar to that of the other models, the UKMO model simulated the largest warming with a value of 5.2°C. However, when it was assumed that the conversion rate from cloud water to precipitation is higher in ice clouds than in water clouds, the sensitivity of the UKMO model decreased to 2.7°C, as a result of cloud cover feedback. When, in addition, the radiative properties of clouds are taken to depend on the prediction of cloud water amount, the resulting cloud optical depth feedback reduces the warming even further, to 1.9°C (Mitchell et al., 1989). Figure 5.6 indicates that the effect of incorporating cloud feedbacks in models is to increase the uncertainty in the simulated temperature sensitivity of the Earth to increased concentrations of greenhouse gases.

These considerable differences among current models, the established uncertainties, and the lack of agreement between present-day simulations and observations are underlined in any intercomparison amongst simulated $2 \times CO_2$ climates. Figures 5.7, 5.8, and 5.9 display the temperature, precipitation, and soil moisture differences which result for the two primary seasons (December/January/February (DJF) and June/July/August (JJA)) when $1 \times CO_2$ simulations are subtracted from their $2 \times CO_2$ counterparts. Results are taken from the three recent models noted with reference to figure 5.5 as having relatively (cf. other AGCMs) high spatial resolution: Geophysical Fluid Dynamics Laboratory (GFDL), Canadian Climate Center (CCC) and the UK Meteorological Office (UKMO) (Mitchell et al., 1989; Houghton et al., 1990). Figure 5.7 shows that these three, relatively high resolution, models simulate a CO_2-induced surface air temperature warming virtually everywhere. In general, the warming is a minimum in the tropics during both seasons, at least over the ocean, and increases towards the winter pole. Figure 5.7 also shows that although there are similarities in the CO_2-induced regional temperature changes simulated by the models, there are marked differences in both their magnitude and seasonality. For example, in central Russia the wintertime (DJF) warming generally increases with latitude in the GFDL model, while the CCC simulation exhibits a warming minimum of ~6°C and the UKMO model shows a warming maximum of >8°C.

Figure 5.8 shows that both positive and negative changes in precipitation rate are simulated by all three models, and that the largest changes generally occur in tropical–subtropical latitudes between 30°S and 30°N. The precipitation changes poleward of these latitudes are generally positive in both seasons over both the ocean and land. However, the CCC model exhibits a much smaller globally averaged increase in precipitation rate (cf. figure 5.5). The models generally simulate precipitation increases of less than 1 mm day^{-1} over the continents during both seasons; the exception being the Indian monsoon in JJA where larger increases occur. Differences abound among these simulations of precipitation rate changes.

The geographical distributions of the CO_2-induced changes in soil moisture over ice-free land simulated by the models for DJF and JJA are presented in figure 5.9. There are few qualitative similarities among the simulated soil moisture changes for DJF. On the other hand, there is more agreement among the simulations of soil moisture changes for JJA, with all three models exhibiting decreased soil moisture over most of the continents except Antarctica.

Numerical climate modellers are still improving the current models and investigating ways of improving the realism of their models. Although there is a broad consensus of agreement, each model differs somewhat from its counterparts in physical construction and in the physical realism of the parameterizations employed, the number and complexity of the feedbacks included and the outcome of increased carbon dioxide levels. Inter-model variability still provides a high margin of uncertainty, not only for temperature changes but also, to an even greater degree, for precipitation and effective moisture.

At least as important as the inter-model differences is the fact that most numerical climate models have, so far, only been applied to the problem of simulating the response to an instantaneously doubled CO_2 amount. That is, in all the three GCM studies discussed above, the CO_2 concentration in the experiment simulation was taken to be constant in time with a value twice that of the control simulation.

The real-world experiment: humanity's transiently increasing burden of greenhouse gases

Since the Industrial Revolution, the atmospheric burden of CO_2 has increased from between 260 and 280 p.p.m.v. to the 1988 value of 350 p.p.m.v. (i.e. a 25 percent increase). This increase has, however, been gradual, not instantaneous. Recently, a small number of GCM simulations have been performed in which the CO_2 concentration in the experiment has been taken to increase with time (i.e. transient experiments). The first of these transient studies was performed with the GISS model by Hansen et al. (1988). They increased the concentrations of CO_2, CH_4, CFCs 11 and 12, and N_2O between 1958 and 1982 following observed increases, and subsequently three scenarios for the future increases

Northern winter (temperature change °C)

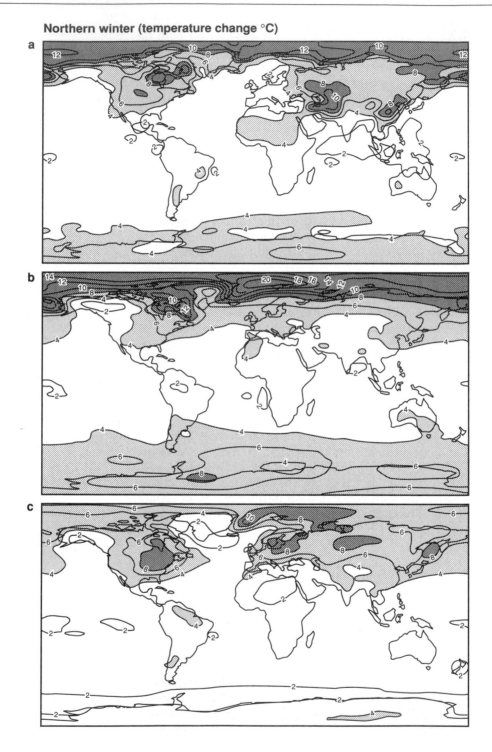

Figure 5.7   Geographical distribution of the surface air temperature change (°C), $2 \times CO_2$ minus $1 \times CO_2$, for DJF (northern winter) and JJA (southern winter) simulated by: (a) the CCC model; (b) the GFDL model;

Southern winter (temperature change °C)

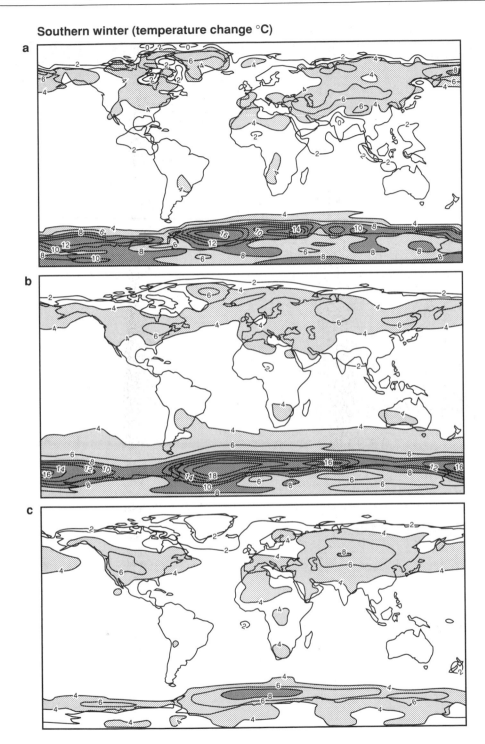

(c) the UKMO model. Stipple indicates temperature increases larger than 4°C and heavy stipple increases greater than 8°C. (After Mitchell et al., 1990.)

Northern winter (precipitation change mm d⁻¹)

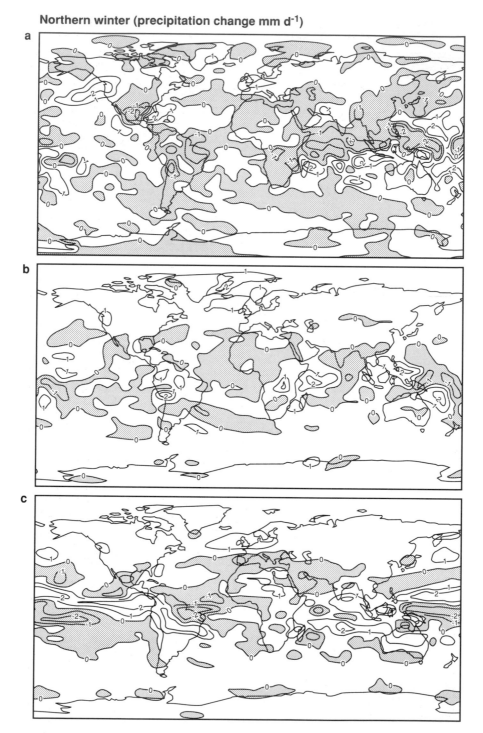

Figure 5.8   Geographical distribution of the precipitation rate change (mm day^{-1}), $2 \times CO_2$ minus $1 \times CO_2$, for DJF (northern winter) and JJA (southern winter) simulated by: (a) the CCC model; (b) the GFDL model; (c) the UKMO model. Stipple indicates areas of precipitation decreases. (After Mitchell et al., 1990.)

Southern winter (precipitation change mm d⁻¹)

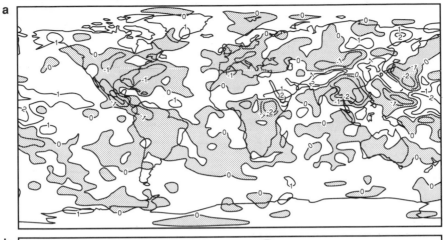

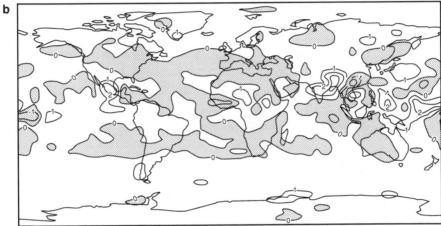

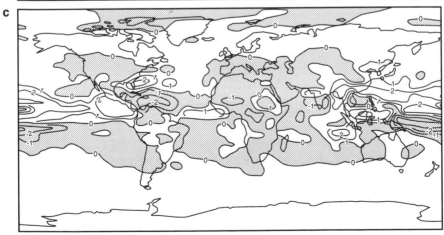

Northern winter (soil moisture change)

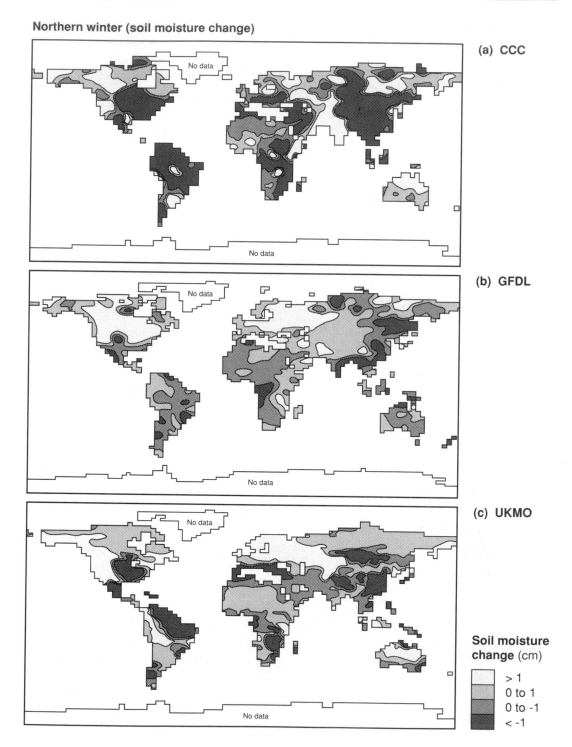

(a) CCC

(b) GFDL

(c) UKMO

Soil moisture change (cm)

> 1
0 to 1
0 to -1
< -1

Figure 5.9   Geographical distribution of the soil water change (cm), 2 × CO$_2$ minus 1 × CO$_2$, for DJF (northern winter) and JJA (southern winter) simulated by: (a) the CCC model; (b) the GFDL model; (c) the

Southern winter (soil moisture change)

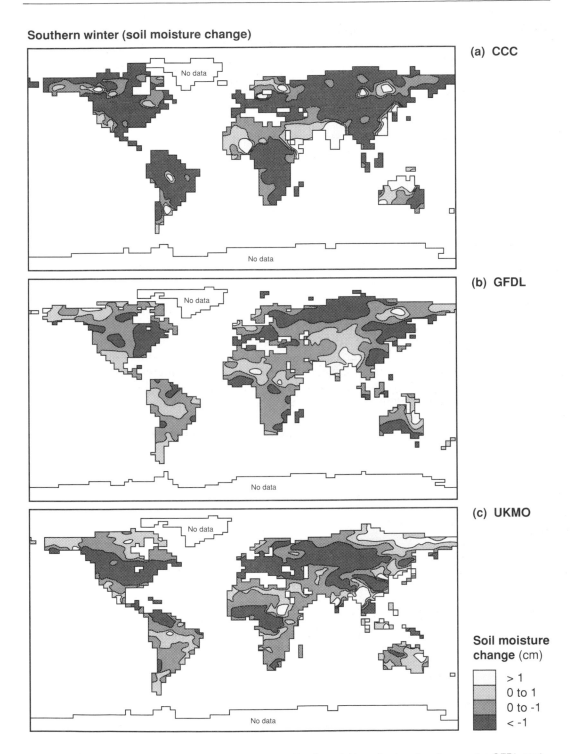

UKMO model. (Note that the CCC model has geographically variable soil capacity whereas the GFDL and UKMO models specify a global value.) (After Mitchell et al., 1990.)

in these gases from 1982 to 2050 were investigated (figures 5.3 and 5.4). In the second experiment, performed with the NCAR model by Washington and Meehl (1989), the CO_2 concentration alone was increased by 1 percent per year over a 30-year period. In a similar experiment with the GFDL model, Stouffer et al. (1989) increased CO_2 concentration by 1 percent each year and analyzed the geographical patterns of temperature differences for the experimental years 61–70.

These experiments are therefore non-equilibrium studies and they differ fundamentally from the equilibrium responses for doubled CO_2 equivalent illustrated in figures 5.6–5.9. They also differ from one another. In the Hansen et al. (1988) simulation, the atmospheric GCM was coupled to a mixed-layer ocean with horizontal heat transport prescribed and vertical heat transport to the deeper ocean predicted, but only by diffusive heat transfer. Washington and Meehl (1989) and Stouffer et al. (1989), however, coupled their atmospheric GCM to an oceanic GCM in which both the horizontal and vertical transports of heat were predicted, the former due to advection and diffusion, and the latter due to these processes as well as convection. Thus the Hansen et al. (1988) experiment might be considered "more complete" in the sense that it used observed increases in all the greenhouse gases between 1958 and 1982 and evaluated the effects of three different future scenarios. On the other hand, the Washington and Meehl (1989) and Stouffer et al. (1989) simulations are, in their turn, also "more complete" in the sense that the ocean circulation as well as that of the atmosphere was calculated. All three simulations and others more recently undertaken are still, therefore, incomplete but are closer to the current real-world experiment than the equilibrium simulations depicted in figures 5.7–5.9. Here I choose one transient experiment, that of Hansen et al. (1988), to compare with the earlier equilibrium simulations.

The time evolution of the global-mean temperature change from 1958 to 2025 simulated in the study by Hansen et al. (1988), presented in figure 5.3, shows that the projected increases in global-mean surface air temperature in 2025 relative to 1958 are about 0.6°C, 1.3°C, and 1.8°C for their hypothesized possible futures A, B, and C, respectively. Hansen et al. (1988) argue that of the three, B is the most probable future projection and hence that the temperature rise of 1.3°C projected from 1989 to 2025 is about 0.8°C larger than the observed temperature rise of 0.5°C from 1880 to 1989.

Accepting Hansen et al.'s (1988) selection of the "most likely" future (while bearing in mind the caveat that numerical climate modellers have little experience upon which to base future projections of greenhouse-gas emissions), we can investigate their model projections of transient temperatures in more detail. The geographical distributions of the July surface air temperature changes relative to July 1958 simulated by Hansen et al. (1988) for their future B are presented in figure 5.4 for the years 1986, 1987,

1990, 2000, 2015, and 2029. This figure shows that in each of these years, there are regions where the temperature changes are negative (i.e. a cooling); this in contrast to the equilibrium surface air temperature changes induced by a CO_2 equivalent doubling, which are everywhere positive (figure 5.7).

Similar results were found in the studies by Washington and Meehl (1989) and Stouffer et al. (1989). The main similarities among these three experiments are: (a) warming at any time is less than the corresponding equilibrium warming; (b) areas of warming are greater at high northern latitudes than in low latitudes but are not zonally uniform; and (c) statistically significant warming is most readily evident over the subtropical oceans. There are also notable differences between the time-dependent experiment using specified ocean transports (Hansen et al., 1988) and the fully coupled ocean plus atmosphere models. These are that the dynamical ocean models predict warming minima in the North Atlantic and, more importantly, a strong minimum or even a slight cooling around Antarctica.

These non-equilibrium (transient) simulation results reveal that earlier characterizations of a greenhouse-gas induced climate change, based on scaling the equilibrium response for a CO_2 equivalent doubling to any smaller level of increase in the equivalent CO_2 concentration, are incorrect. Instead, the actual patterns of climate change are comprised of the greenhouse-gas induced signal superimposed upon the natural variability of climate which exists even in the absence of climatic change (see chapter 4). When the departure of the concentrations of the greenhouse gases from present-day (or pre-industrial) levels is small, the greenhouse-gas induced signal is itself small and is often dominated regionally by the natural variability. This can lead, for example, to negative regional temperature changes (i.e. decreases in surface air temperatures) in individual years, with the locations of these non-warmed regions varying from year to year. Eventually, however, the rise in greenhouse-gas concentrations becomes so large that its signal begins to dominate the natural variability. Only then do the patterns of climate change begin to take on their equilibrium greenhouse-gas induced appearance; in other words, the distributions in figure 5.4 tend, over time, toward those in figure 5.7. Even then the effects of the ocean interactions on the air temperatures and hence the whole climate remain difficult to predict.

Numerical climate modelling is as much an experimental science as field-based geomorphology. The difference is that the nature of the system is explored within a computer (usually a very powerful supercomputer) rather than out of doors. Numerical modelling draws upon theoretical laws, upon analogues of processes that are not fully understood or that cannot be described theoretically at the required scale, and conducts "laboratory" experiments in the form of numerical simulations of the same (or a very similar) phenomenon using many types or configurations of models.

Consensus among model results is seen as a necessary, but not a sufficient, condition for acceptance of the predictions. At present there is worldwide consensus on the direction (an increase) of temperature change resulting from an equilibrium doubling in greenhouse gases over pre-industrial levels. There is also consensus about the intensification of the global hydrological cycle. The magnitude and regional distribution of these increases is, however, dependent upon complex feedback effects which are yet to be fully explored. The gradual (transient) nature of humanity's additional greenhouse further confuses the comparison of numerical climate model projections with observed trends. It is now known that in the real-world transient experiment which humanity is undertaking on Earth, the natural variability of the climate system will make unambiguous detection of a warming trend very difficult in the forthcoming decades.

Further reading

1 An introduction to global-scale numerical climate modelling is given in Henderson-Sellers, A. and McGuffie, K. (1987) *A Climate Modelling Primer*. Chichester: John Wiley and Sons.
2 A detailed description of the development and construction of one particular general circulation climate model is given by Hansen, J., Russell, G., Rind, D., Stone, P., Lacis, A., Lebedeff, S., Ruedy, R. and Travis, L. (1983) Efficient three-dimensional global models for climate studies: models I and II. *Monthly Weather Review*, **111**, 609–22.
3 A review of the state of numerical climate modelling of humanity's additional greenhouse warming is given by Mitchell, J.F.B. (1989) The "greenhouse" effect and climate change. *Reviews of Geophysics*, **27**, 115–39.
4 An overview of the physics and chemistry underlying the theory of humanity's additional greenhouse is given by Ramanathan, V. (1988) The radiative and climatic consequences of the changing atmospheric composition of trace gases. In F.S. Rowland and I.S.A. Isaken (eds), *The Changing Atmosphere*. Chichester: John Wiley and Sons, 159–86.
5 An example of the way in which global numerical climate models can be applied to paleoclimatic simulations is given by Guetter, P.J. and Kutzbach, J.E. (1990) A modified Köppen classification applied to model simulations of glacial and interglacial climates. *Climatic Change*, **16**, 193–215.

PART III

Ice and Ocean

CHAPTER SIX

Global Warming and Periglacial Landscapes

Eduard A. Koster

Introduction

High-latitude, cold-climate regions are recognized to be vital components in shaping global climate and are likely to respond significantly to future climate change. Most of the currently available numerical models of the global climate are now in general agreement that the increasing concentration of CO_2 and other greenhouse gases (like methane, nitrous oxide, CFCs, and ozone) will cause a global warming of several degrees Celsius by the time their overall concentration is equivalent to twice the pre-industrial concentration of atmospheric carbon dioxide (chapter 5, Henderson-Sellers). The boreal and subarctic regions are thought to be particularly sensitive to regional or global changes in the environment (Eddy et al., 1988; Koster and Boer, 1989; IGBP, 1990). The seasons that characterize the climate of temperate regions are exaggerated in northern regions in both duration and amplitude, with extreme summer-to-winter ranges of temperature. Moreover, climate variability is also greater on timescales of decades or longer. Global greenhouse warming is expected to be greatest at high latitudes, and especially winter warming is expected to be much more than the global annual average.

Through several, possibly positive, feedback mechanisms (e.g. sea ice and snow cover versus albedo, northern wetlands as a potential source of CO_2 and CH_4) the initial rise of annual temperatures in high-latitude regions may accelerate the global warming process. Since the installation of the CO_2-monitoring station at Mount Loa (Hawaii) in 1958, and many others that followed, it has become clear that a worldwide increase in atmospheric CO_2 concentration has occurred during the past three decades. However, the seasonal variations in these well-known curves of increasing CO_2 concentration are amplified with increasing latitude, thereby emphasizing the significance of large land masses and their vegetation cover in determining the annual CO_2 cycle; for example, at Point Barrow (Alaska, 71°N) the seasonal amplitude is up to 15 p.p.m. CO_2 whereas at Mount Loa (19.5°N) it is about 6 p.p.m. and at the South Pole only about 2 p.p.m. The CO_2 curves for the Northern Hemisphere all show a marked maximum in May at the end of the winter season and a minimum in September at the end of the growing season; obviously the curve for the South Pole region is out of phase with the curves for the Northern Hemisphere (McBeath, 1984, Boer et al., 1990).

In high-latitude as well as in high-mountain regions important effects of climate change will be through the alteration of freeze–thaw processes and through changes in the occurrence of glacier ice, ground ice, sea ice, and lake ice. Changes in cryospheric processes may have a large range of side-effects on geomorphological events, hydrological conditions, and soil thermal regimes, and thus on vegetation, wildlife habitats, and man-made structures and facilities.

Consequently, the objectives of this chapter are: (a) to define periglacial environments and to outline the areal extent and thermal regime of permafrost; (b) to review the climate variability of northern regions; (c) to emphasize the complex interrelations in the atmosphere–"buffer layer"–permafrost system; (d) to evaluate the significance of northern, especially peatland, regions as sources or sinks of CO_2 and CH_4; (e) to summarize permafrost response to past and future temperature changes; and (f) to describe some geomorphic responses to permafrost degradation and their practical implications.

Definition and areal extent of periglacial environments

According to Washburn (1979) the term periglacial designates non-glacial processes and features of cold climates characterized by intense frost action, regardless of their age or proximity to glaciers. Clearly permafrost, defined as ground that remains at or below 0°C for at least two years (Harris et al., 1988), can be considered as the most diagnostic criterion for periglacial environments. French (1976) defined the periglacial domain as including all areas with a mean annual air temperature of less than +3°C. However useful it is for particular purposes, any such definition remains arbitrary. Permafrost is synonymous with perennially cryotic (frozen if moisture is present) ground as both are only defined on the basis of temperature.

The area of seasonally frozen ground (frost penetration of at least 30 cm once in 10 years) as distinct from permafrost constitutes about 26 percent of the land area of the Northern Hemisphere, assuming that permafrost covers an additional 20–25 percent (table 6.1). A major part of this huge area is designated as discontinuous

permafrost, the southern boundary of which roughly coincides with
a mean annual air temperature of -1 to $-2°C$. Near its southern
boundary it occurs in isolated patches or islands and is sometimes
referred to as sporadic permafrost. Approximately north of the -6
to $-8°C$ isotherm continuous permafrost occurs, but there is no
sharp distinction, or boundary, between the zones of continuous
and discontinuous permafrost. Moreover, an area of approximately
2.3×10^6 km^2, mainly at lower latitudes, is covered by alpine or
mountain permafrost (Péwé, 1983). Most of the alpine permafrost
is discontinuous by definition as it clearly depends very strongly on
highly variable topoclimatic conditions. Approximately 80 percent
of the total permafrost area in China consists of mountain permafrost (1.7×10^6 km^2), usually divided into a lower unstable zone,
a transitional zone and an upper stable zone (Cheng Guodong,
1983). The land surface with a seasonal snow cover (north of
35–40°N) almost doubles that of the permafrost areas during
winter time. Estimates of the areal extent are presented in table 6.1
and are illustrated in figure 6.1. During Pleistocene glacial episodes
the permafrost area was probably twice as extensive as the present-
day extent (Nekrasov, 1984). With the exception of parts of
Antarctica and some high mountain ranges, effectively all periglacial landscapes are located in the Northern Hemisphere. This is
not only because of the absence of land in the Southern Hemisphere
around 60°, but also because periglacial processes such as permafrost tend to be emphasized under a continental climatic regime.

In permafrost areas several types of temperatures are defined,
depending on where they are measured. Combining information by
Ferrians et al. (1969), Washburn (1979), Harris et al. (1988), and
Lachenbruch and Marshall (1986), figure 6.2 has been constructed.
The mean annual air temperature (MAAT) is usually several

Table 6.1   Estimates of areal extent of terrestrial permafrost areas in the Northern Hemisphere, according to
various sources

	Land surface ($\times 10^6$ km^2)	Percentage underlain by permafrost (approx.)	Area ($\times 10^6$ km^2) underlain by permafrost (approx.)
Northern Hemisphere	100.3	20–25	20–25
Continuous permafrost			7.6
Discontinuous permafrost			17.3
Alpine permafrost			2.3
Former USSR	22.4	50	11.2
Canada	9.9	50	4.9
China	9.6	22	2.1
Greenland	2.2	99	2.2
Alaska (USA)	1.5	82	1.2
Land ice and snow cover NH (mean annual range)		4–46	

All data are estimates, which are rather poorly defined and still show large variations in the literature. Offshore permafrost
data are not included.

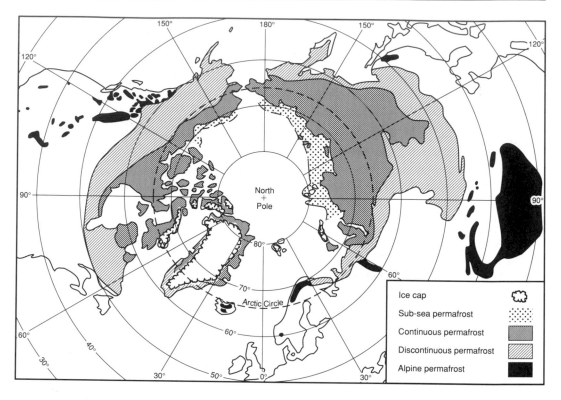

Figure 6.1 Distribution of permafrost in the Northern Hemisphere (after Péwé, 1983).

degrees lower than the mean annual ground temperature (MAGT), the latter being defined as the ground temperature at a depth where temperature fluctuates by less than 0.1°C per year. This level of so-called zero annual amplitude may reach a depth of 20–30 m in extreme cases. Above this depth the ground is subjected to strong seasonal fluctuations. Nevertheless, mean annual ground surface temperature (MAGST) can be deduced by upward extrapolation of the geothermal gradient, provided the measured gradient has achieved equilibrium and there are no recent climatic changes. Fluctuations of the MAGST may reach values of more than 20°C during the year. The mean annual permafrost surface temperature (MAPST) has a limited physical meaning as the position of the base of the active layer may vary from year to year (according to Nicholson, 1978, by as much as 25 percent of the total thickness of the active layer). Moreover, if permafrost is not in equilibrium with present-day conditions, a talik or zone of unfrozen ground between the base of the active layer and the permafrost table may occur. A similar limited meaning obviously applies to the mean (annual) snow surface temperature (M(A)SST). As a reference point for the geothermal regime the MAGT is of most importance. As it does not differ much from the MAGST the ground surface temperature is

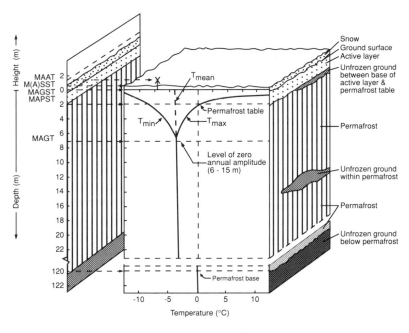

Figure 6.2 Permafrost terminology, typical temperature profile for an area with permafrost, and measurement sites for the different mean annual temperatures. T_{mean} = mean ground temperature; T_{max} = warmest temperature in year; T_{min} = coldest temperature in year. MAAT = mean annual air temperature (measured at standard height); M(A)SST = mean annual snow surface temperature (when present); MAGST = mean annual ground surface temperature; MAPST = mean annual permafrost surface temperature; MAGT = mean annual ground temperature (arbitrarily defined as being the depth where temperature fluctuates by less than 0.1°C per year). The MAGST is assumed to be approx. −4°C, corresponding to a MAAT in the order of −7°C (±2°C). The geothermal gradient (i_g) is assumed to be approx. 1°C/30 m, typical for relatively ice-poor frozen ground with a corresponding relatively low thermal conductivity and therefore a relatively high geothermal gradient (varying according to Washburn, 1979, from 1°C/22 m to 1°C/60 m in permafrost). This leads to a depth (H) of the permafrost base – in this hypothetical case – of approx. 120 m ($H = − \text{MAGST}/i_g$).

also used to characterize surficial permafrost temperatures. There is no direct correspondence between MAGST and MAAT values. This applies especially to the zone of discontinuous permafrost, as the presence or absence of permafrost in that case is strongly determined by local factors. In regions of continuous permafrost local factors are of less importance. Therefore a change in MAAT will lead in due time to a corresponding change in MAGST in those regions. Extrapolation of the geothermal gradient downwards will lead to an approximation of the depth of the permafrost base. Where the geothermal heat flow is constant, the geothermal gradient is inversely proportional to conductivity. The thermal conductivity in its turn strongly varies depending on soil properties and sediment texture and structure. The water or ice content is especially critical. The geothermal gradient in different types of sediment ranges from about 1°C/60 m for sandy, relatively ice-rich material (high conductivity ≈ low gradient ≈ thick permafrost) to about 1°C/22 m for fine-grained, relatively ice-poor material (low conductivity ≈ high gradient ≈ thin permafrost), according to Washburn (1979).

Regional warming

The main equilibrium changes in climate deduced from numerical general circulation models (GCMs) indicate a global average warming at high latitudes that is greater than the global average in winter, increases in precipitation at high latitudes throughout the year, and a diminishing of the area of sea-ice and seasonal snow cover (Houghton et al., 1990). Regional climate scenarios may be generated from GCMs or from historical instrumental data, but both methods have severe limitations (chapter 4, Hulme; Boer et al., 1990). To explore the range of potential changes of climate in order to assess the potential impact of these changes Hulme et al. (1990) have calculated and mapped temperature and precipitation changes for Europe utilizing the information from five independent equilibrium GCM experiments. The $2 \times CO_2$ changes in surface air temperature have been produced from unweighted averages of five model experiments interpolated onto a common grid. The data in table 6.2 clearly indicate that in Europe the largest temperature changes are expected in high northern latitudes, but at the same time reveal that inter-model uncertainty also increases to the north, especially in winter. For precipitation the composite model scenarios suggest an increased winter precipitation over northern Europe, but for the rest of Europe the models are not really consistent in their projections. Information on other climate variables, like minimum and maximum temperatures, spring and autumn frost, rainfall intensity, and snow cover duration, which is particularly relevant for geomorphological and landscape ecological responses, is almost completely lacking.

In general, it is known that the temperature trends during the past century have been of different signs in different parts of the boreal and subarctic regions. Similarly, paleoclimatic evidence based on proxy data clearly indicates that although temperature variations were largest in the high-latitudinal zone, they were also characterized by large regional differences in both timing and magnitude. For instance, Russian scientists reconstructed temperature deviations (as compared to the present-day situation) of +6 to +8°C for northeastern Europe (north of 60°N) and +8 to +12°C for northern Siberia (north of 60°N) during the last (Eemian or Mikulino) Interglacial, whereas the global temperature deviation was only about +2°C. Again, winter temperatures deviated more than summer temperatures. Similar trends seem to apply for the so-called mid-Holocene thermal optimum. Another good example of a paleoclimatic reconstruction, particularly concerning summer temperatures, is given by Nichols (1975) on the basis of pollen stratigraphical studies of peat sections in Arctic Canada. For the

Table 6.2 Average changes in seasonal mean air temperatures in Europe, using five equilibrium GCM experiments ($2 \times CO_2 - 1 \times CO_2$)

	Winter (°C)	Summer (°C)
European boreal and subarctic zone	7–9 (2.5–4.5)	<4 (1–1.5)
Western Europe maritime part	4–6 (1–2)	4 (1–1.5)
Continental parts of Europe	5–7 (1–2)	<4 (1–1.5)
Mediterranean region	3–4 (<1)	3–5 (1–2)

Values in parentheses indicate model-to-model standard deviations (interpretation of temperature maps of Hulme et al., 1990).

Holocene climatic optimum a departure from the modern mean July temperature of up to 4°C was reconstructed, followed by temporary cold fluctuations which took place between 3000 and 2000 BP, around 1500 years ago, and during the Little Ice Age with temperature drops of about 1°C. In the last intervening period during a secondary optimum of the Middle Ages (around AD 850–1200) temperatures were about 1°C higher than today. Temperature fluctuations of the same order of magnitude have also occurred in this century, illustrating once more the large temperature variability in northern regions, especially when looked at on a regional basis. In spite of an increase in atmospheric CO_2 content from 310 to 350 p.p.m. during the past 50 years, the meteorological record from Swedish stations shows a decrease in annual mean temperatures of about 1°C since the 1930s (see Koster and Boer, 1989). This cooling trend is particularly distinct in winter in the northern part of Sweden. Nevertheless, global-mean surface air temperature has increased by 0.3–0.6°C over the past 100 years, with the five global-average warmest years being in the 1980s (Houghton et al., 1990). The variability of the temperature climate makes it doubtful that the temperature signal of climatic change would be detected first in northern areas, in spite of the fact that the largest temperature increases are simulated for these regions.

Interactions of climate and permafrost

The heat exchange interactions of climate and permafrost can be visualized in terms of a so-called "buffer layer" between the atmosphere and the permafrost table (figure 6.3). The vegetation cover, the seasonal snow cover, the organic soil (if present), and the mineral soil, as part of the active layer, are the most important local natural factors affecting permafrost. Figure 6.4 tries to summarize the variability in time and the complex interrelationships of these factors. Naturally this greatly complicates the assessment of the effects of temperature change – let alone changes in precipitation – on permafrost development, as variations in different variables may either enhance or counteract each other (Nieuwenhuijzen and Koster, 1989).

The snow-cover thickness and duration is probably the most important factor owing to its insulating properties. The conductive capacity of dry new snow is 20 percent of that of old compacted snow or dry sand, or 5 percent of that of wet sand or 4 percent of that of ice (Gold and Lachenbruch, 1973). According to Nicholson and Granberg (1973), snow cover variations could explain 70 percent of the variance of ground surface temperatures in their studies near Schefferville, Labrador, Canada. Generally, a snow cover keeps the ground warmer, as it can heat up in summer, but is hampered in cooling down in winter. Goodrich (1982) showed that a doubling of snow cover from 25 to 50 cm could increase the minimum surface temperature by about 7°C and the MAGST by

Figure 6.3 Buffer layer
interactions affecting the
ground thermal regime
(adapted after Luthin and
Guymon, 1974).

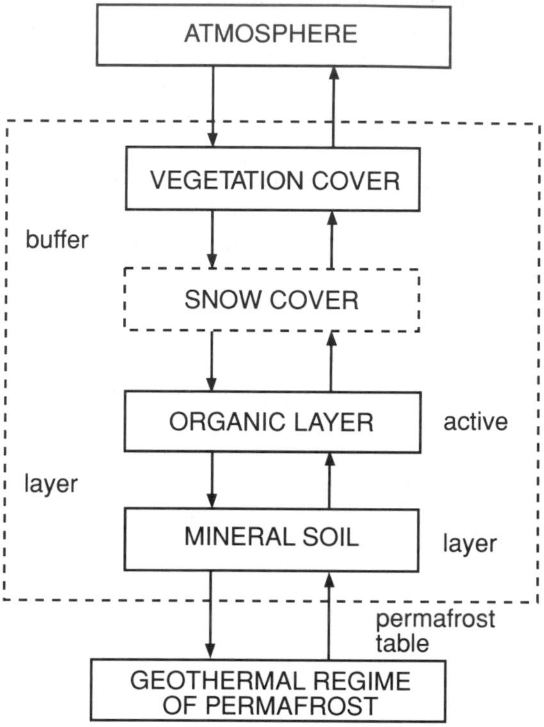

Figure 6.4 Schematic
representation of the
interrelations in the
atmosphere – "buffer
layer" – permafrost
system (after
Nieuwenhuijzen and
Koster, 1989).

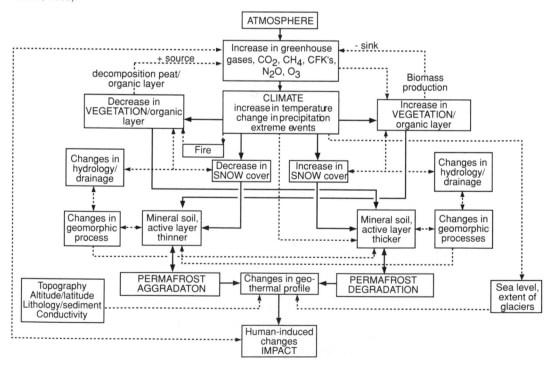

3.5°C. In discontinuous permafrost the snow cover is often the critical factor in determining the presence or absence of permafrost, but even in continuous permafrost it determines MAGST values during winter. The dependency of ground surface temperature on snow depth is illustrated in figure 6.5. However, if the snow cover increases very much it will take longer to melt and therefore less warming will occur in summer, consequently even promoting the occurrence of permafrost. The connection between thickness of snow cover and permafrost occurrence is also nicely revealed by measurements of King (1984, 1986) in the Tarfala region in northern Sweden (MAAT −3.5 to −4°C). Permafrost was regularly found at sites where snow cover in winter was less than 1.1 m, was missing at sites where snow cover thickness reached values between 1.4 and 2.4 m, but occurred again at sites where the thickness increased to more than 2.4 m.

Snow-cover thickness and duration depend to a large extent on the presence and nature of vegetation. Vegetation mainly has an effect upon surface temperatures by shading, thus cooling the

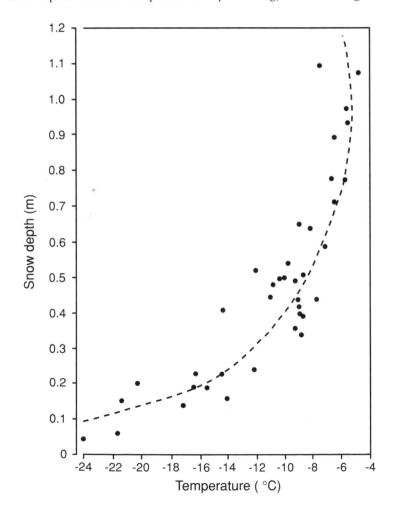

Figure 6.5 Effect of snow depth on ground surface temperature (after Williams and Smith, 1989), including a visually best fitting curve, showing that an increase in snow cover becomes less effective above 50 or 60 cm.

ground (Smith, 1975). Moreover, vegetation prevents radiation back into the sky at night, and the soil is dried by evapotranspiration, which decreases its heat capacity. Although these effects are less than that of the organic layer, they can still be responsible for patches of permafrost occurring under individual trees. The organic layer strongly influences ground temperatures by its differential insulation capacities. The particular influence of organic soil (especially peat) is attributed by Riseborough and Burn (1988) to: (a) the low conductivity of organic soil relative to mineral soil; (b) the effect of seasonal variations in the water content of organic soil on its thermal conductivity; and (c) the seasonal evaporative regime of the surface as controlled by climatic and surface factors. Estimates of the areal extent of peatlands in boreal and tundra regions vary from 2 to 3×10^6 km^2. In summer conductivity of a dry peat layer is extremely low (table 6.3). In winter, however, the organic layer freezes, leading to conductivities several times higher, and consequently the ground cools off. Therefore, in peatlands sporadic permafrost can even occur at MAAT values above 0°C. To a lesser extent the properties of the mineral soil also influence the conductivity (table 6.3). Changes in both snow regime and vegetation will determine the moisture condition of the soil and thereby the thermal conductivity of the materials. The overall effect of all these factors on permafrost differs locally, which means that, for example, on the one hand a gradual change from forest to tundra may lead to permafrost degradation by less insulation through vegetation, but on the other hand the same change may lead to permafrost aggradation through loss of snow cover. Figure 6.4 also illustrates that apart from climatological influences, changes in

Table 6.3 Representative values of the thermal conductivity of soils and their constituents

	Thermal conductivity ($W m^{-1} K^{-1}$)	Thermal conductivity (thawed)	Thermal conductivity (frozen)
Air	0.025		
Organic matter	0.25		
Water (0°C)	0.56		
Ice (0°C)	2.24		
Dry unfrozen peat soil	0.06		
Dry unfrozen clay soil	0.25		
Dry unfrozen sandy soil	0.30		
Wet unfrozen peat soil	0.50		
Wet unfrozen clay soil	1.58		
Wet unfrozen sandy soil	2.20		
Organic soil (high moisture content)		0.40	1.20
Fine-grained soil (medium moisture content)		1.13	2.21
Coarse-grained soil (low moisture content)		2.19	3.01

Source: simplified after Williams and Smith, 1989; Goodrich, 1982

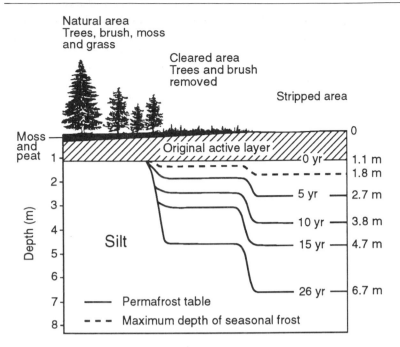

Figure 6.6 Permafrost degradation under different surface treatments over a 26-year period at an experimental site near Fairbanks, Alaska (mean annual temperature −3.3°C) (after Linell, 1973).

hydrology and geomorphology processes affect the aggradation or degradation of permafrost.

Changes in the active layer may occur rather rapidly. This applies especially to human disturbances. Probably the best documented rapid change in active layer development owing to controlled surface modifications of the vegetation cover is that by Linell (1973). Figure 6.6 shows an experimental test site near Fairbanks in an area of discontinuous permafrost, which was originally covered by a dense growth of white and black spruce attaining heights of 9–15 m. All trees and brush were removed from a part of the site and another part was completely stripped off to a depth of about 0.4 m. The strong thermal instability due to surface disturbance and the consequent rapid increase in depth of the permafrost table are clearly proven in this experiment. This was mainly due to a strong increase of summer temperatures. On the other hand, because of lower snow accumulation, ground temperatures in winter decreased, leading to some increase in the depth of seasonal frost.

Permafrost ecosystems

The cold climate, short growing season, and nutrient-poor soils in northern ecosystems have led to adaptations in plants and in soil organisms. In addition, the low rate of decomposition relative to primary production has caused a large accumulation of carbon as dead organic matter during the Holocene (Prudhomme et al., 1984;

Harriss et al., 1985; Sebacher et al., 1986). According to various sources (table 6.4), northern ecosystems contain 400–500 giga-tonnes (Gt) of carbon stored as soil organic matter, compared to 700–800 Gt of atmospheric carbon (Houghton et al., 1990). The areal extent of various northern ecosystems is not clearly defined and estimates of the total carbon storage also vary, but a "best guess" is probably that 30–40 percent of the worldwide soil carbon storage is contained in the taiga and tundra zone. The temperate and boreal forest zone form an especially important carbon sink. Present-day emission of methane, on the other hand, is dominated by northern wetlands, especially between 50 and 70°N, and particularly in western Siberia and east-central Canada, where extensive bog and fen areas occur. Studies by Billings et al. (1982, 1983) have revealed that wet coastal Arctic tundras could become a net carbon dioxide source at higher temperatures. Even slight decreases in water level owing to higher evapotranspiration could strongly accentuate the temperature effect. This effect seems to be stronger than the direct effect of increasing atmospheric CO_2

Table 6.4 Estimates of areal extent of carbon-rich soils and wetlands in northern regions and related sources and sinks of CO_2 and CH_4

Northern regions	Areal extent ($\times 10^6 km^2$)	Emission of CH_4 ($Tg\ year^{-1}$)	Soil carbon storage (Gt)
Global land area	149 (100 NH, 49 SH)		
Boreal (taiga) ecosystems	7.0		
Tundra/alpine ecosystems	6.9		
of which: bogs/fens	2.4		
Total bog areas NH	2.9		
of which: 0–50°N	0.7		
50–66.5°N	2.1		
66.5–90°N	0.1		
Boreal bogs and peatland	1.2		
Tundra wet sedge	1.0		
Tundra tussock	0.9		
Natural wetlands worldwide	5.7	60–140	
of which: Canada	1.3 ⎱		
Former USSR	1.5 ⎰	25	
All biomes			1300
Tundra			200
Northern taiga			270
Northern part of coniferous and temperate broadleaved forests			82
			552 = 42% of total
Worldwide			1400 ± 200
Wet boreal forest			133
Moist boreal forest			49
Boreal desert			20
Tundra			192
			394 = 29% of total

All data are estimates, which are rather poorly defined and still show large variations in the literature (1 Gt = 10^{15} g).
Sources: Bouwman, 1990; Post et al., 1982; Harriss et al., 1985

content on carbon uptake in the tundra. Field experiments by Prudhomme et al. (1984), Hilbert et al. (1987), and Oechel et al. (1987) in tussock tundra sites in Alaska have more or less confirmed these preliminary findings.

These studies on the effects of increasing CO_2 levels indicate that net primary production will increase somewhat, provided that nutrients are not limiting growth. However, in periglacial environments nutrients are usually a limiting factor. On the other hand, if temperature rises and drier conditions prevail, decomposition of organic matter may accelerate, causing more nutrients to become available (Viereck and Van Cleve, 1984). Elevated CO_2 could also cause increased carbon fixation, but that in turn might widen carbon–nutrient ratios to the point that immobilization by microorganisms was substantially increased. That in turn would decrease nutrient availability, increasing plant carbon–nutrient ratios still further, and hence decreasing productivity. The only valid conclusion at present can be that the net results of these interactive effects are fundamental and very poorly known (IGBP, 1990). Increased rates of decomposition with a resulting decreased thermal insulation capacity of organic soils will, if it occurs, certainly lead to permafrost degradation and subsequent thermokarst.

Studies by Harriss et al. (1985) and Sebacher et al. (1986) suggest that warming of Arctic and boreal wetland ecosystems will almost certainly be followed by increased methane emissions, which currently account for the release of about 25–40 megatonnes (Mt) of CH_4 to the atmosphere annually. It is estimated that a 4°C rise in these regions could lead to a 45–65 percent increase in methane release. If warming is accompanied by drying, then there may ultimately be a reduction of methane release, contrary to the possible carbon loss discussed earlier (Peterson et al., 1984; Houghton et al., 1990).

Changes in seasonal snow cover and permafrost will also affect climate through additional feedback mechanisms. Decreasing coverage by snow will result in an increase of the amount of absorbed solar radiation because of low albedo values of natural land surfaces. Furthermore, permafrost degradation could not only alter fluxes of greenhouse gases by oxidation of previously frozen organic materials, but could eventually also lead to methane release from gas hydrates contained within the permafrost (Williams and Smith, 1989). But these timescales are very long and quantitative measurements are as yet unknown (Eddy et al., 1988; Houghton et al., 1990).

Response of permafrost regime to climate change

The uncertainties outlined above notwithstanding, in principle a rise in MAAT and consequently in MAGST will have the following effects. First, the thickness of the active layer will increase. Second, the temperature profile within the permafrost will adjust itself to the new MAGST. The rate with which this happens depends on the thermal conductivity of the permafrost and the ice content of the ground. Eventually permafrost will decrease in thickness. Response times of the active layer, mostly related to local or regional disturbances near the surface, are in the order of years to tens of years, as shown above. But response times of heat diffusion, that is changes in the geothermal gradient in the upper part of permafrost (first hundred meters), are in the order of tens to hundreds of years. Finally, response times with respect to the permafrost base (in some cases of permafrost this is many hundreds of meters) are in the order of hundreds to thousand of years (figure 6.7) (Osterkamp, 1984a, b; Baulin and Danilova, 1984). Moreover, the rate of permafrost response to warming depends on the thermal diffusivity, which is strongly controlled by sediment properties and the water/ice content (Smith, 1988; Anisimov, 1989) (figure 6.8).

On long timescales permafrost distribution has shown a dynamic response to climate changes. Proxy data indicate that during the optimum of the last (Mikulino) Interglacial permafrost had thawed completely over the entire Russian Plain, and that the tundra zone, where permafrost now occurs, was replaced by a forest zone (Velichko and Nechayev, 1984). Compared to the extent during the last glacial maximum, permafrost boundaries have retreated north-

Figure 6.7 Thermal warming model for the continuous permafrost zone. The assumed initial equilibrium temperature profile (t = 0) and the final temperature profile at t = t_x, with a permafrost depth of X_x, are shown with the approximate temperature profiles for t = 10, 100, 600, and 1800 years (after Osterkamp, 1984a, b).

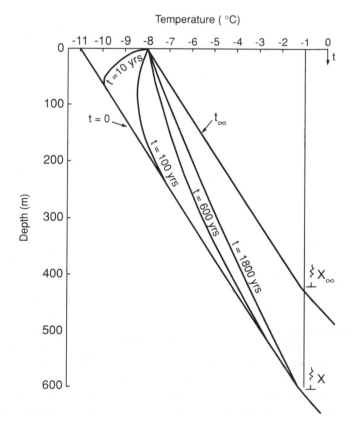

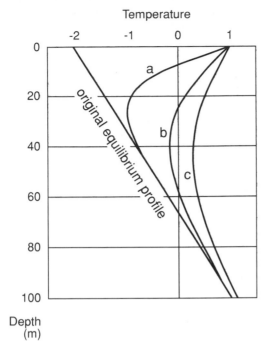

Temperature

original equilibrium profile

Depth
(m)

Figure 6.8 Simulated ground temperature profiles in materials characterized by an increasing thermal diffusivity following a climatic warming (a = low value, ice-rich permafrost; c = high value, unsaturated frozen (sandy) soil; b = intermediate value). The curves show the thermal response of permafrost assuming an increase in the ground surface temperature of 0.5°C per decade over a 60-year period, from an initial value of −2.0°C (after Smith, 1990) (thermal diffusivity is an index of the facility with which a material will undergo temperature change; it is the quotient of thermal conductivity and volumetric heat capacity).

wards by up to 1000 km in West Siberia (E of 75°E) and up to 2000 km in the Russian Plain (E of 30°E), according to Baulin et al. (1984). During the warm optimum of the Holocene the southern limit of discontinuous permafrost in the Russian Arctic was up to 600 km north of its present position. Likewise, vegetation zones and especially the forest/tundra boundary fluctuated by several hundred kilometers in Russia and North America in sympathy with even the short-term climate fluctuations of the late Holocene (Baulin et al., 1984; Nichols, 1975). In the southern part of the discontinuous permafrost zone in Manitoba the southern limit has shifted northwards over the past 150–200 years and the areal extent of permafrost terrain has diminished strongly, according to Thie (1974). Other evidence of relatively recent reactions of permafrost limits to temperature change have been documented by Mackay (1975) for the Mackenzie Valley (Arctic Canada). There is a good probability that MAGT values in this region increased by about 3°C during a recent warm period (late 1800s to the 1940s) and have since decreased by about 1°C. A shift in the continuous–discontinuous permafrost boundary from Fort Norman to Arctic Red River (a distance of about 350 km along the Mackenzie River) has been suggested, resulting from this temperature deviation.

The greatest changes can be expected to occur in those parts of the discontinuous permafrost zone where MAGT values are close to 0°C. Harris (1986) developed a classification of permafrost areas according to their thermal stability. Regions with a MAGT of more than −2°C are classified as unstable. Based upon his maps, Harris

(1986) concludes that a simulated global warming could cause permafrost degradation in at least 40 percent of the permafrost area, resulting in widespread development of thermokarst. Anisimov (1989) constructed a model of heat and water transport in a stratified medium, in conjunction with a paleoclimatic reconstruction method, to simulate changes in the extent of permafrost in Russia. He assumed that permafrost will degrade first in regions dominated by clay and loamy soils with relatively low water/ice content, while sandy sediments with high moisture contents are relatively stable. In this respect it is important to note that the greatest amount of ground ice often occurs near the permafrost table (figure 6.9). This is usually caused by the growth of segregation ice just above the permafrost table, aggradation ice incorporated just below the permafrost table, and wedge ice (Pollard and French, 1980; Harris, 1986; Pissart, 1990). The occurrence of ice-rich ground near the permafrost table improves the thermal stability of the permafrost as it delays the thawing of the ground due to the dissipation of the latent heat of fusion of ice (Washburn, 1979). In the Anisimov model equations, local factors are assumed constant. However, estimates suggest that changes in the "buffer layer" factors may produce the same effect on the permafrost thermal regime as a 1.5–2°C change in MAAT (Smith and Riseborough, 1984). Nevertheless, the model results based on a mean 2°C global warming (assumed to represent a 7–9°C increase in winter and a 4–6°C increase in summer in the latitudinal zone 55–70°N) suggest that in about 50 years the area occupied by continuous permafrost might be 15–20 percent smaller than at present. The vertical movement of the permafrost table is expected to be 0.5–0.8 m year^{-1} during the first year of the model run and to decrease exponentially over time. After 50 years of the model run the maximum depth of the thawed layer was 7 m. The rate of the permafrost boundary retreat was estimated to vary from 10–15 km year^{-1} in western Siberia to 20–25 km year^{-1} in eastern Siberia. Comparable rapid and substantial changes in active layer thickness have been computed by Goodwin et al. (1984) and Osterkamp (1984a, b), for locations along the Dalton Highway from Fairbanks to Prudhoe Bay (Alaska). In these cases maximum percentage changes are expected in soils with a thin active layer under a tundra vegetation. Thaw depth increases of 10–40 percent are simulated for MAAT changes of +3°C, whereas increases of 25–70 percent are simulated for MAAT changes of +6°C; the large variations being dependent on local factors. All model simulations suggest that owing to climate warming the active layer in continuous permafrost will thicken, but that permafrost will remain, in contrast to discontinuous permafrost which might locally disappear completely.

Secular climatic trends are clearly preserved in temperature profiles, mainly due to the slowness of heat diffusion and the retarding effect of latent heat exchange. Cermak (1971) already concluded that the response of ground temperature to climate

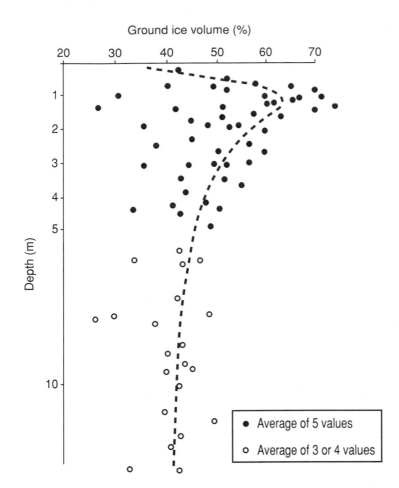

Figure 6.9 Distribution of ground ice volume with depth, Richards Island, Northwest Territories, Canada. The curve represents a visual best fit (after Pollard and French, 1980).

during the recent 150 years has penetrated to depths of about 300 m. Thus, the analysis of permafrost temperature as a function of depth appears to yield an integrated record of air temperature changes in the past. This has been well documented by temperature profiles obtained from boreholes in the Alaskan Arctic Coastal Plain (Lachenbruch et al., 1982, 1988; Lachenbruch and Marshall, 1986). A vast number of the temperature profiles show a distinct curvature toward higher temperature near the surface (figure 6.10). This clearly reflects a systematic trend toward warmer surface temperatures during the past century. The exact onset of warming seems to vary between locations, but they all indicate a warming in the range of 1.5–3°C during the past century, which seems to be in agreement with a similar trend in air temperatures as shown by regional weather records. Some curves even show a superimposed trend to lower temperatures in the uppermost part of the permafrost. This might be caused by the regional air temperature decrease during the past few decades, but it might equally have been caused by engineering disturbance during the drilling operations. Air

Figure 6.10 Measured temperature profiles from oil well locations near Prudhoe Bay (Alaskan Arctic Coastal Plain). Temperature profiles are offset to avoid overlap. The dots represent a small part of the actual measurement points. The shaded region for each curve is the interpretation of a temperature anomaly identified with a recent secular increase in surface temperature. The area above the line of dots on the three right-hand curves represents a superimposed recent cooling. The temperature change (°C) and dates (AD) are given for the best-fitting step-function climate history, reflecting the departure of the upper part of the geotherm from the steady-state linear geotherm (simplified after Lachenbruch et al., 1982, 1988; see also Lachenbruch and Marchall, 1986).

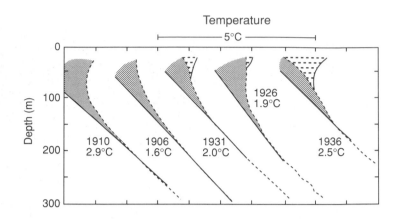

temperatures, averaged throughout the Arctic, seem to indicate a warming from 1880 to about 1940, cooling until about 1970 and warming thereafter.

Geomorphic and economic implications

The most direct consequence of permafrost degradation, particularly of ice-rich ground, is thermokarst. It is defined by Washburn (1979) as karst-like topographic features produced by the melting of ground ice and the subsequent settling or caving of the ground. Although the persistence of permafrost is aided by the occurrence of ice-rich material at shallow depth (plate 6.1), the mechanical consequences of thaw, if degradation sets in, are very serious. Widespread and intense thermokarst development is likely to occur. Apart from climatic influences permafrost degradation may also be induced by geomorphic or vegetational changes. Geomorphic changes, like active layer slope erosion, slumping, soil creep, or deflation, lead to a thinning of the active layer. This will cause a thermal equilibrium whereby thaw of the permafrost table and the associated release of excess ice as water will cause ground subsidence. Similarly, a reduction in the vegetation cover may lead to thermokarst. This form of thermokarst may in its turn promote the formation of ice segregation during winter freeze up, thereby providing additional summer meltwater and consequently further deepening of the active layer. There are many kinds of thermokarst features, including collapsed frost mounds, thaw slumps or ground ice slumps, linear and polygonal troughs, beaded drainage patterns, thaw lakes and alases formed by coalescing individual thermokarst depressions (plate 6.2). All these features are well-known and have been extensively studied in periglacial geomorphology (Washburn, 1979).

A second group of geomorphic features relate to slope stability and are attributed to mass-wasting processes. Various types of thaw slumps, frost creep, gelifluction, and other forms of mass-wasting are conditioned by the mechanical strength of the surficial

Plate 6.1   Ice-rich permafrost and massive ice wedges in organic-rich silt deposits in a former placer gold-mining excavation near Fairbanks, Alaska.

Plate 6.2   View from the top of Ibyuk Pingo (48 m high) to Split Hill Pingo (37 m high) near Tuktoyaktuk, Mackenzie River delta, Northwest Territories, Canada. The pingos are about 1 km apart. Numerous thaw lakes and frost mounds in various stages of aggradation or degradation are found in Mackenzie River delta.

material and are therefore strongly influenced by permafrost degradation.

Due to the development of a thermokarst relief the local drainage patterns will change, and surficial hydrological conditions in general will be drastically changed. With permafrost degradation and a longer summer season, erosion of river banks and lake and reservoir shorelines may increase, which in turn could cause changes in sediment transport as well as river discharge regimes. In its turn, higher discharge of rivers might enhance further permafrost degradation by thermo-erosional retreat of river bluffs (plate 6.3). In combination with sea-level rise, permafrost degradation may also cause an increase in coastal retreat in lowlands (plate 6.4). Moreover, altitudinal shifts in the boundaries of mountain permafrost are likely to occur with initial decreases in slope stability. Most of the permafrost engineering concerns related to warming of permafrost fall specifically into the following categories (Esch and Osterkamp, 1990): (a) warming of permafrost body at depth (increase in creep rate of existing piles, footings and embankment foundations, loss of adfreeze bond support for piling); (b) increases in annual thaw (thaw settlement during summer, increased frost heaving during winter); and (c) development of residual thaw zones (landslide movements and subsidence). In practical terms, increased terrain instability, especially in the first phases of permafrost degradation, would lead to major concerns for the integrity and stability of roads, railways, airstrips, dams, reservoirs, and other foundations in affected areas. Deepening of the active layer would

Plate 6.3 Strong river bank erosion in partly frozen sandy and silty deposits near Ambler along the Kobuk River, northwestern Alaska.

Plate 6.4 Coastal erosion of a region with horizontally banded massive ground ice near Peninsula Point, Mackenzie River delta, Northwest Territories, Canada.

subject foundations to continuing deformations as a result of thaw settlement. Decreases in the amount of ground ice present would lead to decreases in the mechanical strength of the associated soil as well as increases in permeability, both of which have significant consequences for engineering and natural processes. Comprehensive reviews of design procedures, construction methods, and maintenance requirements of structures and facilities in permafrost terrain are given by Harris (1986) and Péwé (1982). The additional factor of a potential climatic change has to be taken into account.

In summary, Harris (1986), French (1987), Williams (1979) and Ferrians et al. (1969) elaborate on the following activities, which are to a certain extent dependent on permafrost conditions: foundation structures for houses and other kinds of buildings (plate 6.5); construction of pads for roads and railways, foundations for bridges; layout of airfields; procedures related to exploratory and seismic surveys, drilling rigs for oil and gas exploration, fluid waste disposal, construction and maintenance of oil and gas pipelines (plate 6.6); placer, open-cast, and underground mining; water supply, waste disposal, and provision of electricity in towns; and agriculture and forestry. Increasingly, human activities are being directed toward the more remote northern regions of the world underlain by permafrost, which influences virtually all aspects of human occupation in these regions. The continued stability of this permafrost in a future warmer Earth can certainly not be taken for granted.

Plate 6.5 Old house damaged by permafrost degradation in Dawson City, Yukon, Canada. Thawing of ice-rich permafrost below the central, formerly heated part of the house, led to differential subsidence.

Further reading

1 A wealth of information concerning periglacial phenomena, permafrost regimes and engineering problems related to permafrost regions can be found in the Proceedings of the International Permafrost Association Conferences; with respect to the issue of global warming significant contributions have appeared in *(Final) Proceedings of the IVth International Conference* (1983/4), Fairbanks, Alaska (National Academy Press, Washington, DC) and in Senneset, K. (ed.) (1988) *Proceedings of the Vth International Conference on Permafrost*, Trondheim, Norway (three volumes). Trondheim: Tapir Publishers.

2 General textbooks by French (1976), Washburn (1979), Clark (1988) and Williams and Smith (1989) still form the most-cited reference works on geocryology and periglacial environments (see bibliography).

3 During the past few years several workshops and symposia have addressed the significance of northern regions in global change, including Vinson, T.S. and Hayley, D.W. (eds) (1990) Global warming and climatic change. Special issue, *Journal of Cold Regions Engineering*, 4/1.

4 A key reference with respect to changing climate and permafrost is that by Lachenbruch, A.H. and Marshall, B.V. (1986) Changing climate: geothermal evidence from permafrost in the Alaskan Arctic. *Science*, 234, 689–96.

Plate 6.6 Part of the 1300 km long Trans-Alaska Pipeline from Prudhoe Bay to Valdez, near Fairbanks. The isolated steel pipe is elevated on vertical support members that are anchored in the underlying permafrost. The zig-zag configuration and the positioning of the pipe on a horizontal brace between the piles is to allow for thermal contraction and expansion as well as earthquake control. The heat exchangers on top of the piles, filled with hydrous gas, serve to keep the ground frozen and firm.

5 A comprehensive review on the management problems concerning foundations for buildings, roads, railways, pipelines, and mining activities in northern regions is found in Harris, S.A. (1986) *The Permafrost Environment*. London: Croom Helm.

CHAPTER SEVEN

Ice Volumes and Climate Change

David Sugden and Nick Hulton

A visitor to this planet in the 1990s would be surprised at the concern with global warming. Viewed on a timescale of hundreds of thousands of years we can count ourselves lucky that ice sheets are unusually restricted in area and have withdrawn from northern Europe and northern North America. For most of the past 2.6 million years there has been more ice on Earth and it is only during interglacials, which have an average length of 10 000–15 000 years, that global conditions have been as warm as they are today. Our present interglacial has already lasted for around 10 000 years and so our extra-terrestrial visitor could well expect us to be concerned with how best to stave off the next ice age. We may yet discover that the rise in greenhouse gases associated with our use of the planet is the best way of preserving our equable present climate for the future.

Role of ice sheets in the global environment

Today, as during the past few million years, ice sheets play a pivotal role in modulating the global environment. Through their growth and decay they change the surface topography and albedo of the Earth and thus influence global temperatures and wind patterns. They lock up water in their mass and change sea level. Their behavior is marked by feedback mechanisms which can amplify small climatic changes, or even introduce large glaciological feedback which may alter the direction of global climatic change itself. Finally, their fluctuations can act as an early warning of change. The break up of the small Wordie Ice Shelf in a relatively warm part of Antarctica might be one such example (Doake and Vaughan, 1991) (plate 7.1).

The most important imprint of ice sheets on the global environment is the 100 000-year ice age climatic cycle. The oxygen isotope

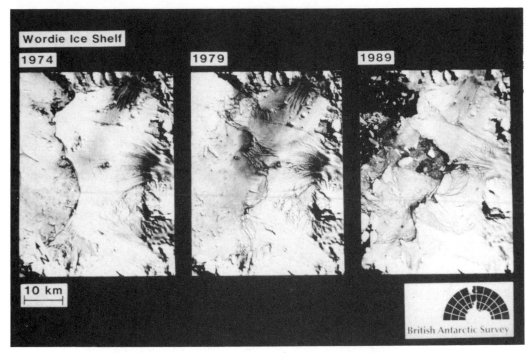

Wordie Ice Shelf
1974   1979   1989
10 km
British Antarctic Survey

ratio in organic remains in sea-floor sediments is an indication of the ice volumes on land (Shackleton and Opdyke, 1973). Dating of the sediments suggests that there have been several cycles of slow but irregular ice-sheet growth followed by sudden terminations as the ice sheets collapse and the world returns to interglacial conditions. The abruptness of the terminations is remarkable. Study of the ice cores in South Greenland points to a rise in temperature of 7°C in only 50 years at the end of the last glaciation 10 700 years ago. This climatic flip was accompanied by a reduction in storms, a 50 percent increase in precipitation and a dramatic decrease in atmospheric dust (Dansgaard et al., 1989).

The extent of the main Arctic and Antarctic ice sheets at the last maximum is shown in figure 7.1. In addition, there were mountain ice caps in South America, the Alps, and the Himalaya. The main features are the greater offshore extent of the existing ice sheets in Greenland and Antarctica and the creation of new ice sheets in North America and Europe, including offshore shelves. The larger ice sheets are, or were, larger than the present United States with altitudes and thicknesses in excess of 4 km.

The cyclic behavior of ice sheets over time is characteristic of a natural system which is hunting for an equilibrium, without achieving stability. The implication is that interglacial conditions are not stable and the global environment moves toward a colder ice age equilibrium. However, before it achieves stability with much more ice than there is at present, it is destabilized and suddenly flips back to interglacial conditions once more. While it is probable that

Plate 7.1 Break-up of Wordie ice shelf on the Antarctic peninsula as recorded in a time series of satellite images (after Doake and Vaughan, 1991).

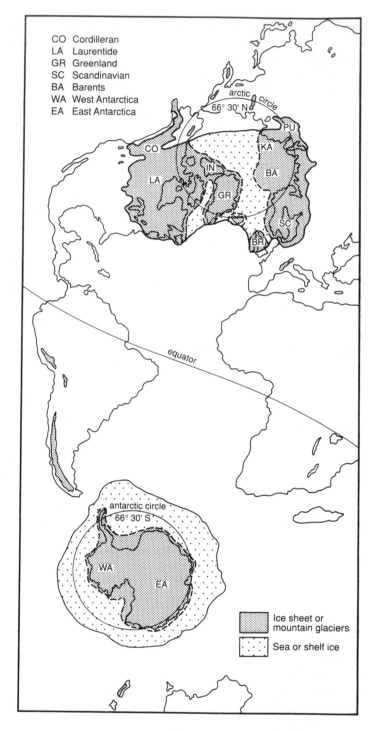

Figure 7.1   The maximum extent of ice sheets during the last ice age (after Denton and Hughes, 1983). Sea ice covers the Arctic Ocean.

this oscillatory pattern is triggered by variations in the Earth's receipt of solar radiation, it is very likely that the growth of the ice sheets themselves accentuates the slide toward an ice age. The increased albedo and the increase in altitude are sufficient to cool the global climate and encourage further ice-sheet growth. Furthermore, there are powerful mechanisms of ice-sheet collapse which can help to terminate ice ages. Calving of icebergs can remove the offshore marine portions of ice sheets within a few hundred years (Mercer, 1978). There is a delicate balance between the weight of the overlying ice and its grounding on the sea floor and quite small sea-level changes may be sufficient to cause the ice to float and to set the calving process in motion (Hughes, 1981).

It is tempting to argue, along with Denton et al. (1986), that it is ice sheets which amplify the small variations in solar radiation caused by orbital parameters. The orbital fluctuations have been recognized since the work of Croll (1867) and Milankovitch (1930), but there is still no clear understanding of the way they cause ice sheets to grow and decay. The two shorter cycles involve no change in the total amount of radiation reaching the earth. Rather, they simply redistribute the existing solar radiation between the hemispheres. This implies that the contrasting geographies of the two hemispheres must be involved in any theory of ice-sheet growth and decay. This being so, it is important to understand the geographical constraints on the behavior of individual ice sheets.

Response of existing ice sheets to warming

Our present lack of understanding is illustrated by the very different views about the state of health of existing ice sheets and their possible response to global warming. The Antarctic ice sheet, representing a sea level rise of 65 m if it melted, consists of two contrasting components (table 7.1). The East Antarctic ice sheet is

Table 7.1   Some physical characteristics of ice on Earth

	Antarctica (grounded ice)	Greenland	Glaciers & small ice caps
Area (10^6 km^2)	11.97	1.68	0.55
Volume (10^6 km^3 ice)	29.33	2.95	0.11
Mean thickness (m)	2488	1575	200
Mean elevation (m)	2000	2080	–
Equivalent sea level (m)	65	7	0.35
Accumulation (10^{12} kg year^{-1})	2200	535	–
Ablation (10^{12} kg year^{-1})	<10	280	–
Calving (10^{12} kg year^{-1})	2200	255	–
Mean equilibrium-line altitude (m)	–	950	0–6300
Mass turnover time (year)	~15 000	~5000	50–1000

Source: after Warrick and Oerlemans, 1990

based on a coherent continental base and forms a broad dome over 4 km in altitude which is drained around the margins by a series of radial ice streams (figure 7.2a and plate 7.2). The ice streams may lead into rock-bound glacial troughs, particularly in the Transantarctic Mountains, where they are known as outlet glaciers. The West Antarctic ice sheet, which comprises three contiguous domes

Figure 7.2 (a) The topography of the existing Antarctic ice sheet.
(b) The former reduced East Antarctic ice sheet postulated on the basis of marine fossils in Sirius till in the Transantarctic Mountains (after Webb et al., 1984).

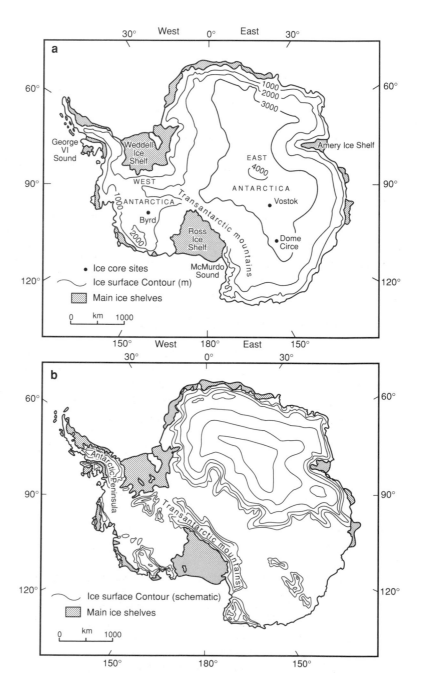

Plate 7.2   Oblique air photograph of an Antarctic ice stream.

rising to approximately 2000 m, is in contrast grounded in many areas on a bedrock floor below sea level. It too is drained by ice streams, many of which flow into ice shelves, floating sheets of ice which occupy coastal embayments and flow seawards at velocities of several hundred meters per year. The Ross Ice Shelf, which occupies the Ross embayment, is as large as France and is fed by outlet glaciers from both East and West Antarctica.

For some time it has been thought that the East Antarctic ice sheet formed in the Tertiary period and that it is an exceedingly stable feature of the global environment. Originally based on the deep-sea isotopic record (Kennett, 1977), but now amplified by geological evidence (Denton et al., 1984) and glaciological considerations (Robin, 1988), it seems possible that ice began to build up in East Antarctica some 50–60 million years ago and formed a major ice sheet by 38–18 million years ago (Sugden, 1987). The feedback effect caused by its own mass lowered Antarctic temperatures so far that it has survived any subsequent warm periods intact. The existence of unbroken cold polar desert conditions for at least four million years in ice-free areas of the Dry Valleys is a recent discovery which is in full agreement with such a hypothesis (Marchant and Denton, 1990).

The discovery of marine diatoms in the Sirius formation, a glacial till found throughout the Transantarctic Mountains, represents a challenge to the view of long-term stability. The diatoms encompass ages of 2.5–3.0 million years and it is argued that they were originally growing in marine embayments in East Antarctica at that time (Webb and Harwood, 1987; Barrett, 1991). The implication is that the ice sheet must have been much reduced in size and that it has exhibited unstable behavior in the relatively recent past (figure 7.2b). The implication is also that conditions warmed up sufficiently to remove an ice sheet from what is currently the coldest part of the globe, a conclusion apparently backed up by the presence of tree fossils in the same deposits in areas which now have mean annual temperatures lower than −20°C.

The contrast between the two views is dramatic. Both are based on field interpretations of geomorphic or geologic evidence and yet they have fundamentally different implications for the behavior of the East Antarctic ice sheet during warmer climates. The former interpretation implies that there is no risk of removing the East Antarctic ice sheet under any likely scenario of future global warming. The other implies that the ice sheet is unstable and could become much smaller in a future warm period and contribute many tens of meters to sea-level rise.

There is uncertainty about the West Antarctic ice sheet, which is grounded below sea level in many places. Following Mercer (1978), Hughes (1981), and van der Veen (1987), it is possible that the ice sheet is buttressed by the surrounding ice shelves and that a small rise in sea level or an increase in the melting rates beneath the ice shelves could trigger ice calving and the collapse of the ice sheet. Such an event would raise global sea level by 5–7 m. Such collapse is known in other areas from the Quaternary record, for example the final loss of the North American ice sheet over Hudson Bay between 8000 and 7000 years ago (Bryson et al., 1969). In the case of West Antarctica, the scenario depends critically on the shape of the subglacial topography beneath the ice sheet and particularly the existence of suitable calving bays. If the calving ice edge retreats into deeper water, then the rate of calving will increase, leading to progressive collapse. At present we simply do not know the shape of the bed in sufficient detail to know if collapse is imminent for the ice sheet as a whole or even for one drainage basin. In the absence of such information, there has been a tendency to assume that the risk to the ice sheet in a warmer world has been overstated and to assume a slight thinning owing to increased bottom melting and sea level rise (Oerlemans, 1989; Houghton et al., 1990).

The response of the Greenland ice sheet to any warming is also uncertain (Rech, 1989). On the basis of the retreat of outlet glaciers in West Greenland over the past century (Weidick, 1984), which has coincided with a marked warming, it seems fair to expect more melting and a decrease in the size of the ice sheet. On the other hand, satellite observations show that in the years 1978–86 the ice sheet in the warmer south increased in thickness (Zwally, 1989). An indication of the uncertain response of the ice sheet to warmer conditions may be made by analogy with the warm interval of the last interglacial. Koerner (1989) argued on the basis of ice characteristics in the bottom of ice cores that the ice sheet had melted, at least in the south. But geomorphic and geological evidence tends to support the view that the ice sheet survived the interglacial (Funder, 1989).

Considerable agreement concerns the behavior of mountain glaciers and ice caps in a warmer world. They only contain a small part of the ice mass on Earth (table 7.1), but there is good evidence around the world for their retreat since the cooling of the Little Ice Age between about AD 1590 and 1850 (Grove, 1988) (see figure 4.11). The most complete study was made by Meier (1984), who documented the loss of ice mass and calculated that they contributed 2–4 cm of sea-level rise in the past 100 years. Because this period coincided with warming, it is reasonable to expect further warming to cause further retreat of mountain glaciers (plate 7.2). As small mountain glaciers are likely to respond most quickly to climatic change they are likely to be major contributors to the initial phase of any global warming.

Problems of prediction

It is clear from the above that our understanding of the behavior of the largest ice masses is very poor. In part this reflects the inaccessibility of the polar ice sheets, their large size, and the sparseness of observations on which to make calculations of their mass balance. However, the problem runs deeper than this. The ice sheets are part of a complex system with links with the atmosphere, oceans, biosphere, and lithosphere. In order to understand how they may respond to forcing, for example a warming of 3°C, it is necessary to understand the relative importance of the feedbacks between all components of the system. Ideally, one can aim at a type of global model akin to the atmospheric general circulation models, based on fundamental physical principles, but which includes ice, oceans, biosphere, and lithosphere. The trouble is that ice sheets grow and decay in a non-linear way with thresholds of behavior which are very sensitive to initial conditions and slight perturbations. The difference between a stable marine ice sheet and one that is unstable may simply be too dependent on detailed topography and bed conditions to be modelled with confidence. If this is so, then the system as a whole with many such thresholds may be impossible to model, at least in a deterministic way.

Following Saltzman (1985), perhaps we should concentrate on using field evidence to develop and constrain simpler models, subject, of course, to them satisfying current knowledge of basic physical principles. In time it may be possible to combine different models and make some probability statements about particular predictions. The first step is to develop a model of each ice sheet, which takes into account its unique climate and topography, and is able to predict the growth and decay of the ice sheet through the course of a glacial cycle. These predictions can then be compared with the past record of change obtained from field evidence. The matching of the predicted and actual behavior provides the opportunity to refine the model and to gain some idea of its sensitivity to different climatic assumptions. Once the model is able to simulate known behavior effectively, then, and only then, can one expect it to provide sensible predictions for the future.

The most complete field record of the morphology and behavior of individual ice sheets comes from terrestrial geomorphological and geological evidence. It should be possible to reconstruct the ebb and flow of a margin, the ice thicknesses at different times, former subglacial thermal conditions and ice margin conditions. All this evidence is at the relatively high resolution of individual landforms and thus, if it is to be employed effectively, we need to develop a model at a sufficiently high resolution to be able to predict the location of an ice sheet margin at a particular place at a particular time. This suggests that at the very minimum the model must be able to take into account features as small as ice streams or individual valleys.

One important implication of this approach is the importance attached to field studies. Rather than representing an activity long on description and short on theory, fieldwork now becomes central to the development of theory. The types of question, the optimum locations and the types of field evidence required are directed by the need to understand ice-sheet behavior. In return the fieldwork provides the basis on which to reject certain hypotheses and to rediscover or develop others.

Understanding ice sheet behavior

The rest of this chapter explores a coupled modelling/field approach for ice sheets. We discuss the glaciological underpinning of a high-resolution ice sheet model for Greenland, test its performance against the record of past change, and then make some predictions of how the ice sheet is likely to respond to warming of 3 and 6°C.

Glaciological mechanisms

Ice sheets may be expected to respond to warming in three main ways: by warming the ice, which affects the flow characteristics; by changes in surface mass balance, which increases or reduces the ice volume; and by changes in sea level or sea temperature, which affect the amount of ice lost by calving.

Ice which accumulates at the surface of an ice sheet may take many tens of thousands of years to pass through the system. The surface ice accumulates close to the mean annual temperature and is influenced in its flow characteristics as a result. Cold ice is stiffer than warm ice. After a sudden warming, warmer ice accumulating at the surface would gradually replace colder ice in the system. Eventually this softer ice would lead to a lower ice sheet with a reduced volume and thus contribute to a rise in sea level. In practice, however, the response time for this adjustment is so long and the effect so small that it is not significant on a human timescale. It has been calculated that there is still abundant cold ice in the Greenland ice sheet dating from the last glacial maximum ca.18 000 years ago, and that its eventual disappearance will only affect ice-surface altitudes and volumes by a few percent (Dahl-Jensen and Johnsen, 1986). The effect of warming on the high East Antarctic ice sheet at Vostok has been calculated to lower the surface by some 30 m, but it would take 100 000 years.

Changes in the surface mass balance are critical to the response of an ice sheet to warming. The mass balance describes the total gain or loss of mass to an ice sheet and is a measure of the total accumulation minus the total ablation. The surface mass balance describes the direct link with climate on the surface of the ice sheet and measures snow accumulation minus the amount lost by melting (and/or sublimation in conditions where it is too cold for melting). The altitude at which accumulation balances ablation is called the equilibrium line altitude. It may be likened to the snow line in temperate climates. Superficially, it sounds sensible that more warming will lead to more melting and thus a reduction in the volume of an ice mass. But it is not so simple in the case of ice sheets. Figure 7.3 shows how ablation increases sharply from environments where the mean annual temperature is $-12°C$ to values of over 4 m per year where the mean is near $0°C$. In cold conditions, accumulation is related to the ability of air to hold water vapor. Thus it only measures a few centimeters per year at temperatures below $-20°C$ and rises as temperatures approach zero. In cold dry parts of the interior of ice sheets the mass balance is limited by the snowfall. The implications of this are that the change in mass balance related to warming depends on where you are. In environments colder than about $-10°C$ warming will increase the mass balance, while in warmer environments, increased melting will reduce the mass balance.

There are important implications for ice sheets. Ice sheets such as East Antarctica, where most temperatures are below $-10°C$, are likely to increase in size with modest warming. So also will

Figure 7.3 (a) The relationship of ablation and accumulation to annual surface temperature for an ice sheet. (b) The dependence of net annual mass balance on temperature changes from positive to negative, which complicates the response of ice sheets to climate change. Schematic diagram from Warrick and Oerlemans (1990).

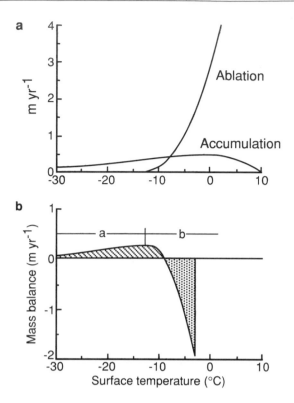

northern sectors in Greenland. But southerly sectors of Greenland are likely to be reduced by increased melting. An extra complication relates to mass balance changes with altitude. In South Greenland, melting could lower the marginal areas at the same time that the high interior is thickening owing to increased snowfall. The result is to steepen the surface slope and to increase flow velocities, which then counteract the thickening of the interior. This example is a good illustration of the feedback links which make prediction so difficult.

Iceberg calving is an ablation mechanism whose importance justifies separate treatment. There is a strong and exponential link between the rate of calving and water depth (figure 7.4). The relationship is so sensitive that an increase in water depth at an ice sheet margin can trigger iceberg calving, increase flow toward the margin, and thus lower the surface of the ice sheet interior (plate 7.3). This is the mechanism thought to have caused the high rates of retreat of marine ice margins documented in the Quaternary record, such as the disintegration of the Laurentide ice sheet over Hudson Bay. The water depth could be increased by the worldwide rise in sea level associated with the thermal expansion of the oceans in warmer conditions, or locally by tectonic or isostatic subsidence of the Earth's crust at an ice margin. Alternatively, increased bottom melting at a margin could thin the ice and cause it to float and flow faster.

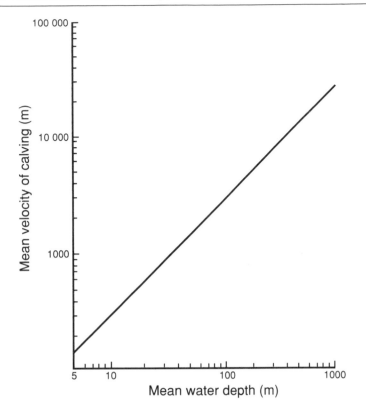

Figure 7.4 The relationship between glacier calving and water depth based on observations in Alaska by Brown et al. (1982).

Plate 7.3 Marine margin of Antarctic ice sheet, showing annually banded layers of ice.

Modelling the Greenland ice sheet

We have developed a high resolution numerical model which can
trace the evolution of an ice sheet over time. It has the particular
advantage that it can simulate non-linear ice sheet behavior, such as
rapid collapse or growth of one sector, in response to topographic
variations. The model, which builds on the work of Mahaffy
(1976) and Budd and Smith (1981), was developed by A. J. Payne
(Payne et al., 1989). It depends on the continuity equation for ice
solved over a finite difference grid. Average ice velocities are
calculated for idealized columns of ice on the basis of surface slope
and thickness, using separate terms to describe ice deformation and
ice sliding. The surface mass balance is calculated for each column
and added to the ice flux from adjacent columns (figure 7.5). This is
then represented as a change in ice thickness in the next iteration of
the model. Iterations for all the columns in an ice sheet are carried
out about every two years of simulated time. The model is linked to
the lithosphere via isostasy, which takes into account the ice load,
and the whole is related to eustatic sea level. The output of the
model is a series of maps of ice extent, thickness, and velocity at any
stage of a model run.

Figure 7.5 The
calculations in the ice
sheet model are carried
out for idealized columns
with iterations every two
years of simulated time.
Change in ice thickness
over time (dZ/dt) is the
sum of the surface
balance, b and the ice
flux to other parts of the
system, $V(\overline{V}. Z)$, where $\overline{V}$
is the vertically averaged
column velocity defined
as $\overline{V} = V_{shear} + V_{sliding}$.
This can be thought of as
the transfer of ice
between adjacent ice
columns.

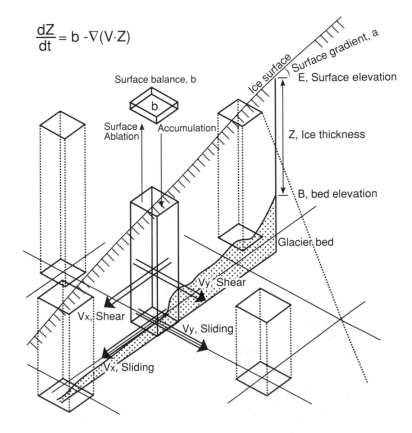

$$\frac{dZ}{dt} = b - \nabla(V \cdot Z)$$

If the long-term processes controlling ice sheets are to be correctly simulated, it is crucial to describe the mass balance over the ice sheet correctly. It does not matter how sophisticated the modelling of ice flow is if the modelled ice sheet simply melts away because the mass balance is incorrect. Moreover, by allowing the simulated climate over the ice sheet to change, the modelled ice sheet can be made to grow and shrink. The mass balance is difficult to model because the atmospheric processes acting above the ice sheet and at its surface are much more complex than the relatively simple physics controlling ice sheet flow.

There are a number of techniques that can be used to overcome these problems and yet ensure that the model is simulating mass balance realistically over the long term, while introducing global and local climatic factors. The present ice sheet in Greenland provides data on the real mass balance which can be used to constrain the model. The most important atmospheric variable is temperature. It varies with latitude and altitude and also with distance from the oceanic heat sinks. Temperature effects are relatively simple to model and are included in the modelling of both accumulation and surface ablation.

Surface ablation rates depend upon the heat energy balance at the surface. This is affected not only by the atmospheric and ice surface temperatures but also by the sun's insolation and by atmospheric and surface albedo. It is possible to derive characteristic surface ablation rates for given latitudes which account for differences in insolation and seasonal length and which are averaged for differences in albedo. Budd and Smith (1981) give averaged measured values for Northern Hemisphere surface ablation rates and these were used to simulate values in Greenland model. Surface ablation is a non-linear function of latitude and altitude.

Accumulation is a little harder to simulate. The Greenland model relies on a predictive statistical correlation between accumulation on the one hand and surface temperature and surface slope on the other. The known pattern of variability over the ice sheet was used to calibrate the relationship. Surface temperature is a good predictor because it strongly controls the amount of moisture the atmosphere can hold. Accumulation is a function of latitude, altitude, and also distance from the ocean.

Changing the accumulation and surface ablation patterns over time is done by running the equations with a forced temperature change. In the model, this alters the entire distribution of both mass balance elements. Areas which previously had net gain in ice mass might demonstrate net loss and vice versa. Thus, some areas of the modelled ice sheet can be made to shrink and others made to expand as the ice model continually tries to keep up with the changing surface mass balance. The surface balance can evolve in response to the changing elevation of the ice-sheet surface because mass balance is related to altitude. This replicates the feedback known to exist in the real world.

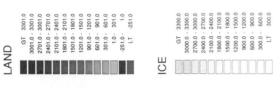

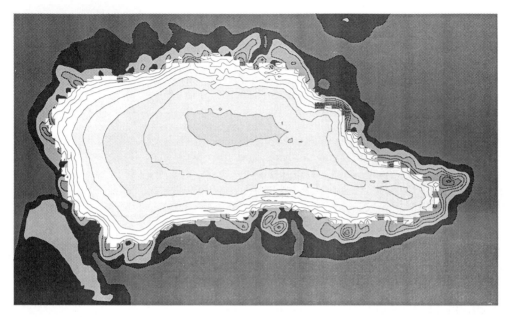

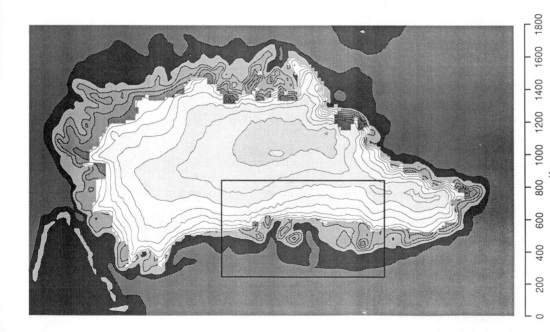

Figure 7.6 The surface topography of the Greenland ice sheet (left) compared to a modelled Greenland ice sheet (right) which is in equilibrium with present climate. The good fit suggests that the model is able to replicate the real ice sheet effectively. The box shows the location of figure 7.8.

Even with such seemingly simple models of the ice sheet it is possible to produce good simulations of the present ice sheet, and to simulate glacial change over the past and into the future. The present ice sheet gives us a very good test of the model. If we assume that the present ice sheet is in equilibrium with present climate, then running the model with the present climate and topography should

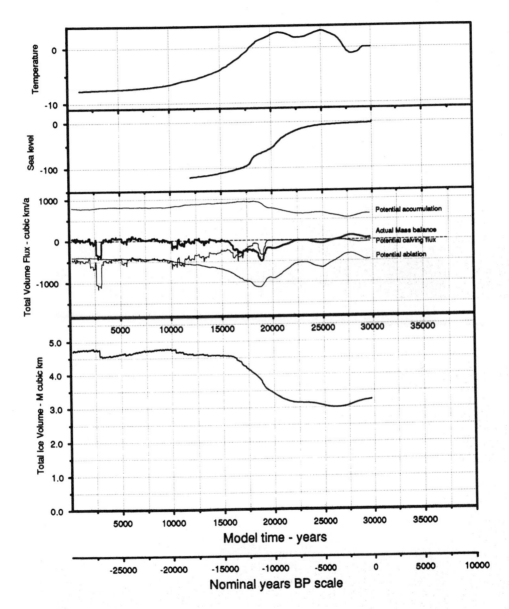

Figure 7.7 Modelled change in the Greenland ice volumes during the last phase of global deglaciation. The model is driven by the temperature record derived from the Camp Century ice core and by change in sea level. The mass balance changes which result are shown in the middle, while the ice volume response is shown below. Time is years BP.

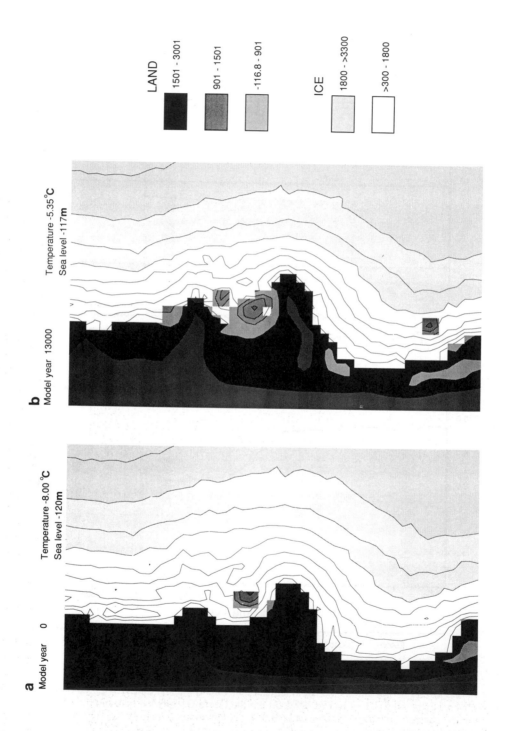

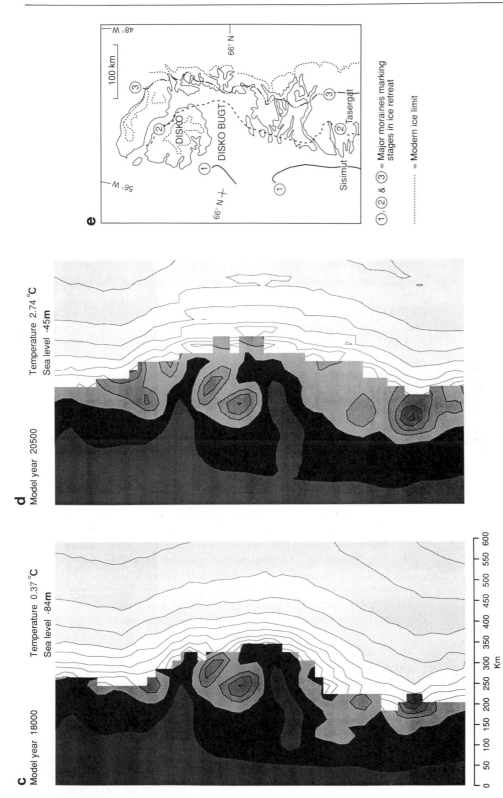

Figure 7.8 Modelled stages of deglaciation compared with the geological record of the margin changes in the Disko Island Area, West Greenland. The maps (a–d) show the changes from the maximum (model year 0; 30 000 BP), the early stages of deglaciation (model year 13 000, 17 000 BP) and during the process of rapid deglaciation (model year 18 000, 12 000 BP and model year 20 500, 9500 BP). The known retreat stages for the same area as portrayed by Funder (1989) are shown in (e) and reveal a similar pattern.

show little change. To a large extent this happens (figure 7.6). In fact the present ice sheet grows a little, which concurs with other research that suggests the ice sheet is actually in equilibrium with a slightly cooler climate than today (Oerlemans, 1991). Even if we alter the mass balance values by only 10 percent a very different model ice sheet results. This demonstrates that the present ice sheet is sensitive to changes in mass balance and also that the simulated mass-balance values are relatively good. Likewise, simulations of the glacial maximum state fit very well with the geological and geomorphological interpretations of the shape and size of the ice sheet at the last glacial maximum.

It is possible to force the model by altering temperature as well as sea level. The Camp Century ice core in Greenland gives us a record of temperature fluctuations since the end of the last glacial maximum (Dansgaard et al., 1971). In addition, evidence from former shorelines on coral reefs (Fairbanks, 1989) gives an indication of the change in global (eustatic) sea levels over the same period. Using these two records, the model was run to simulate the behavior of the ice sheet during the last global deglaciation. Figure 7.7 shows the change in total ice volume in comparison to the changing temperature and sea-level curves. In general, the volume change lags behind the forcing curves but the greatest changes follow the strongest forcing. Rising sea levels are significant in the early part of the deglacial sequence when ice loss by calving is more important because the ice sheet still has an extensive marine fringe.

It is possible to look in some detail at the model predictions for deglaciation. Figure 7.8 gives a detailed view of the modelled ice sheet at three stages during deglaciation, together with a map of major moraine systems associated with the retreating position of the ice margin. These major moraine limits are widely believed to be related to periods of standstill or readvance during deglaciation (Weidick and Ten Brink, 1974; Kelly, 1980) but it is difficult to ascertain whether this hiccup in overall retreat has climatological or topographical causes. The model can give some pointers as to why the margin pulls back in the way it does. The temperature signal driving the model is smooth and has no reversals, and yet the modelled ice margin still has positions where it lingers, and other areas where it retreats rapidly. For instance, the deep marine trough south of Disko Island causes rapid calving, but over areas of high ground like Sukkertoppen, the ice margin retreats more slowly because ablation is reduced. The model helps us understand the relationship between global climate change and its effect on small areas or sectors of the ice sheet which have particular topographies and which are part of a larger ice mass whose other parts may be responding differently. It clarifies why certain kinds of glacial behavior occur in certain locations within an overarching pattern of continuous global change. This is important, because the evidence for change obtained from geomorphology is gained initially at a local scale. By comparing the model predictions at a number of locations it is possible to see how the record of change in different

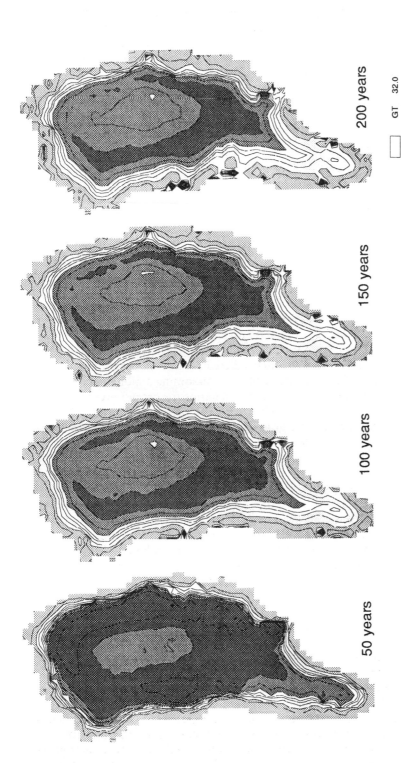

Figure 7.9 Numerically modelled change in the thickness (m) of the Greenland ice sheet at 50-year intervals for the next 200 years, assuming a stepped global warming of 6°C. The four maps represent the change in thickness after 50, 100, 150, and 200 years. Note the thickening of the high center and the thinning of the lower altitude fringe, especially in the south.

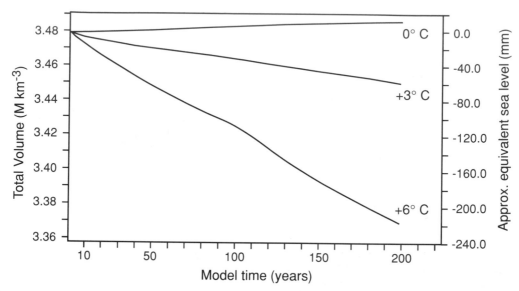

Figure 7.10 The ice volume loss from Greenland associated with a warming of 3 and 6°C and its effect on global sea level rise.

areas might be interrelated and what the broader causes for interlinked patterns of behavior might be.

In addition, the model can make predictions about future behavior. Figure 7.9 shows predicted ice thickness change over 50 year intervals for the next 200 years, assuming a global warming of 6°C. This is at the higher end of scenarios for greenhouse warming. The general pattern is for the ice to thin around the edge and to thicken in the middle; gradients steepen and the ice flows faster. The model gives a value of ice volume loss over this period as being equivalent to a rise in global sea level of 22 cm (figure 7.10). Such predictions have to be treated with caution. This one assumes that the ice sheet is in equilibrium at present and that the warming is instantaneous and constant. Over thousands of years with this amount of warming the model predicts that the ice sheet would melt completely. The model is not intended to make the most accurate or definitive prediction about future behavior but rather to demonstrate what ice sheet response would be anticipated for given certain magnitudes of climatic change and also to show what to look for if we are to observe change as it ensues.

Other ice sheets

The thrust of this chapter is that individual ice sheets respond in a distinctive way to a change in climate. There is not space to consider the response of other ice sheets in detail, but it is useful to hint at tendencies. Earlier in this chapter it was suggested that ice sheets in cold dry areas will increase in mass under warmer conditions. The Antarctic ice sheet is in such an environment and one might expect such an outcome if there is global warming. Fortuin and Oerlemans (1990) used regression analysis to establish

Table 7.2 Factors contributing to projected sea level rise (cm) 1985–2030

	Thermal expan- sion	Mountain glaciers	Greenland	Antarctica	Total
High	14.9	10.3	3.7	0.0	28.9
Best estimate	10.1	7.0	1.8	−0.6	18.3
Low	6.8	2.3	0.5	−0.8	8.7

Source: after Warrick and Oerlemans, 1990

the link between temperature and accumulation on the ice sheet surface. The predicted effect of a 1°C rise in temperature is an increase in volume and a slight fall in world sea level. Interesting confirmation of this comes from two field studies. In the Dry Valleys, Taylor Glacier, an outlet glacier of the East Antarctic Ice Sheet, was thicker in the past on several occasions. An intensive dating program on the moraines reflecting this thickening shows that it occurred during recent interglacials and the Pliocene (Denton et al., 1989). It was thinner during full glacials. This carries the implication that the outlet glacier increased its volume in warmer times and reduced its volume in colder periods. Further support for this relationship comes from recent observations on the Antarctic Peninusla. Over the past 30 years temperatures have risen by almost 2°C and have been accompanied by a 25 percent increase in snow accumulation (Peel and Mulvaney, 1988).

These Antarctic considerations ignore the possible instability of the West Antarctic ice sheet. Although it is possible that it may grow in size during a period of warming in response to changing surface mass balance, the sea-level link is poorly understood. While this remains the case and until the sub-ice topography is known in detail, we are not be able to rule out the possibility that it might begin a catastrophic collapse by calving, either in one drainage area or as a whole. In short, both accumulation and loss through calving are likely to increase under a warmer climate.

With the possible exception of the instability of the West Antarctic ice sheet, the main thrust of this chapter agrees closely with the Intergovernmental Report on Climate Change (1990). Table 7.2 shows the sea-level predictions made in that report. The sign and the amplitude of change are similar to those argued here, with about half of the projected sea-level rise being accounted for by ice volume changes. In the initial stages, at least, most of this melting will affect mountain glaciers rather than large ice sheets.

Conclusions

This chapter has tried to show that ice sheets are a key component of the global environmental system, with profound implications for climate and for sea-level change over time. The system is so complex, with many poorly understood feedback loops and sensitive thresholds, that it may not be possible to predict the response to human-induced changes for many years. A particular problem is the way topography, which is unique beneath a particular ice sheet, influences the dynamics of growth and decay. In this chapter we suggest that global warming will cause the Greenland ice sheet to be reduced in volume as melting in the south more than compensates for any increase of mass in the north. Probably, the East Antarctic ice sheet will grow slightly in a period of warming. The behavior of the West Antarctic ice sheet is difficult to predict with confidence, as it could respond in an unstable way. These differences in behavior illustrate the need to understand the characteristics of particular ice sheets in order to make broader global generalizations.

Further reading

Doake, C.S.M. (1987) Antarctic ice and rocks. In D.W.H. Walton (ed.), *Antarctic Science*. Cambridge: Cambridge University Press, 140–92.

Mercer, J.H. (1978) West Antarctic ice sheet and CO_2 effect. A threat of disaster. *Nature*, 271, 321–5.

Payne, A.J., Sugden, D.E. and Clapperton, C.M. (1989) Modelling the growth and decay of the Antarctic Peninsula ice sheet. *Quaternary Research*, 31, 119–34.

Sugden, D.E. (1982) *Arctic and Antarctic*. Oxford: Blackwell.

Warwick, R. and Oerlemans, J. (1990) Sea level rise. In J.T. Houghton, G.J. Jenkins and J.J. Ephraums (eds), *Climate Change*, IPCC Scientific Assessment. Cambridge: Cambridge University Press, 257–81.

CHAPTER EIGHT

Sea-level Response to Climate

Michael J. Tooley

Introduction

Stephen Schneider (1989) has noted that "sea-level rise is undoubtedly the most dramatic and visible effect of global warming into the greenhouse century." Sea-level rise will impact on all the coastlines of the world, and the effects will be particularly disruptive to coastlines formed of unconsolidated sediments backed by coastal lowlands and to the human populations that inhabit them. In UNEP's (1990) report on *The State of the Marine Environment*, attention was drawn to the fact that "half the world's population dwells in coastal regions which are already under great demographic pressure, and exposed to pollution, flooding, land subsidence and compaction, and to the effects of upland water diversion. A rise in sea-level would have its most severe effects in low-lying coastal regions, beaches and wetland."

These two quotations illustrate the dilemma faced by decision-makers planning for coastal change and in the coastal lowlands of the world. But while there is undoubtedly a link between sea-level rise and climatic change any changes in sea level will arise from a complex interaction of factors, of which some are natural and others are human-induced. The impacts of sea-level change will be global but their magnitude will vary on a regional and local scale.

Climate affects sea level in a variety of ways. Changes in wind regime may affect the incidence and magnitude of extreme water levels associated with storm surges. Temperature rise will result in higher sea level due to steric effects, that is, the expansion of surface water masses as their temperature rises. Temperature rise will also result in the wastage of high latitude and high altitude glaciers and ice caps. In the latter two cases, changes in the volume of water in the ocean basins result in what are known as eustatic changes in sea level. Over long geological timescales, eustatic or worldwide changes in sea level may also arise from changes in the capacity of the ocean basins caused by basin subsidence and sedimentation and lead to a fall and rise in sea level respectively.

Kuenen (1950) and Fairbridge (1961) elaborated on the eustatic mechanisms and identified three main causes:

1   Changing the shape and, hence, the capacity of the ocean basins by tectonic activity (tectono-eustasy), sedimentation (sedimento-eustasy), and water load (hydro-isostasy).

2   Changing the geohydrologic balance, or the ratio of water in the ocean basins to that on land. Glacio-eustasy is the dominant process, but secular changes in groundwater volume will also have an impact on the ratio, and both are climatically controlled.

3   Changing the total volume of water, by the addition of juvenile water from volcanic activity or by the loss of water during sedimentation. Steric changes are included here, and are manifest by the expansion or contraction of water masses resulting from changes in ocean water density as temperature and salinity changes.

The altitude at which sea level intersects the land is further complicated by vertical land movements arising from isostatic adjustments. Isostasy is a condition of equilibrium within the crust of the Earth which will be disturbed by the addition or removal of a load such as ice, water, or sediment. The depression of the crust in one area will be compensated by uplift elsewhere, as material flows in the transition zone between crust and mantle. Maximum depression occurred over Hudson Bay and the Gulf of Bothnia. The forebulge in the Northern Hemisphere during the maximum of the last glaciation extended 1000 km south of the Laurentide ice margin in North America and 800 km south of the Fenno-Scandinavian ice margin (Fairbridge, 1983). Following deglaciation rebound occurred in areas of ice loading and the collapsing forebulge migrated toward the former ice centers; uplift is continuing in both ice centers, and in Fenno-Scandinavia, since 13 000 BP, total uplift has been 830 m, whereas subsidence in the forebulge area has been 170 m (Mörner, 1980).

Fairbridge and Jelgersma (1990) have demonstrated the interrelationships between climatic change and sea level (figure 8.1) and drawn attention to the multiple forcings of sea-level changes, of which climate is only one.

Kidson and Heyworth (1979) demonstrated that the sea surface was far from "level" at many different spatial scales. The "topography" of the ocean surface is called the geoid. It has been defined by Pugh (1990) as the equipotential surface that would be assumed to be the sea surface in the absence of tides, water density variations, currents, and atmospheric effects. No horizontal forces disturb the surface, and it is the force of gravity that acts on it. Hence, the distribution of density of materials in the crust and mantle of the Earth results in a marked "topography" of the ocean surface as measured from a single fixed datum, which is the geometric center

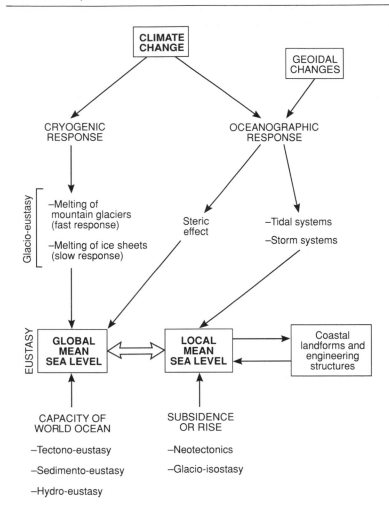

Figure 8.1   A diagram to show the interrelationships between climatic change and sea level with an indication of causative and feedback systems.

of the Earth. The surface deviates positively and negatively from the constructed ellipsoid of revolution of the Earth (King-Hele et al., 1980), comparable in shape to an oblate spheroid. The distribution of geoid relief (figure 8.2) is based, with refinements, on satellite altimetry from the SEASAT mission of 1978. On Goddard Earth Model 10^3 (Marsh and Martin, 1982) the range of deviations from the ellipsoid is −104 m south of India to +74 m over Irian Jaya and Papua New Guinea. The distribution of geoidal highs and lows is meridional and appears to be associated with the global pattern of positive and negative gravity anomalies, respectively, and therefore density variations (Fifield, 1984). Climatic change manifest by glacial and interglacial conditions will have an effect on geoid relief not only through isostatic processes, but also through the addition and loss of ice mass.

Dynamic sea-level changes and glacio-eustasy, however, are the two climatically forced variables that have the greatest impact on

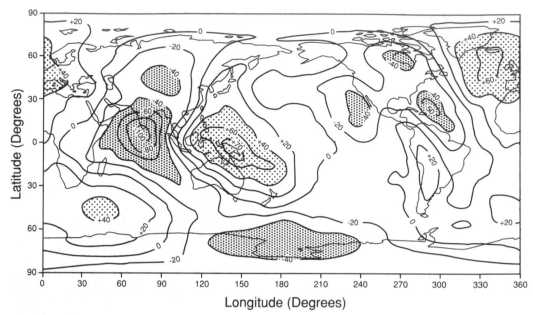

Figure 8.2 A map of the geoid surface topography derived from altimeter data from the NASA SEASAT mission of 1978. The long wavelength components of the geoid reveal undulations with a maximum altitude of about 180 m (Marsh and Martin, 1982; Fifield, 1984).

the continental shelves, coastlines and coastal lowlands of the Earth.

Dynamic sea-level changes

Changes of ocean level arising from meteorological, hydrological, and oceanographic processes may result in extreme water levels that erode natural and artifical sea defences and lead to extensive flooding of coastal lowlands (plate 8.1).

The distribution of atmospheric pressure elicits a response in ocean levels. Beneath the persistent subtropical anticyclonic cells, water level is depressed, whereas it rises beneath low pressure cells. The effects of pressure change and associated wind stress on water surfaces are difficult to separate, but, in theory, an increase in atmospheric pressure of 1 millibar will elicit a fall in sea level of 1 cm.

Both pressure changes and wind stresses operate in the North Sea, described by Heaps (1983) as "a splendid sea for storm surges," to produce the infamous surges of 1953 and 1978, during the former of which, when up to 3 m was added to predicted tidal levels, more than 2100 people lost their lives in the United Kingdom and The Netherlands (von Ufford, 1953; Steers, 1953).

In the Irish Sea, storm surges and extreme water levels also occur, of which the Towyn flood of February 1990 is one of the most recent manifestations (Englefield et al., 1990; Roe, 1991). During the last week of February, an intense low pressure cell (951 mb) crossed the north Irish Sea basin. Winds veered from southwest to

Plate 8.1 The Thames tidal barrage, London. Similar costly sea defence structures may be necessary in the future along other low lying coastlines to combat rising sea levels.

northwest and gusts up to 83 knots were recorded. Wave heights reached 4.3 m and an extreme water level (still water level plus wave height) of +11.1 m OD was recorded (OD = Ordnance Datum, the average value of mean sea level at Newlyn, Cornwall, England, 1915–21). The parapet on the sea wall at Towyn had a crest at +8.4 m OD. This was overtopped, and 467 m of the sea wall breached. The threshold of the breach was at +4.0 m OD. Flood waters rose to +5.5 m OD in Towyn and pushed south some 2 km: more than 10 km² were inundated. More than 5000 people were evacuated and 2800 properties (1290 of the housing stock) were damaged. Over one million pounds (US $1.5 million) was expended in six days to plug the breach.

Such losses pale into insignificance when compared to the loss of life and damage to property in the lowlands around the Bay of Bengal. Here, tropical cyclones travelling northwards into the Bay with associated onshore winds elevate water levels up to 9 m above predicted levels, as happened during the "deadliest tropical cyclone in history" in 1970 (Frank and Hussain, 1971), when more than 200 000 people lost their lives. The surge of May 1991 in the Bay of Bengal and its impact on the geologically subsiding Ganges delta around Chittagong was greater than that of 1970 and confirms the susceptibility of this area to devastating storms.

In the Northern Hemisphere, there is a long documentary record of storm surges and coastal inundations (Gottschalk, 1971, 1975,

1977; Lamb, 1980; Tooley, 1985). Lamb (in Pfister, 1990) shows a rise in the frequency of severe storms in the North Sea from AD 1550 to 1700, followed by a decline to AD 1750, a second peak in AD 1800 and a third in the late 1980s. Both overlapping this record and extending it back into the early Holocene (past 10 000 years) is the natural record of storms preserved in beach ridges (Fairbridge, 1987). In the English Channel, the flinty shingle ridges of Dungeness are an archive of storm surge incidence and severity, while among the rich documentary sources are records of the great thirteenth-century storms (Eddison and Green, 1988).

The associations of these storms with periods of climatic change have been explored by Fairbridge (1989: "crescendo events") and by Fairbridge and Jelgersma (1990). The culminating stages of the Medieval Warm Period were associated with ferocious storms in northwest Europe as the jet stream and frontal storm tracks shifted south, whereas the peak of the Little Ice Age was also associated with storminess in northwest Europe and polar front storms.

The storm surges in the North Sea in 1953 and the Irish Sea in 1990 left no gross sedimentary signature (Tooley, 1979; Tooley and Jelgersma, 1992). The consequence of the submarine Storegga slide off the Norwegian coast ca.7000 years ago, however, was the generation of a tsunami (seismic wave) in the North Sea and the deposition of a sand layer up to 74 cm thick in coastal sediments of eastern Scotland (Long et al., 1989). The sedimentary record of both extreme events such as this and gross changes in sea-level due to glacio-eustatic processes is preserved in continental shelf sediments and the unconsolidated sediments of the coastal lowlands of the world, some of which have undergone subsidence whereas others have experienced uplift.

The record of sea-level changes

The coastal lowlands and continental shelves of the world contain an archive of sea-level signatures from which both sea-level movements and vertical land movements can be determined. Rates and directions can be calculated, and together with proxy oceanographic and climatic data provide the surest foundation for outlining future sea-level changes and their impacts.

The imprint of sea level on the land is both erosional and depositional. The former can provide important altitudinal controls, but is difficult to date, particularly if inheritance plays a part in coastal history and evolution. Baulig (1935) had linked the ideas of erosional surfaces to eustatic sea levels and concluded that they were superior to litho- and bio-stratigraphic methods as a basis for subdivision and correlation (see discussion in Tooley, 1987). In fact, a combination of morphological mapping, levelling, stratigraphic and micropaleontological analyses (as exemplified by Haggart, 1987; Firth and Haggart, 1989) yields a database from which sea-level and land-level changes can be calculated. For between-site and within-site correlations an objective methodology is required, and this has been developed (Shennan et al., 1983) and applied to both middle and low latitude coasts (e.g. Shennan, 1982, 1986a, b; Haggart, 1987; Ireland, 1987). A battery of techniques is available for studying sea-level changes, and many of the techniques have been drawn together and described in van de Plassche (1986).

Two records of sea-level change are described here: one from the humid, temperate mid-latitudes of England's Lancashire coast (53–54°N) and one from the humid, subtropical low latitudes of the Brazilian coast (23°S).

The Lancashire coast, United Kingdom

The Lancashire coast, from the south shore of Morecambe Bay to the north shore of Liverpool Bay, is formed of unconsolidated sediments, and no solid rocks outcrop. The area has been affected by repetitive glaciations and emplaced tills up to 77 m thick have been laid down in the Fylde, east of Blackpool. Basins and valleys in the till, opening into the Irish Sea and affected both directly and indirectly by relative sea-level changes, have been filled with alternating organic and inorganic sediments of terrestrial, telmatic, limnic, and marine origin, attaining thicknesses of 20 m (Tooley, 1978). Although the area has been affected by glaciation, Shennan (1989) has demonstrated that there has been no isostatic uplift for the past 5000 years in south Lancashire (between the Mersey and Ribble estuaries), whereas imperceptible uplift of +0.10 mm year^{-1} has occurred for the past 5000 years in north Lancashire (between the Ribble estuary and south shore of Morecambe Bay) and +0.35 mm year^{-1} in Morecambe Bay.

Lytham, on the south coast of the Fylde, has the most complete record of sea-level changes from present-day land sites, spanning the period from 8575 ± 105 to 1370 ± 85 radiocarbon years BP (where present is AD 1950). The record comes from several sites within two basins in this area, of which the easterly of the two, Nancy's Bay, was one of four inlets on the coasts of southwest Lancashire and the Fylde that remained open until the recent historic past. Downholland inlet was occupied by the sea from ca.8500 until ca.4000 years ago, when it became choked by blown sand; Martin inlet was affected by the sea from ca.7000 years ago until AD 700 (Tooley, 1985); Lytham Hall Park inlet from ca.5000 to 800 years ago when it also became choked by blown sand; and Nancy's Bay from ca.7600 BP until AD 1741 when it was enclosed and reclaimed by the construction of a new sea embankment.

The stratigraphic record (figure 16 in Tooley, 1978; figure 1.4 in Tooley, 1981) is based on a series of boreholes and excavations close to the central axis of the inlet. The ground altitudes of the reclaimed inlet range from +1.5 to +4.1 m OD, rising significantly seaward. The maximum thickness of Holocene marine sediments is about 18 m, and boulder-clay was encountered at Dock Bridge at about −13 m OD (Wray and Cope, 1948).

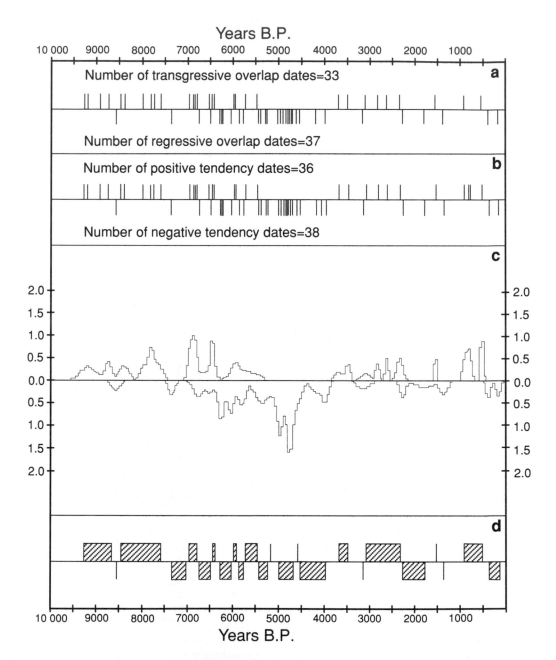

Figure 8.3 Stages in the construction of the sea-level chronology for northwest England. (a) Distribution of 70 radiocarbon dates from transgressive and regressive overlaps; (b) distribution of 74 radiocarbon dates showing positive or negative tendencies of sea-level movement; (c) combined frequency histograms of the 74 radiocarbon dates in which two standard deviations are accommodated; (d) a partial chronology of tendencies of sea-level movement. Boxes or bars above the line are positive tendencies (coastline retreat) and below the line are negative tendencies (coastline advance) (Shennan et al., 1983).

The sediments comprise marine silts, clays, sands, and, locally, pebble beds, intercalated with *in situ* peats of different origin — some are dominantly limnic laid down in brackish water or freshwater, whereas others are composed of the roots of herbaceous plants such as the common reed (*Phragmites*) and of woody detritus. Interruptions in marine sedimentation are shown by thin peat horizons dated to 7605 ± 85 and 6885 ± 80 radiocarbon years BP and confirmed by pollen analysis, and a thick peat horizon that accumulated about 6000 years ago. This peat was overlapped by marine clays and silts with abundant pebbles locally, which represent the upper tidal flat and beach deposits of the inlet before its reclamation in the eighteenth century. At more seaward sites ground altitudes rise from 1.5 to 4.1 m OD, and the stratigraphy displays a younger sequence of marine clays and silts intercalating peat, dated between 6290 ± 85 and 2330 ± 65 BP.

One of the thickest and highest peat beds on the Lancashire coast accumulated between 5775 ± 85 and 2330 ± 65 years BP, and both the stratigraphic record and the pollen record indicate that neither sedimentation nor the local and regional vegetation was uniform. At the base, monocotyledonous *turfa* with *Phragmites* rhizomes and the pollen of saltmarsh taxa, such as Chenopodiaceae and *Artemisia*, give way to freshwater conditions with *Typha angustifolia*, *Lemna*, and *Cladium mariscus*.

A sharp boundary separates the organic deposit from the overlying coarsely laminated clayey, silty sands, with gravel and pebble beds indicating either an erosional contact or the sudden arrival of a sand and shingle bank at the site, or both. This event happened shortly after 2330 ± 65 years BP and above this level a rich diatom assemblage with *Coscinodiscus*, *Diploneis*, and *Navicula* indicates fully open marine conditions.

These data have contributed transgressive (marine sediments overlying terrestrial sediments) overlap and regressive (terrestrial sediments overlying marine sediments) overlap dates from which a chronology of tendencies of sea level movement in northwest England (figure 8.3) has been derived (Shennan et al., 1983; Tooley, 1982). This clearly shows that there have been 12 periods of positive sea level tendency and 12 periods of negative tendency, and associated coastline retreat and advance during the past 9000 years.

The Rio de Janeiro coast, Brazil

The coast and coastal lowlands of Rio de Janeiro state have been described by Dansereau (1947), Muehe (1979), and Ireland (1987). The faulted escarpment of the Brazilian Shield with monadnock outliers of crystalline rock, such as Pão de Açúcar (the Sugar Loaf), is fronted by a narrow, discontinuous coastal plain with barrier beaches (*restingas*) and lagoons (*lagoas*). There are 20 lagoons, ranging in size from 1 to 200 km², from 0.5 to 3.0 m in depth, and with catchments ranging in extent from 270 km² (Marica) to 20 km² (Lagoa de Rodrigo Freitas). Crest altitudes of the *restingas* belong to two distinct altitudinal populations – one at about 4 m and the other at about 12 m (Muehe, 1979). The evidence of sea-level changes on this coast comprises prehistoric middens (*sambaquis*: Fairbridge, 1976; Kneip and Pallestrini, 1982), the sedimentary infill of lagoons and the encrusting tubes of gastropod Vermetids on crystalline rocks.

Ireland (1987) has summarized the evidence for sea-level changes on the Rio de Janeiro coast based on a consideration of the sedimentary sequence in the Lagoa de Itaipú, the Lagoa de Padre, and at Itaipú-Açu. At the first, where engineering borehole records had indicated a maximum infill of unconsolidated sediments of 17.7 m, 17 boreholes were put down from east to west (parallel to the coast) across the eastern telmatic zone of the lagoon. The maximum thickness of unconsolidated sediments was proved to be −8.7 m (Imbituba Datum), and comprised silt with a trace of sand. It was overlaid, as in all the boreholes, by a dark reddish brown, well-humified, woody, and herbaceous peat of variable thickness (0.45–1.85 m) and altitudinal range and capped in many boreholes by charcoal. This is overlaid in turn by silt and a second organic deposit, passing up locally into a *limus*, the altitudinal limits of which vary. This, in turn, is overlaid by another clay and silt, interrupted between −1 and −2 m (Imbituba Datum) by another peat, capped by a minerogenic horizon, before the surface peat is reached.

A diatom diagram from borehole 33/1 (figure 8.4) shows marine or brackish water conditions at all levels analyzed with variable freshwater components (Ireland, 1987). The radiocarbon assays

Figure 8.4 The stratigraphy, diatoms, and radiocarbon dates from site 33/1 in the Lagoa de Itaipú (Ireland, 1987).

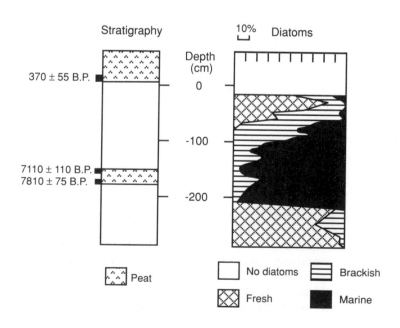

indicate two distinct chronostratigraphic units: the older dates are >35 000 BP, indicating an interglacial or interstadial age, whereas the younger ones are interglacial (Holocene) in age. The marine and brackish water diatoms of interstadial age indicate evidence of either a high interstadial sea level between −4 and −5 m (Imbituba Datum) between 35 300 and 42 500 radiocarbon years or *Sabkha* conditions in inland lagoons. Thom (1973), although critical, marshalled evidence for a climatic amelioration about 35 000 years ago (in northwest Europe equivalent to the Hengeloo Interstadial; Jelgersma et al., 1979) and a higher sea level, and Cronin (1983) showed a sea-level peak at this time.

Another line of evidence has been used in Brazil to determine former sea-level altitudes, and this is the age and latitude of encrusting tubes of gastropod Vermetids (Delibrias and Laborel, 1971). Two genera of the family Vermetidae (*Dendropoma* and *Petaloconchus*) occupy a very narrow range of the intertidal zone between low water of neap tides and low water of spring tides. In Brazil near Recife they live 0.4–0.9 m below mid-tide levels; in Bermuda 0.2–0.3 m and in the Mediterranean 0.1 m (Laborel, 1986). Where they grew under sheltered conditions, fossil Vermitid casts can indicate a former sea-level position with an accuracy of ± 0.1 m compared with ± 0.5 m for salt marsh and brackish reedswamp *turfas* in humid temperate coastal paleoenvironments. Employing the evidence from fossil Vermetids on the coast of Rio de Janeiro state, Delibrias and Laborel (1971) were able to demonstrate a sea level higher than present and falling about 4000 years ago, when the stratigraphic record shows no apparent sea-level changes.

Sea-level changes

It has been estimated that over the past 100 years global mean sea level has been rising by between 0.5 and 3.0 mm year^{-1} (Warrick and Oerlemans, 1990). Barnett (1988, in Warrick and Oerlemans, 1990) analyzed data from 155 tidegauge stations and established a rate of 1.15 mm year^{-1} for the period 1880–1986 (figure 8.5b), but with an apparent acceleration from 1910 to 1.7 mm year^{-1}. A comparison with the global-mean, combined land-air and sea-surface temperatures (figure 8.5a) shows some similarities. But Pirazzoli (1989) has cautioned against the acceptance of these rates of sea-level rise and argued that if there is a eustatic signal it is weak and obscured by tectonic, glacio-isostatic, oceanographic, anthropogenic, and atmospheric effects, with a value of 0.4–0.6 mm year^{-1}. Such low rates of sea-level rise lie well below those predicted for the next century and those that occurred during the past 140 000 years, including the major rise of sea level at the end of the last glaciation.

There are many estimates of future sea-level rise driven by predicted increases in the amounts of radiatively active gases in the

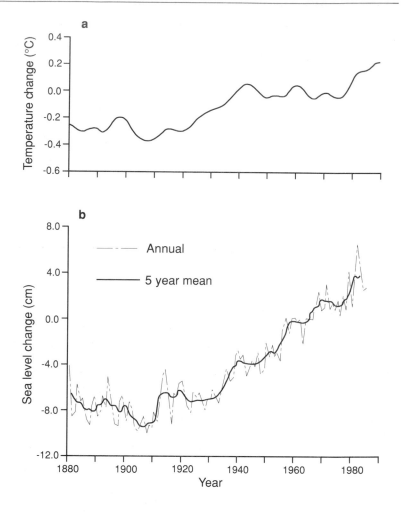

Figure 8.5 Sea-level and temperature variations since AD 1880: (a) global mean combined land-air and sea-surface temperatures, relative to the average for AD 1951–80 (Houghton et al., 1990); (b) global mean sea level rise, relative to the average for AD 1951–70. The dashed line connects annual mean values and the solid line is the 5-year running mean (Barnett, 1988, in Warrick and Oerlemans, 1990).

atmosphere, such as CO_2, CH_4, CFCs, H_2O, and N_2O, and consequential global temperature rise. Hoffman (1984) estimated a range by AD 2100 from 56.2 cm (conservative estimate) to 345 cm (high estimate), yielding rates of from 4.8 to 29.7 mm year^{-1}. Warrick and Oerlemans (1990) estimated for the "business-as-usual" scenario a range from 31 to 110 cm, yielding rates of from 2.8 to 10.0 mm year^{-1} (figure 8.6). Both these estimates ignore the impact of the distribution of density of Earth materials on the new geoid topography following a partial deglaciation and return of meltwater to the ocean basins. Clark and Primus (1987) demonstrated the consequences for geoid topography of melting part of the Greenland and Antarctic ice caps (figure 8.7) to liberate sufficient water to raise sea level globally by 100 cm. In the case of a partial Antarctic deglaciation, sea level would rise 111.3 cm in the Thames estuary and 120 cm along the Japanese coast, whereas there would be a fall of sea level of 210.8 cm along the Filchner-Ronne coast of Antarctica due to a loss of mass of the ice cap. This variability, to which local factors such as tidal changes, subsidence

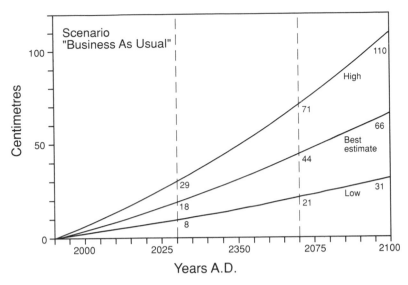

Figure 8.6 Sea-level rise predicted to result from business-as-usual (scenario A) assumptions of the IPCC (Houghton et al., 1990), showing the best estimate and the range. For AD 2030, the estimated rise is 18 cm with a range of 8–29 cm. The major contribution to sea-level rise is estimated to be thermal expansion (10.1 cm) with meltwater from mountain glaciers contributing 7.0 cm. Greenland contributes 1.8 cm, but because of increased precipitation Antarctica contributes negatively (−0.6 cm) (Warrick and Oerlemans, 1990).

rates, and hydro-isostasy would be added, will require consideration to assess the local and regional impacts of sea-level rise.

Such variability and uncertainty exists when considering sea-level index points and sea-level curves from the past. It has been a common practice to plot sea-level variates on age–altitude graphs and to fit sea-level curves by eye (for example, Fairbridge, 1961, 1976; Jelgersma, 1961; Jelgersma et al., 1979; Tooley, 1974, 1978; Devoy, 1979). While this practice continues (for example, National

Figure 8.7 A map to show the amount of sea-level rise caused by the melting of part of the Antarctic ice sheet sufficient to raise sea level by 100 cm. Contours are in centimeters (Clark and Primus, 1987).

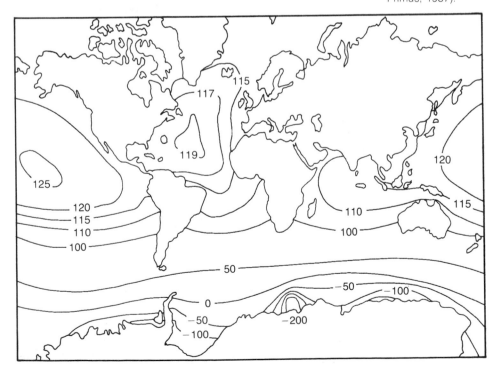

Research Council, 1987; Fairbanks, 1989) the recognition of all the age and altitudinal errors to which sea-level index points are subject has led to the use of error boxes, error bands, or ellipses for each variate (Kidson, 1982). Godwin (1940) initiated this practice, but it appears to have been abandoned after van Straaten's (1954) work and was not revived until the 1970s (Akeroyd, 1972; Thom and Chappell, 1975; Louwe Kooijmans, 1974; Preuss, 1979). Shennan (1982, 1986a, b) has done much to address the problem of the errors associated with sea-level variates, and demonstrated that a sea-level band (figure 8.8) represents more accurately the range of age and altitudinal errors that affect individual sea-level index points. The actual course of sea-level change (probably the paleo-mean high water mark of spring tides on macrotidal coasts) lies within the enveloping band.

Determining rates of sea-level change in the past by employing banded sea-level curves will yield a range of values, but this has rarely been undertaken. For northwest England, it has been demonstrated (Tooley, 1989) that rates of sea-level rise, associated with the catastrophic melting of the Laurentide ice cap about 7800 radiocarbon years ago, ranged from 34 to 44 mm year^{-1}. The measured rise of sea level (ignoring consolidation) in west Lanca-shire was $\geq$ 7.3 m over a period of time from 70 to 360 radiocar-

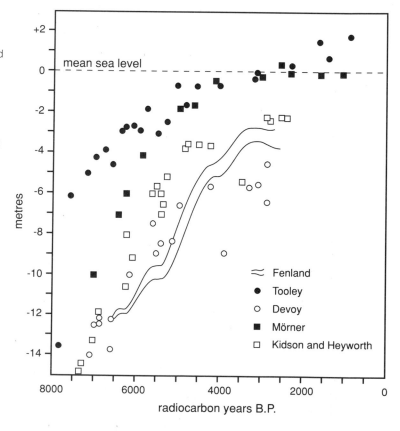

Figure 8.8 The Fenland sea-level band with altitude and age errors accommodated compared with sea-level variates from northwest England (Tooley, 1978), the Thames estuary (Devoy, 1979), the Bristol Channel (Kidson and Heyworth, 1973), and the west coast of Sweden (Mörner, 1969). All values are related to the present local mean sea level (Shennan, 1986b). The dispersion of the regional sea-level data gives an indication of land movements, relative to Mörner's (1976) "regional eustatic" sea-level curve for northwest Europe.

bon years (one standard deviation) or within 505 radiocarbon years (two standard deviations). Such rates of sea-level rise are not exceptional. Cronin (1983) reviewed the evidence and concluded that sea level has risen and fallen at rates ranging from 10 to 30 mm year^{-1}. In this interglacial, rapid rates of rise associated with the melting of the Laurentide ice cap have come from Denmark (Petersen, 1981) and the German Bight (Ludwig et al., 1981). For the last interglacial, Zagwijn (1983) and Streif (1990) present evidence for a rise in sea level of 40 mm year^{-1} during pollen zones IIIa and early IIIb.

The rise of sea level since the maximum of the last glaciation has been established by Fairbanks (1989) from the Bahamas (figure 8.9) and although the data are presented without age and altitudinal errors (see Chappell and Polach, 1991), Fairbanks identifies two periods of "exceedingly rapid" sea-level rise: 42 mm year^{-1} some 12 000^{14}C years ago, and 28 mm year^{-1} 9500^{14}C years ago. These enhanced rates of sea-level rise are associated with peak discharge rates of 14 000 and 9500 km^3 year^{-1} from the ice cap over North America. The liberation of meltwater in such large quantities has been questioned, but Shaw (1989) has argued that the catastrophic release of meltwater forming one drumlin field in Canada liberated 84 000 km^3 of water that may have raised sea level by 23 cm in a few weeks. The simultaneous melting and

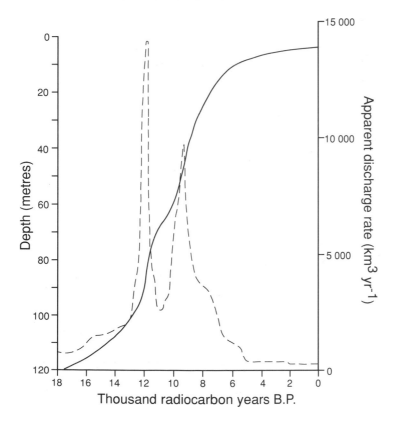

Figure 8.9 The Caribbean sea-level curve (continuous line) based on data from Barbados before 6400 BP, and the two periods of enhanced sea-level rise (12 000 and 9500 radiocarbon years BP), when sea level was rising at 42 mm year^{-1} and 28 mm year^{-1}, are associated with two glacial meltwater discharge peaks (dashed line) calculated from the curve (after Fairbanks, 1989). The maximum depression of sea level during the last glacial maximum about 18 000 ^{14}C years BP was 121 ± 5 m below present levels.

catastrophic discharge of water from other drumlin fields in the Northern Hemisphere, associated with climatic amelioration, could be sufficient to raise sea level by several meters in a few years.

Conclusions

The Quaternary period has been characterized by marked climatic and sea-level changes. The consequences of the former have been a series of glacial and interglacial stages that have affected the volume of water in the ocean basins by interrupting the hydrological cycle. This climatically forced effect on seawater volume has resulted in a change of sea level, the measured magnitude of which on the Bahamas has been 120 ± 5 m in the past 18 000 years. The return of water to the ocean basins does not appear to have been steady, but was achieved by the addition periodically of considerable volumes of cold freshwater, elevating sea level at rates up to 44 mm year^{-1}, disrupting the biological productivity and thermal structure of surface water masses, and affecting regional climates.

Land uplift due to isostatic recovery following deglaciation has resulted in raised beaches, and continues at the present in the main centers of ice loading at rates exceeding projected sea-level rise. The effects of the distribution of Earth materials in the crust and mantle of the Earth dictate that a rise (or fall) of sea level will be neither of the same order nor of the same direction everywhere. The geoid topography will vary. In addition, there will be local effects, such as a change in the tidal amplitude and a consequential change in the strength of tidal currents and the capacity to move a bed load.

Correcting sea-level index points on age–altitude graphs to accommodate regional and local effects demands the use of error-banded sea-level curves. Many of the sea level prognostications for the next century ignore these considerations, and greater attention should be directed toward clarifying the links between rapid rates of sea-level rise and climatic change, and elaborating dynamic sea-level changes, particularly extreme water levels during storm surges. Coker et al. (1989) have calculated that for a 20 cm rise of sea level only, there would be an increase in storm surges in Eastern England equalling or exceeding the danger levels specified by the Storm Tide Warning Service from 112 occasions to 334 occasions for the period 1972/3 to 1988/9. Recorded extreme water levels already intersect parts of the British coast at altitudes up to +7.5 m OD, and to calculate this return period requires a longer time series of records than the natural archives in the coastal lowlands can provide.

Further reading

Devoy, R.J.N. (ed.) (1987) *Sea Surface Studies: a Global View*. London: Croom Helm.

Lisitzin, E. (1974) *Sea Level Changes*. Amsterdam: Elsevier.

van de Plassche, O. (ed.) (1986) *Sea-level Research: a Manual for the Collection and Evaluation of Data*. Norwich: Geo Books.

Pugh, D.T. (1987) *Tides, Surges and Mean Sea-level*. Chichester: John Wiley and Sons.

Smith, D.E. and Dawson, A.G. (eds) (1983) *Shorelines and Isostasy*. London: Academic Press.

Tooley, M.J. and Jelgersma, S. (eds) (1992) *Impacts of Sea-level Rise on European Coastal Lowlands*. Oxford: Basil Blackwell.

Tooley, M.J. and Shennan, I. (eds) (1987) *Sea-level Changes*. Oxford: Basil Blackwell.

CHAPTER NINE

Tropical Coral Islands – an Uncertain Future?

Tom Spencer

Introduction

Beyond the continental shelves, the islands associated with oceanic lithosphere are geologically young and transient features in cycles of ocean growth and destruction. These islands show a remarkable dichotomy of form between high, volcanic islands on the one hand and low islands composed of carbonate sediments on the other (plate 9.1 and plate 9.2). It was Charles Darwin who first argued that these very different island types are linked by a common history. In the reef seas, coral growth around an extinct volcano will initially produce a fringing reef in contact with the shore. As the volcano subsides the coral is able to grow upwards to sea level: the result is the development of a barrier reef island, with a lagoon separating the reef from the volcano. Ultimately, the central volcano disappears completely below sea level, leaving a ring of upward-growing coral around a central lagoon. Darwin's theory was successfully confirmed by deep drilling in the Pacific Ocean basin in the 1950s: at Enewetak atoll, for example, over 1200 m of near-surface limestones, indicative of continued subsidence, were found to overlie 50 million year old basalts. Further deep drilling and seismology at other Pacific Ocean sites has revealed similar island histories, with an average subsidence rate of 2.3 cm per thousand years (Stoddart, 1973). Plate tectonic theory now provides a mechanism for this process. New oceanic lithosphere formed at the mid-ocean spreading ridges is hot and buoyant but as it moves away from the ridges it cools, thickens, and sinks, taking passenger volcanic islands below sea level. This basic pattern is, however, complicated by regional patterns of plate re-heating and loading by new volcanoes, which can result in the uplift of sea-level atolls and their transformation into raised reef islands. By

Plate 9.1   'Tonga Vava'u group: sand cays are vulnerable to sea-level rise.

Plate 9.2   'Tonga Vava'u group: will coral growth keep pace with sea-level rise and protect reef islands?

comparison with plate interiors, plate-marginal locations are characterized by thin fringing reefs around volcanoes associated with tectonically active subduction zones (see Rosen, 1982; Scoffin and Dixon, 1983 for reviews).

Of these island types, it is the coral atolls, supporting islands of unconsolidated sands only a few meters above sea level, which are thought to be particularly vulnerable to sea-level rise consequent upon global warming. There has been considerable concern over the future of nation states (e.g. Tuvalu, the Marshall Islands and the Maldives) composed entirely of such islands. However, scenarios of future global environmental change which predict the catastrophic flooding, loss of island area, and passive removal to wave base on rapid sea-level rise need reassessment in the light of two factors. First, there has been a downwards revision of projected rates of future, greenhouse-effect induced sea-level rise. Second, studies of coral reef geomorphology show sensitive and complex responses to environmental change by both individual corals and whole-reef systems.

Changing views on changing sea level

Changes in global average sea level, on the 1–100 year timescale, are largely a function of changes in the temperature of ocean water and variations in the terrestrial ice volume held within both major ice sheets and smaller glaciers and ice caps. Following any global warming, sea-level rise results from both the thermal expansion of seawater and the increased freshwater supply from enhanced ice melt. The relative contribution of these two factors is, however, subject to debate and difficult to unravel from neotectonic and oceanographic signals in existing tidegauge records (Bryant, 1987; Peltier and Tushingham, 1989; Pirazzoli, 1989; Tooley, chapter 8). Thermal expansion models, on an assumed global warming of 3–5°C by 2050 from equivalent $2 \times$ atmospheric CO_2, suggest a complete sea level response of +51 cm. However, the response rate is relatively slow (e.g. 38 percent of total sea-level rise being achieved after 50 years, compared to 72 percent of comparable rise towards equilibrium by atmospheric temperatures; Mikolajewicz et al., 1990) and recent modelling has suggested a thermal effect sea-level rise contribution of 7–14 cm, with a best estimate of 10 cm, by 2030 (Wigley and Raper, 1987; Warrick and Oerlemans, 1990). When enhanced ice-melt effects are factored in, a total sea level rise of 9–29 cm, with a best estimate of 18 cm, by 2030 has been predicted (Warrick and Oerlemans, 1990), rising to 44 cm by 2070 and 66 cm by 2100 (figure 9.1a), although with regional disparities as a result of ocean circulation systems (Mikolajewicz et al., 1990). Thus it is now becoming apparent (Stewart et al., 1990) that these predictions are considerable downward revisions on some earlier standards for maximum sea-level rise (e.g 345–367 cm by 2100: Hoffman et al., 1983, 1986), which may have underestimated the

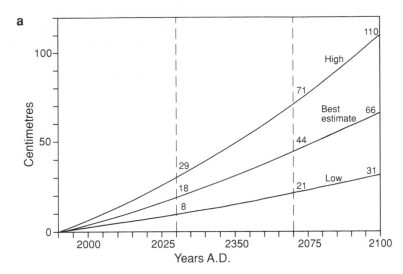

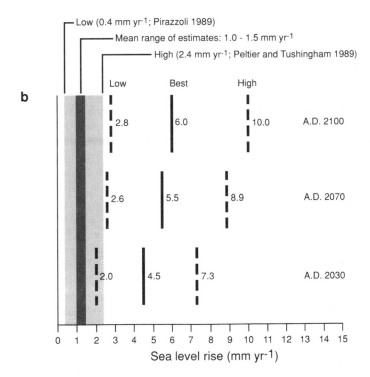

Figure 9.1 (a) Projected sea-level rise using IPCC "business as usual" scenario which envisages negligible controls on greenhouse gas emissions (after Warrick and Oerlemans, 1990). (b) Converted to rates of sea-level rise and compared to observed historical rates of sea-level rise (for sources see text).

complexities, and thus overestimated the magnitude, of the ice-melt component (Meier, 1990).

With these revisions, scenarios of future global environmental change which simply predict catastrophic inundation and erosion of low-lying coral islands on rapid sea-level rise are likely to be inadequate. Rather, coral reef environments will be modified in the future by the interaction between more gradual sea-level rise and

large-scale ocean–atmosphere processes (such as El Niño reorganizations and tropical storm patterns), sediment budget changes, and
local anthropogenic effects. As coral reefs are bio-constructional
systems, any change in environmental factors within tropical
circulation systems can be expected to result in a landform response. Thus any discussion of future change on reef islands should
include a strong *geomorphological* component. Furthermore, the
importance of the interactions mentioned above will require an
understanding of the changing magnitude and frequency of controlling processes and of the leads and lags between different processes
and the response of reef island morphology. Thus the *rate* of future
sea-level rise, rather than its magnitude, becomes of critical importance.

Despite this change of emphasis, it should not be assumed that
morphological modifications on reef islands are likely to be trivial,
for future rates of sea-level rise are likely to be of the order of 3–6
times those experienced in the recent past (figure 9.1b). How will
coral reefs respond to the acceleration in this forcing function to
rates of about 4.5–6.0 mm year^{-1}?

The nature of coral growth

The Scleractinia, or stony corals, are a highly successful class of
anthozoans (which includes the sea anemones) with a rigid exoske-
leton composed of calcium carbonate (plate 9.3). Although corals
are widely distributed over most latitudes, the reef-building, or
hermatypic, corals are restricted to warm (sea-surface temperature
> 20°C in the coldest month of the year) and shallow (low tide level
to < 100 m) water within the tropics and subtropics. They are
organized into two major realms of contrasting diversity, the
Atlantic (with > 20 genera at its Caribbean core) and the Indo-
Pacific (> 70 genera at western Pacific centre; Rosen, 1984). At a
community level, the composition and distribution of framework-
building hermatypes, and associated coralline algae, is determined
both by environmental factors such as light availability (and thus
water depth), tidal range, wave action, and salinity, and by
biological interactions both between coral species and between
corals and other organisms, including predators and parasites. The
biological and topographic complexity of coral reefs results from
their high sensitivity to these controls (e.g. corals can grow twice as
fast on a sunny day as on a cloudy one; Goreau and Goreau, 1959)
and their ability, because of their colonial organization (a coral
colony is composed of a number of genetically identical polyps), to
adapt their growth rapidly to changing environmental conditions.

There are several ways of assessing the growth potential of reef
corals. Quantitative studies of the growth rates of individual corals
show that massive, hemispherical corals typically exhibit rates of
radial expansion of 4–20 mm year^{-1}, averaging 10–12 mm year^{-1}
(Buddemeier and Kinzie, 1976). Open-structured branching corals,

Plate 9.3 Hard coral
fragments (foreground)
support soft corals and
other reef organisms.

such as *Acropora* spp. (e.g. staghorn and elkhorn corals), show much more rapid rates of branch-tip extension of 100 mm year^{-1}, and occasionally 200 mm year^{-1}. However, when these rates are collapsed down to the equivalent growth of a solid structure then their rates of radial expansion equal those of the hemispherical corals (Maragos, 1978). However, such rates of growth relate only to individual colonies; clearly estimates of reef growth by this method require the measurement of growth rate and cover/ abundance characteristics for all reef-builders in a biologically diverse reef community. The only detailed assessment along these lines, by Stearn, Scoffin and co-workers (Stearn and Scoffin, 1977; Stearn et al., 1977; Scoffin et al., 1980) for a fringing reef in Barbados, produced a net balance of 15 $CaCO_3$ m^{-2} year^{-1}. Assuming a reef porosity of 50 percent, this production was equivalent to a vertical accretion rate of 11 mm year^{-1} (Budde-meier and Smith, 1988). An alternative approach, which avoids over-concentration on corals alone, estimates the rate at which $CaCO_3$ production in the reef community depletes the overlying water column. This cannot easily be measured directly but the depletion of total alkalinity offers an appropriate surrogate (for method see Smith and Kinsey, 1978). The "alkalinity depression" technique typically identifies three modal rates of $CaCO_3$ production (figure 9.2a): a "fast" rate of 10 kg $CaCO_3$ m^{-2} year^{-1}, characteristic of high cover in coral thickets; an "intermediate" rate of 4 kg $CaCO_3$ m^{-2} year^{-1} typical of reef flats on reef ocean margins; and a "slow" rate of 0.8 kg $CaCO_3$ m^{-2} year^{-1} associated with large expanses of sand/rubble lagoon floors (Kinsey, 1983). If 1–2 percent of the reef area calcifies at the fast rate, 4–8 percent at the intermediate rate and 90–95 percent at the slow rate, then coral reefs produce an average of 1.0–1.2 kg $CaCO_3$ m^{-2} year^{-1} (Smith, 1978), equivalent to a vertical growth rate of 0.8–0.9 mm year^{-1}.

These estimates of reef calcification from fluctuations in water chemistry are based on short-term experiments over limited areas of reef substrate and deal only with *in situ* accretion and solutional loss. Of much greater value is the estimation of reef growth rates from ^{14}C dating control on vertical sequences of reef carbonates recovered by coring. Importantly, this approach records sediment retention and accumulation, not simply initial production. It also provides a time-averaged accretion rate by allowing for the post-depositional removal and localized redeposition of coral debris by storms and hurricanes. Table 9.1 shows how rates of reef growth derived from the Holocene core record can be classified by plate-tectonic setting. Although highly variable, the high rates of coral growth are particularly associated with mid-plate fringing and barrier reefs and at some continental shelf and subduction zone settings. Mid-ocean atolls show lower rates of reef accretion, typically 1–3 mm year^{-1}, but reaching 5–8 mm year^{-1}.

Figure 9.2c compares these rates of former reef growth with the predicted rates of future sea-level rise. Clearly, in general, reefs should be able to keep pace with rising sea levels over the next

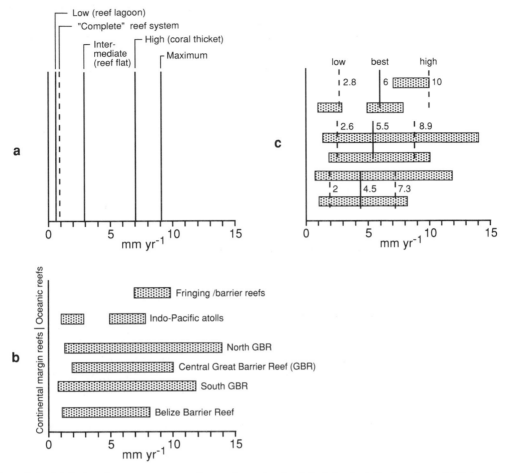

Figure 9.2   (a) Rates of contemporary reef accretion using the "alkalinity depletion" method. (b) Rates of Holocene reef growth from radiometrically dated drillcores in different tectonic settings (for sources see text). (c) Rates of projected sea level rise compared to rates of Holocene reef growth.

Table 9.1   Rates of reef growth from radiocarbon dating of vertical sequences of reef carbonates recovered by reef-top drilling

Plate-tectonic setting	Locations	Rate of coral growth (mm year⁻¹)	Source
Plate-margin, subduction zone	US Virgin Islands	1–2	Adey, 1978
Plate-margin, continental shelf	Atlantic Panama	0.3–12	Graus et al., 1985
	Alacran, Gulf of Mexico	12	Macintyre et al., 1979
	Belize barrier reef	1.1–8.3 (max. 12)	Macintyre et al., 1981
	N. Great Barrier Reef	1.3–14.2	Marshall and Davies, 1982;
	C. Great Barrier Reef	1.9–10.2	Davies et al., 1985
	S. Great Barrier Reef	0.6–8.3	
Mid-plate fringing reefs and barrier reefs	Reunion, Indian Ocean; Society Islands and Hawaiian Islands, Pacific Ocean	7–10 (range 1.6–50)	Easton and Olson, 1976 Montaggioni, 1988 Grigg, 1988
Mid-plate, coral atolls	Enewetak and Bikini atolls, Central Pacific Ocean	1–2 (max. 10)	Tracey and Ladd, 1974 Tracey in Adey, 1978
	Mururoa atoll, S. Pacific Ocean	2–3	Labeyrie et al., 1969
	Tarawa atoll, Central Pacific Ocean	5–8	Marshall and Jacobsen, 1985

century but the variation in the rate of growth within reef type groupings has been very wide in the past. This variation can be explained by different reef growth strategies in response to sea-level rise.

Coral growth strategies and response to future sea-level rise

Three growth strategies have been identified (Neumann and Macintyre, 1985; Davies and Montaggioni, 1985): the "keep-up" strategy, related to a reef's ability to keep pace with sea-level rise; the "catch-up" strategy characteristic of reefs initially drowned by a transgression but then able to recover and regain sea level by rapid vertical accretion of coral; and the "give-up" strategy, used to describe reefs terminally affected by initial drowning. In the last case, platforms drowned by rapidly rising sea level find themselves in water too deep for adequate upward growth once sea level rise has slowed. The highest rates of calcification occur at water depths of 5–10 m and the rate of reef accretion at 30 m depth in Jamaica has been shown to be only 20 percent that at 10 m (Dustan, 1975). The critical depth for Holocene drowning appears to have been about 20 m in the Caribbean (Adey and Burke, 1977) and 30–40 m in the Pacific Ocean (Grigg and Epp, 1989), although with great sub-regional variation. Shallower platforms showing aborted reef growth may have been the victims of freshwater and sediment inputs from coastal drainage systems (e.g. Neumann and Macintyre's (1985) "reefs shot in the back by their own lagoons") or of excess nutrients released from flooded continental shelf soils (Hallock and Schlager, 1986).

Figure 9.3 indicates that the three growth strategies possess characteristic geological facies signatures in the drill core record and thus it is possible to match this signature to the rate of sea-level rise in the early Holocene. Table 9.2 shows that coral reefs appear to "give up" when rates of sea-level rise exceed 20 mm year^{-1}. This threshold is now well established from the dating of aborted coral reef ridges off the island of Barbados, which appear to have failed in two phases of very rapid sea-level rise of 37 mm year^{-1} at 13 500 years BP (when sea level was at about −90 m), and of 25 mm year^{-1} at 11 000 years BP (about −60 m). These rises resulted from the addition of deglaciation meltwater pulses of more than a million cubic kilometers of continental ice per 100 years (Fairbanks, 1989; Bard et al., 1990). In the context of island futures, however, the boundary between "catch-up" and "keep-up" reefs is also of importance because island destruction may take place before reef response to sea-level rise can be initiated. This threshold appears to fall at a rate of sea-level rise of about 8–10 mm year^{-1}, which starts to approach some of the higher predictions for future sea-level rise (figure 9.1b).

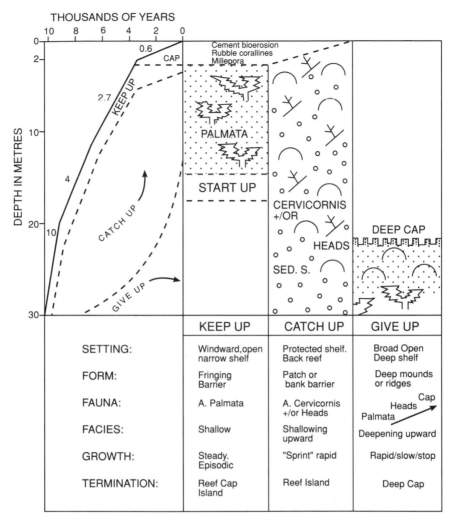

Figure 9.3 Summary of growth histories and other characteristics associated with keep-up, catch-up, and give-up reefs (after Neumann and Macintyre, 1985). The curve on the left shows sea-level rise, with units in meters per thousand years.

Near the keep-up/catch-up threshold, recent sea-level history and the island morphology inherited from this history becomes potentially significant. The varying regional importance of the eustatic, glacio-isostatic and hydro-isostatic components of Holocene sea level change has led to very different sea-level histories in the Atlantic and Indo-Pacific reef provinces (Stoddart, 1990; Woodroffe, 1990). Thus the Florida sea-level record is one of a progressive slowing down of sea-level rise through the post-glacial transgression with present sea level having been reached very recently, whereas the Indo-Pacific record is one of higher than present sea levels being achieved in the past 6000 years (figure 9.4).

Table 9.2   Early Holocene sea-level rise and "keep-up," "catch-up," and "give-up" modes of coral reef growth response

Reef location	Sea-level rise (mm year⁻¹)	Time period (thousand years BP)	Reef response
Bahamas/Florida/ St Croix/Panama	5–6	9.0–7.0	Keep-up
Reunion Is.	8	7.7–5.5	Keep-up
Oahu, Hawaii	8	9.0–6.0	Keep-up
Central Great Barrier Reef	10	to 7.5	Catch-up
Society Is.	15	9.0–6.0	Keep-up (windward reef) Catch-up (leeward reef)
Barbados	25	~13.5	Give-up
Barbados	37	~11.0	Give-up

Source: see text

Figure 9.4   Minimum sea-level curve from *in situ*, framework-building *Acropora palmata* coral from the tropical western Atlantic (after Lighty et al., 1982), and theoretical sea-level curves (after Nakada and Lambeck, 1987) for the S. Pacific "far field" with observed sea levels (closed circles) from the northwest Tuamotu archipelago (after Pirazzoli and Montaggioni, 1986).

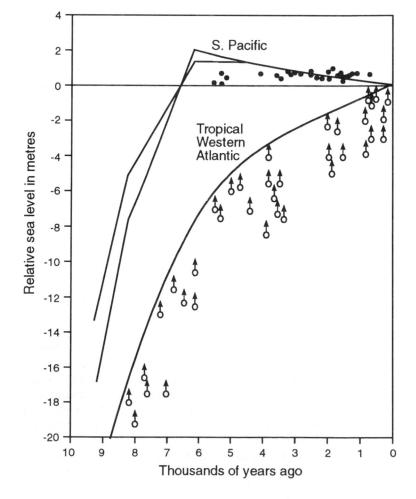

In the latter region, therefore, reef islands will be protected in the initial stages of sea-level rise by fossil reef flats, cemented beachrock ramparts and lithified boulder beach conglomerates adjusted to the former +1 m high stand (Woodroffe et al., 1990). On reef flats where growth is currently constrained by low tide exposure, rejuvenation of coral growth will take place (Hopley and Kinsey, 1988). By comparison, Caribbean reef islands will be more vulnerable to sea-level rise, as reefs have only just reached sea level in this region and protective reef flats and other structures are absent.

Stress in reef systems in a warmer world

The above analysis assumes that the past is a good analogue for what may happen in the reef seas in the next century. One important proviso, however, is that sea-level rise in the future will start from the base of a much warmer ocean than the immediate postglacial. Two confounding factors need to be taken into account: first, a potential increase in *physical* stress on reef corals from increased intensities and frequencies of tropical storms; second, a possible rise in *physiological* stress in the coral ecosystem from higher ocean temperatures. Both these effects can be evaluated, to a degree, by studying reef behavior during quasi-periodic reorganizations of the ocean–atmosphere system in the Pacific Ocean – El Niño–Southern Oscillation (ENSO) events – when warm surface waters replace relatively cool upwelling water in the eastern and central Pacific. The 1982–3 event is of particular value in this regard, being probably the strongest warming of the equatorial Pacific this century (e.g. Halpern et al., 1983).

Reefs and storms

Coral reefs will not keep pace with future sea-level rise if vertical accretion is removed by increased storm attack. However, as Stoddart (1990) points out, coral rubble ridges built from storm debris may increase island areas and provide suitable substrates from which coral growth to higher sea levels may take place. The role of storms in the sediment budget and development of reef islands very much depends upon the timescale over which their impact is viewed. Not surprisingly, therefore, the long-term effect of changing storm and storm-associated surges on reef islands is difficult to assess (although see Bayliss-Smith, 1988, for an imaginative model). In the short-term, however, mechanical studies (Hernandez-Avila et al., 1977) have shown that hurricane wave forces are capable of fracturing and transporting cobble-sized fragments of coral, with the rapid-growing, finely branched corals being particularly susceptible to breakage. Thus Hurricane Allen (1980) reduced the living area of acroporid corals by 99 percent on the north coast of Jamaica (Woodley et al., 1981). One factor controlling levels of reef modification is storm intensity: field studies have shown that cyclones with wind speeds of 120–150 km h^{-1} typically produce limited damage which can be repaired in under 10 years, whereas severe hurricanes with wind speeds in excess of 200 km h^{-1} are capable of reef destruction which may require over 50 years for full regrowth (Stoddart, 1985).

Simple models of the cyclone energy cycle permit the establishment of relationships between storm intensity (as measured by minimum sustainable central pressure), windspeed, and sea-surface temperature (SST) (Emanuel, 1987). As the latent heat content of air is a strongly exponential function of temperature, an increase in SST of 3°C yields a 30–40 percent increase in the maximum fall in central pressure. Maximum windspeed increases as the square root of the pressure fall and there is a strong step function in potential maximum windspeeds for North Atlantic storms above a SST of 28°C (Holland et al., 1988). Models of cyclone incidence which incorporate sea-surface warming of 2–5°C with 2 × atmospheric CO_2 show, first, reduced recurrence intervals for storms of particular central pressure. Thus, for example, Love (1988) has shown for Darwin, North Australia, how this magnitude of warming may transform the 950 mb central pressure cyclone from a 1-in-100 year to a 1-in-40 year event (figure 9.5). The implications of this increase in storm frequency are considered further below. Second, ocean warming is likely to lead to an extension of the area affected by cyclones to areas with a normally low storm incidence at present (plate 9.4). Figure 9.6 shows the changing ocean areas for storms of a given minimum central pressure on predicted warming (950 mb = common, ≤ 880 mb = exceptionally severe storm, using criteria of Barry and Chorley, 1987): the changes in risk are clear, particularly for coral ecosystems in the Caribbean and on the reef

Figure 9.5 Gumbel plot of cyclone recurrence interval against central pressure for Darwin, north Australia. Solid line = current climate, dashed line = scenario of increased cyclone intensity (after Love, 1988). The changed recurrence for cyclones with a central pressure of 950 mb is indicated.

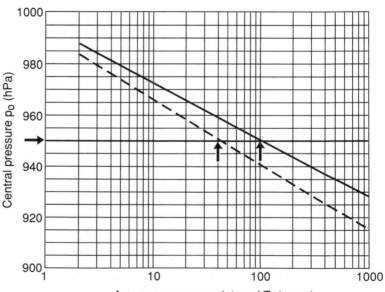

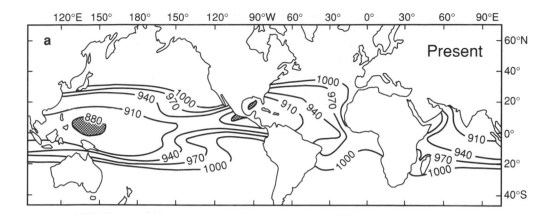

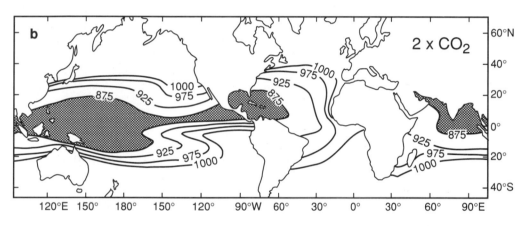

and mangrove coasts of southern India and the Bay of Bengal. The consequences of such events on reef structures and luxuriant coral growth unused to storm impacts can be evaluated by reference to the 1982–3 ENSO episode in French Polynesia. This area of the South Pacific, not affected by severe storms since 1906, was ravaged by six cyclones in a 5-month period. On Tikehau, north-west Tuamotu archipelago, these cyclones were responsible for 50–80 percent coral destruction to a depth of 20 m, with relocation of debris both onshore in subaerial boulder ramparts and to ocean depths by avalanching down the steep atoll margin (Harmelin-Vivien and Laboute, 1986).

Changes in storm frequency *per se* may also be important. For the Caribbean region, Hubbard (1988) has argued that coral reef morphology is controlled by the interaction of wave energy and tropical storm frequency. Thus it appears that high storm frequencies maintain wave-swept pavements in areas of low wave energy (e.g. Bahamas) and promote the dominance of coralline algal ridges in the areas of high wave energy in the eastern Caribbean (figure 9.7). Thus the characteristic *Acropora palmata* reef crests of the

Figure 9.6 (a) Minimum attainable surface central pressure (mb) in September at present day. (b) Minimum sustainable pressures using sea-surface temperature increases predicted by GISS general circulation model II averaged over five Augusts in simulations with 2 × atmospheric CO_2 (after Emanuel, 1986, 1987). Area below 880/875 mb is shaded.

Plate 9.4 Cyclone Winifred, which had a central pressure of 955 mb and wind speeds up to 250 km h^{-1}, hitting the Queensland coast in February 1986. Tropical storms such as this may become more frequent as a result of global warming.

Caribbean may be replaced by coralline–algal dominated structures if storm frequencies and intensities increase. Alternatively, there is the fear that reefs may not be replaced by hard structures at all but by soft substrates, dominated by fast-growing fleshy algae, particularly where high storm frequencies interact with ongoing disease, predation, and the loss of algal grazers, as was the case in the Caribbean in the 1980s with the mass mortality of the herbivorous urchin *Diadema antillarum* (e.g. Jamaica: Knowlton et al., 1990).

The frequency of occurrence of storms relates to factors controlling atmospheric circulation systems as well as ocean heat sources

Figure 9.7 Wave energy (open circles = sites with > 30 years wave data, wave roses = regional energy trends) and storm incidence as a control on reef character in the Caribbean (after Hubbard, 1988).

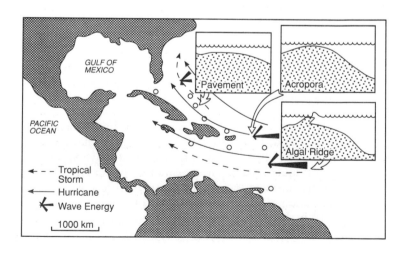

and thus cannot be easily predicted from projected changes in sea-surface temperature. Nevertheless, historical evidence suggests that tropical storm frequencies fell well below a frequency of 5 per year (compared to a recent frequency of 9 per year) in the period of the "Little Ice Age" (Wendland, 1977) and in the last century. The relatively cold sea-surface temperatures centered on 1910 and the subsequent warming to the 1940s (Folland et al., 1984) appear to be reflected in the changing incidence of storm frequency in the Caribbean (Spencer and Douglas, 1985). Interestingly, the warm temperatures of the 1980s (Jones et al., 1986) were accompanied by a series of severe storms in the western Atlantic, culminating in the record minimum sea-level pressure of Hurricane Gilbert (Willoughby et al., 1989). In the South Pacific, Fiji has experienced a rise in hurricane-force cyclone frequencies from 3.1 events per decade for the period 1941–80 to 11.4 per decade 1981–9. A similar pattern of increased activity is also apparent for the Tuvalu region (Nunn, 1990a) and severe storms have been recorded since 1986 in the Solomon Islands and Vanuatu (Nunn, 1990b). Furthermore, average annual cyclone frequency appears to have accelerated in the recent record from the Southwest Pacific (1940–55 = 5.4 ± 1.5 cyclones per year, 1956–78 = 8.0 ± 2.4 cyclones per year; Grant, 1981). Clearly, however, only ongoing records will reveal if all these changes are sustained or transient.

Reefs and high temperatures

As well as growth increments being removed by storm activity, growth itself may be inhibited by high water temperatures. Classic experiments over 50 years ago (e.g. Yonge and Nicholls, 1931) established that corals respond to thermal stress by "bleaching," the loss of pigmentation being the result of the expulsion from the host of the zooxanthellae, which may account for half the tissue mass. The symbiotic zooxanthellae play a vital role in several aspects of coral physiology and their expulsion is accompanied by reduced carbon fixation and skeletal growth (Coles and Jokiel, 1977, 1978; Hudson, 1981). Thus, for example, at Discovery Bay, Jamaica, bleached colonies of the framework-building coral *Montastrea annularis* virtually ceased to grow during the 1987 Caribbean bleaching episode (Goreau and MacFarlane, 1990). Several weeks of temperature elevation of 1–2°C may produce bleaching, although with low mortality and rapid (1–2 month) recovery of pigment in cooler months (Lasker et al., 1987; Jaap, 1985); by comparison, field temperatures of 4°C above normal, sustained for perhaps only a few days, can produce mass bleaching followed by 90–95 percent coral mortality (Glynn, 1984; Harriott, 1985). Renewed recruitment from the plankton and coral cover recovery after such events exceeds 5 years (Jokiel and Coles, 1990; Brown and Suharsono, 1990). Mass bleaching events across the Pacific in 1983 can be attributed to the seawater warming resulting from the prolonged 1982–3 ENSO event. In the Java Sea, mean seawater temperatures rose by 2–3°C with 80–90 percent of corals dying on reef flats after massive bleaching (Brown and Suharsono, 1990), while in the eastern Pacific coral mortality on bleaching ranged from 51 (southwest Costa Rica) to 97 percent (Galapagos Islands) on sea-surface warming of 2–4°C above normal ambient temperatures (Glynn et al., 1988). As coral reef development in the eastern Pacific is characteristically poor – with reef thicknesses of only 1–8 m despite high skeletal growth rates – it appears that it is these warming events which act as a strong brake on reef development, rather than the low temperatures associated with upwelling as previously supposed (table 9.3; Glynn and D'Croz, 1990).

Given these established relationships, the appearance of complex, multiphase bleaching episodes outside ENSO-influenced areas (e.g. Great Barrier Reef: Fisk and Done, 1985), culminating in mass bleaching throughout the Caribbean in 1987 (Williams et al., 1987; Williams and Bunkley-Williams, 1988) and 1989 (Goreau, 1990), have led reef scientists to speculate as to whether these events

Table 9.3 Response of *Pocillopora damicornis* to temperature stress at Saboga Island Reef, Pearl Islands, Gulf of Panama

Oceanography	*Coral condition*		
	Normal (%)	*Bleached (%)*	*Dead (%)*
Upwelling – cooling (March 19, 1985)	62.4	27.2	10.4
El Niño – warming (September 12, 1983)	9.7	21.8	68.5

Coral mortality was significantly higher after El Niño warming than after subsequent episodes of unusually intense cool upwellings
Source: after Glynn and D'Croz, 1990

Figure 9.8 Air and sea surface temperature trends for Oahu, Hawaiian archipelago (after Nullet, 1989). Maximum summer temperatures (after Jokiel and Coles, 1990) now start to approach the thermal threshold for Hawaiian corals.

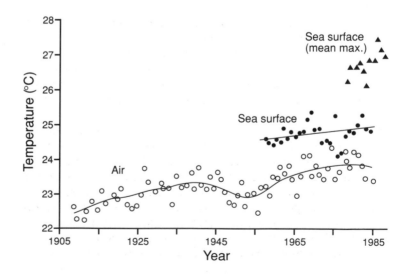

represent the first strong signal of the greenhouse effect in the shallow marine biosphere. Jokiel and Coles (1990), for example, have drawn attention to a prolonged warming trend in Hawaiian waters bringing corals close to their thermal thresholds of 28–29°C (figure 9.8); indeed, summer heating of reef flats and warm meso-scale ocean eddies in the lee of the islands of Maui and Hawaii produced localized bleaching from 0 to −20 m in 1986, 1987, and 1988.

It would be a mistake to see bleaching as simply temperature-controlled. There appear to be synergistic relationships between bleaching, temperature, and high light levels (e.g. Hoegh-Guldberg and Smith, 1989), and bleaching also appears to be induced by a range of other causative factors, including subaerial exposure, freshwater dilution, turbidity and sedimentation, calm water, and coral infections (Williams and Bunkley-Williams, 1990). Thus bleaching may be symptomatic of wider anthropogenic stresses on the shallow marine environment.

Reefs, humankind and environmental change

There are large-scale interactions between coral reefs and other tropical shallow marine ecosystems and their hinterlands. Reef structures dissipate ocean wave energy and create reef flats and lagoons which permit the opportunistic growth of seagrasses and mangroves. By acting as sinks for terrestrially derived sediments and nutrients, mangrove forests and seagrass beds trap and bind sediments, reduce sediment concentrations in the water column, and intercept freshwater runoff, thereby promoting the growth of corals offshore. Thus coral density on reefs in the Ryukyu Islands is strongly influenced by the presence or absence of a natural shoreline forest (Kuhlmann, 1985). The combination of increased sediment supply from deforestation and watershed clearance in the tropics and the removal of coastal margin sediment traps with agricultural, urban, and industrial development pressures is thus of enormous consequence to continuing reef health. A particularly well documented case was the decline in lagoon coral communities and the corresponding explosive growth of the smothering green algae *Dictyosphaeria cavernosa* with urbanization, dredging and filling, and massive sewage inputs to Kaneohe Bay, Oahu, in the 1960s and 1970s; encouragingly, coral recovery has followed the termination of sewage discharges in 1978 (Maragos et al., 1985). However, deterioration of coral reefs continues to be reported from the northwest (Muzik, 1985) and south (Dahl, 1985) Pacific and the Caribbean (Rogers, 1985). Coral reefs may be caught, therefore, in a tightening vice whose jaws are on the one hand a secular warming trend and on the other a decreasing resilience to environmental impacts with environmental deterioration (figure 9.9).

Figure 9.9 Model of the
causes of worldwide
coral-reef bleaching.
Interaction of long-term
trends, temporary
warming events, and
seasonal variations in
temperature yields
potential for multi-phase
bleaching episodes (after
Williams and
Bunkley-Williams, 1990).

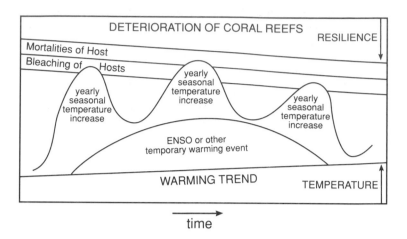

time

Conclusion

If one ceases to see coral reefs as only passive structures to be drowned in time (and where socio-economic impacts can be simply determined through GIS packages which utilize a critical contour height overlay), it becomes necessary to ask questions about the dynamic functioning of reef systems and their potential response to environmental change. What, for example, are the key parameters of sea-level change as a forcing function for reef response? Is it change in low tide level? Or changing wave energy? Or variations in temperature and ocean circulation? What is the relative importance of the changing magnitude and frequency of "average events" as opposed to the changing incidence of storms or mass bleachings? And how is this signal, whatever its form may be, translated into a morphological change in the reef system? How – and how quickly – does reef calcification, sedimentation, and lithification change on sea-level rise (and how do changes in reef community structure, function, and metabolism condition these geological processes)? While the past record of coral growth offers some useful pointers to future reef behavior, we must remain aware of the difference in boundary conditions from the early Holocene and of the anthropogenically stressed nature of many modern reef systems. While there is no shortage of concern over the future of atoll island states in the reef seas, scientific research into the kinds of basic questions outlined above is needed if we are to begin to understand what the future really holds for these dynamic and environmentally sensitive systems.

Further reading

A special section on "sea-level rise as a global geomorphic issue" in volume 14 (1990) of *Progress in Physical Geography* describes the past sea level record in more detail than this contribution and usefully extends these arguments to saltmarsh and mangrove environments. D. R. Stoddart deals specifically with coral reefs. Good introductions to coral reef geomorphology are provided by A. Guilcher's *Coral Reef Geomorphology* (1990, Chichester, J. Wiley) and D. Hopley's *The Geomorphology of the Great Barrier Reef* (1982, New York, Wiley-Interscience), which is a more wide-ranging text than the title suggests. For a beautifully illustrated guide to reefs as geological structures, see H. W. Menard's *Islands* (1986, New York, W. H. Freeman (Scientific American Library)). For an ecological approach, stunning photography, and taxonomic detail, consult J. E. N. Veron's *Corals of Australia and the Indo-Pacific* (1986, North Ryde, Angus and Robertson).

PART IV

The Hydrological System

CHAPTER TEN

Surface Water Acidification

Richard W. Battarbee

Introduction

Two simple claims were made over 20 years ago in Sweden by Svante Odén (1968). The first was that "acid rain" was causing lakes to acidify and fish to die. The second was that the increased acidity in precipitation was derived from long range cross-frontier transport of air pollutants, especially sulfur dioxide, from Britain and other heavily industrialized countries in Europe.

Ironically, but not surprisingly, given Britain's long history of industry, the first recognition that gas emissions from fossil fuel combustion caused environmental damage occurred in Britain. In the mid-nineteenth century Angus Smith (1852) commented extensively on the acidity of the atmosphere and its effect on materials, buildings, and vegetation in British towns and cities, and he is credited with the first use of the term "acid rain." A century later Eville Gorham (plate 10.1) pointed out that rainfall acidity in the English Lake District was significantly influenced by air masses passing over industrial regions upwind, and that acid deposition could cause changes in the ecology of upland lakes (Gorham, 1958).

Despite the accuracy and importance of his observations, Gorham's work was not pursued, perhaps because Britain was preoccupied in the 1960s with the human health issue of improving air quality in its towns and cities. And in the 1970s the claims made by Odén and other Scandinavian environmental scientists, that Britain's large power stations, built partly to relieve urban pollution in the UK, were causing acidified precipitation in Norway and Sweden, were not believed (plate 10.2). Nevertheless, the claims were fully justified and it was established by 1980, from air quality monitoring within Europe, that significant transboundary air pollution did, and does, occur. Import/export balances between countries in pollutants such as sulfur dioxide (SO_2) and nitrogen oxides

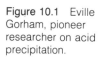

Figure 10.1 Eville Gorham, pioneer researcher on acid precipitation.

Plate 10.2 Industrial emissions of sulfur dioxide and other gases, especially from power stations, are now known to have significantly increased the atmospheric deposition of acids over Europe and eastern North America.

(NO_x) are now calculated annually (e.g. EMEP, 1991), and, although countries argue over the figures, the concept of trans-boundary pollution has been accepted.

There has been a greater controversy over the extent to which acid deposition has caused environmental damage. In the case of surface water acidification the debate has been over the importance of acid deposition relative to other possible alternative causes, such as land-use change and natural factors (Battarbee et al., 1990; Renberg and Battarbee, 1990). However, the importance of acid deposition as the major cause of surface water acidification has also now been accepted, again justifying Odén's claims.

This chapter, consequently, seeks to explain how acid deposition affects surface water chemistry and biology, and it outlines the results of recent research concerned with acidification reversal.

Natural acidity of precipitation and surface waters

Water weakly dissociates into H^+ and OH^- ions, and because the amount of H^+ in pure water is 10^{-7} gram-ions per liter, pure water has a pH of 7. Rain, even clean rain, is not pure water since it contains naturally occurring acids and bases and reacts with gases in the atmosphere, especially with CO_2 to form carbonic acid (H_2CO_3).

At equilibrium, "clean rain" has a pH of 5.6. This theoretical value is rarely attained, as contamination by alkaline dusts, for example over the North American prairies (Barrie and Hales, 1984), can raise pH, and contact with naturally occurring sulfur and nitrogen gases from volcanic activity and biological sources can significantly depress pH. Galloway et al. (1982) give background values of about pH 5.0 for many remote sites in the Southern Hemisphere. Consequently many authors take this value as the background pre-industrial acidity of rain over Europe.

The natural pH of streams and lakes is only partly influenced by the pH of precipitation. More important are the ion exchange reactions that take place in the soil solution as rain water moves through the watershed. The crucial process is the production of carbonic acid from the decomposition of organic matter and from mineral weathering, and its dissociation into H^+ and HCO_3^- in the soil solution. The H^+ ions are then exchanged for base cations (mainly Ca^{2+} and Mg^{2+}) and leave the soil with bicarbonate anions (alkalinity) in the groundwater and runoff. The groundwater has a high partial pressure of CO_2 and an increase in water pH occurs as the CO_2 out-gasses when the water emerges in streams and lakes and re-equilibrates with the external atmosphere.

The pH of clean (non-acidified) surface waters is therefore controlled by these processes, and, in regions where bedrock weathers slowly and soils have a naturally low base saturation, such waters typically have pH values from 5.5 to 7.0. Although acidic, these values are substantially higher than those for the pH of incident precipitation, and this is due to the alkalinity generated in the soil. Unpolluted waters with pH < 5.5 are rare except where organic acids from peaty soils or strong acids from naturally occurring sulfides in the catchment are important.

Acid deposition and its effect on surface waters

Strong acidity in the atmosphere is derived principally from the burning of coal and oil. The sulfur dioxide (SO_2) and nitrogen oxides (NO_x) produced in this way are either deposited close to source (as dry deposition) or, after oxidation to SO_4^{2-} and NO_3^- respectively, deposited at greater distances from the source (as wet deposition) (figure 10.1).

In upland regions occult deposition, in which wind-driven mist droplets are impacted onto leaf and branch surfaces, is also important (Dollard et al., 1983). The total acid deposition received at any point on the ground is, therefore, a combination of sulfur and nitrogen deposition from these different processes. It is usually expressed as kEq H^+ ha^{-1} $year^{-1}$. Because the deposited nitrogen in most regions becomes incorporated in terrestrial biomass or in soil organic matter, sulfur rather than nitrogen is the main cause of surface water acidification. However, in some cases where soils have become saturated with nitrate, nitrate lost to drainage waters can also contribute to stream and lake acidity.

The impact of acid deposition on surface water chemistry varies enormously according to the original alkalinity of the water (see above). The resulting alkalinity of waters influenced by acid deposition is then expressed as:

Figure 10.1 Schematic diagram showing emission, transport, transformation, and deposition of wet and dry acid deposition (from Crane and Cocks, 1987).

$$Alk = [Ca^{2+} + Mg^{2+}] - [SO_4^{2-} + NO_3^-]$$
$$= [HCO_3^-] - [H^+] - [Al^{3+}]$$

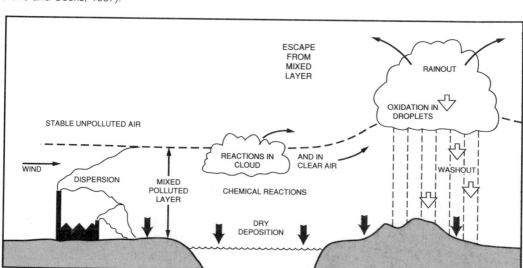

In waters with low original alkalinity (< 100 µEq l^{-1}) the addition of the strong acid anions (SO_4^{2-}, NO_3^-) leads to a reduction in bicarbonate (loss of alkalinity), and increases in H^+ (pH decrease) and Al^{3+}. Base cation (Ca^{2+} and Mg^{2+}) values either remain constant or increase depending on the cation-exchange capacity of the soils.

The variable response of surface waters to acid deposition can be illustrated by comparing the modern chemistry of streams with differing catchment characteristics but similar acid deposition. Table 10.1 shows an example from the Galloway region of Scotland (figure 10.2), and compares the chemistry of bulk precipitation and selected streams. The precipitation has a mean pH of 4.4, high sulfate values (98 µEq l^{-1}), high sea salts (Na^+ and Cl^-), low calcium (18 µEq l^{-1}) and no aluminum. The stream chemistries are arranged in increasing order of calcium. The Dargall Lane and Grannoch 3 are both extremely acidic (pH 4.35–4.9) with low calcium values (40–65 µEq l^{-1}), while the Burnfoot stream has pH 6.7 and a calcium of 165 µEq l^{-1}, and Barlay a pH of 8.0 and calcium of 1050 µEq l^{-1}.

Barlay is situated in a catchment with base-rich soils. The runoff water is consequently high in base cations and considerable alkalinity is generated both by mineral weathering and by respiration of the soil microbial community. Acidity from acid precipitation is neutralized by catchment soils as the H^+ in precipitation is exchanged for base cations. Alkalinity in the water remains high and, except for the sea-salts Na^+ and Cl^- there is no similarity between the stream and precipitation chemistry. In contrast, Grannoch 3 is situated on hard granitic bedrock. Weathering occurs only slowly, the soils are naturally acidic and peaty, and the exchange surfaces have very low base saturation; that is, most of the exchange sites are occupied by the acid cations H^+ and Al^{3+}. The addition of sulfate causes H^+ and Al^{3+} rather than base cations to be exchanged and transported out to surface waters, pH is not dissimilar from that of the precipitation, and Al^{3+} has reached levels that are toxic to trout. The Burnfoot stream has an intermediate chemistry with just enough alkalinity to neutralize acid deposition.

Table 10.1 Water chemistry analyses for selected sites in the Galloway district, southwest Scotland

	Altitude (at sampling site) (m)	Forest cover (%)	pH	Na+	K+	Mg2+	Ca2+	Cl-	SO4²-	NO3-	Alk	Al
Galloway precipitation chemistry												
Cairnsmore of Fleet	–	–	4.42	118	7	28	18	130	78	11	–	–
Galloway stream chemistry												
Dargall Lane	255	0	4.9	122	12	58	40	142	119	10	0	126
Grannoch 3	235	70	4.35	170	5	50	65	186	182	8	0	307
Burnfoot	100	50	6.7	180	8	84	165	232	196	44	38	36
Barlay	40	0	8.0	170	9	320	1050	220	220	51	1273	<10

Ion concentrations are as µEq l^{-1} or µg l^{-1} (Al).
Data from Harriman et al., 1987

Figure 10.2 Location of
streams in Galloway with
differing chemistry.
Shaded area shows
granitic bedrock (see
table 10.1) (from Harriman
et al., 1987, reprinted by
permission of Kluwer
Academic Publishers).

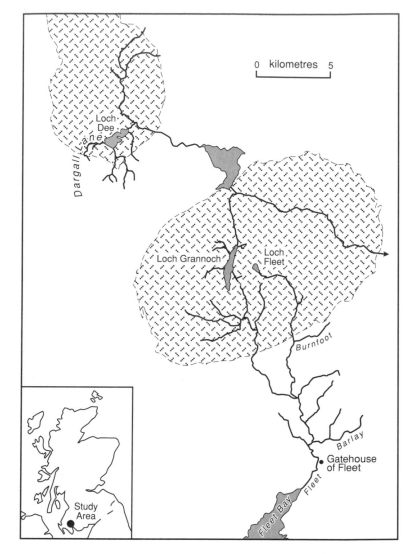

Figure 10.2 Location of streams in Galloway with differing chemistry. Shaded area shows granitic bedrock (see table 10.1) (from Harriman et al., 1987, reprinted by permission of Kluwer Academic Publishers).

The main difference between these three streams is related to the original alkalinity, or neutralizing capacity, of the catchment soils. However, comparison between the Dargall Lane chemistry and Grannoch 3 (table 10.1) also indicates a secondary factor related to afforestation. Grannoch 3, with 70 percent afforestation, has a lower pH and a much higher aluminum value than Dargall Lane, despite its somewhat higher calcium. The greater than expected acidity at Grannoch can be attributed to the forest, where sulfur deposition is enhanced by the scavenging effect of the forest canopy (Harriman and Morrison, 1982), an effect that leads to the higher sulfate values in the water.

In summary, the key processes involved in the acidification of streams and lakes are the deposition of acidity, principally sulfur, the exchange of cations and anions, the weathering of primary soil minerals, the generation of alkalinity in soils, and modifications to these processes caused by catchment vegetation, especially forestry. More detailed variations between catchments in time and space are also related to microgeological and hydrological differences, the frequency and intensity of sea-salt episodes, and a range of other random and non-random factors.

Evidence of recent acidification

Modern measurements of stream and lake water chemistry show the presence of highly acidic waters in regions of high acid deposition. However, this correlation is not an adequate basis for concluding that recent acidification has taken place. More direct evidence for a *change* in water quality is required. In a limited number of cases historical records have provided useful information on past chemistry (Watt et al., 1979; Sutcliffe et al., 1982; Sutcliffe, 1983). Unfortunately such data are rare, usually lack temporal and methodological continuity, have insufficient antiquity, or suffer from problems of seasonal and inter-annual variation (Battarbee, 1984).

In the absence of high-quality, long-term records, evidence of acidification can be based either on empirical chemical models (the "Henriksen" approach) or on diatom studies of lake sediments (the paleolimnological approach).

The Henriksen approach

It is clear from the discussion above that waters acidified by acid deposition differ chemically in a number of ways from non-acidified waters; in particular, they have lower pH and alkalinity and higher sulfate for any given level of base cations. Consequently, an index of acidification (AI) of a water body is the difference between the original pre-acidification alkalinity (A_o) and the current alkalinity (A_t):

$$AI = A_o - A_t \tag{1}$$

A_t is measured and values for A_o can be derived empirically as alkalinity in non-acidified waters is directly related to non-marine base cation concentration. Henriksen (1979) used data from 13 lakes worldwide, that were assumed to be non-acidified, to generate a linear regression model for A_o:

$$A_o = 0.93 \, ([Ca^{2+}]^* + [Mg^{2+}]^*) - 14 \tag{2}$$

where all concentrations are in $\mu Eq \, l^{-1}$ and an asterisk indicates sea-salt correction.

Acidification, then, is equivalent to the loss of alkalinity (1). If the acidification is caused by acid deposition then the loss of alkalinity (the acidification index) will be balanced by an increase

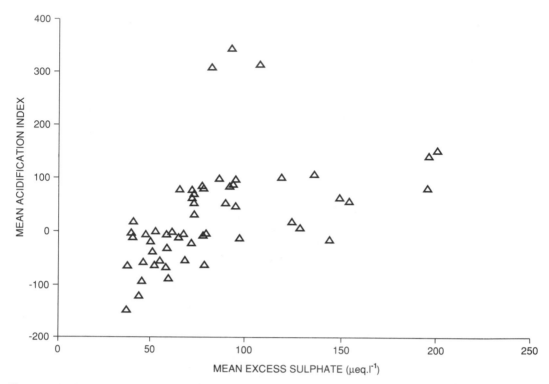

Figure 10.3 Henriksen acidification index (AI) plotted against excess sulfate for surface waters in the UK (from AWRG, 1988).

in sulfate. In a study of Scottish waters, Wells et al. (1986) showed that there was indeed a very good agreement between non-marine sulfate concentration in the water and the acidification index ($n = 92$, $r = 0.87$). Analysis of a fuller data set for a wider range of sites from all parts of the UK produced a somewhat weaker correlation (figure 10.3).

Although this method is not applicable to all water bodies, such as ones with significant amounts of dissolved organic carbon, and although it gives no indication of the timing and rate of acidification, it has been used extensively and provides an effective and rapid means of assessing the acidification status of large numbers of lakes and streams.

The paleolimnological approach

The paleolimnological approach is a much slower technique for assessing acidification than the Henriksen method but it has the advantage of being less ambiguous, is suitable for all waters, and provides information on both the timing and rate of acidification. The diatom microfossil record is particularly useful. Diatoms are algae (class Bacillariophyta) with siliceous cell walls (plate 10.3). They are excellent indicators of water chemistry and changes in the composition of fossil assemblages can be used to reconstruct changes in water pH.

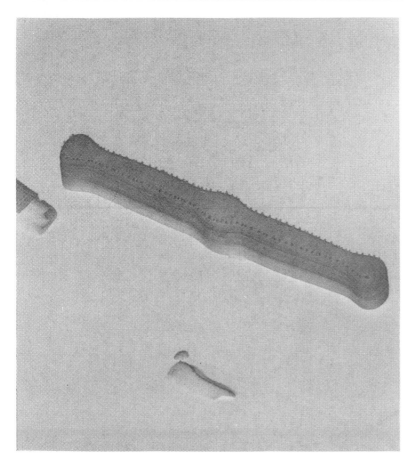

Plate 10.3 Scanning electron micrograph of *Tabellaria quadriseptata* (about 60 μm in length), a diatom which can survive in strongly acid water. Along with other acid tolerant diatoms, this species increases in the uppermost part of sediment cores from recently acidified lakes.

The most recent statistical approach for pH reconstruction is based on calculations of the pH optima of individual species using a weighted averaging technique (Birks et al., 1990a, b). Figure 10.4 shows the relationship between pH inferred from diatom weighted-averaging and pH measured electrometrically for a training set of 171 lakes used in the Surface Waters Acidification Programme (Battarbee et al., 1990; Stevenson et al., 1991).

An example of the use of this approach is seen in figure 10.5, which shows a diatom diagram from a core from the Round Loch of Glenhead in Galloway, southwest Scotland (Jones et al., 1989) (plate 10.4). The lake has a present pH of about 4.8. A small number of indigenous brown trout (*Salmo trutta*) are still present in the lake but some adjacent lakes are fishless. The lake and its catchment are situated entirely on granitic bedrock and the catchment has mainly peaty soils and a moorland vegetation which is grazed by sheep and periodically burned.

The core has been dated by the ^{210}Pb method (Appleby et al., 1986) and the proportion of the main diatom species at successive sediment levels has been calculated. Figure 10.5 shows that there

Figure 10.4 Observed
pH versus diatom-inferred
pH for 171 lakes in
Norway, Sweden and the
UK (from Stevenson et al.,
1991).

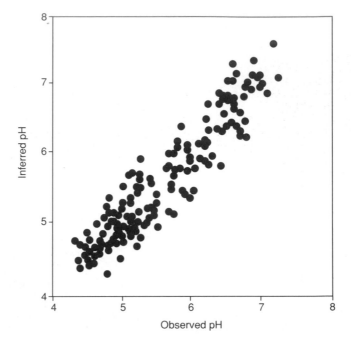

Figure 10.5 Changes in
diatom assemblages, and
reconstructed pH for a
sediment core from the
Round Loch of Glenhead.
Dates are interpolated
from ^{210}Pb dating
(adapted from Jones et
al., 1989; Birks et al.,
1990a).

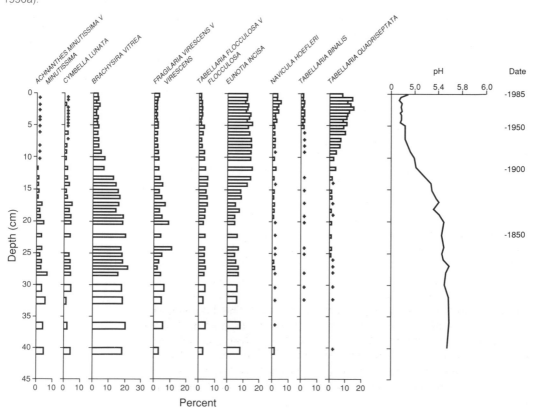

Plate 10.4   Round Loch of Glenhead, in Galloway, southwest Scotland. Notwithstanding its isolation, this lake has been acidified by almost 1 pH unit since the Industrial Revolution through atmospheric pollution.

was little change in the diatom flora of this lake until the mid-nineteenth century, when diatoms characteristic of circumneutral water, such as *Brachysira vitrea*, began to decline and be replaced by more acidophilous species. Acidification continued through the twentieth century and by the early 1980s the diatom flora of the lake was dominated only by acid tolerant taxa such as *Tabellaria quadriseptata* and *T. binalis*. The reconstructed pH curve since about 1700 (figure 10.5) shows a stable pH of about 5.4 until about 1850, followed by a reduction of almost one pH unit in the following 130 years to 1980.

Similar work has been carried out at many sites within the UK and in Europe and North America (Battarbee and Charles, 1987; Charles et al., 1989). In both continents, sites with clear recent acidification trends similar to the one from the Round Loch of Glenhead occur in areas of high sulfur deposition. Lakes with low calcium waters in the western USA remain unaffected, as do those in northwest Europe where sulfur deposition is low.

The dose–response relationship between sulfur deposition and acidification is clearly seen in the UK. Figure 10.6 shows the water calcium and total sulfur deposition relationships for all sites with diatom-based pH reconstructions. The data show that all the acidified lakes fall below a line which relates acidification to a ratio of calcium to sulfur of approximately 60 : 1. This ratio can then be used to predict acidification at sites where core records are not available. It can also be used to calculate "critical sulfur loads" (see below).

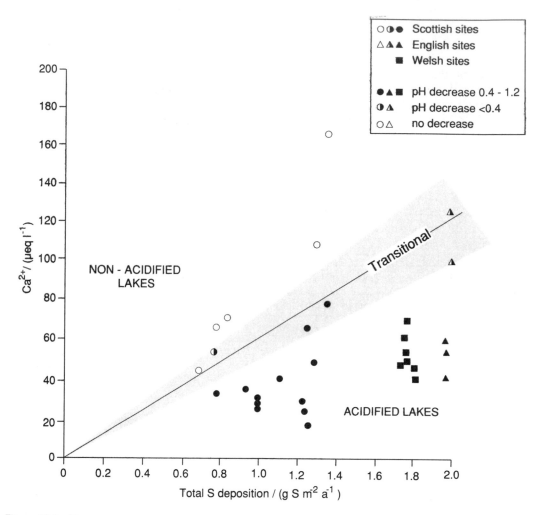

Figure 10.6 The relationship between lake-water calcium concentration (as a measure of sensitivity) and sulfur deposition for lakes in the United Kingdom. The status of each site (acidified, non-acidified, transitional) is based on pH reconstruction from diatom analysis (from Battarbee, 1990).

Biological consequences of surface water acidification

The chemical changes brought about by acid deposition, such as pH decrease, loss of alkalinity, and increase in toxic metals, especially aluminum, have caused major changes in the structure and function of soft-water aquatic ecosystems. Shifts in algal, aquatic macrophyte, zooplankton, and benthic invertebrate populations are well documented (Muniz, 1991; Morrison and Battarbee, 1988). At higher trophic levels there is evidence in the UK for the acidification of natterjack toad (*Bufo calamita*) breeding ponds (Beebee et al., 1990) and the loss of dipper (*Cinclus cinclus*) populations (Ormerod et al., 1986; Tyler and Ormerod, 1988). But the major concern in all countries has been over the loss of fish populations.

The first evidence for decline in salmonid fish populations in association with acid waters was presented for Norway by Jensen

and Snekvik (1972) and for Sweden by Almer et al. (1974). In
Norway annual statistical data for salmon from 1876 were used to
show that, for the rivers in the south of the country, salmon (*Salmo
salar*) catches declined progressively through the twentieth century,
to the extent that these rivers are now almost entirely fishless
(Leivestad et al., 1976). Brown trout (*S. trutta*) populations in small
lakes in the south of Norway have also declined dramatically. A
survey of 1679 lakes in 1974 and 1975 in the southernmost
counties of Norway showed that brown trout were absent or very
sparse in over half the lakes and that the proportion of sites with
fish decreased with increasing acidity. A further questionnaire
survey of local landowners and fishermen showed that the main
increase in the number of fishless lakes occurred in the decades
following 1940 (Sevaldrud et al., 1980) (figure 10.7).

In the UK the first evidence of fish losses came from the Loch Ard
area in Scotland, where Harriman and Morrison (1982) showed
that brown trout populations were sparse in moorland streams and
absent from afforested streams. In a parallel study of 26 streams
and 22 lochs in Galloway, fishless streams and lochs were found in
moorland as well as afforested regions (Harriman et al., 1987).

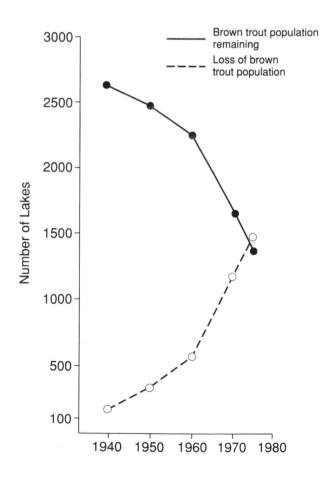

Brown trout population remaining

Loss of brown trout population

Figure 10.7 Loss of fish populations in southern Norway (from Sevaldrup et al., 1980).

a.

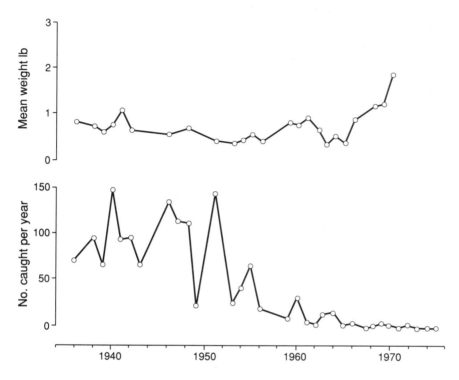

b.

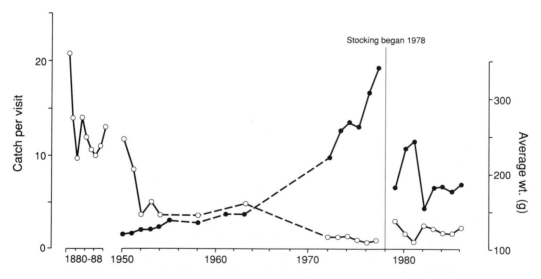

Figure 10.8 Changes in fish catch and weight for (a) Loch Fleet (from Brown and Turnpenny, 1988) and (b) Lyn Conwy (from Milner and Varallo, 1990, reprinted by permission of Kluwer Academic Publishers).

Analysis of the data suggested that the presence or absence of brown trout could not be related to factors such as access, presence of suitable spawning grounds, food availability, or overfishing, but was related to acid deposition, exacerbated in some cases by afforestation (Harriman et al., 1987). A subsequent study by Maitland et al. (1988), selecting sensitive lakes with granitic catchments throughout Scotland, confirmed the Harriman et al. findings for Galloway, and showed that Galloway was the most affected region of Scotland.

In Wales, similar surveys have been carried out. While the number of sites which are fishless is small and the regional picture has been complicated by stocking, it is clear that there have been substantial declines in populations in the most sensitive parts of the country and especially at sites with afforested catchments. In a study of 37 sensitive lakes in Gwynedd, Milner and Varallo (1990) found that between 8 and 21 sites had suffered declines that could have been acid-water related, and, by extrapolation to Wales as a whole, 35–92 sites could be affected.

Many of the data used to infer fishery decline in acid waters have been based on questionnaire surveys about fishery quality (usually presence or absence) through time. At a few sites, however, historical catch statistics are available (Harriman et al., 1987; Brown and Turnpenny, 1988; Milner and Varallo, 1990). These data show not only declines in population size but also an increase in average weight as recruitment failure progressively eliminates younger age classes. In the case of Loch Fleet (figure 10.8a), stocking to boost the falling population was unsuccessful and the lake became fishless in 1976. In Llyn Conwy (figure 10.8b), stocking in 1978 was partially successful and the trend toward increasingly large fish was reversed. A small fishery including some indigenous stock has been maintained.

Sulfur reduction and critical loads

The clear linkage between acid deposition and surface-water acidification suggests that the most obvious method for controlling and reversing acidification is an elimination or reduction of acid deposition, especially over regions sensitive to acidification. Even before the scientific basis for this linkage was fully established and accepted by all countries, Scandinavian countries had already persuaded the UNECE to set up a Convention on Long Range Transboundary Air Pollution (LRTAP) and promote protocols for reductions in both sulfur and nitrogen. The ensuing sulfur protocol, which encouraged member states to reduce national emissions of SO_2 by a minimum of 30 percent by 1993 based on a start year of 1980, was adopted in 1985. Membership of the "30 percent club" expanded relatively rapidly but a number of net sulfur-exporting countries, most notably the United Kingdom, failed to join. Nevertheless, the UK conceded the importance of SO_2 reductions in 1986 and began the planned introduction of flue gas desulfurization (FGD) to a number of large British power stations. More recently the UK has signed the EC Directive for large combustion plants, which requires a 60 percent decline in emissions from these sources by 2003, based on the 1980 datum. Power companies are planning to build new gas-fired power stations and import coal with a low sulfur content in order to reduce emissions to meet this target.

Despite the political success of the "30 percent club" this "across-the-board" approach to emission reduction has inadequacies, partly because reductions need to be much higher than 30 percent to cause any significant improvement, but mainly because reductions need to be more specifically targeted at regions and countries with large emission sources. It is also sensible to optimize reduction to gain the most benefit to damaged regions. For example, it might be more important to remove almost all emissions from power stations directly upwind of acidified regions and have lower reductions for power stations whose emissions largely fall in the sea or on land with high neutralizing capacity.

For this reason a "critical loads" concept has been introduced by the UNECE as a replacement for the sulfur protocol (30 percent club). The most commonly used general definition of a critical load is that of Nilsson and Grennfelt (1988): "The highest deposition of acidifying compounds that will not cause chemical changes leading to long term harmful effects on ecosystem structure and function." The concept for freshwaters can be illustrated as in figure 10.9, where a chemical or biological effect is related to an increased acid (S) loading over time. If the critical load is less than the measured S load at any site then the critical load is exceeded. Maps of exceedance within and between countries can be drawn up and used for policy making and the fixing of target loads (see below).

The Nilsson and Grennfelt definition is quite ambiguous, mainly due to the subjectivity required to judge a "long-term harmful effect." The definition can be improved if the term "harmful" is omitted. Then, with reference to figure 10.9, the critical load is exceeded when the first change in the aquatic ecosystem that can be related to acid deposition occurs (figure 10.9, point a). Additional critical values can, in theory, be defined as loads increase and progressively less sensitive organisms in the ecosystem decline or disappear (figure 10.9, points b, c). Target loads (figure 10.9, point T) may then be set according to the need to protect any selected species or groups of species, or according to government expediency.

In the UK, critical loads are being set using a variety of methods including the diatom model (Battarbee, 1990; figure 10.6 above), the steady-state chemical model (Henriksen et al., 1990), and the MAGIC model (see below). The diatom model is most suited to

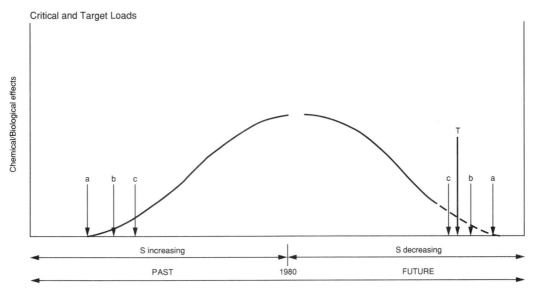

Critical and Target Loads

setting the initial critical load (figure 10.9, point a) as diatoms are among the most sensitive organisms to acidification in freshwaters. Other models are used to calculate the critical load required to maintain alkalinity at levels needed to protect individual organisms, such as brown trout.

Whichever critical load is selected, the extent to which one or more water bodies are receiving higher loads (exceeding the critical load) can be simply calculated by subtracting the critical load from the measured (or modelled) acid load for sulfur at individual sites. Although the calculation is simple there will be errors in the calculation of both critical and measured loads, and the maps that are now becoming available using these techniques should at this stage be regarded as a first approximation.

Figure 10.10 shows the exceedance map for Scotland based on the diatom model. The value for each 10 km grid square is based on the chemistry of only one water sample and on sulfur deposition values extrapolated from a low density network of monitoring stations. Nevertheless, the general pattern of exceedances, especially in the center and south of the country, conform to expectations based on earlier studies of many of these regions (Flower et al., 1987, 1988; Battarbee et al., 1988b; Harriman et al., 1987). In the northwest, somewhat unexpectedly, sites exceeding the critical load appear on the map. Prior research suggests that there are few or no problems of acidification in this region, and it is more likely that the critical load exceedance calculations are in error, due to problems with estimating sea salts in the models and/or to over-estimates of sulfur deposition.

Figure 10.9 Schematic diagram illustrating the critical and target load concepts. Point a refers to the diatom-based point of change and represents the critical load for the ecosystem. Points b and c represent the critical loads for given target organisms. T represents a future target load.

Critical load exceedance (diatom model)

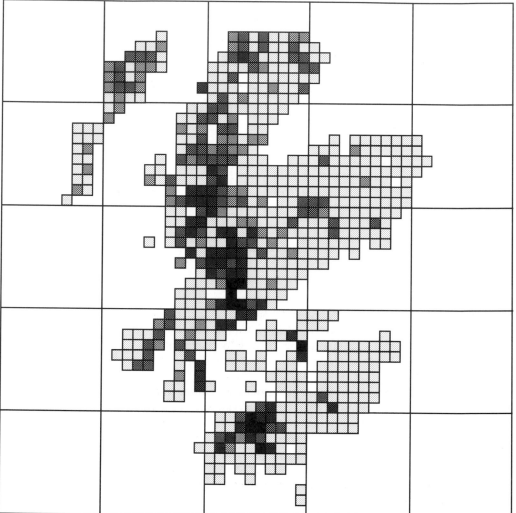

■ > 1.0 kEq H+ ha -1 yr-1

■ 0.5 - 1.0 kEq H+ ha-1 yr-1

▨ 0.2 -0.5 kEq H+ ha-1 yr-1

▨ 0.0 - 0.2 kEq H+ ha-1 yr-1

☐ Not exceeded

Figure 10.10   Critical load exceedance map for Scottish freshwaters based on the diatom model. The exceedance values for the 10 km grid squares are based on the chemical analysis of a single sample from the most sensitive standing water (headwater streams in some cases) in each square. Some values for the northwest of Scotland where surface water chemistry is strongly influenced by sea salts are thought to be erroneous (see text).

Reversibility

The purpose of critical load exceedance maps is to show the distribution of "environmental damage" both at the present and in the future; for example, sites that still exceed the critical load when the target load is reached. Such future exceedance maps assume, however, that lakes and streams will respond without any lag to deposition reduction or, in other words, that acidification and its effects are readily reversible.

In theory the response of lakes to reductions in acid deposition will vary according to the sensitivity of catchment soils. In high base saturation soils a decrease in SO_4 in runoff water will be accompanied primarily by reduction in base cation output with little change in pH. However, at sites with low base saturation a similar reduction in sulfate runoff will lead to a reduction in the acid cations Al^{3+} and H^+ (leading to an increase in pH). From these considerations it can be predicted that the most acidified lakes are the ones most likely to show a rapid initial improvement, although a complete recovery could take a very long time as primary weathering is required to regenerate soil bases.

The extent to which individual sites might respond to sulfur reduction can be assessed by the use of process-driven dynamic models. The most frequently used model is MAGIC (Model of Acidification of Groundwaters in Catchments) which simulates the physical and chemical processes that occur in response to sulfur loading on catchments (Cosby et al., 1985). The model can be set up for any individual site given a knowledge of rainfall and runoff quantity and quality, sulfur deposition history, and soil characteristics (especially bulk density, depth, cation exchange capacity, sulfate adsorption rates and base saturation, temperature, and carbon dioxide levels). Once set up, the model is run to reconstruct past acidification trends and to predict future pH and water chemistry according to specified future sulfur deposition scenarios.

The output of MAGIC can be validated by comparing model-based pH reconstructions with diatom-based reconstructions (Jenkins et al., 1990). For the Round Loch of Glenhead there is close agreement between these two independent methods, giving confidence in the use of the model to predict future changes. Figure 10.11 takes three different future scenarios: (a) a reduction in sulfur deposition of 60 percent by 2003 and then constant levels; (b) a reduction of 60 percent by 2003, and a further reduction to 80 percent by 2020 and then constant levels; and (c) a reduction of 60 percent by 2003 and a further reduction to 100 percent by 2020. The last scenario is almost impossible to achieve but it is of interest in that it indicates the time required for complete recovery. The first scenario approximates present European policies, assuming that a 60 percent reduction in emissions is translated into a similar reduction in deposition. The intermediate scenario assumes a further commitment to sulfur reduction in Europe that might come

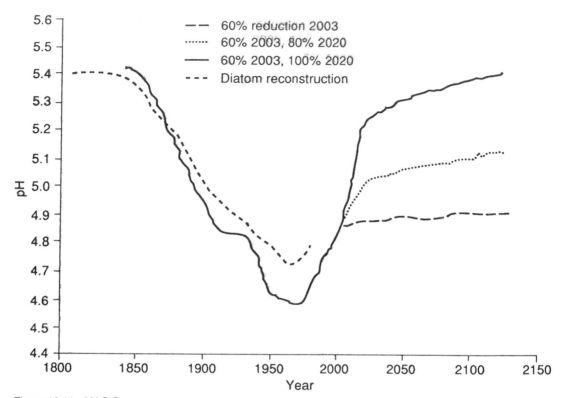

Figure 10.11   MAGIC model pH hindcasts and forecasts for the Round Loch of Glenhead. The pH hindcast is compared to diatom-based reconstruction for the site. Forecasts are for: (a) a reduction in S deposition by 2003 of 60 percent and then constant levels; (b) a reduction of 60 percent by 2003, a further reduction to 80 percent by 2020, and then constant levels; (c) as (b) but with 100 percent reduction from 2020 (pH reconstruction from Flower and Battarbee, 1983, MAGIC data from Jenkins, personal communication).

about following the adoption of a new UNECE policy based on critical loads concepts. According to MAGIC the acidification of the Round Loch of Glenhead should be reversible, and initially at least the reversal should be in line with reductions in sulfur deposition, although a full recovery might take a long time.

Because model forecasts are inherently uncertain it is of considerable interest to examine the responses of such acidified surface waters to real decreases in acid deposition. In southwest Scotland sulfur deposition has already declined by about 40 percent from its maximum in the 1970s and, as predicted by the model, pH has increased from a mean of 4.7 in 1979 to a mean of 4.9 in 1990, reflected by a slight shift in the diatom assemblages in sediment cores sampled in 1990 (Allott, 1991; Allott et al., 1992).

Even more striking have been the changes at Loch Enoch, a lake close to the Round Loch of Glenhead, but at higher altitude, and with lower calcium and pH. Loch Enoch lost its brown trout population in the 1950s (Harriman et al., 1987) and diatom data (Flower et al., 1987) show it to have been strongly acidified beginning about 1850, with a pH of 4.4 in 1979 when first direct measurements were made. Since then, following declines in sulfur deposition (figure 10.12a) the sum of strong acids (SSA) has decreased and pH has shown a sustained increase to 4.8 (figure 10.12b).

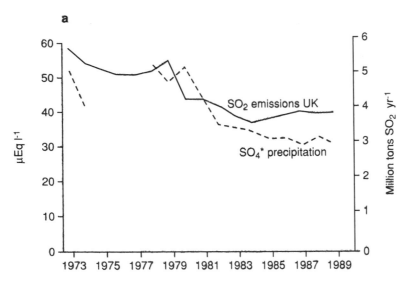

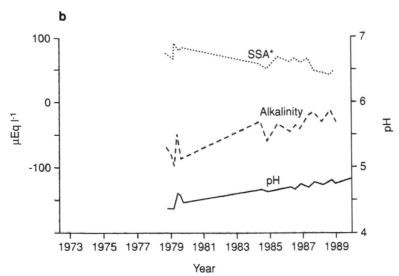

Figure 10.12   Recent changes: (a) in UK sulfur dioxide emissions, and non-marine sulfate deposition at Eskdalemuir, south Scotland; and (b) in sum of strong acids (SSA), alkalinity and pH for Loch Enoch, Galloway, southwest Scotland (from Harriman, personal communication).

In Norway the question of reversibility has been tackled using large-scale catchment experiments (Wright and Hauhs, 1991). In 1984 the RAIN (Reversing Acidification in Norway) project constructed a large transparent roof over an acidified stream catchment to remove incident acid precipitation. In the experiment in the Risdalsheia catchment the precipitation was collected from the roof, cleaned to remove sulfate and nitrate and then sprayed below the roof onto the forest canopy. The response of the outflow stream chemistry was predicted before the experiment began using the MAGIC model. Figure 10.13 shows that the declines in sulfate, nitrate, and base cations and rise in alkalinity were rapid and close to the MAGIC model predictions during the 1984–90 period of the

Figure 10.13 Results from the RAIN project experiment at Risdalsheia, southern Norway after exclusion of incident acid deposition: (a) declines in sulfate and nitrate; (b) decline in the sum of base cations (SBC); (c) increase in alkalinity. Changes simulated by MAGIC are also shown compared with observations for the past, and as predictions for the future following re-introduction of acid deposition (from Wright and Hauhs, 1991).

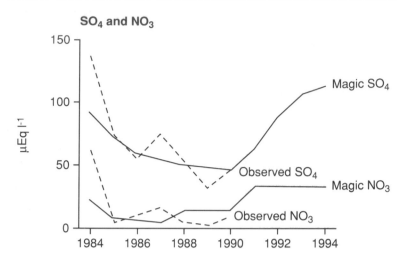

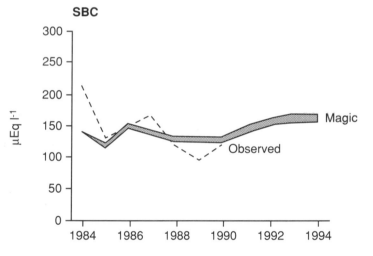

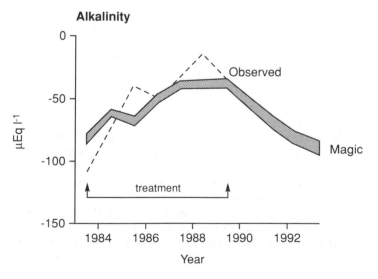

experiment. Predicted re-acidification of the catchment is also shown for the next few years as the experiment is reversed and ambient precipitation allowed to fall on the catchment again.

Biological recovery

Most studies of reversibility have been concerned with the chemical response of streams and lakes to a reduction in acid deposition. Of prime importance, however, is the biological response. As pH increases organisms still existing in the stream or lake should recover their population sizes, and rapid colonizers, such as algae and many invertebrates, might be expected to reinvade. On the other hand, fish populations may be slow to recover unless helped by stocking.

The extent to which the pre-acidification ecological balance can be restored is largely unknown. The problem is exacerbated because there is little information on the pre-acidification biology of acidified streams and lakes. For lakes the sediment record of diatoms, cladocera, and chironomids can be useful in identifying previous biological conditions. Otherwise non-acidified sensitive streams and lakes in non-polluted regions are needed to provide analogue situations. Such regions are rare as the effects of acid deposition are so widespread in Europe. However, parts of north-west Scotland and central Norway still have relatively pristine streams and lakes that can be used as reference or control sites to assess the speed and direction of biological recovery (Battarbee et al., 1989a).

The desirability of complete restoration may be debated. The conservation ethic suggests that whole ecosystem recovery is most important. For economic or recreation purposes, on the other hand, it is often the survival and recovery of fish, especially brown trout, populations that is paramount. In these cases acidity can be mitigated by liming to maintain an artificially high pH. In Sweden wholesale liming of acid waters has been adopted as a national strategy to protect fish stocks (plate 10.5). Other countries have taken a much more cautious approach and in the UK only experimental liming has been carried out (Burns et al., 1984; Underwood et al., 1987; Howells and Brown, 1987). Lime can be applied either directly to the lake surface or to the lake catchment. The former is simpler and quicker, but its effectiveness is limited by the water residence time of the lake. In the UK, where most lakes have residence times in weeks or months, liming needs to be frequently repeated. Catchment liming is more long-lasting, but liming upland catchments damages terrestrial ecosystems.

The most comprehensive experiment in liming in the UK was carried out by the former Central Electricity Generating Board (CEGB) on Loch Fleet (Howells and Brown, 1987), a site which had lost its brown trout population in the 1960s (see figure 10.8a). Lime was applied to various sub-catchments in 1986 (362 tonnes) and 1987 (83 tonnes). The pH of the loch water rose from 4.5 to

Plate 10.5 Liming of acidified lakes has been used as a short-term measure to protect fish stocks, as here in Sweden, but a permanent solution to acidification of freshwaters must come from a reduction in sulfur emissions.

over 6.0 in a few weeks and this was followed by rapid changes in the diatom flora (figure 10.14). This was accompanied by a threefold increase in calcium and a steep drop in labile monomeric aluminum, providing a water chemistry suitable for brown trout survival. Fish were successfully reintroduced to the loch in 1987 and a self-sustaining population was established. On the negative side, bryophyte communities in the catchment have been killed by liming and inflow streams have excessive pH levels. A comparison of the diatom flora after liming with the pre-acidification flora (Anderson et al., 1986) shows that the lake is now substantially different from its condition at any previous time (Cameron, 1990).

Because of these negative ecological consequences, and the need for expensive repeated applications, liming is a limited management technique, useful in maintaining or reintroducing fish populations at selected sites where recreational angling is more important than nature conservation, but not a substitute for emission reduction.

Monitoring and future issues

Now that the role of acid deposition in the process of surface-water acidification is largely understood many countries have diverted their efforts away from intensive studies of individual problem sites

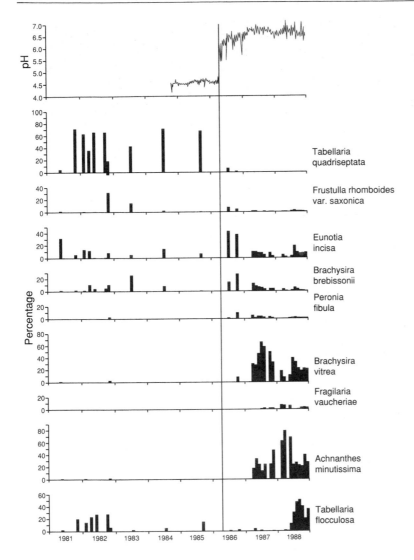

Figure 10.14 Response
of pH and epilithic diatom
communities to liming in
Loch Fleet, Galloway
(from Cameron, 1990).

to more extensive programs of long-term monitoring. Such programs are designed to assess the response of acidified aquatic ecosystems to planned reductions in sulfur emissions, and can be used to assess and validate the critical load models. In the United Kingdom a monitoring network of 22 sites (figure 10.15) was established in 1988 (Patrick et al., 1991). This network includes sensitive streams and lakes in areas of high and low acid deposition and at each site key chemical and biological variables are measured or surveyed at frequent intervals, usually quarterly or annually. In addition some stream sites are monitored continuously for pH, conductivity, and flow in order to detect trends in the frequency and intensity of acid episodes.

Because most acid deposition in the UK is derived from UK emission sources rather than from overseas sources the sites in the

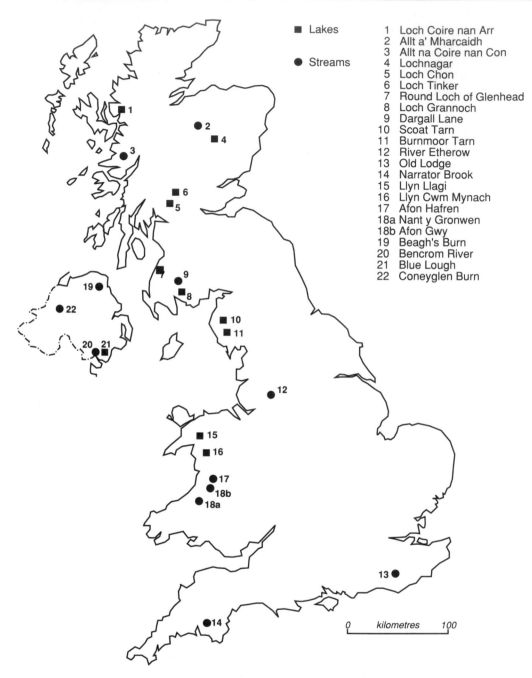

■ Lakes

● Streams

1 Loch Coire nan Arr
2 Allt a' Mharcaidh
3 Allt na Coire nan Con
4 Lochnagar
5 Loch Chon
6 Loch Tinker
7 Round Loch of Glenhead
8 Loch Grannoch
9 Dargall Lane
10 Scoat Tarn
11 Burnmoor Tarn
12 River Etherow
13 Old Lodge
14 Narrator Brook
15 Llyn Llagi
16 Llyn Cwm Mynach
17 Afon Hafren
18a Nant y Gronwen
18b Afon Gwy
19 Beagh's Burn
20 Bencrom River
21 Blue Lough
22 Coneyglen Burn

0   kilometres   100

Figure 10.15 Location of sites in the UK Acid Waters Monitoring Network (from Patrick et al., 1991).

monitoring network should respond to changes in UK emissions. Although a decline in emissions of approximately 40 percent occurred between 1970 and 1985 (see figure 10.12a) there has been no decline since the establishment of the monitoring network, and further declines are not expected until the second half of the 1990s. For this reason the monitoring network is not likely to show any clear indication of improvement until towards the end of this century. Moreover, any trend related to decreases in sulfur deposition may be offset by changing trends in other factors such as nitrogen deposition and climate change. Nitrogen deposition from acid rain is usually taken up by the biomass in the terrestrial ecosystem. However, there is increasing concern in some countries that normally nitrogen-deficient soils are becoming "saturated" with nitrogen, which is leaching through to surface waters, causing acidification. This is particularly of concern in Germany in areas of forest dieback, where damaged forests are less efficient at using available nitrogen (Hauhs, 1990). Nitrate concentration in surface waters in Norway has also increased (Henriksen et al., 1988), offsetting decreases in SO_4 concentrations.

A final concern is over the response of processes involving both sulfur and nitrogen compounds to climate change, especially the extent to which both nitrogen and sulfur currently stored in soil organic matter will be released to water courses if temperatures rise, as predicted. Increases in nitrogen loss are particularly likely and this could not only increase surface water acidification but also cause eutrophication of nitrogen-limited water bodies, especially coastal areas (Wright and Hauhs, 1991) (figure 10.16).

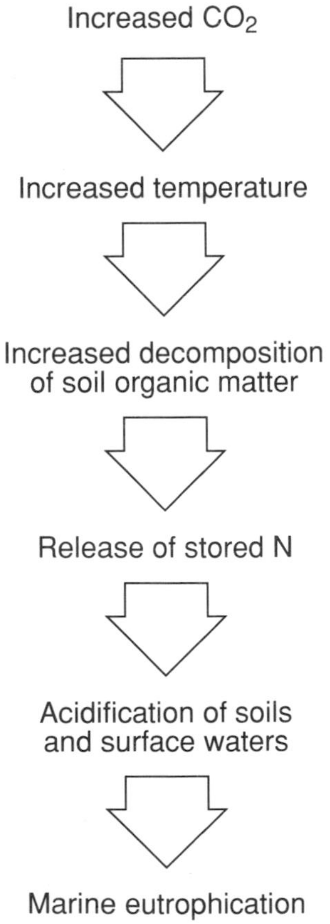

Increased CO$_2$

Increased temperature

Increased decomposition
of soil organic matter

Release of stored N

Acidification of soils
and surface waters

Marine eutrophication

Figure 10.16 Possible effects of climate change on nitrogen release from terrestrial ecosystems (from Wright and Hauhs, 1991).

Conclusion

Although acid waters occur naturally and changes in catchment land management can, in some situations, increase the acidity of surface waters, it has been clearly established that the overwhelming cause of surface-water acidification is acid deposition derived from fossil fuel combustion. Lake sediment studies have clearly shown that acidification in Europe and North America is a post-1800 phenomenon, that it only occurs in areas of high sulfur deposition, and that it is accompanied by increasing concentrations of air pollutants, including fly-ash particles from power stations. The ecological changes brought about by surface water acidification include major changes in the species composition of algal, higher plant, and invertebrate communities, and, in the worst cases, the entire loss of fish populations.

For Europe as a whole, sulfur emissions are no longer rising, and in some countries there has been a decline. So far these changes have been sufficient to prevent further acidification but, except for limited cases, they have been insufficient to cause much improvement. Initial monitoring and modelling data, however, suggest that a rapid improvement in conditions would follow if significant reductions in deposition occurred. In the longer term a recovery depends on the extent to which the critical loads calculated for specific affected sites and regions can be attained, but a complete recovery for many sites would be delayed until primary weathering restored the base saturation of soils. Recovery also depends on trends in atmospheric nitrogen deposition and on the influence of changing climate over the coming decades. Separating the respective individual and synergistic impacts of these various influences is an important task for environmental scientists that requires not only a developing understanding of the key processes but also a commitment to long-term environmental monitoring.

I am grateful to many colleagues, especially members of the ECRC, for many of the data presented here. Alan Jenkins provided the Round Loch of Glenhead MAGIC forecasts and Ron Harriman made chemical data from Galloway streams and lakes available. Our work on lake acidification has been funded by the Central Electricity Generating Board, The Department of Environment, The Royal Society of London, and the Natural Environment Research Council.

Further reading

Ashmore, M.H., Bell, N.J.B. and Garretty, C. (eds) (1988) *Acid Rain and Britain's Natural Ecosystems*. London: Centre for Environmental Technology, Imperial College.

Barrie, L.A. and Hales, J.M. (1984) The spatial distributions of precipitation acidity and major ion wet deposition in North America during 1980. *Tellus*, 36B, 333–55.

Battarbee, R.W. (1990) The causes of lake acidification with special reference to the role of acid deposition. *Philosophical Transactions of the Royal Society, London*, B 327, 339–47.

Battarbee, R. W. and Charles, D.F. (1987) The use of diatom assemblages in lake sediments as a means of assessing the timing, trends, and causes of lake acidification. *Progress in Physical Geography*, 11, 552–80.

Henriksen, A., Lien, L., Traaen, T.S., Sevaldrud, I.S. and Brakke, D.F. (1988) Lake acidification in Norway – present and predicted status. *Ambio*, 17, 259–66.

Mason, B.J. (1992) *Acid Rain: Its Causes and Effects on Inland Waters.* Oxford: Clarendon Press.

Wright, R.F. and Hauhs, M. (1991) Reversibility of acidification: soils and surface waters. *Proceedings of the Royal Society of Edinburgh*, 97B, 169–91.

CHAPTER ELEVEN

Reconstructing the History of Soil Erosion

John Dearing

An eroding planet

The way of the Earth's surface is to erode. This state of affairs is predicted by the second law of thermodynamics, used to drive W. M. Davis's model of landform evolution, and is vividly demonstrated in descriptive accounts of erosion from as early as the writings of Plato, through to Steinbeck's *Grapes of Wrath*. During recent years the impact of its effects on human society in deforested and dried-out Ethiopia has been witnessed by millions through the television medium, and has impressed upon us that stable relationships between our use of land and the soil can never be taken for granted.

Such concerns immediately lead us to take stock of what we know about soil erosion: its measured rate, its causes, and its consequences. And apart from visual and descriptive records there is a good deal of information. For instance, the United States has lost one-third of its topsoil in the past 200 years (Pimental et al., 1976) and crop productivity losses of 3–5 percent are expected over the next century in parts of the Cornbelt (Napier, 1990). In England, one field on the South Downs recently lost 10 percent of all its soil in a single erosion event (Boardman, 1987). And the impact of erosion is not restricted to the land which loses soil. Off-farm effects in the form of silted-up reservoirs, polluted streams and flooding have been reckoned to cost more than $6 billion in the United States in the early 1980s (quoted by Napier, 1990). In fact, in an environmental league table of quantity of information amassed, soil erosion and its effects rank fairly high. The Universal Soil Loss Equation, a statistical model developed in the USA and used widely to predict soil erosion on farmed slopes, utilizes the equivalent of more than 10 000 years of data taken from instrumented field plots. But the ability of this model and others to predict the magnitude of erosion under a given set of circumstances is far from perfect, and certainly inadequate to predict with confidence the impact of climatic change on soil erosion, whether it be direct, or indirect as its effect filters down to the users of land.

This chapter is concerned with one methodology by which our understanding of erosion processes may be improved, a methodology that embraces the problem of how to cope with erosion processes, which can range in their rates of activity from the unmeasurably small through to the catastrophic. In order to highlight the need for this methodology, let us take the case of soil erosion on the South Downs during the 1980s (plate 11.1). John Boardman and his colleagues have systematically monitored the rilling and gullying over a large area of the Downs, and have been able to make associations between soil losses, daily rainfall totals, and farming operations (Boardman, 1990). They show that the most important land-use factor has been the switch, since the 1970s, to growing winter cereals, which leaves the soil surface bare and vulnerable to erosive rainfall during late fall and winter. Importantly, the studies have pinpointed the longer-term risk of the present farming system: many slopes are losing soil at such a fast rate that profitable crop yields are already threatened. In short, this set of studies has provided some of the most detailed and convincing evidence for the on-farm and off-farm impacts of agriculture in Britain today, and has contributed greatly to our understanding of what causes soil erosion to take place on the chalk downlands. What then are the questions or problems that are not addressed or covered by this approach?

Plate 11.1 Rill and gully erosion on the South Downs. These chalk hills were among the first areas to be settled by farming peoples in England, and they have suffered from anthropogenic erosion for much of the last 5000 years, a process which continues today.

The relationships that have been established between soil erosion, climate, and land use are restricted to the period of the investigation, about 10 years. Will those relationships still be good for future years in which the climate or the mode of land use is of a type not experienced during that 10 years? In other words, with what level of confidence can we apply our knowledge to future environmental settings and events? Is 10 years a sufficiently long period to witness all the important processes and thresholds? The role of rilling is now understood, perhaps, but what about chalk solution or the breakdown of soil aggregates? Questions such as these can be levelled at any study which uses data from a relatively short time period, and make us aware of how few continuously monitored records of enviromental processes that cover more than a decade exist.

In the UK, climatologists lead the way in having recourse to long-term data, with annual records of rainfall stretching back to 1725, and monthly mean temperature records back to 1659 (see Hulme, Chapter 4). These records are hardly matched by hydrological data for sediment carried by streams and rivers, where few data sets extend back more than a handful of years. Soil scientists fare slightly better, with records of wheat yield from the Rothamsted Experimental Station in Harpenden going back to 1843, but unfortunately they are accompanied by little information about rates of soil loss. Elsewhere, there are many instances of soil erosion events being recorded in documents, particularly in requests for rent or tax relief and in official enquiries on rural conditions from as early as 1574 (Vogt, 1990), but as with most documentary records the information is usually dominated by extreme events and does not provide a continuous record at any one location. The documentary data are therefore historically interesting rather than scientifically useful.

The task then is to find ways of obtaining data about erosion over timescales longer than a decade, data which are accurate enough to complement and extend the many shorter records derived from contemporary measurements. This involves the scientific *reconstruction* of erosion records rather than their direct measurement.

Sediments and sinks

On bare cultivated fields it is commonplace to see the extent of eroding hillcrests effectively mapped by the pale subsoil which has been brought close to the surface by ploughing. At the bottom edge of the field, a drop in height between the upslope and downslope sides of a field boundary may testify to the long-term build-up of eroded soil, and the presence of rills on a slope provides the process link between the two. With features such as these most landscapes show that they possess interlinking elements which may be eroding, transporting, or sedimenting. The key idea in reconstructing ero-

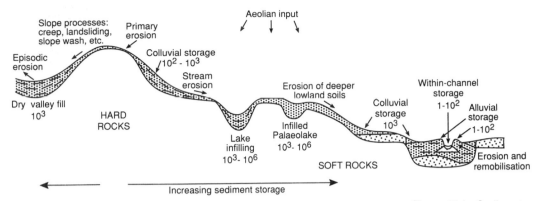

Figure 11.1 Sediment sinks and stores in the natural landscape, with typical timescales in years over which the sediment is immobilized (after Brown, 1987).

sion is to exploit these links and to study those landscape elements where eroded soil, or sediment, is temporarily stored: the natural environmental sinks. Figure 11.1 shows that there are basically three kinds of terrestrial sinks — alluvial, colluvial, and limnic — representing accumulations of eroded sediment on timescales varying anywhere between decades and thousands of years.

The following sections briefly look at these sinks and, through examples, show the approaches to obtaining erosion records and the quality of them. It will become clear that the links between source and sinks have so far concentrated on processes involving flowing water because the opportunities to reconstruct histories of wind erosion or mass movements are limited.

Alluvium and colluvium

In the British Isles there is evidence for widespread water erosion, with soil banked up against field boundaries, and colluvium and alluvium deposited on valley floors (R. Evans, 1990). Measurement of the depth, volume, and age of such deposits has utilized many techniques, from simple mapping of the spatial extent of these deposits, through to radio-isotope, geochemical, and archeological analyses of cores and vertical sections. Evans's paper describes a survey of seventy-four 1 : 25 000 soil maps of England and Wales and their soil descriptions which allowed him to estimate the amounts of stored colluvium and alluvium and the equivalent soil surface lowering in the surrounding drainage basins. Over large parts of England and Wales 40–250 mm of soil has been lost, probably since Bronze Age times, equivalent to soil erosion rates of 12–75 tonnes per square kilometer per year ($t \ km^{-2} \ year^{-1}$) over the past 5000 years. These figures are for whole drainage basins ranging in area from less than one to several thousand hectares and lumped over different timescales. They consequently hide local variations which are apparent when large river catchments are analyzed in detail. In the drainage basins of the Severn and Wye rivers, Brown (1987) has extended the mapping approach to

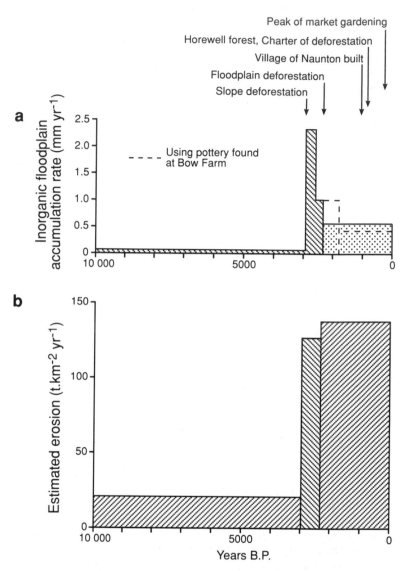

Figur 11.2   (a) Rates of floodplain accumulation and (b) estimated erosion rates for the Ripple Brook catchment, a sub-catchment of the River Severn, England. Also shown are historical and vegetation records from the area. Note the links between early deforestation and the rise followed by a decline in floodplain accumulation rates (after Brown and Barber, 1985).

include field studies, pollen analysis and radiocarbon (^{14}C) dating and gives estimates of historical erosion rates of between 2 and 140 t km^{-2} year^{-1} depending upon the size of the sub-catchment, the soil and landscape, its land use history and timescale. Figure 11.2 shows that for one of the lowland areas there was a dramatic increase in sedimentation during the late Bronze Age and early Iron Age, some 200–300 years after deforestation, suggesting that cultivation caused progressive, rather than immediate, soil deterioration.

Bell's work (1983, 1986) on valley sediments shows widespread evidence for post-glacial erosion on the chalk landscapes of southern England. There, archeological finds such as pottery fragments have been used to date colluvial sequences which show erosion starting as early as the Neolithic, with the first evidence of major colluviation in early Bronze Age times. Crude calculations of soil depth losses from drainage basins on the South Downs range from 2.5 to 24 cm over varying timescales, but these values may in reality be much higher because of losses of sediment out of the chalk valleys. As much as 30–40 cm of soil may have been lost since the first phases of deforestation, a loss which could have been felt by farmers in terms of reduced yields as early as Romano-British times. If enough vertical sequences through colluvium are studied it may be possible to estimate the total amount of soil which has been lost from the valley sides. For instance, Smyth and Jennings (1990) used borings and pollen analysis to estimate the total volume of colluvium stored on the floodplain of the Coombe Haven in East Sussex since the Iron Age to be 1.16 million m^3.

In semi-arid Mediterranean environments, archeological investigations have long pointed to phases of soil instability in the past, a feature which some believe to be the result of climatic change acting upon a sensitive ecosystem (plate 11.2). Recently van Andel et al.

Plate 11.2 Many Mediterranean landscapes, like this one in Morocco, show evidence of past and present erosion, but debate continues about the relative roles of human and natural factors in determining this.

(1990) have considered the question of whether anthropogenic impacts or climatic change were the prime factor. They summarize evidence for soil erosion episodes from studies of alluvial sequences in Greece, and conclude that there is little evidence for climatic links with erosion. The landscape was stable prior to and during the first spread of settlement and farming during the Final Neolithic and Early Bronze Age (4500–3500 BC), and major erosion did not occur for a further 1000 years.

The practice of terracing, introduced ca.2000–1000 BC, led to low rates of erosion and sedimentation, although erosion rates appear to have accelerated during the centuries either side of the BC/AD boundary, as they did in southern Italy. Over the past 1500 years stream aggradation has been episodic and localized, suggesting that human interference rather than climatic change was the principal cause (plate 11.3). Overall they regard climate and sea-level change as being of minor importance in the rate of alluviation in Greece, with the erosion record one of long periods of stability interspersed with relatively short-lived erosion episodes lasting up to a few 1000 years. They estimate that erosion due to the anthropogenic impact has led to no more than an average of 40 cm total surface erosion, which is similar to maximum figures calculated for the English chalk downlands.

Plate 11.3 A dissected alluvial fan and terrace in Caucasia, reflecting alternating phases of stream aggradation and erosion.

In some areas sedimentation is a more recent phenomenon, and one technique in particular has allowed scientists to monitor and map the net movement of topsoil in fields over just the past 30 years. Caesium-137 (^{137}Cs) is an artificial radio-isotope produced primarily through bomb tests since the 1950s which rains out of the atmosphere onto the soil surface. It is tightly adsorbed to fine particles and generally stays in the upper part of the soil profile. Any loss or addition of soil by erosion processes will alter the total amount of the radio-isotope from the expected value, such that total measurements can be used to assess the degree of erosion or deposition. In colluvial and alluvial deposits, ^{137}Cs may be used as a marker for topsoil material laid down since 1950, in much the same way as pottery fragments in older deposits. For instance, cores of floodplain sediment, 1.5 m in length, sampled in the River Start valley in South Devon are believed by Owens (1990) to be only 100 years old, and ^{137}Cs analyses show that the upper 40 cm has been deposited in the past 30 years. Owens's intensive sampling of a small floodplain coupled with measurements of wet and dry sediment density allowed him to express the total sediment volume as a range of mean erosion rates from the surrounding valley sides of 80–150 t km^{-2} year^{-1}. These rates are most likely attributable to higher levels of erosion during the post-war period of agricultural intensification.

Overall, the translation of alluvial and colluvial sediment storage into erosion records is not without its problems. Sediment dates are often uncertain, giving rise to calculations of erosion rates which are averaged over long periods of time, often with large error terms. In the case of alluvial deposits, complex responses of streams to changing conditions lead to spatially variable effects in the same drainage basin, which underlines the need for high-density sampling. And for both types of sink there is an unmeasurable loss of sediment which means that calculations are usually minimum values. Brown (1987), for instance, concluded for the River Severn that as much sediment went into floodplain stores as was lost via suspended sediment transport in the river. Therefore the ideal sink should trap all the sediment and provide a continuous record of sedimentation which can be studied at any timescale. The environmental sinks that come closest to fulfilling these requirements of an environmental tape-recorder are lake and reservoir basins.

Lakes and sediments

The limnologist John Mackereth (1966) suggested that "lake sediments comprise a suite of eroded soils from the catchment." A quarter of a century later we can see that, despite its simplistic overtones, this statement has proved to be true in many studies (plate 11.4). These studies use many different techniques, depending upon the type of lake and its sediments, the age of the sediments and the purpose of the study. In general there are common steps of

Plate 11.4 Lakes act as natural receptacles for materials eroded from their catchments, and by coring and analyzing lake-bottom sediments (as here at Groby Pool in the English Midlands) it is possible to reconstruct the erosional history of the landscape.

sampling, analysis, dating, and expressing the data. Sediments may be sampled at a single location or from several points, and some studies use echo-sounding equipment to measure the thickness of sediment. In reservoirs it is commonplace to estimate the sediment volumes from the difference between the original and present bathymetries. Fresh lake sediments may comprise "eroded soil" but they also contain the remains of aquatic plants and animals, and chemical precipitations from the lake water. It is therefore necessary to analyze the sediment in order to isolate those components which have come from the drainage basin. Working out how much catchment-derived material has come into the lake over a certain period of time requires the sediment to be dated. Annual layers or laminations, like tree-rings, are some of nature's built-in clocks, but unfortunately their presence is rare. Instead, natural radio-isotopes existing in the sediment, such as radiocarbon and lead-210 (^{210}Pb), whose decay rates are known, are routinely used to construct a curve of sediment depth versus sediment age. This curve is then used to calculate accumulation rates between known dates in a sediment core. Where large numbers of cores have been sampled these are usually correlated with the dated core on the basis of some simple property, like density. Finally the amount of catchment-derived material in the lake is usually expressed as a dry mass unit arriving at the lake per unit time, sometimes as a sediment yield from the drainage basin area in t km^{-2} $year^{-1}$.

The first successful attempt to reconstruct erosion rates from lake sediments was by Davis (1976) working at Frains Lake in Michigan (figure 11.3). The reconstructed records of sediment yield and vegetation (derived from pollen analysis) from ca.1800 onwards show clearly the erosional impact of woodland clearance by the earliest settlers when erosion rates increased 30–80 times over pre-settlement rates. Figure 11.3b shows how this type of record, detailed and well-dated, can be used to quantify the steady-state or stationary values and reaction and recovery rates to change, and, in this case, to define vegetation removal as the important factor in taking the system over its stability threshold. If we were concerned about the erosional impact of proposed woodland clearance in this type of landscape today, this study tells us what to expect over subsequent decades: an early peak in erosion followed by a new steady state after 30–40 years, with erosion rates some 10 times higher than under the present undisturbed woodland.

Since the mid-1970s a growing number of studies of reservoir and lake sediments in other landscapes have produced similar records of erosion rates or sediment yields over the past two centuries. This is a timescale which neatly combines the merits of

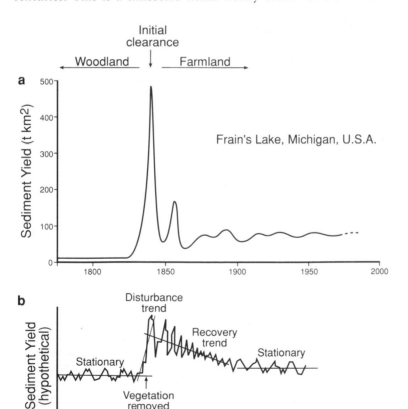

Figure 11.3 (a) The historical record of sediment yield at Frains Lake, Michigan since AD 1800, showing the rapid rise in erosion at the time of woodland clearance and the subsequent erosion level caused by agriculture (Davis, 1976). (b) The Frains Lake sediment yield data annotated to show how the record can be defined in terms of different states and responses in the sediment system.

the ^{210}Pb dating technique and the period in which human population growth and development have stressed natural environments more than at any other time. We are now beginning to obtain a clear view of the long-term erosional impact of a variety of environmental disturbances, particularly those which have occurred in North America and northwest Europe. The erosional impact of recent quarrying and mining, and lowland agriculture, in Britain are exemplified in figures 11.4 and 11.5. For the agriculture example, the idea of an experimental control has been used whereby the record from a disturbed site is compared to one from an adjacent or local undisturbed site. Clearly, these kinds of environmental disturbance have produced an erosional response, with sediment yields (where calculated) rising from less than 10 to about 35 t km^{-2} year^{-1} in the modern cultivated catchment and from less than 10 to about 15–40 t km^{-2} year^{-1} in the mined upland catchments. Foster et al. (1990) review these and other

Figure 11.4 Sediment yields through phases of mining in Wales, at Llyn Peris (after Dearing et al., 1981) and Llyn Geirionydd (after Dearing, 1992).

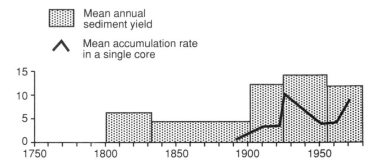

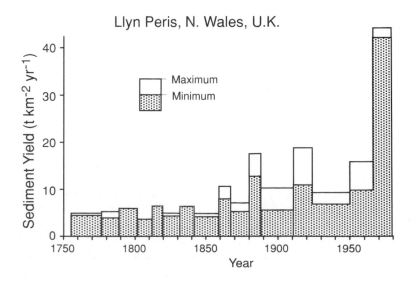

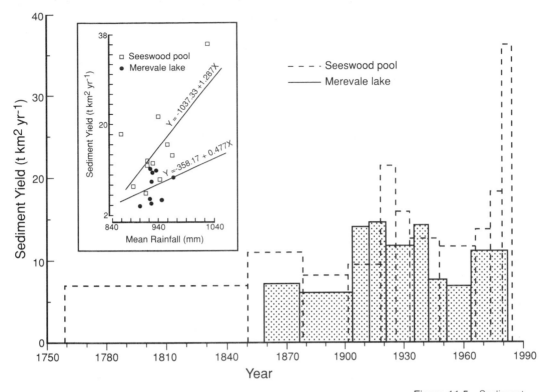

records and show that the erosional response to any disturbance is normally smaller than that seen in Frains Lake, being often less than a 10 times increase. They also demonstrate that the magnitude of sediment yield is strongly related to the catchment area, with sediment yields per unit area increasing as catchment size declines. This suggests that there are no straightforward relationships between the type of disturbance and the magnitude of sediment yields because, as discussed above, the storage of eroded soil in colluvium and alluvium is important in virtually all catchments. It is now clear from Frains Lake that the record produced there may only be a good analogue for the effects of clearance at other sites if they possess similar physical and hydrological attributes.

In some parts of the world, a 200-year record is too short to observe the true natural background value of sediment yield or the effects of early anthropogenic impact or climatic change. Although 200 years is old for a reservoir, it is young for a lake and many lake basins have accumulated several meters of sediment over 10 000 years in previously glaciated regions and for far longer elsewhere. In Poland, recent studies of sediments from Lake Goscienz show an amazing record of over 10 000 annual laminations but such an accurate and precise chronology is very rare, and for this reason it is more practicable to reconstruct the trend, rather than the magnitude, of sediment yield through time (from accumulation rates in single long cores). Figure 11.6 shows Holocene records of erosion

Figure 11.5 Sediment yields at two lowland reservoir-catchments in Midland England showing the difference between the agricultural catchment of Seeswood Pool and the wooded catchment of Merevale Lake. The inset shows the relationships between the sediment yields and annual mean rainfall (after Dearing and Foster, 1987).

Figure 11.6 Generalized records of Holocene erosion from northwest Europe, North America and tropical forest areas. (a) European records based on data of sediment yield and erosion indicators in Dearing (1991), Dearing et al. (1987), Digerfeldt (1972), Pennington (1981), Pennington et al. (1972), Renberg and Segerström (1981), Snowball and Thompson (1990). (b) North American records of sediment yield from Frains Lake, Michigan (Davis, 1976), sediment accumulation in Mirror Lake, New Hampshire (Likens and Davis, 1975), and loss-on-ignition data from Lake Hope Simpson, Labrador (Engstrom and Hansen, 1985). (c) Record of SiO_2 accumulation in Lake Quexil, Guatemala (Deevey et al., 1979) and sediment yields in the Lake Egari catchment, Papua New Guinea (Oldfield et al., 1985).

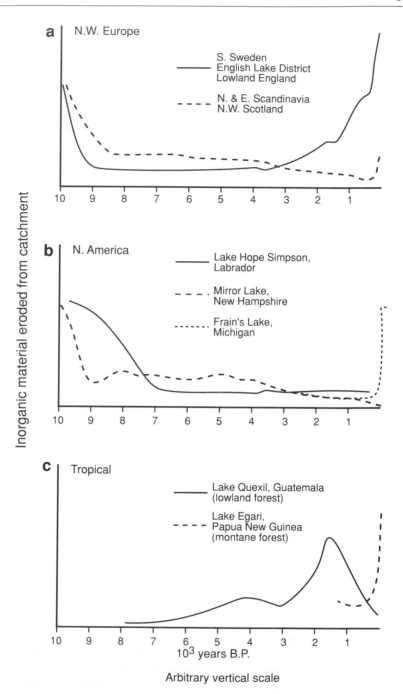

from different global regions. The records in figure 11.6a are generalized from sediment yield and sedimentation rate data from northwest Europe, comparing the trends in erosion between landscapes where human impact has been large, and where it has been largely absent. In the "undisturbed" record the highest erosion rates occurred as a result of intense erosional activity during the periglacial phase at the end of the late-glacial period. As the climate improved the landscapes became vegetated and erosion rates declined. Throughout the past 9000–8000 years the erosion rates have remained roughly constant or have declined, except in the past couple of centuries when afforestation or local construction has led to more erosion. In contrast, where a landscape has a long history of agriculture, erosion rates have generally increased since the time of initial deforestation, some 5000–2000 years ago. The characteristic pattern of erosion since then is a stepwise increase which apparently reflects the many phases and revolutions of agricultural activity and technology. Comparisons of sediment yield between pre-agricultural times and today show typical increases of five to ten-fold. In much of North America the history of agriculture is short, extending back only two or three hundred years. In figure 11.6b the records from three lakes suggest that before the pioneer farmers the long-term erosion rate trend was similar to that of northwest Europe, with high early rates lasting longer in cooler Labrador than in more southerly New Hampshire. The spectacular rise in erosion following deforestation is taken from the Frains Lake study in Michigan, shown in figure 11.3. Data from other regions of the world are sparse, but figure 11.6c shows two records from tropical forest which both show the strong impact of the forest disturbance. In Guatemala, the dramatic rise and decline of the Mayan civilization is reflected in the erosion record. Deforestation, agriculture, and construction raised erosion rates by 30 times over the pre-disturbance level. In the mountains of Papua-New Guinea the sediment records suggest that the effect of disturbance was slight before explorers and colonists introduced new crops, like the sweet potato, and new tools, like metal axes, to the native people. The subsequent agricultural intensification led to a five times increase in erosion.

So far erosion has been considered simply as an amount of soil or sediment, but in terms of understanding its causes this is only half the story. It is important to determine where the sediment has come from; in many catchment studies it is necessary to confirm that the eroded sediment is soil from slopes and not sediment from other sources. In lake studies there have been a number of attempts to identify the changes in sediment source through time. Most convenient in temperate zones is the division of sources into surface and subsurface soil horizons. Except where gullying is operating, this division neatly brackets together geomorphological processes, such as fluvial scouring and stream channel incision, and, through land management concerns, surface soil erosion and its control. Distinguishing between topsoil and subsoil is not easy, but one property shows promise. The mineralogy of soils can be described by its magnetic properties. Many igneous and metamorphic rocks from which soils are derived contain magnetic minerals, such as magnetite. But in soils there is also a second source of these minerals, which grow in the organic-rich topsoils of freely draining

Figure 11.7   10 000 year record of sediment yield and source at Lake Busjösjö, southern Sweden. The sediment source record is based on magnetic properties of soils and sediments, where high values indicate unweathered subsoil and low values indicate surface soil (after Dearing et al., 1990).

soils. Magnetometers are used to measure the concentration of these and other minerals in order to define material of either topsoil or subsoil origin. Figure 11.7 shows a 10 000-year record of sediment yield from a Swedish site, bearing all the hallmarks of a lowland agricultural landscape with high levels of erosion in early Holocene times and progressively higher levels from 3000 years ago. In addition, figure 11.7 shows data for the source of the sediment, where high values indicate a dominance of subsoil rather than topsoil reaching the lake. Comparison of the two records indicates that prior to 3000 years ago the sediment was topsoil-dominated. Over the past 3000 years the source has changed

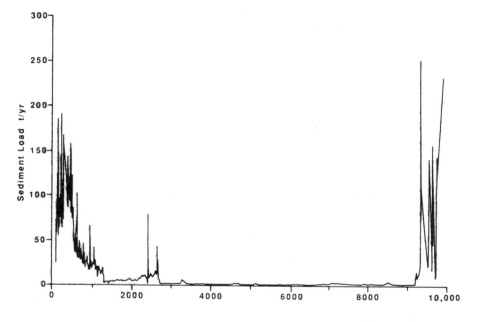

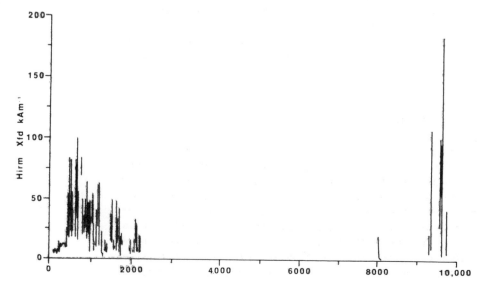

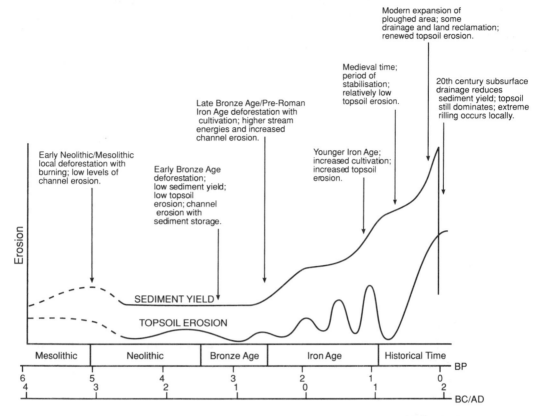

Figure 11.8 A generalized model of erosional intensity and sources since the Mesolithic in southern Scania, Sweden (after Dearing, 1991).

dramatically, especially in the past 500 years when there has been a switch from subsoil to topsoil as the major source. In this region of southern Sweden a number of studies like this have allowed some generalization of the erosional impact and processes over the past few thousand years (figure 11.8). The records suggest that in this area losses of topsoil have only become important during modern times, and indeed the present farming regime is highly intensive and, it has been argued, inappropriate for the environment and its climate.

Putting the records to work

With values and trends of erosion, the reconstructed records have, in themselves, provided much insight into how the geomorphology of our landscapes has developed. They are now being used to confirm previously held views about the impact of early deforestation and farming, the nature of erosion during the last periglacial phase in the temperate zone, and the magnitude of erosion caused by modern farming methods compared with traditional systems. But these are not the only uses of the records. More exciting and challenging objectives lie with trying to model the workings of environmental processes over long timescales. This chapter ends, therefore, by describing how the records might be "put to work" in alternative ways in order that we may be able to answer the "what if?" questions of the future.

One objective of geomorphologists has been to view the development of landscapes and the operation of processes at different timescales; the phrase "the present as the key to the past" is often used to justify monitoring contemporary processes. Lake sediments perhaps offer one of few ways of linking the present with the past, and some studies have actually married together contemporary observations with reconstructed data at the same site. Figure 11.9, for instance, shows how contemporary and reconstructed data have been combined to produce sediment budgets for the present and past agricultural landscapes around Seeswood Pool in Central England (Foster et al., 1988). The present landscape, half of which is now under intensive cultivation, shows a much higher loss of sediment than occurred in the older landscape, but the reasons for this are not as simple as one might think. Close observation of the two flow diagrams reveals that the catchment which is losing most sediment today is the one which has remained under pasture! In fact the most destructive process is the damage to stream banks and surface soil ("poaching") caused by large numbers of cattle. Comparison of the two budgets shows us that in this kind of environment the amount of sediment leaving the catchment is directly related to the stability of the channels and the hedge zone which borders them; rilling may be taking place on some of the

Figure 11.9 Sediment budgets for the Seeswood Pool sub-catchments (southern stream and northern stream), in Central England based on contemporary monitoring of erosion and sediment sources and lake-based sediment yields and sources. Shown here are the budgets for 1765–1853, during which both the sub-catchments were predominantly under improved pasture with cattle and sheep grazing, and for 1986–7 when the northern sub-catchment remained under pasture but the southern sub-catchment was completely cultivated (after Foster et al., 1988).

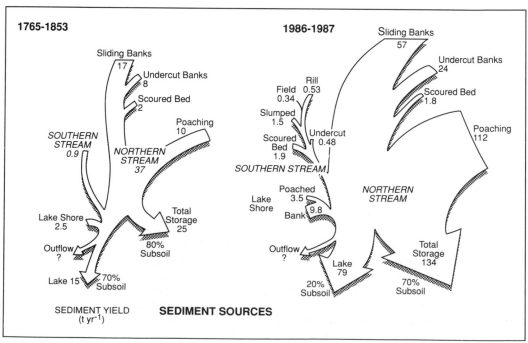

cultivated slopes but very little topsoil moves from the fields into the streams unless the channel edge is disturbed. Only the availability of the older sediment budget allows us to be so certain that the channel edge is this landscape's Achilles' heel. At the same site the lake sediment data have been used to explain the role of climate in affecting the annual sediment yield. Figure 11.5 (inset) shows that there is a positive and statistically significant relationship between the mean annual rainfall and annual sediment yield at Seeswood Pool over the past 200 years, though the relationship is not significant at the local wooded Merevale site. We can therefore set up a number of hypotheses which in summary would say that although the management of the Seeswood Pool catchment is important in controlling how sediment becomes available to erosive forces, over a timescale of decades the actual quantity of sediment removed is at least partly a function of annual rainfall or, more realistically, the size and number of flood events in the channels.

This example illustrates the major problem of using reconstructed records in statistical models: that success relies completely on the availability of accurate records of other processes, like rainfall. Unfortunately, as we have already seen, these are not always available. With the pressing need to predict the environmental response to climatic change, it will require a different approach in the case of reconstructed erosion data. One approach, for instance, will be to create unit hydrographs from contemporary measurements for different conditions of land use and climate which are known to have existed in recent centuries. The hydrographs could be calibrated against sediment movements, and allowed to "run" in order to predict the amount of sediment removed from channels. The predicted sequence of sediment yield could then be compared with the reconstructed record from the same area to see how sensitive the landscape was to changes in climate or land use.

Models like these are yet to come, but already we can say quite a lot about how climate has affected the long records of sedimentation in alluvial, colluvial, and limnic sinks. The first important point is that too few data exist to make confident generalizations, especially in those parts of the world where it is logical to expect a greater degree of sensitivity to climatic change, like the Mediterranean. But even there the evidence from alluvial and colluvial sequences for climatically related erosion is not as clear as once thought (van Andel et al., 1990). This is supported by the limited information from lake studies in that region, from Morocco (Flower et al., 1984) and Italy (Hutchinson et al., 1970), which points to recent drainage schemes and early road construction as the triggers for erosion. In more northerly climes, better dating of alluvial fills shows that within different periods the fills may not be synchronous and that local land-use factors played a more significant role than climate (Bell, 1982). In contrast, what is clear from contemporary studies almost everywhere is that extreme climatic events can and do produce strong erosional responses, both on slopes and in river channels, and we have to assume that the same was true for the past. We could argue from this that climatic change will only dramatically affect erosion directly if it becomes more extreme than it is at present, breaking down thresholds of soil, slope, and channel stability more often. Alternatively, we could argue that should the climate deteriorate, in agricultural terms especially, then a change in land-use management will impose an indirect response to climatic change. For instance, there is some evidence in the reconstructed records of erosion from Sweden that during the so-called "Little Ice Age" the wetter and cooler conditions actually led to reduced erosion because farming became less viable and less intensive (Dearing et al., 1987). Ideally we need to study records of erosion through historical periods of climatic warming which represent analogues to predictions for the next century. The closest analogue (though not for land use!) is probably the Medieval Warm Epoch which reached its peak in northwest Europe between AD 1150 and 1300 (Lamb, 1984), but a review of sediment yield records during these times in upland Britain and Scandinavia is not particularly informative. The dating control for that period is generally poor, which means that the data represent long periods of time. It was a time of Scandinavian migration and overall a period of agricultural expansion and population growth throughout northwest Europe. These facts alone could be expected to lead to increased erosion, and to some extent this may be borne out in figures 11.7 and 11.8, where the data indicate an acceleration of sediment yields at this time. But the direct role of a warm climate is not clear. Its role is also unclear in higher cooler latitudes, where disturbance and land use change have tended to be minimal. In Renberg and Segerström's (1981) study of laminated sediments from central Sweden there is the crucial advantage of an annual chronology (figure 11.10). Although the organic records indicate that temperature levels fluctuated enough to affect the aquatic

Figure 11.10 Records of organic and inorganic accumulation from the sediments in Lake Kassjön, central Sweden over the past 6200 years. The inorganic accumulations mainly represent material from the catchment. The organic material probably originated in the lake water and may be related to temperature changes (after Renberg and Segerström, 1981).

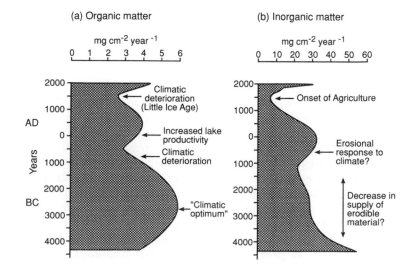

productivity of the lake, the record of sediment from the catchment shows an overall declining trend as seen elsewhere in relatively undisturbed catchments (figure 11.6a). The peak accumulation of inorganic sediment around 2000 years ago has been suggested to have registered the effects of a colder climate with increased snowfall and flooding, but there is no evidence that a Medieval Warm Epoch led to higher rates of erosion there. There is a great need for more highly resolved studies of sediment accumulation in all types of environment. But looking at the overall trend of sediment yields from lake studies which have been done so far, there is little correlation between them and reconstructed records of climate: in the majority of records the land-use factor appears to be paramount.

Further reading

van Andel, T.H., Zangger, E. and Demitrack, A. (1990) Land use and soil erosion in prehistoric and historical Greece. *Journal of Field Archaeology*, 17, 379–96.

Dearing, J.A. (1990) Lake sediment records of erosional processes. *Hydrobiologia*, 214, 99–106 (also published separately as Smith, J.P. et al. (eds), *Environmental History and Paleolimnology*. Kluwer Academic).

Evans, R. (1990) Soil erosion: its impact on the English and Welsh landscape since woodland clearance. In J. Boardman, I.D.L. Foster and J.A. Dearing (eds), *Soil Erosion on Agricultural Land*. New York: Wiley, 231–54.

Foster, I.D.L., Dearing, J.A., Grew, R. and Orend, K. (1990) The sedimentary data base: an appraisal of lake and reservoir based studies of sediment yield. *Erosion Transport and Deposition Processes (Proc. Jerusalem Workshop)*. IAHS Publication 189, 119–43.

Oldfield, F. and Clark, R.L. (1990) Lake sediment-based studies of soil erosion. In J. Boardman, I.D.L. Foster and J.A. Dearing (eds) *Soil Erosion on Agricultural Land*. New York: Wiley, 201–28.

CHAPTER TWELVE

Large-scale River Regulation

Geoff Petts

Introduction

Recent concern over human impacts on the environment has tended to focus on climatic change, desertification, destruction of tropical rain forests, and pollution. Yet large-scale water projects such as dams, reservoirs, and inter-basin transfers are among the most dramatic and extensive ways in which our environment has been, and continues to be, transformed by human action. Water running to the sea is perceived as a lost resource, floods are viewed as major hazards, and wetlands are seen as wastelands (Cosgrove and Petts, 1990). River regulation, involving the redistribution of water in time and space, is a key concept in socio-economic development. To achieve water and food security, to develop drylands, and to prevent desertification and drought are primary aims for many countries (Beaumont, 1989). A second key concept is ecological sustainability. Yet the ecology of rivers and their floodplains is dependent on the natural hydrological regime, and its related biochemical and geomorphological dynamics. River regulation has led to the widespread degradation of ecological resources. Conflict between the two concepts of river regulation and sustainable development has become a focus for international concern.

History of river regulation

The earliest civilizations were established along major rivers, among them the Nile, Tigris–Euphrates, and Indus. It is on these rivers that the history of water projects and hydraulic engineering began about 5000 years ago (Smith, 1971). Natural seasonal flow variations were used to supply water to agricultural land. Such floodplain agriculture was supplemented by elaborate gravity-fed irrigation systems, which made large-scale agriculture possible.

Unpredictable river flows, especially low flow years, constrained agricultural expansion and water-storage dams were constructed to regulate the natural flow variations. The earliest known dam was built at Sadd el Kafara in Egypt before 2759 BC. By 2000 BC irrigation agriculture was widespread and highly developed, notably in China, where by the Qin Dynasty (about 250 BC) hydraulic engineering, including river channelization projects for navigation and flood control, was also well developed. However, until about AD 1750 the intensity of river regulation remained low (Petts et al., 1989a), the effects being local and the natural systems being frequently "reset" by large floods that at least partly destroyed the regulation schemes.

The era of the mega-project in water development opened in the second half of the nineteenth century (Cosgrove and Petts, 1990), and an explosive growth in water projects worldwide has characterized the past 60 years. The construction of large dams, many over 200 m high, and the planned exploitation of the water resource potential of entire drainage basins, became symbols of social advancement and the power of technology to control nature. The 221 m high Hoover Dam on the Colorado River, USA, and its 37×10^9 m^3 reservoir Lake Mead, constructed in 1935, provided one model for water development: the creation of massive impoundments. A second model was provided by the Tennessee Valley Authority in the USA, formed in 1933. Multi-purpose dams were grouped within an entire river basin in order to exploit the full potential of the catchment's water resources (water supply and hydro-electricity) and navigation, while at the same time achieving control over flooding. Today the TVA has 32 large dams and 14 small ones to provide local flood relief and local water supply. The Aluminum Company of America has another six large dams in the basin.

River regulation in the 1990s

Rivers form corridors through the landscape and their physical and biological characteristics tend to vary in a more or less predictable way from headwaters to mouth. Upland streams are strongly influenced by hillslope processes and these low-order basins supply water, sediment, organic matter, and nutrients to the middle reaches. Progressive downstream changes of flow, water quality, and channel morphology give changes of biota a strong longitudinal dimension. This has been described by freshwater ecologists as the "river continuum concept" (Vannote et al., 1980), but lowland reaches of large rivers tend to be dominated by lateral exchanges between the river and the floodplain (Amoros and Petts, 1993). These important interactions occur in two ways. First, exchanges and associated faunal migrations occur during overbank flooding, as described by the flood-pulse concept (Junk et al., 1989). Second, exchanges take place between the groundwater and surface waters through the channel bed and through floodplain springs (see *Regulated Rivers*, 1992, p. 7.1). Dams and reservoirs interrupt both longitudinal and lateral transfers – trapping sediment, changing temperature and water-quality regimes, regulating floods, and inhibiting species dispersal.

Today, virtually every river in Europe and North America has been regulated, and even small streams have been incorporated into regional water resource plans. On the other continents progress toward river regulation is moving rapidly, such that few rivers remain that are not affected by regulation in some way. Croome et al. (1976) suggested that by the year 2000 about 66 percent of the world's total stream-flow will be controlled by dams.

Dams

A large dam project can be defined as one that meets one or more of the following criteria: the dam is higher than 150 m, the dam volume is more than $15\ 000 \times 10^3\ \text{m}^3$, the reservoir storage capacity is greater than $25\ 000 \times 10^6\ \text{m}^3$, or the hydro-plant has a total generation capacity of greater than 1000 MW. Together, the 410 large projects constructed, under construction, or planned would impound up to 4674 km³ of runoff. An analysis of the global distribution of these large projects (figure 12.1) shows that Europe

Figure 12.1 World distribution of large water projects (data from Mermel, 1990).

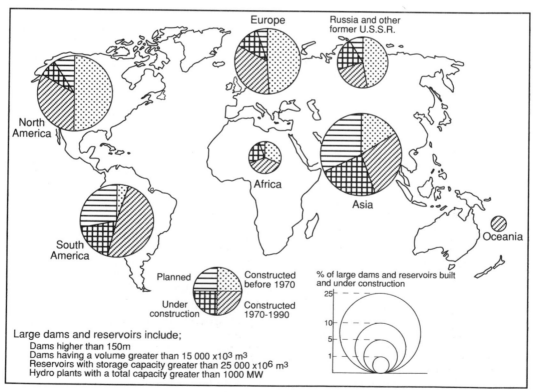

and North America have more than half of those projects already constructed.

Lists of the world's highest dams and largest capacity reservoirs are given in tables 12.1 and 12.2 respectively. Hoover Dam is now nineteenth in the world list of highest structures, which includes two 300 m high embankments on the River Vakhsh in Siberia. Lake Mead now ranks thirty-first in the list of largest capacity reservoirs. Excluding Lake Victoria (Owen Falls Project), the major volume of which is natural, the largest reservoirs, both created in the 1950s, have maximum capacities of over 180 000 000 m^3.

In addition to the large projects, there are many thousands of dams more than 15 m high (there were already more than 12 000 in 1971: Fels and Keller, 1973). Although there is a relatively small number of projects in Africa, the Nile, Zambezi, Niger, Volta, and Orange–Vaal basins are all significantly influenced by dams; of the continent's major rivers only the Zaire (Congo) remains largely uncontrolled. Currently, the most rapid growth in dam construction is taking place in Asia and South America (Petts, 1990), where the main driving force is hydro-electric power development. The Parana – an international river shared by Brazil and Argentina – is the major focus of dam building and the core of development programs in an area of rapid population growth (figure 12.2).

Table 12.1 The world's highest dams

Rank	River	Country	Dam	Year completed	Height (m)
1	Vakhsh	Russia	Rogun*	1990	300
2	Vakhsh	Russia	Nurek*	1980	300
3	Dixence	Switzerland	Grande Dixence	1961	285
4	Inguri	Russia	Inguri	1980	272
5	Vajont	Italy	Vajont	1961	262
6	Grijalva	Mexico	Chicoasen*	1980	261
7	Bhagirathi	India	Tehri*	1995	261
8	Tons	India	Kishau*	1995	253
9	Guavio	Colombia	Guavio*	1989	246
10	Yalongjiang	China	Erlan	u/c	245
11	Yenisei	Russia	Sayano-Shushensk	u/c	245
12	Columbia	Canada	Mica*	1973	242
13	Bata	Colombia	Chivor*	1975	237
14	Dranse de Bagnes	Switzerland	Mauvoisin	1957	237
15	Humuya	Honduras	El Cajon	1985	234
16	Sulak	Russia[a]	Chirkey	1978	233
17	Feather	USA	Oroville*	1968	230
18	Sutlej	India	Bhakra	1963	226
19	Colorado	USA	Hoover	1936	221
20=	Piva	Bosnia	Mrantinje	1976	220
20=	Verzasca	Switzerland	Contra	1965	220
22	Clearwater	USA	Dworshak	1973	219
23	Colorado	USA	Glen Canyon	1966	216
24	Naryn	Kirgizia	Toktogul	1978	215

Those marked* are earth embankments, the others are concrete dams.
[a]Caucasia. u/c, uncertain.
Source: after Mermel, 1990

Table 12.2 The world's largest capacity reservoirs

Rank	River	Country	Reservoir	Year completed	Reservoir volume ($m^3 \times 10^3$)
1	Nile	Uganda	Owen Falls[a]	1954	2 700 000
2	Dnieper	Ukraine	Kakhovskaya	1955	182 000
3	Zambezi	Zimbabwe/Zambia	Kariba	1959	180 600
4	Angara	Russia	Bratsk	1964	169 270
5	Nile	Egypt	Aswan High	1970	168 900
6	Volta	Ghana	Akosombo	1965	148 000
7	Manicouagan	Canada	Daniel Johnson	1968	141 852
8	Caroni	Venezuela	Guri	1986	138 000
9	Yenisei	Russia	Krasnovarsk	1967	73 300
10	Peace	Canada	Bennett W.A.C.	1967	70 309
11	Zeya	Russia	Zeya	1978	68 400
12	Zambezi	Mozambique	Cabora Bassa	1974	63 000
13	La Grande	Canada	La Grande 2	1978	61 715
14	Parana	Argentina	Chapeton	(1998)	60 600
15	La Grande	Canada	La Grande 3	1981	60 020
16	Angara	Russia	Ust-Llim	1977	59 300
17	Angara	Russia	Boguchany	1989	58 200
18	Volga	Russia	Kuibyshev (Volga-VAI Lenin)	1955	58 000
19	Tocantins	Brazil	Serra da Mesa (São Felix)	(1993)	54 000
20	Caniapiscau	Canada	Caniapiscau	1981	53 800

[a]Major part of lake volume (Lake Victoria) is natural. The dam added 270 km³ of storage to existing capacity of lake.
Source: after Mermel, 1990

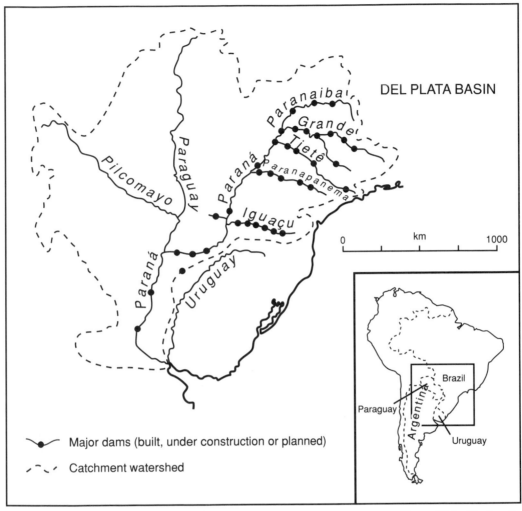

DEL PLATA BASIN

Major dams (built, under construction or planned)

Catchment watershed

Figure 12.2 The location of major dams in the Parana River basin, South America.

Worldwide, reservoirs have drowned about 1 000 000 km² of land. In the USA alone, 25 000 km of fast-flowing main rivers have been converted to stillwater lakes. The nation's major rivers, the Mississippi, Colorado, and Columbia, for example, are completely regulated by a combination of large dams and small weirs; on the Columbia–Snake system, reservoirs have replaced 822 km of river habitats. Flow regulation has had marked environmental impacts, both direct and indirect, on the river and floodplain systems downstream of the dams. For example, the area of floodplain woodland in USA has been reduced by 80 percent (Swift, 1984), reflecting the combined influence of flood regulation, channelization, and land drainage.

River fisheries have been severely degraded. Salmonid and floodplain fisheries have been particularly affected. For example, Atlantic salmon (*Salmo salar*) disappeared from the Dordogne

River, France, soon after the first dams were built (Décamps et al., 1979) and the valuable Acipenseridae (sturgeon family) have been seriously impacted by dams on the Volga, Don, and Caucasian rivers in eastern Europe (Zhadin and Gerd, 1963). In the Columbia River, between 1911 and 1972 the annual catch of Pacific salmon (*Oncorhynchus* spp.) diminished from 22 440 t to less than 700 t as a result of impoundment (Trefethen, 1972). On the Missouri River, the disappearance of some species and an 80 percent decline in fish catches was related to the marked reduction in floodplain area following regulation (Whitley and Campbell, 1974). Tropical fisheries are a particularly important component of indigenous cultures; they are dependent on floodplain and marginal habitats (Welcomme, 1979) and regulation can cause dramatic impacts (Petrere, 1989; Petts, 1990).

In some cases the influence of river regulation extends to the coastal zone. Regulation of the River Nile by the Aswan High Dam, located about 1000 km inland from the sea, has caused considerable changes of the hydrographic conditions over the continental shelf in the southeast Mediterranean (Din, 1977). The intrusion of saline water into the delta or lower river has been a consequence of regulation on several major rivers, including the Zambezi, Dnieper, and Orinoco.

Inter-basin transfers

The integrity of the drainage basin, with the watershed forming a barrier to water, sediment and nutrient transfers, species migration and dispersion, is fundamental to the concept of the catchment ecosystem (some would argue that at the scale under consideration here, the term "ecobiome" is more appropriate than "ecosystem"). Anomalies to this norm receive special attention. Some anomalies are topographical, others are paleogeographical. Today, for example, the Orinoco and Amazon river systems are linked by the Casiquiare river in the swamps of the Guayanian Shield and, consequently, many fish species are common to both systems (Lowe-McConnell, 1986). Similar situations existed elsewhere in the past. Thus, during the Quaternary, the Nile was linked to Lake Chad and the Niger and Zaire rivers. Increased precipitation in equatorial Africa, the Ethiopian Plateau and over most of the Sahara caused massive increases in the areas of African lakes (see chapter 3, by Street-Perrott and Roberts) and this allowed faunal interchanges between Central Africa, the Mediterranean area, and Asia (Dumont, 1986).

Artificial inter-basin connections have been created to increase water yields to reservoirs, especially for hydro-electric power production, or directly to demand centers, such as cities and areas of extensive irrigated agriculture. The diversion of runoff from a number of small headwater catchments into a reservoir to increase the reliable yield of the scheme has been a common practice. For example, the Grande Dixence Scheme, completed in 1965, involves

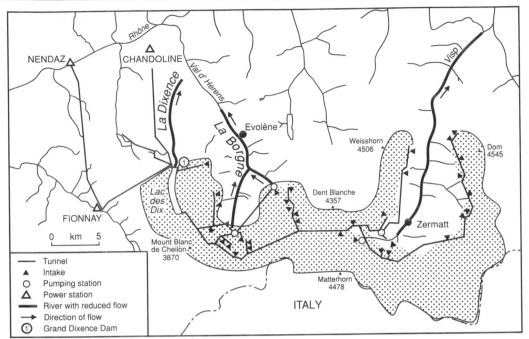

N/A

90 km of tunnel connecting intakes on 34 mountain and pro-glacial streams, extending the effective catchment of the Lac des Dix to 357 km^2 (figure 12.3, plate 12.1). The annual runoff to the reservoir has been increased to about 420×10^6 m^3, yielding 55 m^3 s^{-1} to generate 1680 GWh of electricity, but the meltwater discharge of three alpine rivers has been drastically reduced.

Figure 12.3 A major Alpine basin development: the Grande Dixence scheme, Valais, Switzerland. The stippled area shows the contributing high altitude catchments.

Plate 12.1 The 285 m high Dixence dam in Switzerland (the ten-story hotel at the base provides a scale). The resulting reservoir is fed by pumped glacial meltwater from an area of over 350 km^2.

Larger-scale schemes include the La Grande Rivière Hydroelectric Complex in Quebec (figure 12.4). The effective catchment area is $176\,810$ km^2, of which 43 percent is outside the natural La Grande catchment. The 100 kilometer lower reach of the La Grande Rivière now experiences increased flows: the average discharge of 3400 m^3 s^{-1} includes 845 m^3 s^{-1} transferred 150 km from the Eastmain river to the south and 790 m^3 s^{-1} transferred 230 km from the Caniapiscau river to the east. About 1000 km of river now experiences reduced flows and in the most extreme case at the mouth of the Eastmain the average flow reduction is 90 percent. The reservoirs flood an area of $13\,341$ km^2, of which about 85 percent is submerged land systems. The total planned generation capacity is 24 496 MW and the potential annual output is 128.9 TWh.

Artificial connections can cause accidental introductions of species across natural biogeographical divides. For example, in South Africa, five fish species passed through the 82 km long tunnel between the Orange River and the Great Fish River where they successfuly established (Cambray and Jubb, 1977). Furthermore, diversions reduce flows in the supply river and increase flows in the receiving stream; the resulting hydrological and biogeochemical changes may significantly alter the ecology of both systems. Thus,

Figure 12.4   A major Boreal basin development: La Grande Rivière hydro-electric complex, Quebec, Canada.

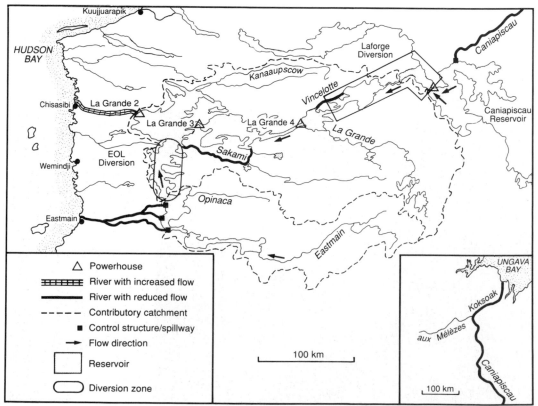

for example, in the Great Fish River, since the opening of the diversion tunnel there have been pest outbreaks of the mammal-biting blackfly *Simulium chutteri*, causing severe stock disruption and damage to sheep, cattle, and goats (O'Keefe and De Moor, 1988).

Case studies

The conflicts between water development and ecological sustainability may be illustrated by discussion of two examples. First, the damming of the Colorado (plate 12.2) demonstrates how even the world's mighty rivers can be dramatically transformed into a tamed system of placid lakes. Consumptive uses have caused the river virtually to dry up in its lower reaches. Second, although shelved in November 1985, the Siberian Rivers Diversion Project remains a good example of the potential scale of future water projects and highlights many important environmental issues. The need for additional water resources in rapidly developing areas, such as the semi-arid region of Central Asia, seems certain to increase the pressure to divert "wasted" Siberian rivers southwards.

The Colorado River

Water demand in the arid southwest USA has resulted in the construction of hundreds of dams to regulate virtually the entire

Plate 12.2 Despite its image as a wild river, the Colorado is now dammed over much of its course, and as a result of water consumption is almost dry in its lower reaches.

Colorado River (Turner and Karpiscak, 1980; Adams and Lamarra, 1983; Stanford and Ward, 1986a; Graf, 1987). Between 1922 and 1948 legislation apportioned the flow of the Colorado River between the seven basin states, with California, Colorado, and Arizona receiving the largest shares. In an international treaty of 1944, Mexico was assured 10 percent of the river's annual flow. The legislation of 1922 paved the way for construction of works to control, regulate, and utilize the river. The allocations were based on river flow data for the first two decades of the twentieth century. Subsequently flows declined to an average about 20 percent less than these values, and today there are more legal rights to water in the river than there is real water!

The water projects

In American culture, the Colorado River symbolizes the romantic West: the frontier and the wilderness beyond. Today, however, the wild river has been dramatically transformed into a series of desert lakes (figure 12.5). The Colorado River drains an area of

Figure 12.5 Water development within the Colorado River basin, USA (boxed area shown in figure 12.6a).

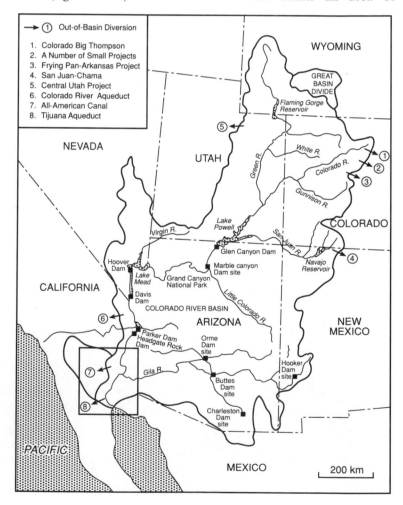

630 00 km^2, equal to 8 percent of the United States, and has a length of 2320 km. Although the river drains the southern Rocky Mountains and rises at an altitude of about 3100 m, the discharge is low, reflecting the semi-arid nature of the basin. Runoff is only 55 mm year^{-1} but flow is highly seasonal and flood flows, following the spring snow melt, exceed 5000 m^3 s^{-1}. The reservoirs of the Colorado basin vary in area from less than 1 hectare in headwater reaches to more than 650 km^2 on the main river. Nineteen high dams alone store an amount of water equal to almost five times the river's mean annual flow. Runoff during wet years can be stored for use in dry years. One consequence of this is that less than 1 percent of the river's virgin flow reaches its delta and along its lowest 100 km the once mighty river today flows only intermittently (figure 12.6).

a

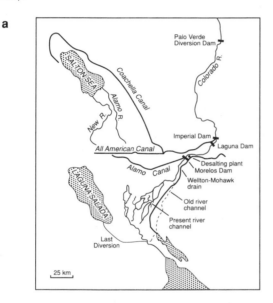

Figure 12.6 (a) Lower part of Colorado River; (b) declining twentieth century water flows of the Colorado; (c) similar decline in sediment yield (from Graf, 1987).

b

c

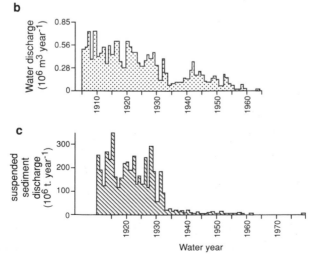

Plate 12.3 The Glen
Canyon dam on the
Colorado River.

Use of the waters of the Colorado river began in the late
nineteenth century with diversions to irrigate the desert areas of
southern California. The completion of Hoover Dam in 1935
provided the first major regulation of floods and harnessed the
spring snowmelt for supply. Subsequently, a series of dams were
constructed on the main river (figure 12.5 and plate 12.3). On the
main river, only the reach above Lake Powell to the Gunnison River
confluence retains a near pre-regulation flow regime. Elsewhere,
the few important free-flowing reaches are highly regulated and the
only significant free-flowing tributaries in the entire basin are the
Little Colorado and White rivers.

Impacts
The ecology of the pristine river evolved in harmony with the spring
floods and was adjusted to the long periods of low flows during late
summer and autumn. Turbid floodwaters renewed the vast desert
marshes, cued spawning of riverine fish, and encouraged the
growth of riparian plants (Stanford and Ward, 1986a). Within this
system evolved a fish fauna with one of the highest rates of
endemism of any river basin in North America (Miller, 1958;
Minckley and Deacon, 1968).

Twenty-five percent of the river has been converted to lentic
(standing water) habitats by impoundment, the backwaters and
marshes have been isolated from the river, and the thermal regime
and trophic structure of the remaining reaches have been changed
by releases from the reservoirs. The effect of Glen Canyon Dam on
the river through Glen Canyon is illustrated in table 12.3. Most of

Table 12.3 The Colorado at Lees Ferry below Glen Canyon Dam (Lake Powell)

	Pre-dam	Post-dam
Flows (cm)		
Minimum	20	130
95 percentile flow	102	156
Mean annual flood	2434	764
10-year flood	3481	849
Temperature (°C)		
Minimum	4	7
Maximum	30	10
Sediment load		
(t × 10^6 year^{-1})	140	20

Source: after Turner and Karpiscak, 1980; Carothers and Johnson, 1983

the sediment load is trapped within the 290 km long Lake Powell. The water releases from the reservoir are controlled by the power turbines and water is drawn from a depth of about 60 m in the reservoir, that is from the cold, deeper waters of the hypolimnion. The river is characterized by a regulated flow regime, clear water, increased salinities, reduced temperature fluctuations, stabilized channel, profuse algal growths (mainly *Cladophora* spp.), and lower diversity zoobenthos. Many endemic fish have been pushed near to extinction; of the 32 native freshwater fish, one is extinct and 15 occur in sparse populations (Stanford and Ward, 1986b).

Habitat changes have had major impacts on the biota. Of equal importance has been the introduction of exotic species to exploit the new (particularly cold water) habitats created by regulation. In the Glen Canyon area, for example, rainbow trout (*Salmo gairdneri*) was introduced in 1964 and has been sustained in an ongoing stocking program. Rainbow trout is now the most common species and has resulted in arguably the best trout fishery in North America (Carothers and Dolan, 1982). Stanford and Ward (1986b) identify 50 exotic fish that are established within the basin and have been especially successful in reservoirs and heavily regulated reaches. Competition, the spread of introduced alien diseases, and hybridization between native and exotic species, have contributed to the precipitous declines among the indigenous fish.

Through Grand Canyon, vegetation has established in response to the stabilized flow regime, and the once barren flood zone has been transformed into a narrow riparian woodland with an associated fauna of rodents, reptiles, amphibians, and small birds. Beaver (*Castor*) is common and appears to have increased in number, as have several species of lizard. The dense thickets comprise not only the native willows but also the exotic salt cedar (*Tamarix chinensis*). This was introduced into the southwest USA in the mid-eighteenth century (Robinson, 1965) and has spread along the riparian zone of the regulated river, replacing the native cottonwood–willow communities (*Populus fremontii–Salix goodingii*). Although highest densities of terrestrial vertebrates occur along the river margin (Carothers and Johnson, 1983), the salt cedar thickets support fewer birds than most native riparian vegetation and are considered to be of relatively low value for wildlife habitat and recreation (Swift, 1984).

The irrigated lands of the lower basin have expanded to more than 2000 km^2. Imperial Valley is the largest expanse of irrigated agriculture in the Western Hemisphere (Pillsbury, 1981) and uses most of the remaining river flow via diversions at Imperial Dam. However, one consequence of hydrological changes has been to raise solute concentrations in the Lower Colorado from about 380 mg l^{-1} before regulation to over 825 mg l^{-1}. This has been related to dissolution via irrigation and impoundment, diversions of low salinity headwaters, flow depletion via evapotranspiration on irrigated land, and concentration via reservoir evaporation (Pillsbury, 1981). The heavy salt load of 9–10 million tonnes per year cost users $113 million per year in 1982 and is expected to rise to $267 million annually by 2010 if the salinization cannot be controlled (El Ashry et al., 1985).

Twenty million people depend on the Colorado River for water and power (Graf, 1987). Although the landscape along the Colorado still retains some wilderness characteristics, the river has been transformed into a highly managed, but ecologically degraded, system. The lessons from the Colorado and many other rivers in North America and Europe are clear, but the increasing demands of society for more water – for domestic supply, irrigation, and hydro-electric power – not least in the American "Sunbelt" (Beaumont, 1989), means that water for environmental conservation is given low priority.

The Trans-Siberian Transfer

Russia and the rest of the former Soviet Union have enormous water reserves that in 1985 exceeded the annual consumption by more than 15 times (Voropaev and Velikanov, 1985). However, 75 percent of the population live in areas that contain only 16 percent of the available freshwater. Particular problems exist in arid Central Asian regions such as Kazakhstan, where irrigation is the main water user. Voropaev and Velikanov estimate the potential area of irrigable land is about 100 million ha, of which only 7 million ha is currently irrigated. Furthermore, there is rapid growth among the Muslim population, averaging 2.5–3 percent annually. It is predicted that by the end of the century the annual water deficit in this region will amount to 20 km^3.

The declining water resources are reflected most dramatically by the falling levels of the Aral Sea, a saline non-outlet lake. Since 1960, water level has fallen by 15 m as a result of water impoundment and irrigation use of runoff from the main feeding rivers, the Amu Darya and Syr Darya (Williams and Aladin, 1991; see also Gerasimov and Gindin, 1977; Hollis, 1978; Micklin, 1988). The

Plate 12.4   The Aral Sea in Central Asia, whose surface area has been reduced dramatically during the twentieth century because of water abstraction, mostly for irrigation, along the Syr-Darya and Amu-Darya. The rotting hulks of boats testify to the demise of a once-thriving fishing industry.

volume of the lake has declined from over 1000 km^2 to less than 350 km^2 and the shoreline has receded by 120 km in places (plate 12.4). The water is also becoming more saline, rising from 10 mg l^{-1} in 1960 to 30 mg l^{-1} in 1990. Shrinkage of the sea from 68 000 km^2 to less than 40 000 km^2 has already led to the degradation of the river deltas and salt-laden dust storms have further denuded the surrounding area, threatening both agriculture and human health. Of the once healthy fishery, only four species remain and these now face extinction (Williams and Aladin, 1991). The number of nesting bird species has been reduced by nearly 50 percent and mammals have declined from 70 to 30 species (Kotlyakov, 1991). Predictions for the year 2000 by Williams and Aladin (1991) suggest that the lake level could fall by another 10 m, its area will decline to less than 25 000 km^2 and its volume to 162 km^3, its average depth will be no more than 7 m, and salinity levels could reach 50 mg l^{-1}.

Large-scale, long-distance water transfers have been considered since the 1930s as a solution to the regional water deficiency in the

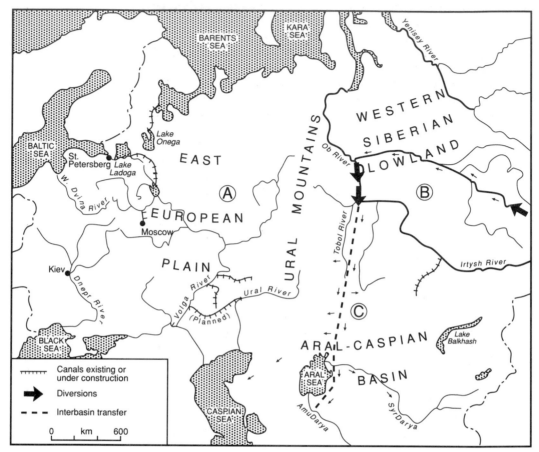

Figure 12.7 The Siberian River Diversion Project.

southern drylands, and have been reviewed by Micklin (1986). Economically available groundwater resources amount to about 10 km^3 per year, while desalination of drainage water from irrigated areas is feasible, but expensive. Voropaev and Velikanov (1985) considered that a long-distance water transfer was the only radical measure which could be regarded as feasible from both socio-economic *and* ecological viewpoints. One central component of the water redistribution plan formulated in the 1970s was the Siberian Rivers Diversion Project.

The great Siberian rivers discharge northwards into the Arctic Ocean and thus away from the main centers of water demand (figure 12.7). The proposals involve the north-to-south transfer of water from the River Ob to the Aral Sea basin, a distance of more than 2500 km. Micklin (1986) reports the proposals as three phases, the final phase providing a total transfer of 200 km^3 each year.

The transfers

The proposed first and second phases have received most attention. These would provide 25–27 km^3 and 60 km^3, respectively, of water to support industrial and domestic use and to irrigate an additional 4.5 million ha (3 million ha in Central Asia). Optimistic estimates suggest that the first phase alone could generate a massive increase in agricultual output: about 17 million tonnes of grain, about 30 million tonnes of other produce, and 45 million tonnes of fodder.

Water would be abstracted from the Ob and lower reaches of the Irtysh rivers, with the main intake below the confluence. A series of reservoirs, canals, and pumping stations would be required to convey the water over the Ob–Irtysh and Syr–Darya divides, providing a lift of 113 m, with gravity flow over the remainder of the route. The scheme has four components. For the first, between the Ob and the Tobol–Irtysh confluence, there are two alternatives. The first involves a 316 km canal along the west bank of the Irtysh. The alternative is to use four dams to reverse the flow of the Irtysh along this 651 km section. The second component is a 1500 km canal lifting the water to the catchment divide. Here the water will be stored in the Tegiz reservoir, the third component, having a capacity of 5.4 km^3 and a surface area of 485 km^2. The reservoir would regulate the flow to the Aral basin so as to meet irrigation requirements. The fourth component is the gravity canal from the Tegiz reservoir to the Amu Darya. The canal sections would have widths of between 110 and 220 m and depths varying between 12 and 16 m, providing a capacity of 1400 m^3 s^{-1} through the section below Tegiz reservoir.

Environmental impacts
The transfer canal would be equivalent to a sizable river. The route traverses a wide variety of geographical conditions: taiga, forest-steppe, steppe, semi-desert. Along the canal, which would be unlined, seepage would influence a zone mostly from 1 to 10 km wide, but occasionally up to 20 km wide, causing water-logging, accompanied by soil degradation. If the west-bank canal was constructed from the Ob to Tobolsk, an area of about 450 km^2 would be swamped; agricultural land along the 400 km long canal between Tobolsk and the Tegiz reservoir would be partially lost; southward from the forest-steppe zone, the formation of swamps would be accompanied by soil salinization and productivity would decline. Through the Turgay trough at the Ob basin–Syr Darya basin divide, ground and surface water resources feeding several important lakes would be intercepted, causing shrinkage and salinization of the lakes. The Turgay Lakes may be severely affected, reducing the waterfowl population, which in 1983 amounted to nearly one million (Voropaev and Velikanov, 1985).

In addition to the above, the breakdown of the biogeographic divide could cause the spread of opisthorchiasis (liver-fluke disease) and diphyllobothriasis (fish tapeworm disease) along the entire canal route (Voronov et al., 1983). The intermediate hosts of these diseases, freshwater snails (*Bithynia leachi*) and crustaceans (*Cyclops*) respectively, are common in northern waters and could invade the southern areas where they are virtually absent.

During phase 1 the annual diversion of 25 km^3 would be in addition to the rising consumptive water use from the Ob, which is estimated to reach 25 km^3 by the turn of the century. The net withdrawal would be about 20 percent of the river flow during dry years. With the implementations of the phase 2 proposals the reduction in flow during dry years would be about 30 percent, reaching up to 55 percent at Belogor'ye, immediately below the diversion point. The greatest reductions would occur during the autumn.

Water abstraction would lead to lower river levels, especially during the spring–summer flood, by between 1.5 and 2.5 m during phase 1. Consequently, the area of the Ob floodplains would be reduced by 2760 km^2, about 30 percent, from 8920 km^2 to 6160 km^2. The duration of inundation would be reduced by from 10 to 20 days and groundwater levels would be reduced by 0.3 to 0.6 m. During phase 2 the floodplain area would be reduced further to 5890 km^2. Loss of floodplain area is important for three reasons in particular. First, the ecological diversity and productivity of the floodplain will be reduced, affecting many species of fauna and flora, including 200 species of birds (plate 12.5). Second, the Ob

Plate 12.5 Floodplain woodland along a Siberian river. Diversion of these rivers southwards might solve the water shortage in the Aral Sea catchment, but is likely to have a negative environmental impact on ecosystems in the Taiga and the Arctic.

floodplain provides a rich source of fodder with a potential area of 2.5–5.0 million ha and a potential annual production of 3.5–7.0 million t. To compensate for the loss of production following regulation, other floodplain areas could be put to more intensive agriculture but this may lead to other adverse ecological effects.

Third, there would be significant losses in fish productivity. Currently, there is an important fishery in the Ob–Irtysh basin. Of the 55 species, 20 have commercial value, including salmon and sturgeon. Average catch during the 1970s was over 40 000 t, about 55 percent of the Siberian catch. The productivity of the river depends mainly on the vastness of the floodplains and the duration and predictability of the flood season. Inundated floodplains and floodplain lakes provide spawning grounds and fattening areas for juvenile fish that feed on the abundant phytoplankton. It has been estimated that water diversion could reduce fish catch by 20 percent.

Additional changes of freshwater ecology, including impacts on the fishery, would result from changes of water quality below the diversions. At Belogor'ye the summer water temperature would be reduced by 0.2 to 0.7°C and ice cover would form three to nine days earlier; in the lower reaches freezing would be advanced by between 14 and 19 days. Beneath the ice cover, the water becomes anoxic and the period of oxygen-deficient conditions would be extended. Reduced flows would cause the mixing zone of saline and freshwater, represented by water salinity of $1 \, g \, l^{-1}$, to move outward by one degree longitude (110 km).

The Ob Gulf, a large river estuary of 40 000 km^3, would experience a reduction of heat flow owing to the diversion of warm, southern waters, and to greater diversions in summer. The heat input, estimated to be 3.217×10^{15} kilocalories per year, would be reduced by 15 and 30 percent for phases 1 and 2 diversions, respectively. The effect would be to increase the thickness of the ice cover by 10 percent and to delay the spring ice break-up by 8–10 days in the southern part of the Ob Gulf.

The estuary opens into the Kara Sea, which receives the inflow of the Ob and several other large rivers, together yielding 1300 km^3 year^{-1}. Although the phase 1 and 2 diversions from the Ob alone represent only a very small proportion of the total inflow, potential diversions from the four major rivers feeding the Kara Sea (Ob, Yenesei, Divina, and Pechora) could increase to 300 km^3 year^{-1} (Semtner, 1984): about 8 percent of all runoff to the Arctic Ocean. Concern has been expressed about the effect of the long-distance water transfers on the behavior of the Arctic Ocean and its sea ice, and then on climate. Indeed, Golden and Amfitheotrof (1982) concluded that the diversion could drastically alter climate not only in Russia but throughout the Northern Hemisphere.

Sea-ice formation in polar regions is promoted by low-density freshwater at the ocean surface. The ice inhibits the transfer of heat and moisture from the ocean to the atmosphere, and reflects incoming solar radiation, producing an extremely cold and dry climate. If reduced inflows of freshwater caused the Arctic icepack to shrink, the result would be a warmer, wetter climate in polar regions. Models of the potential effects of water diversion on circulation in the Arctic Ocean and macro-scale variations of climate (e.g. Semtner, 1984) in fact show remarkably little impact of the diversions, with only minor effects on ice extent. However, Holt et al. (1984) conclude that the diversions could noticeably affect the Arctic ice cover, especially in the Barents Sea throughout much of the period from late winter to early summer, and the eastern Kara Sea from July and September. Micklin (1986) reports former Soviet research suggesting that the phase 1 and 2 diversions would increase the extent of summer ice by 6 and 9.5 percent respectively. Differences between studies reflect the limitations of the databases used, especially the spatial and temporal resolution of the data. While the scientific community remains divided on the likely effects of the Siberian diversions, the level of concern reflects the potential global effects of long-distance water transfers.

Prospect

The history of water resources development is characterized by a progression of technological achievements. The first phase involved water storage, flood control, and land drainage. The modern era began with electricity generation, and has continued with rising demands for energy, especially hydro-electric power. The third stage in this development is the mitigation of the environmental problems created by "modernization." Indeed, the conflict between development and ecological sustainability has arisen because of a failure to recognize that development is more than a technological problem: there are also important environmental and social issues (Cosgrove and Petts, 1990).

A model of potential development is presented in figure 12.8. The problems of river management in less developed and developing countries relate fundamentally to (a) the problem of population growth and (b) resource consumption. Continued rapid increase in population is inevitable because of the demographic structure of the populations, about 50 percent being under the age of 16. For example, across Latin America as a whole, population growth is 2.4 percent (UN, 1987) but the distribution is very uneven. Nearly 70 percent of Latin America's population live in urban centers. Only about 40 percent of this urban population has access to sewers and some 90 percent of the collected waste-water is discharged to receiving waters untreated (Bartone, 1990). Of the major rivers, the Tiete and Grande rivers – headwater tributaries of the Parana – are severely affected. Within the Upper Parana basin, pollution problems may be made critical by flow regulation (see figure 12.2), which severely reduces the capacity of a river to flush, dilute and assimilate domestic, industrial, and agricultural wastes.

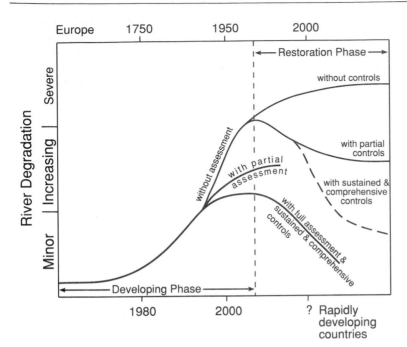

Figure 12.8 Model for river degradation and restoration in Europe and in rapidly developing countries in relation to level of controls and environmental assessment (based on Meybeck and Helmer, 1989).

Although rapid and continuing urban development in drylands is placing a particular strain on water distribution systems (Beaumont, 1989), hydro-power development is the main driving force behind dam building. Worldwide only 13.5 percent of the technically feasible hydro-electric power potential is currently exploited. One scenario for developing countries requires 565 000 MW of installed hydro-power capacity compared with an actual level of 154 000 MW in 1986 (Goldemberg et al., 1988). In Latin America, an increase in per capita resources demand to that in the United States or Western Europe would cause a massive, perhaps ten-fold, increase in energy and materials use.

Opportunities exist to improve the efficiency of water use, but in developing countries a major expansion of energy production will be required nonetheless. Water projects which have gestation periods of more than 15 years in countries with high population growth will have to proceed, because the risks of not harnessing these resources in a timely manner may be too high. The inescapable conclusion is that for the next 50 years, at least, large dams and reservoirs will remain a most attractive option to achieve energy and food security.

In many cases environmental degradation has not been caused by engineering works *per se* but by poor planning and management. Water projects have often lacked the necessary coordination and administration, and there has been an excessive reliance on large-scale engineering without the necessary assessments of social and environmental issues. Such assessments are required, first, when considering alternative locations for dams and, second, when developing management scenarios to mitigate environmental impacts. A range of management tools are being developed for this latter purpose (Gore and Petts, 1989; Petts et al., 1989b; Petts, 1990), including flow control, water-quality control, structural design and maintenance, biological controls, and regulations on human activities.

The post-modernist vision has concern for the quality of life, focuses on investment in the environment, and is expressed by symbols of social advancement: landscape restoration and the protection of wilderness areas (Cosgrove and Petts, 1990). Indeed, it could be argued that some rivers, at least, should be preserved in their pristine state – although this would mean conserving entire drainage basins! Sustainable development requires large water engineering projects but the goals of these projects and their operation must be radically revised. Large dams and reservoirs must be developed sparingly and as part of a comprehensive catchment, national, or regional plan for integrated land and water management. Furthermore, secondary regulation measures to mitigate impacts must be a requirement for all schemes.

World maps of water resources and population density show contrasting patterns and imaginative schemes have been proposed for water redistribution. Falkenmark (1989), for example, has suggested that the problem of water supply in Africa might be solved by inter-basin transfers from the Zaire. The electricity needs of Venezuela could be met by converting the 1000 km long Caroni River into a series of artificial lakes drowning an area of over 9000 km^2 (Anon., 1989). The environmental degradation resulting from river regulation projects throughout this century has led many to conclude that large-scale dam projects encapsulate every imaginable social, environmental, and economic folly. The establishment of effective management structures to administer and coordinate developments giving due regard to long-term environmental needs could form the basis for environmentally sound water projects for the twenty-first century.

Further reading

Amoros, C. and Petts, G.E. (eds) (1993) *Fluvial Hydrosystems*. London: Chapman and Hall.

Beaumont, P. (1989) *Environmental Management and Development in Drylands*. London: Routledge.

Falkenmark, M. (1989) The massive water scarcity now threatening Africa – why isn't it being addressed? *Ambio*, 18, 112–18.

Petts, G.E., Moller, H. and Roux, A.L. (eds) (1989) *Historical Changes of Large Alluvial Rivers: Western Europe*. Chichester: Wiley.

Welcomme, R. (1979) *Fisheries Ecology of Floodplain Rivers*. London: Longman.

Williams, W.D. and Aladin, N.V. (1991) The Aral Sea: recent limnological changes and their conservation significance. *Aquatic Conservation*, 1, 3–23.

PART V

The Tropics

CHAPTER THIRTEEN

Savanna Landscapes and Global Environmental Change

Philip Stott

Savannas, or, in the old spelling, "savannahs," are vegetation formations of the seasonal tropics and subtropics that are co-dominated by various mixtures of woody species, namely trees and shrubs, and grasses (Huntley, 1982). Despite media obsessions with the tropical moist forests, it is actually savannas which dominate the southern continents, covering 65 percent of Africa, 60 percent of Australia, and 45 percent of South America (figure 13.1). In the northern continents, savannas are also widespread, occupying 10 percent of the Indian subcontinent and mainland South East Asia, and pushing north into Mexico and Texas (Küchler, 1964; Archer, 1989; Stott, 1990; Yadava, 1990). The majority of people in the tropics, over one-fifth of the world's population, live in savannas, which are a core part of the Monsoon lands that, overall, support more than 50 percent of the global village. For pastoralists and farmers who inhabit the developing countries of the South, the

Figure 13.1 The world distribution of savanna lands. (Adapted from Deshmukh, 1986, figure 8.1, p. 224.)

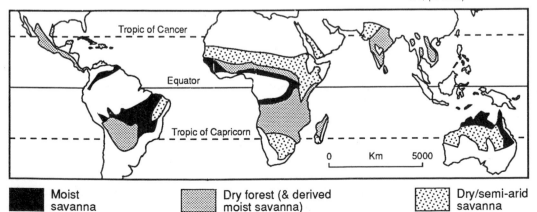

Moist savanna

Dry forest (& derived moist savanna)

Dry/semi-arid savanna

savannas are thus a far more significant landscape element than the tropical moist forests. Accordingly, the relationships between global environmental change and the savannas are especially important for the people of the South, although these relationships have received considerably less attention from conservationists and ecologists in the North (Stott, 1991). An increased understanding of savanna ecology, in the light of global environmental changes, is thus vital for the long-term survival of so many people, who are already threatened with landscape degradation, drought, famine, and disease.

Savanna ecology

In savanna ecosystems, the majority of the grass species involved belong to the C_4 group of photosynthetic plants. In such plants, the rate of photosynthesis continues to increase with the intensity of "photosynthetically active radiation" (PAR), instead of the carbon dioxide (CO_2) uptake curve flattening off to a plateau, as is more normal in C_3 species (Begon et al., 1990, pp. 81–3). In C_3 plants, the carbon dioxide is fixed initially as a three-carbon compound, phosphoglyceric acid; by contrast, in C_4 plants, the carbon dioxide is fixed as a four-carbon compound, oxaloacetic acid (Walker and Edwards, 1983; Deshmukh, 1986, pp. 24–6). C_4 plants are the most efficient photosynthetically where light conditions are maximized, so that their optimum temperature for CO_2 fixation lies between 30 and 45°C, their photosynthetic rate under optimal conditions is between 40 and 80 mg CO_2 dm^{-2} h^{-1} (3 to 4 times that of C_3 species), and their light saturation is 100 percent. This means that they are ideally adapted to hot, bright, and fairly dry environments, precisely the conditions which obtain in the open savanna grasslands of the seasonal tropics. However, there is some evidence that this competitive advantage in hot, dry regions may be significantly diminished as a consequence of predicted "greenhouse" effects, such as increased atmospheric CO_2, because C_4 species appear to show little photosynthetic response to elevated CO_2 (Woolhouse, 1990). Nevertheless, dry-matter production may still be enhanced by an improvement of the plant-water status, through the reduction of stomatal aperture which usually accompanies the effects of high CO_2 concentrations (Squire, 1990).

Although C_4 grasses form an essential ground component of nearly all savannas, the woody element of shrubs and trees varies considerably, both in the degree of its co-dominance with the grasses and in the mix of deciduous and evergreen species (see the four contrasting landscapes shown in plates 13.1 to 13.4). On a world scale, savannas range in structure from virtually treeless grasslands, with a canopy cover of less than 2 percent, such as the mitchell (*Astrebla* spp.) grasslands of Queensland and the Serengeti Plains of Tanzania, to savanna woodlands or forests, like the *miombo* (*Brachystegia–Julbernardia* woodlands) of southern

Plate 13.2 A baobab (*Adansonia digitata*) savanna, Tarangire Game Reserve, northern Tanzania.

Plate 13.4 A *Burkea africana* savanna on infertile, sandy soil, Nylsvley, South Africa.

Plate 13.1 A eucalypt savanna woodland in the "top end" of the Northern Territory, Australia.

Plate 13.3 An *Acacia* savanna, Tarangire Game Reserve, northern Tanzania.

Africa, the eucalypt savannas of Northern Australia, and the dry deciduous dipterocarp forests of mainland Southeast Asia, in which the canopy cover, when in full leaf, attains 50–75 percent (Belsky, 1990; Burrows et al., 1990; Mott et al., 1985; Stott, 1990). This marked variation in landscape appearance partly reflects the fact that savannas form under an equally wide range of tropical macroclimates, from moist savannas with a mean annual rainfall of 1000–2000 mm and a dry season of up to 5 months, through dry savannas with a mean of 500–1000 mm and a dry season of up to 7.5 months, to semi-arid formations, in which the mean annual rainfall is as low as 250–500 mm and the dry season as long as 10 months of the year (Harris, 1980). At one end of their distributional spectrum, savannas can abut directly onto tropical semi-evergreen rain forest, while at the other end, they grade into semi-desert and desert (figure 13.2) (Whittaker, 1975). This extremely wide climatic tolerance helps to explain the global significance of savannas, which occupy just under one-third of the world's land surface (Solbrig, 1990; Werner, 1991).

Savannas are of particular significance not only because they are home to a large number of people in the tropics, but also for a range of mainly theoretical reasons. First, it must be recognized that the "core" savannas are ecosystems in their own right, and that they are not simply secondary forms derived from forest communities (Stott, 1989). These "core" savannas probably cover more of the tropics than do the primary forests, and their conservation and

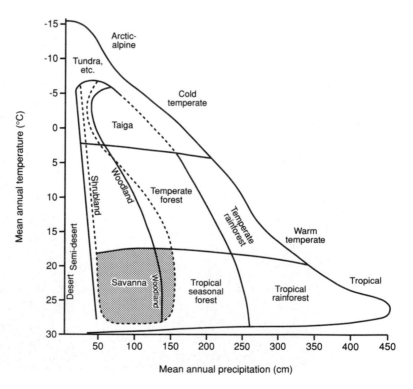

Figure 13.2 The ecological position of savannas in relation to the major terrestrial biomes of the earth, based on temperature and rainfall gradients. (Modified from Whittaker, 1975, and after Belsky, 1990, figure 1.)

correct management depend on detailed research into their own peculiar ecologies and ethnoecologies. The results of such research often necessitate a totally different assessment of the role of various ecological stresses from the more general and widespread interpretations based on the study of temperate and forest ecosystems. The ecology of fire is a case in point (plate 13.5) (Stott, 1988 a, b; Goldammer, 1990). In the savannas, both natural fire, from lightning-strike, and human-induced fire are often entirely integral to the maintenance of the ecosystem, whereas in forests this ecological force is usually perceived as fundamentally inimical to the functioning of the system. In many savannas, it would be impossible to maintain the economic productivity of the grass stratum without the use of fire as a prescribed tool, whereas in forests, by contrast, fire is largely regarded as destructive and, of course, a contributor to the release of "greenhouse" gases. Savanna ecologists, and the people of the South who manage the savannas, think of fire, correctly applied, as a useful force, even a friend; forest ecologists automatically see it as a threat and an enemy. To say that all fire in the tropics is "bad" is simply to reassert the hegemony of "forest ecology," and thus the continuing hegemony of the North over the South (Stott, 1991). Such differing attitudes to fire and global environmental change are not, however, easily reconciled, particularly as the amounts of CO_2 and other trace gases emitted by savanna fires are about three times larger than those emitted from deforestation in the tropics (Wei Min Hao et al., 1990, pp. 452–60).

Plate 13.5 A headfire in the dry deciduous dipterocarp forests (savanna forests) of mainland southeast Asia: Khao Nang-Ram, Huay Kha Khaeng Wildlife Sanctuary, Uthai thani, Western Thailand.

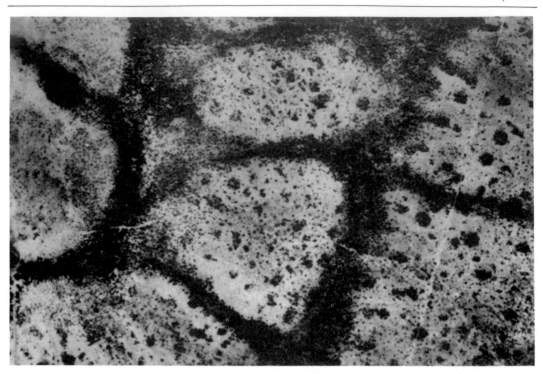

Plate 13.6   An aerial
photograph of the
woody/grass mosaic at La
Copita Research Area in
southern Texas, USA. The
woody clusters are
formed around a honey
mesquite (*Prosopis
glandulosa* var.
glandulosa) nucleus.

Second, savannas are of further theoretical significance because
of their unique mixture of woody species and grasses (plate 13.6).
Most ecological models developed so far can cope well with
ecosystems that are dominated either by trees and shrubs or by a
graminoid layer, but they are less successful with those co-
dominated by both at the same time. This problem presents an
interesting challenge to scientists wishing to model savanna landscapes in relation to global environmental change (Solbrig, 1991).
Predicted global environmental changes could have a significant
effect on the mixture of trees and grass, thus affecting not only the
form of savannas, but also their productivity from the human point
of view. Similarly, changes in the tree/grass ratio, brought about by
the human management of savannas through fire and grazing,
could, in turn, have important feedback effects on the actual
direction of global environmental change itself. The need to
develop suitable savanna models, which relate widely to climate,
hydrology, and geomorphology at all spatial scales, is thus an
urgent necessity. Because savannas lie in the sensitive intermediate
zone between the tropical forests and the deserts, it is extremely
likely that they will respond significantly to global changes, both at
the local level in terms of their species composition, and at the
higher spatial scales in terms of their overall form and structure, but
particularly at the forest–savanna boundary. Monitoring the
changes at this boundary, either in detail through botanical and
ecological surveys, or, in more general terms, by satellite imagery
and aerial photography, will be essential (Kaufman et al., 1990;
Furley et al., 1992).

In essence, therefore, savanna landscapes are of enormous significance for global environmental change, first because they actually support the majority of people who live in the tropics, and second because their geographical location and distinctive form as a tree/grass ecosystem makes them of particular theoretical value.

A hierarchy of determinants

It is now recognized that the savanna form is generally governed by the intricate interplay of five key ecological factors, namely plant available moisture (PAM), plant available nutrients (PAN), fire, herbivory, and major anthropogenic events (Frost et al., 1986; Walker, 1987; Walker and Menaut, 1988; Solbrig, 1990, 1991). The initial determinants of biomass are seen to be PAM and PAN, whereas fire and herbivory are regarded as secondary modifiers of this primary productivity. Anthropogenic events, such as human settlement or its abandonment, and even relatively sudden and intensive global environmental changes caused by human action, can trigger a whole savanna system into a new mode of functioning, so that an essentially historical factor continues to determine the basic form of the savanna over a long period of time. Blackmore et al. (1990) provide an excellent example of such an ecological "memory," demonstrating the close correlation between the occurrence of nutrient-enriched patches, supporting *Acacia tortilis*, within a more general nutrient-poor *Burkea africana* savanna and the former presence of Iron Age settlements at Nylsvley in South Africa. Levels of Ca, Mg, K, and PO_4 in the soil under the *Acacia* communities were found to be 10–100 times higher than those under the *Burkea* savanna, and order-of-magnitude calculations indicated that the activities within an Iron Age settlement would be sufficient to account for the nutrient accumulations observed. Iron Age artifacts were only found within the *Acacia tortilis* patches.

As Solbrig (1991, pp. 18–28) demonstrates, it is extremely helpful to construct a hierarchy of control for these five key savanna determinants. Initially, PAM and PAN were regarded by savanna scientists as equal in significance, and this led to the development of the concept of the PAM/AN plane (the plant available moisture/(plant) available nutrients plane). By this it is hypothesized that the functional properties of all savanna plants, including their ecophysiological, morphological, and demographic patterns of response to environmental variability, especially seasonal variations in water supply, can be located within a plane defined by the axes of PAM and PAN. This PAM/AN plane then allows a functional classification of savannas based on abiotic characteristics, which in turn facilitates the comparison of both savanna sites throughout the world (figure 13.3) and savanna functional groups or lifeforms, such as the disposition of evergreen and deciduous woody elements within the plane (figure 13.4). By its nature, however, the PAM/AN plane intrinsically assumes that PAM and PAN are of equal importance, whereas experience in

Figure 13.3 The
disposition of East African
savanna types within the
PAM/AN plane. (After
Frost et al., 1986; Belsky,
1990, figure 5.)

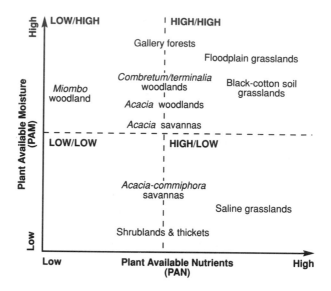

Figure 13.4 The
suggested phenological
and physiognomic
disposition of South
American savanna tree
types within the PAM/AN
plane. (After Walker and
Menaut, 1988, figure 2.1;
and Stott, 1991.)

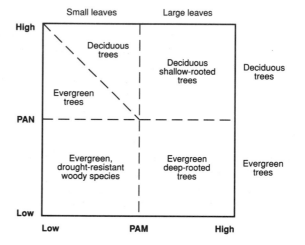

actually using the concept increasingly indicates that PAN is nearly
always constrained by PAM, and that, in many instances, PAN
exercises no significant control at all. PAN is thus lower in the
hierarchy of determinants than PAM.

Figure 13.5 presents one possible hierachy of savanna determinants, including all the five key factors of PAM, PAN, fire (F),
herbivory (H), and historical anthropogenic events (A). This hierarchy differs slightly from that presented by Solbrig (1991, figure
9), in that historical anthropogenic events (A) are placed at the very
top of the hierarchy, rather than in the third level, where they
would be of co-significance with fire (F) and herbivory (H). Solbrig
(1991), however, admits of this possibility in his discussion. Each
level of the hierarchy has one or more holons. A holon is simply a
subsystem which interacts frequently and strongly with other

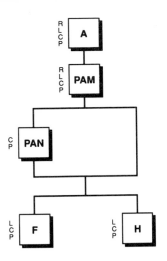

Figure 13.5 Suggested hierarchy of savanna determinants. A = catastrophic anthropogenic factors (historical events); F = fire, both natural and human-induced; H = herbivory, both natural and human pastoral systems; PAM = plant available moisture; PAN = plant available nutrients; P = the patch level; C = the catena level; L = the landscape level; R = the regional level. See text for full explanation. (Modified from Solbrig, 1991, figure 9.)

holons at the same level of the hierarchy. For example, herbivory (H) and fire (F) are holons which influence each other considerably, the one often precluding or limiting the other, and both affecting each other in more subtle ways, through such factors as nutrient release and forage quality (e.g. Winter, 1987). Fire (F) includes both natural and human-induced fires, while herbivory (H) involves not only the role played by large indigenous mammal browsers and grazers, but also the impact of introduced large exotic herbivorous mammals (Freeland, 1990), herbivory by invertebrates (Andersen and Lonsdale, 1990), and the part played by human pastoral economies (McKeon et al., 1990).

Holons at one level of the hierarchy are constrained by those above, whereas the rates of ecological processes become slower as one moves up the hierarchy. Figure 13.6 supplements and helps to interpret the core hierachy diagram (figure 13.5), in that it illustrates the fundamental relationships between the biological

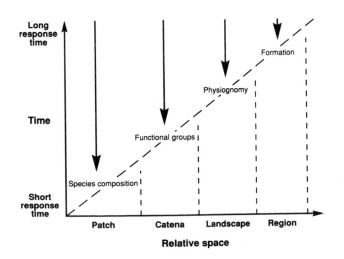

Figure 13.6 A diagrammatic conceptual framework for evaluating savanna hierarchy determinants at various spatial and temporal scales. The horizontal axis represents the relative spatial scales; the vertical axis, temporal response scales. The appropriate level of biological analysis is indicated on the diagonal. See text for full explanation. (After Solbrig, 1991, figure 1.)

characteristics of savannas (species composition, functional groups or lifeforms, physiognomy, and formation type) and both temporal and spatial response scales. In spatial terms, savannas have always been studied at one of four basic scales, namely the patch (P; a local area with a relatively homogeneous species composition), the catena (C; a linear topographic linking of such patches), the landscape (L; an areal jigsaw of patches, bounded by clear topographic limits, such as those of a drainage basin), and the region (R; a distinctive geographical entity or *pays*, like the Llanos of Venezuela). The hierarchy diagram (figure 13.5) accordingly indicates at which of these spatial scales the different holons in the hierarchy are likely to function. Figure 13.6, on the other hand, indicates which biological attribute is most appropriate for the characterization of each of these spatial units. At the patch level (P), for example, species composition will be the main indicator, and this will, in general, have a shorter response time to changes mediated through the hierarchy. The mechanisms of vegetation change suggest that plant diversity will be altered before the vegetation switches to a new, discrete type (Schlesinger et al., 1990; Woodward and Rochefort, 1991, p. 8). At the landscape level (L), in contrast, the overall physiognomy of the vegetation will be more signficiant, and this will have a much slower time response to change. Global environmental changes will thus affect savannas in different ways at different spatial and temporal scales.

Each holon, or subsystem of the hierarchy, is itself susceptible to further hierarchical analysis. The hierarchy of PAM determination is given in figure 13.7. This illustrates two important points. First, at the catena level of the subsystem hierarchy, PAM will be mediated in at least two ways. In topographical situations where groundwater emerges at the surface of the catena for some of the year, PAM will be controlled essentially by fluctuations in the water table. Elsewhere, however, where the water table is always well below the rooting zone of the catena surface, infiltration capacity

Figure 13.7 Suggested hierarchy of measurable variables for PAM in savannas, at different spatial scales. See text for full explanation. (After Solbrig, 1991, figure 10.)

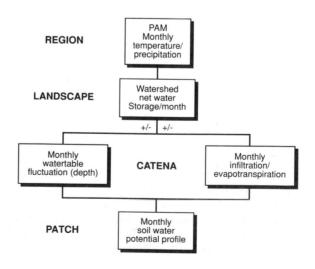

and evapotranspiration will be the significant factors involved. Second, and more importantly, the diagram demonstrates the imperative of finding an easily measurable variable through which PAM is mediated at each level of the hierarchy. At the patch level (P), the monthly soil water potential profile has been selected, with other grosser, but scale-appropriate, measures for the higher levels of the hierarchy. It is worth noting that the measures chosen fit in well with the concerns of other relevant disciplines, especially climatology, hydrology, and geomorphology. It is through such agreed measures that the biology of the savannas may be better linked to our wider understanding of global environmental change, and, more particularly, to the models of global environmental change developed in other disciplines, such as general circulation models (GCMs) of the atmosphere. Moreover, if all savanna scientists can agree on a standard measure for each level of the PAM subsystem hierarchy, this will greatly facilitate both world savanna comparisons and savanna ecosystem modelling. Solbrig (1991) already presents an agreed list of variables for the standard measurement of PAM and PAN when they are employed to define the PAM/AN plane; further lists should now be drawn up for the other key factors, and for PAM and PAN at different hierarchical levels.

Global environmental changes will thus be mediated *down* to savannas through the hierarchy of determinants outlined in figure 13.5, but at different time responses and through different biological attributes at each spatial scale. Research measures to analyze the impact of global environmental change will thus vary according to the scale under consideration. Unfortunately, many biologists have refused to work at anything but the individual plant or patch scale (P), and this has tended to hinder our understanding of savanna changes at the higher spatial scales (Allan, 1991). Similarly, changes in savannas will, in their turn, influence global environmental change *upwards* through the hierarchy, linking with geomorphological, hydrological, and climatological processes at the points, and in the ways, indicated. The modelling of savanna biology in terms of PAM, PAN, fire, herbivory, and other anthropogenic factors is thus a vital next concern, so that we can begin to answer the fundamental questions: how will predicted global environmental changes affect different savannas; and, how will savanna processes feed back to influence global enviromental changes?

Modelling the changes in savannas

In modelling the global environmental changes to which savannas will respond, interest focuses primarily on the relationships between greenhouse-gas emissions and changing climate, particularly precipitation, temperature, evapotranspiration, and wind speeds at the regional and biome levels. However, current models of global environmental change, such as those GCMs that attempt to predict future climates resulting from alterations in the emission of greenhouse gases, "diverge in the degree, sign and spatial distribution of the predicted changes in both temperature and precipitation" (Woodward and Rochefort, 1991, p. 7; see also Houghton et al., 1990). Figure 13.8 illustrates, as just one example, the different predictions for the geographical distribution of the rate of precipitation change (mm day^{-1}), during the wet season of December to February, for the savanna regions of Australia, based on five competing GCM simulations, in which CO_2 is instantaneously doubled. As Henderson-Sellers (this volume, and 1991, pp. 57, 62)

Figure 13.8 The predicted geographical distribution of the change in precipitation rate (mm day^{-1}) following $2 \times CO_2$ minus $1 \times CO_2$, during December–February, for Australia, simulated by five different models. Note the marked lack of correlation between the scenarios presented. (After Henderson-Sellers, 1991, figure 4a.)

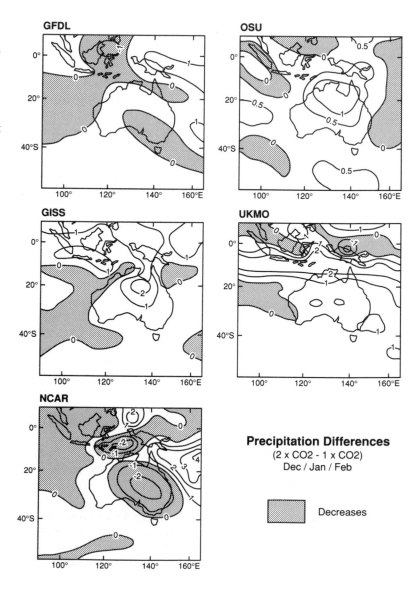

comments, these equilibrium simulations exhibit a "complete lack of regional predictive skill," and they fail to take into account that CO_2 actually increases gradually through time, for which we require transient studies, such as that of Washington and Meehl (1989). Unfortunately, therefore, there are still large uncertainties in all our estimates of global change, especially at the regional or biome levels, because our basic understanding of the relevant physics remains incomplete, because our limitations in computing power slow the progess of model development, and because our actual predictions of greenhouse-gas emissions remain approximate (Rowntree, 1990).

Moffatt (1991), however, has tried to assess the ecological impacts of the greenhouse effect on the savanna woodlands of Northern Territory, Australia, drawing on the preliminary regional results of the CSIRO4 atmospheric global simulation model (CSIRO, 1990). Predictions that net primary production will be augmented in the northern part of the Territory, and that increased monsoonal precipitation will favor the development of pasture legumes and cattle raising, must be set against the fact that the soils of the region are nutrient poor and often limiting in elements such as phosphorus, so that the new possibilities may prove a false hope, in the end actually encouraging overgrazing and severe soil erosion. Moreover, the assumed rise in sea level brought about through global warming may also cause significant coastal inundation of the northern portion of the World Heritage site of Kakadu National Park, facilitating the spread of mangrove communities, dominated by *Rhizophora* spp., into areas at present carrying stringybark (*Eucalyptus tetrodonta*) woodland. Yet even here it is possible that the formation of natural levées will eventually occur, compensating for the loss of land and resulting in the respread of low-lying vegetation. Moffatt's regional analysis thus illustrates all too vividly the essentially speculative and qualitative nature of our understanding of the possible ecological impacts of global climate changes on savanna biomes (see also Werner, 1988).

Nevertheless, there is some consensus that CO_2 concentrations in the atmosphere will increase to about 600 p.p.m. over the next 50 years. Although predicting the actual climatic repercussions of this rise is, as we have just seen, rather hazardous, some comment must be made on the direct physiological influences of the CO_2 increase, particularly in respect of the C_4 savanna grasses (Beadle et al., 1985; Woolhouse, 1990, pp. 35–8). The potential effect of the increasing CO_2 concentration on the rate of photosynthesis must be seen in terms of the competitive suppression of the oxygenation reaction catalyzed by the carboxylating enzyme, RuBPCase (ribulose 1,5-biphosphate carboxylase). CO_2 enhancement will have its greatest effect under hot, dry climates, where leaf temperatures and stomatal closure lead to elevated O_2/CO_2 ratios at the site of CO_2 fixation. In C_4 plants, however, there already exists a CO_2-concentrating mechanism, the C_4 cycle, which is not susceptible to

O_2 competition, and in which the RuBPCase is protected by a locally elevated concentration of CO_2. It is not surprising, therefore, that C_4 plants show little response to elevated CO_2. This means that their competitive advantage may be reduced under increased atmospheric CO_2, having significant effects on both the composition and the productivity of the grass stratum in the tropical savannas. But yet again, there is much uncertainty about this simple prediction, because the single variable advantage may be offset or transcended by significant changes in the other environmental factors concerned with dry-matter production, particularly temperature and precipitation (Squire, 1990).

If there are problems in developing and interpreting the general global and regional models of environmental change, it has to be admitted that savanna-specific modelling is still in its infancy. Nevertheless, as Solbrig (1991, pp. 29–36) indicates, there are a range of models for savannas which merit attention and further refinement. Basically, these models fall into one of three categories: individual plant models, large-scale process-response models, or ecosystem production models. The first type is well exemplified by the "anlaytical model of grass–woody plant interactions" developed by Walker and Noy-Meir (1982). This was designed primarily to test the hypothesis that the coexistence of woody plants and grasses in savannas is a consequence of their separate use of the topsoil and subsoil mixture. The model is limited by its assumption of equilibrium conditions, which are never really attained in a savanna, but the model does provide a useful view of equilibrium behavior in savannas under a wide range of climatic conditions.

STEP (the South Turkana Ecosystem Project), developed by Coughenour, Swift and Ellis, on the other hand, is a good example of a large-scale process-response model. The aim of this project, which is a spatially explicit regional model, superimposed on a geographical information system (GIS), is to track the biomass dynamics of trees and herbaceous plants, herbivory by livestock, livestock production for milk, and the interactive variation of these elements through time, in relation to precipitation and drought. CENTURY, by contrast, is an ecosystem production model. This was developed by Parton and others, originally to simulate the dynamics of C, N, P, and S in uncultivated and cultivated grassland soils. The savanna version of the model integrates elements of the original with those of its extension, which simulates the dynamics of the same nutrients in forest ecosystems. The difficulty of dealing with both the woody and the grass element is one of the perennial problems in all savanna modelling (see Solbrig, 1991, for further details of all these models).

The main significance of such models is not, however, their intrinsic, theoretical value, but rather their potential importance in formulating applied plans for the human management of savanna ecosystems, particularly in the light of the possible effects of climate change on secondary production and land use.

The human dimension

Many of the theoretical models being developed for savanna ecosystems can provide the basis for useful resource system planning models (RSPM) and decision support systems (DSS) (Stott, 1991, p. 25; Stuth et al., 1990). For example, the large-scale process-response model, PYRO, developed by Mentis in South Africa (Solbrig, 1991), is a DSS designed to help savanna park administrators with their management of both wildfires and two types of prescribed burn, namely security burns to protect people and property and ecological burns to maintain biotic diversity. Similarly, SHRUBKILL, produced by Ludwig (1988, 1990), is a microcomputer-based advisory program with three key modules, namely BURNTIME to provide advice on the time of prescribed fire, BURNWAYS on how to carry out burns safely and effectively, and BURNECON to compute the cost benefits of such a prescribed burn. This program has been designed specifically for on-site decision-making in relation to the crucial problems of controlling shrub/grass ratios in the savannas of Northern Australia. The package comes with an explanatory booklet, containing illustrations of different woody/grass landscape mixtures, so that pastoralists and farmers can visually compare their own tall grass pastures with a range of savanna types.

Such decision support systems (DSS) will inevitably become increasingly important in the light of global environmental change, as savanna farmers and pastoralists seek to adapt their lands to the new conditions. The potential effects of climate change on secondary savanna production and on land use are summarized in figure 13.9. Because so many people depend for their livelihood on savanna grasslands, the development of these advice packages, based, of course, on a sound theoretical understanding of savanna ecology, must be regarded as a top priority in the next few years. In fact, it is entirely arguable that much of the money at present being directed to the so-called hard science aspects of global environmental change modelling, such as GCMs, would really be far better used at the local and regional scales to aid directly human responses and management. After all, as already pointed out, our large-scale general models remain totally inadequate, and, in attempting to control the direction of global changes, we may, in the end, trigger further totally unpredicted changes. In contrast, aiding the human response to change, whatever that change may be, rather than attempting to manage the direction of change itself, may be a wiser course to follow. Until the present generation, this has always been the way in which human populations have responded to environmental change, which, of course, is not a new phenomenon. The advice systems employed, however, must be culturally relevant, and, for many of the poorer parts of the world, they will have to be presented in ways that can be easily afforded

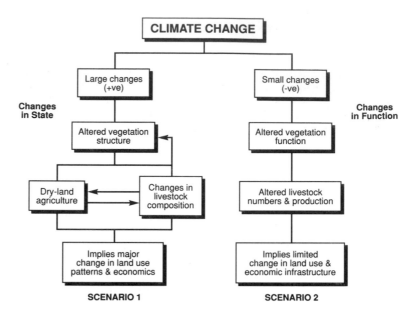

Figure 13.9 The potential effects of climate change on secondary production and land use in savannas. (After Solbrig, 1991, figure 12.)

and understood. DSS should not, therefore, always be computer-based, but should be available in many different formats and locally relevant packages, such as drama, puppetry, picture books, religious tracts, film, or video.

Finally, it must be remembered that savanna ecosystems have themselves long been a significant contributor to the never-ending story of global environmental change. Even before the impact of humans, fire, through lightning strike, friction, and refraction, was a natural stress in savanna lands. Some 1.4 million years ago, hominids began the intensification of this stress (Gowlett et al., 1981), which was then further intensified by the spread of human populations, some 50 000 years ago (Stott, 1988a). Today, savanna fires contribute 75 percent of all the CO_2 emitted to the atmosphere through biomass burning in the tropics (Wei Min Hao et al., 1990, p. 456). In the Brazilian *cerradao*, for example, the entire humid savanna is fired once every two years, while 75 percent of the humid savannas of Africa burn annually, ensuring that 85 percent of the CO_2 emitted from tropical Africa is derived from its savannas. Many ecologists and conservationists thus regard biomass burning in the savannas as a major contributor to global warming. But, in reality, the issue is much more complex, because the savannas have always burned. What we need to know is the exact nature of the *recent* increase in burning over the long-term historical levels. And would this recent increase really be significant at all if it were not being added to the historical contribution over the past 150 years or so of the industrialized countries of the North, who have yet to pay the real price of their development?

In many ways, our future management of the savanna landscapes of Australia, South East Asia, India, Africa, and Latin America will indicate all too starkly whether we are succeeding in our *human* response to global environmental change.

I am especially grateful to Professors Brian Walker, Otto Solbrig, Patricia Werner, and David Wigston for many hours of valuable discussion on savanna ecology over the past few years. I am also grateful to IUBS for sponsoring my participation in the workshop "Savannas and Global Change," held at Harvard Forest, Petersham, Massachusetts, during October 1990.

Further reading

Deshmukh, I. (1986) *Ecology and Tropical Biology*. Oxford: Blackwell Scientific.

Furley, P.A., Proctor, J. and Ratter, J.A. (eds) (1992) *Nature and Dynamics of Forest–Savanna Boundaries*. London: Chapman and Hall.

Goldammer, J.G. (ed.) (1990) *Fire in the Tropical Biota: Ecosystem Processes and Global Challenges* (Ecological Studies **84**). Berlin: Springer.

Harris, D.R. (ed.) (1980) *Human Ecology in Savanna Environments*. London: Academic Press.

Huntley, B.J. and Walker, B.H. (eds) (1982) *Ecology of Tropical Savannas*. Berlin: Springer.

Jackson, M., Ford-Lloyd, B.V. and Parry, M. L. (eds) (1990) *Climatic Change and Plant Genetic Resources*. London and New York: Belhaven.

Solbrig, O. (ed.) (1991) Savanna modelling for global change. *Biology International* (Special Issue), **24**, 1–47.

Stott, P. (1991) Recent trends in the ecology and management of the world's savanna formations. *Progress in Physical Geography*, **15**, 18–28.

Walker, B.H. and Menaut, J.-C. (eds) (1988) *Research Procedure and Experimental Design for Savanna Ecology and Management* (RSSD Australia Publication 1). Melbourne: CSIRO for IUBS/UnescoMAB.

Werner, P.A. (ed.) (1991) *Savanna Ecology and Management: Australian Perspectives and Intercontinental Comparisons* (RSSD Publication 5). Oxford: Blackwell Scientific.

CHAPTER FOURTEEN

Tropical Moist Forests: Transformation or Conservation?

Peter A. Furley

Introduction

The conflict between conservation and development in tropical forests lies at the heart of the argument concerning our management of the Earth's resources. On the one hand, it is starkly clear that continuous clearance or major disturbance of the forests leads inexorably to impoverishment and extinction of life forms and essentially disinherits future generations. In the longer term, such squandering of our birthright is both inexcusable and self-destructive. On the other hand, it is estimated that some 200 million tribal people (Goodland, 1982) or possibly up to 500 million people in total (Myers, 1990) live either within or directly upon tropical forests and that many millions of desperately poor peasants, often lacking most means of sustenance, live in close proximity. To deny them a share in the forest resources condemns them to perpetual poverty. The object of this chapter is to examine the intermediate possibilities between attempts at conservation of all or most tropical moist forests and their virtual elimination, in the belief that the only realistic path forward is a series of compromise solutions.

Increasing concern over the accelerating pace of disturbance and destruction has progressed far beyond emotive response to endangered species. The value of the forest cover is perceived in many different ways. In the developed world, the *environmental* and *ecological* role of tropical forest is frequently emphasized, in its influence upon regional and to some extent world climates, in its ability to maintain soils capable of supporting the highest biomass productivities, in its protection of water courses, and in its biological richness (plate 14.1). The richest biome, the moist forests, contains over 50 percent of the world's biota in only 7 percent of

Plate 14.1 Undisturbed Amazonian rainforest like this contains the greatest biodiversity of all the world's ecosystems.

the Earth's surface (Wilson, 1988). Within tropical countries themselves, *productive capacity* is often considered more important, providing food, fodder for domesticated animals, timber, fuel, and a myriad uses from medicine to shelter, with increasing understanding and sympathy for the fact that the forests are home to millions of people with distinct social and cultural traditions but usually little influence on the environmental changes occurring around them. Discussion of potential for development is intricately bound up with issues of conservation and dynamic change. All such discussions are bedevilled by a lack of accurate information, particularly on the long-term effects of changes in landcover. The

Table 14.1 Tropical forest formations

A *Tropical evergreen forests*: Rainforests composed mainly of evergreens with little or no bud protection against drying or cold

1 *Lowland forest*: multilayered structure with many trees exceeding 30 m in height; discontinuous ground herbaceous layer

2 *Mountain forest*: tree sizes markedly reduced and few species exceed 30 m; abundant undergrowth, often represented by tree ferns, small palms, vascular and other epiphytes; ground layer rich in herbs and bryophytes

3 *Cloud forest*: closed structure with numerous gaps and liana thickets; trees often gnarled, rarely over 20 m; burdened with epiphytes, lianas and herbaceous climbers; extensive ground coverage of mosses, liverworts, broad-leaved herbs and herbaceous ferns

4 *Alluvial forest*: multilayered closed structure with numerous gaps, distinctive herbaceous undergrowth and palms; many buttresses and stilt roots; can be subdivided into (a) riparian forests occupying the frequently flooded narrow strip along river banks, (b) occasionally-flooded forests which can contain emergent trees over 50 m with buttresses and numerous climbers, and (c) seasonally waterlogged forests with an uneven tree cover where water stagnates for several months

5 *Swamp forest*: tree species limited and usually less than 30 m high; stilt roots and air-breathing roots common; located in depressions and smaller valley bottoms with permanently excessive soil moisture and poor soil aeration

6 *Peat forest*: stands generally less than 20 m high with a few, slow-growing, broad-leaved trees or palms; few ground plants – mostly ferns; air-breathing roots and stilt roots common; found on nutrient-poor soils covered by an accumulating organic layer

B *Tropical and sub-tropical seasonal forests*: dominantly evergreen and leaf-exchanging trees with some bud protection. Upper canopy species reach around 40 m and are moderately drought-resistant with some leaf shedding common during the dry season. Transitional forest between A and C

1 *Lowland forest*: the most widespread subtype

2 *Montane forest*: fewer tree ferns than the evergreen equivalent

3 *Dry sub-alpine forest*: dense stands of sclerophytic trees reaching around 10 m; few ground herbs or shrubs; lichens the most common epiphytes

C *Tropical or sub-tropical semi-deciduous forests*: most upper layer trees are drought resistant and shed leaves fairly regularly in the dry season. Some emergents reach 40 m; middle layer pygmy trees and undergrowth shrubs are normally evergreen and may have sclerophyllous leaves. Most trees have some form of bud protection

1 *Lowland forest*: multilayered structure with marked emergents (some 30 m); undergrowth of tree seedlings, saplings and woody shrubs; practically no epiphytes; both short-cycle and perennial herbaceous lianas; many grasses

2 *Montane forest*: similar structure to other montane and some cloud forests but with smaller trees

D *Sub-tropical evergreen forests*: multilayered stands with similar divisions to A but less vigorous and with more shrubs in the understory. Pronounced temperature differences between the seasons. Localized stands in Taiwan and northern Australia

E *Mangrove forests*: Halophilous (salt-adapted) forests of intertidal zones in the tropics and sub-tropics; single layer trees up to 30 m with evergreen and sclerophyllous leaves and various kinds of stilt roots, peg roots, knee and other air-breathing adaptations; epiphytes and ground herbs rare; a few algae, bryophytes and lichens attached to aerial roots and lower stems

Source: after Furley and Newey, 1983

emphasis of this chapter will be upon tropical moist lowland forests (table 14.1) and especially the largest area still remaining – the Neotropical or New World forests. These forests are defined, for the present purposes, as dominantly tree formations characterized by constantly high temperatures (lowest month around 18°C or above), with nearly continuous or seasonal rainfall (usually over 1000 mm year^{-1}) and, in most cases, a low temperature differential between the coldest and the warmest months.

Many of the most endangered and disturbed forests are among the least known and publicized. The dry forests, for instance, have been characterized as the most endangered tropical system (Janzen, 1988). The scattered deciduous forests stretching NE–SW across South America, from the dry *caatinga* of Brazil to the *chaco* of northern Argentina and Paraguay, have been heavily utilized for centuries on account of their more favorable soils; they are now mostly restricted to inaccessible and unusable patches. Equally, the long corridor of Atlantic forest down the coast of Brazil survives today only in isolated mountainous remnants with little hope of re-connection, and only about 7 percent of the forest remains that inspired Darwin in 1832. Madagascar, with one of the world's most distinctive faunas and floras, has lost all but about 1 percent (figure 14.1a).

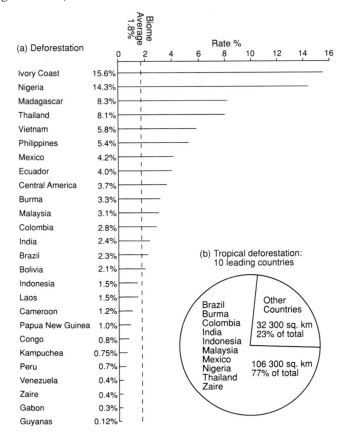

Figure 14.1 Tropical deforestation: (a) rate of loss in different countries; (b) total area of moist forest loss (after Myers, 1989).

Rates of tropical forest destruction and conversion

Projections for deforestation give figures of less than 30 years to 100 years or more for the life of contemporary forests. The stark message is the same: at current rates of depletion, re-afforestation, and conservation, there will be very little left to pass onto generations two or three times removed from the present.

The most recent and acceptable figures at a comparative, global scale (Myers, 1989) have been criticized as exaggerations in respect of some countries because of the methodology involved. Whereas early calculations tended to be underestimates, the onset of satellite image analysis has tended to exaggerate. As remote sensing is by far the most rapid and effective means of assessment, and as satellite images produce a high frequency of scenes and are therefore capable of monitoring change, they have emerged as the most appropriate tool for analysis. However, the quickest and most frequently used sensor, the AVHRR based on NOAA weather satellites, has been shown to be subject to considerable error (e.g. Setzer et al., 1988). The AVHRR (advanced very high resolution radiometer) has a coarse spatial resolution with pixel size 1.1 km square, compared to 80 m square for the earlier LANDSAT MSS, 30 m square for LANDSAT TM and 20 m square for SPOT (see chapter 2, Haines-Young). The AVHRR survey works by picking up heat from a thermal sensor, and it has been demonstrated that when as little as half of a pixel shows burning forest, the whole pixel is recorded as burnt (Fearnside, 1990) (plate 14.2). Since these early estimates, various corrections have been applied (such as Malingreau and Tucker, 1988) and it is now believed that the figures are reasonably reliable (Cross, 1990).

A greater resolution is obtained from Earth resource satellites such as LANDSAT TM or SPOT. With these systems, areas are less frequently scanned but the vegetation can be viewed in greater detail. Indeed, the amount of information generated is so great that the problem is often one of data handling. If photographic prints are taken off the monitor and one date simply overlain on another manually, it is possible to monitor large areas fairly effectively if tediously, as has been demonstrated by the Brazilian forest surveys reported in Tardin and da Cunha (1990). In a recent study, 222 LANDSAT TM images, covering virtually all of Brazil's Amazonia Legal, have been interpreted on colour composites of bands 3, 4, and 5 at a scale of 1:250 000. The figures for total deforestation work out at a level of 400 000 km^2 ± 5 percent (August 1989). The average yearly deforestation figure for dominantly tree formations was calculated at 21 218 km^2 ± 10 percent. This compares with the figures, now widely accepted as exaggerations, produced by the World Resources Institute (1990) of 80 000 km^2, or Myers's (1989) figure of 50 000 km^2, both of which were based on AVHRR data for 1987, when there also happened to be a peak of burning (Fearnside et al., 1990).

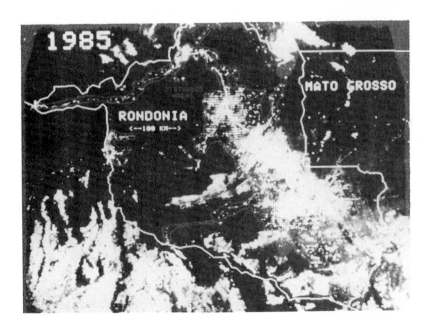

Plate 14.2 AVHRR satellite images show the explosive increase in the area of cleared forest since 1980 in the Brazilian province of Rondonia. Even at this scale it can be seen how clearance follows the main road highways.

Using computer scanning and geographical information systems, it is possible to make more accurate calculations for smaller areas. For example, using LANDSAT TM in northern Roraima, at the edge of the agricultural frontier in a remote part of Brazil, a 2–4 percent loss in the forest cover was seen between 1978 and 1985 and, even more a cause for concern, a 20–25 percent loss in the valuable gallery forests of the *várzeas* or floodplains (Dargie and Furley, 1993). Radar has also been used for forest surveys. The pioneering work of Projeto Radam in Brazil provided black and white mosaics at a scale of 1:250 000 as a basis for natural resource surveys, using SLAR (side looking airborne radar). Most recently a combination of radar and thermal scanners (SAR, synthetic aperture radar) has been able to distinguish subtle vegetational differences and produce hard copy color prints for field checking (TREE, the Tropical Rainforest Ecology Experiment).

Such data are, however, only available for certain of the world's moist forests and some very important areas, for instance Zaire, have a relatively poor quality of information. The overall global data quality is therefore extremely varied. Myers (1989) reported that the tropical forests lost 142 200 km^2 in 1989, an area approximately equivalent to Austria, Belgium, and Switzerland. Assuming the area remaining forested is 8 million km^2 and the rate of loss is about 1.8 percent per annum, then a crude extrapolation gives a terminal life-span of around 38 years. The loss figure calculated for 1979 was 75 000 km^2 and so there has been a 90 percent increase in the yearly loss over a decade. Consequently, although the total *amount* so far destroyed may be one-third of the maximum pre-human intervention level (arguably Latin America and Asia have lost two-fifths and Africa over half), it is the accelerating *rate* of destruction that is causing such anxiety. Whether these figures are accurate to within 5, 10, or even 25 percent is not really of much consequence in view of the apparent inevitability of the forest removal.

The pattern of destruction varies over the tropics. West Africa comes out with the worst loss rates (Ivory Coast and Nigeria) and much of the magnificent forest of West Africa has already been lost. Madagascar has suffered irreparable damage, as have countries which have suffered political and military struggles, such as Thailand and Vietnam (figure 14.1a). Ten countries account for over three-quarters of the total area deforested (figure 14.1b).

While the pattern of destruction is reasonably well known, the present issue concerns the amount of forest remaining. Three countries, one from each of the main tropical regions, account for well over 50 percent of the forest that is left (figure 14.2). Brazil,

Figure 14.2 Distribution of remaining tropical moist forests (after Myers, 1989).

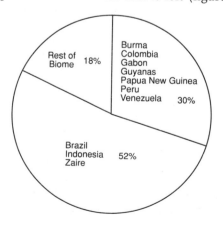

Indonesia, and Zaire have very large areas of intact forest despite intense and growing pressures. Most of the forests in these countries and in the next seven countries, which account for a further 30 percent of the remaining stock, are located in isolated areas. They remain, not so much by design, as by inaccessibility.

Ecological and environmental consequences

In an assessment of global environmental change, the impact of tropical forest destruction is high on the list of priorities demanding action. Although biological considerations tend to dominate the discussion, other environmental arguments are also compelling.

The principal case for restraining disturbance to the forest lies in their biological diversity. Wilson (1988) quotes a global figure of 1.4 million species for all types of living organisms which have been scientifically described. However, the likely figure, based on extrapolations from invertebrates and particularly insects, may be much higher. Erwin (1988) suggests that the species list for tropical moist forests alone may be 30 million and some speculations have lifted this to 50 million. All the other biomes combined may only contain 2.5 million species (Myers, 1988a).

Species richness varies geographically. Costa Rica, for example, is believed to have a biodiversity equal to the whole of West Africa, and Brunei plus Sarawak, which are half the size of Britain, have around 2500 tree species compared to Britain's 35 indigenous species. A 0.8 km^2 patch of the Rio Palenque Biological Station in coastal Ecuador has been shown to have the extraordinary figure of 1033 plant species (Gentry, 1982). High levels of endemism occur in distinct localities or regions, as illustrated in Soulé (1986), including highly threatened areas such as Madagascar, Western Ecuador, and peninsular Malaysia. It has been suggested that within the ten major areas of exceptional biodiversity, covering only 0.2 percent of the Earth's surface and only 3.5 percent of tropical forest, there are 17 000 plant species and at least 350 000 animals, all of which are threatened with extinction within the next decade (HMSO, 1991). The genetic, medical, food, and other resources are considered under the arguments for conservation.

The impact of deforestation on climate has been the subject of considerable discussion. This has been related principally to global warming and local or regional effects on water balance. At one time, it was thought that the burning of tropical forests contributed large amounts of carbon dioxide to the atmosphere and thereby global warming. The release of carbon dixoide, methane, nitrous oxide, and possibly other greenhouse gases, certainly occurs with forest burning but the amounts are not comparable to, for example, vehicle and industrial emissions in the developed world, although they may add about 30 percent of anthropogenic carbon emissions and 1.5–4 percent nitrous oxides (plate 14.3). In total this represents less than 20 percent of all emissions of radiative forcing gases.

Plate 14.3 Large-scale burning of tropical moist forest releases carbon stocks into the atmosphere, but its contribution to the greenhouse effect is smaller than that resulting from industrial emissions in more developed economies.

Deforestation increases the surface albedo by around 10 percent (HMSO, 1991), reduces roughness, and increases runoff. The interactive effects on the atmosphere have been mostly assessed through global circulation models or GCMs (Dickinson and Henderson-Sellers, 1988; Lean and Warrilow, 1989; Houghton, 1990b). Increases in albedo cool the surface but this is counteracted by other processes making the surface more arid, especially during the dry season. The net effect is a possible warming by 2°C. The increased albedo also influences rainfall, with a decrease of 2–4 percent for every 1 percent increase in albedo (Rowntree, 1991). However, there are many variations and many factors affect precipitation. In Amazonia, the westerly moving "cells" of convective rainfall are believed to lose moisture as they travel from the Atlantic inland, and Salati et al. (1986) and Salati (1987) found that about 50 percent of the moisture was generated by evapotranspiration and that, therefore, complete deforestation would reduce the rainfall at the next convective cell by a half. Decreases in rainfall are likely to be accompanied by decreased cloud cover and the greater photosynthetic energy could have a significant impact on plant composition and growth. It should be remembered that climate has been continuously changing through time and that the forest edge has been retreating and advancing (Furley et al., 1992).

The process of forest clearance itself may severely affect the environment and has been said to be "the most crucial step in affecting the future productivity of farming systems" (Cochrane and Sanchez, 1982). Clearance may range from small-scale manual cutting of selected trees to large-scale mechanized removal involving destruction of the soil surface and organic litter (Lal et al., 1986). On the whole, the greater the efficiency of biomass removal the more long term are the problems of recuperation. Manual cutting and burning generate a flush of nutrients from the ash which is depleted in periods from a few months to several years (figure 14.3). During the year after clearance, there are usually large volatization losses of nitrogen and sulfur, soil organic matter decreases, pH increases, and aluminum levels (especially any toxicity) decrease. There may also be a drastic reduction in the microbial population with its valuable nutrient scavenging role. Mechanical clearance, quite apart from its drastic effect on the biotic components, physically affects soil down to 20 cm or more (Nortcliff and Dias, 1988). Compaction leads to increased bulk density and lower aeration, often leading to local impermeability which can reduce seed germination or successful seedling transplantation. In Cameroon, for example, Ngeh et al. (submitted)

Figure 14.3 Effect of cutting and burning on soil properties (after Jordan, 1987). ex = exchangeable; sat = saturated; CEC = cation exchange capacity.

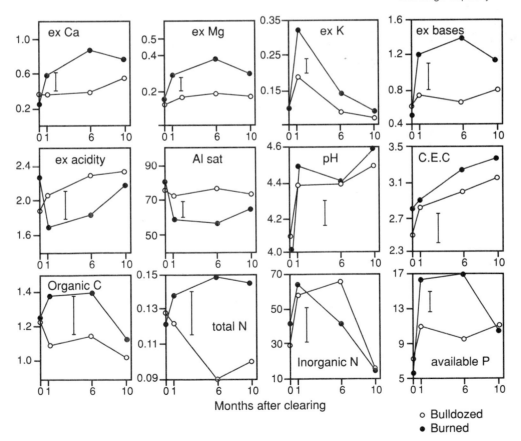

Months after clearing

○ Bulldozed
● Burned

Figure 14.4 Soil compaction following mechanization (after Ngeh et al., submitted).

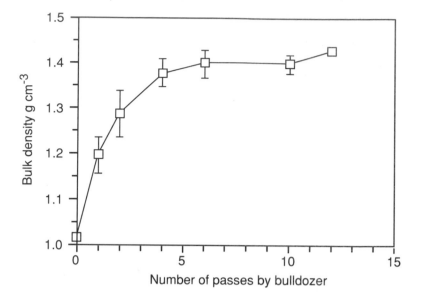

showed that for different management treatments, plots with clear felling and bulldozing had the most damaging soil effect, interplanting methods which left a number of standing trees had less damage, and manual clearance leaving a large number of trees with much of the forest structure intact showed very little damage. Only a few bulldozer passes were sufficient to cause compaction damage (figure 14.4).

The impact of forest clearance on soils can be grouped into sub-aerial and nutrient cycling effects. Since the micro-climate of mature tropical forest is well stratified, with drier, lighter, and hotter conditions in the upper vegetation strata, and frequently saturated humidities, low light levels, and reduced temperatures near the ground, it is obvious that deforestation will have dramatic consequences (figure 14.5). Clearance will stimulate the germination of light-sensitive plants and may destroy more susceptible

Figure 14.5 Micro-climatic differences between gaps and forest.

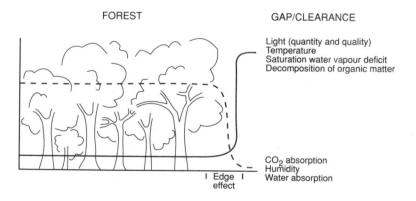

seeds. Surface soils heat up and become desiccated, organic matter is oxidized more swiftly, and the surface is physically buffeted by wind and rain. Infiltration capacities are exceeded, leading to runoff and erosion. Soil erosion losses have been reported many thousand times greater than in neighboring undisturbed forest (Salati and Vose, 1983; see also chapter 15, Douglas). This is accentuated by the construction of access roads (Hamilton and Pearce, 1987). Vulnerability to erosion is more critical in sloping areas. Much of the Amazon Basin, for example, is extremely level and not considered to have much erosion risk (less than 8 percent of the land is sloping; Nicholaides et al., 1984), but the most fertile igneous areas have steep slopes and are highly erodible (Furley, 1990).

Nutrient cycling mechanisms may be completely destroyed or, at best, so severely disturbed that tree recovery is long delayed. Nutrient scavenging by the biological system as a whole takes place above ground as well as at the ground surface and in the topsoil (Jordan, 1985). Within a mature tropical forest, the nutrient conservation mechanisms are finely tuned and extremely effective. Radioactive tracers have indicated that very low quantities of nutrients escape through to ground water (Stark and Jordan, 1978), with the role of mycorrhizae (fungi) being especially important in oligotrophic soils (St John, 1985).

Tree removal also affects the hydrology of an area. Uncontrolled flooding from the removal of critical upslope catchment forests is reported throughout the tropics. Runoff differences of 2 m^3 ha^{-1} $month^{-1}$ in undisturbed forest to 180 m^3 ha^{-1} $month^{-1}$ have been reported (Hamilton and Pearce, 1989). The interference in river flow may also result in a reduction in atmospheric water vapor. The presence of forest in a catchment may be likened to a sponge, absorbing water and releasing it slowly to soil or streams or as vapor to the atmosphere.

Contemporary pressures on tropical forests and sustainability

Although forest disturbance can be traced back to human beginnings, a relationship nowhere more vivid than in Africa, the scale and intensity of current forest conversion has a relatively recent origin. Difficulties of access, the inhospitable living environment for pioneer settlers, and the problems of obtaining a permanent source of livelihood provided barriers to penetration until the 1950s and 1960s. These barriers have only been overcome with the advent of mechanization, chain saws, modern road building techniques, and, above all, political pressures aimed at opening up what was often perceived as late as the early 1970s to be a cornucopia of untapped resources (see, for example, Bourne, 1978; Goodland and Irwin, 1975).

Agricultural developments

Agricultural and particularly pastoral expansion has been frequently cited as the principal factor in tropical deforestation. Under indigenous and smallholder cultivation systems, clearances tend to be small scale, occur over short time periods, and involve tree and shrub as well as ground crops, which often mimic the structure of the forest. Such disturbances supported low densities of population with relatively little perceptible change in the forest except for species composition. With larger, more permanent clearances and the introduction of monocultures, the intensity and destructiveness of agriculture has increased (table 14.2).

Table 14.2 Characteristics of various land utilization types in tropical rain forest areas

Land utilization type	Nature of produce	Type of ecosystem	Long-term prospects
1 Nature conservation	n	o	s
2 Protection of environment	n	o/d	s
3 Recreation and tourism	n	o/d	s
4 Gene pool conservation	e	o	s
5 Gathering and hunting	e	o/m	s
6 Selective commercial exploitation of natural forest	e	m	s/i/t
7 Wood production from managed natural forest	e	m	s
8 Shifting cultivation, aboriginal	e	o/m	s
Shifting cultivation, established	e	d	s
Shifting cultivation, guided	e	d	s
Shifting cultivation, deteriorated	e	d	i
Shifting cultivation, transmigratory	e	d	t
9 Gardening	e	d/m	s
10 Non-wood production from ligneous species in plantations	e	d	s
11 Wood production from plantations	e	d	s/i
12 Pasture	e	d	s/i
13 Permanent cropping (no trees), rain fed	e	d	s/i
14 Permanent cropping (no trees), irrigated	e	d	s
15 Clear cutting for integral harvesting of wood	e	d	t
16 Zero use	nil	n.a.	n.a.

d, derived ecosystem, i.e. artificial system not resembling original ecosystem in species composition and structure;
e, produce by extractive exploitation;
i, production not sustained (decreasing);
m, modified ecosystem, resembling original ecosystems in structure and species composition;
n, produce not tangible;
n.a., not applicable;
nil, no produce;
o, original ecosystem, eventually imperceptibly modified;
s, sustained production;
t, transitory production system.
Source: after Boerboom and Jonkers, 1987

(1) *Shifting cultivation* accounts for between an estimated 45 percent (FAO, 1982) and 60 percent (Myers, 1989) of all deforestation, varying by region from around 70 percent in Africa and nearly 50 percent in tropical Asia to 35 percent in tropical America. Migratory cultivation has been practiced for thousands of years and is sustainable, providing that a sufficiently long fallow period is possible (20–100 years) to allow for the recovery of soil fertility. In contrast, monitoring of logging sites indicates that regeneration of mature forest could take centuries (Wilson, 1988). The stability and sustained production of traditional agricultural methods is partly ascribed to crop densities (Merrick, 1990). During the fallow period, secondary plant succession accumulates nutrients and organic matter and helps to restore soil structure and aeration. The damage to nutrient cycling mechanisms, notably mycorrhizal associations, is also restored gradually. Unfortunately, as in several parts of West Africa, land pressures resulting from rapidly increasing population have reduced the fallow period to under 10 years and this process, if continued over a long period, progressively reduces both biological diversity and soil resources. It has been shown that shifting cultivation, while ecologically and pedologically sound, is a prodigal consumer of land and can only maintain populations at a subsistence level. Families tend to become marginalized in terms of economic, social, political, and institutional status and have little scope for improvement. Without significantly greater and radical attempts from governments and international development agencies, there is unlikely to be a solution to the problem of continuous migratory forest depletion.

(2) *Smallholder agriculture* varies in size of holding and agronomic practice but is essentially the same throughout the tropics. The system is small-scale, multi-crop, family-based, and semipermanent to permanent, producing only small surpluses which are either exchanged or marketed locally. Traditional smallholdings tend to be ecologically diverse, conserving genetic resources (Gliessman, 1990; Oldfield and Alcorn, 1991). Despite this, the smallholder units have been shown to be unsustainable without continuous capital inputs and credit availability, which has not usually been forthcoming. Many smallholders are relatively recent immigrants, often from totally different environments and with little idea of the appropriate skills or access to the resources necessary to succeed.

Smallholder farming dominates the conversion of tropical forest and can be subdivided into spontaneous and traditional or government/private colonization. The first group consists, for example, of the home gardens typical of intensive agricultural occupation and dense rural settlement of parts of southeast Asia (Marten, 1990). The spontaneous type of smallholding is exemplified by the recent land occupation which is occurring in many parts of tropical America, where a relatively impoverished and often ill-adapted small-scale version of "Western" agriculture is implanted, usually against a background of land tenure problems. An estimated 60 percent of all rural Brazilian smallholders are landless, while in other areas of South America, such as Peru and Ecuador, the figure rises to 75 percent (HMSO, 1991). This problem of uneconomic size of holding and lack of stable land tenure is found alongside the phenomenon of very large holdings. In Latin America, 1 percent of the landowners possess something like 40 percent of the arable land. The history of land colonization schemes is littered with failures (see examples in Barbira-Scazzochio, 1979; Moran, 1990; Hemming, 1985; Fearnside, 1987). Governments have frequently been assisted by international agencies in the financing of such developments, for example the World Bank Polonoroueste Project in northwest Brazil. The Indonesian Transmigration Scheme, the largest resettlement scheme in the world, is moving millions of people from the densely populated islands, such as Java and Bali, to the less populated parts of Sumatra, Sulawesi, Irian Jaya (Western New Guinea), and Kalimantan (Borneo). World Bank and Asian Development Bank finance has assisted in this threat to over 3 million hectares of relatively undisturbed forest, which includes, as in other examples throughout the tropical forests, indigenous tribal areas (Secrett, 1986).

(3) *Commercial agriculture and agroforestry* have been neither widespread nor on the whole successful in tropical moist forest. The devastating effect of land clearance, the rapidly impoverished soils, the vulnerability to erosion and rain damage, and the risk of pests and diseases, among a number of other adverse factors, have reduced the area of forest converted to this extreme form. Annual (or short-cycle) crops tend to be less adapted and successful than perennial (or long-cycle) crops. A few specialist short-cycle crops have succeeded, however, notably paddy rice in seasonally flooded or high water-table conditions and particularly in the wet lowlands of tropical Asia. Commercial crops generally require external markets to be grown on a large scale (e.g. cassava in Thailand, soybean in Brazil and sugar cane in the Caribbean). Perennial crops which have been perceived by agronomists to hold promise for development have generally been trees or shrubs (e.g. cacoa, coffee, rubber, and some of the spices such as black pepper or climbers such as guarana).

There are powerful arguments for encouraging perennials, particularly trees, as opposed to short-cycle crops, for biological and environmental reasons as well as food. Agroforestry combines the two, although the emphasis has been on smallholder rather than commercial development until recently (table 14.3). A range of agroforestry techniques has been tried, aiming to increase productivity whilst maintaining conservation. The benefits that are normally perceived are given in table 14.4 (MacDicken and Vergara, 1990; Nair, 1990). The tree canopy provides shade and diminishes desiccation and oxidation at the organic ground surface.

Table 14.3 Agroforestry technologies

1 *Mainly agrosilvicultural (trees with crops)*
  Rotational:
    Shifting cultivation
    Improved tree fallow
    Taungya
  Spatial mixed:
    Trees on cropland
    Plantation crop combinations
    Multistory tree gardens
  Spatial zoned:
    Hedgerow intercropping (barrier, hedges, alley cropping) (also
    agrosilvopastoral)
    Boundary planting
    Trees on erosion-control structures
    Windbreaks and shelterbelts (also silvopastoral)
    Biomass transfer

2 *Mainly or partly silvopastoral (trees with pastures and livestock)*
  Spatial mixed:
    Trees on rangeland or pastures
    Plantation crops with pastures
  Spatial zoned:
    Living fences
    Fodder banks

3 *Tree component predominant (see also taungya)*
  Woodlots with multipurpose management
  Reclamation forestry leading to multiple use

4 *Other components present*
  Entomoforestry (trees with insects)
  Aquaforestry (trees with fisheries)

Source: after Young, 1989

Table 14.4 Assumed benefits of agroforestry practice

1 Control of soil and sediment erosion
2 Maintenance of soil organic matter and soil fertility
3 Maintenance of favorable soil physical properties
4 Encourages an effective rooting system
5 Nitrogen fertilization through growth of N-fixing trees and shrubs
6 Nutrient trapping and accumulation from atmospheric and lower soil
   horizon sources
7 Encouragement of "closed" nutrient cycling
8 Mitigates surface acidity and helps to neutralize subsoil aluminum toxity
9 Augments soil water availability by reducing wind desiccation, absorbing
   atmospheric and ground water and maintaining surface and topsoil
   humidities
10 Useful component of soil rehabilitation

Source: after Young, 1989

The direct effect of precipitation is also minimized, although removal of undercanopy trees has been shown to increase the size and impact of droplets (Brand, 1988). Nutrient losses are lessened and nutrient recycling mechanisms can be preserved (Ingram,

1990). The value of the crop frequently justifies the cost of fertilizer and other capital requirements. However, one of the problems of developing commercial stands is the increased susceptibility to pests and diseases. In the wild, the trees tend to be scattered throughout the forest as individuals, perhaps an evolutionary trait resulting from selective pressure. The better performance of many species when they are transported outside their place of origin is well known. An example is rubber, which has a history of plantation failure in Brazil, notably the Ford Motor Company plantation along the Amazon Tapajos River. Plantations form an extreme case. In some instances, the complex ecological relationships with pollinators or dispersing agents make it very difficult to cultivate trees in this way. An example is the Brazil nut, *Bertholletia excelsa*, where bee pollinators are required whose life cycle demands the presence of a specific orchid only found in undisturbed forest conditions.

Ranching

Although livestock rearing is commonplace throughout the humid tropics, the large-scale clearance of land for cattle rearing is a problem of particular importance in Latin America (plate 14.4). It has been estimated that between 75 (Repetto and Gillis, 1988) and 85 percent (Hecht and Cockburn, 1990) of forest conversion in the Brazilian Legal Amazon can be attributed to the expansion of cattle pastures. Most of this can be linked directly to government policies in enlarging the road network and in the provision of credits and fiscal incentives, such as tax remissions. Much of the disturbance has been a result of land speculation and attempts by wealthy investors to obtain credits and tax relief. International demand for beef has led to an expansion of livestock elsewhere in Latin America. Mahar (1979) reported that pasture land prices in northern Mato Grosso rose 38 percent annually through the early part of the 1970s, even allowing for inflation. In Guatemala, 2.2 percent of the population owns 70 percent of the agricultural land, most of which is used for ranching (Myers, 1985).

Ranching makes a major impact on the environment. Primarily, it provides an exceedingly poor replacement system, converting one of the highest productivities and most biologically diverse eco-systems into what is frequently monospecific grassland. Ranchland often carries no more than one animal per hectare and it has been argued that the cost of raising cattle rarely meets the selling price. Landowners are more than compensated by artificial credits and increased land prices. Second, it has been shown that pastures deteriorate rapidly. They dry out quickly, suffer soil compaction, and lose fertility, especially available phosphorus (Hecht 1981; Hecht et al., 1988), sustaining little more than 5–10 years of production without expensive and probably uneconomic capital input. After about 7 years, the already low carrying capacity may drop to as little as 0.3 head per hectare (Sternberg, 1975). Over 20

Plate 14.4 Livestock on cleared forest, Maraba, Brazil, with smoke rising in the background from new burning. The resulting grazing is often poor in quality, and will typically support only one or two head of cattle per hectare. Note the presence of termite mounds, which invade the deforested land.

percent of all established pastures from forested land in the Brazilian Amazon have been abandoned (Serrão and Toledo, 1988). The end result is a weed-infested scrubby secondary growth which may persist for many years where there are poor soils and a pronounced dry season.

Forestry and tree plantations

Logging is only one aspect of disturbance to the natural forest though it frequently attracts the most publicity. Its destructive capacity is undoubted; something like 40 percent of the forest may be cleared for logging trails and access alone (Nicholson, quoted in HMSO, 1991). Forest management is less developed in the Neo-tropical forests than in tropical Asia and Africa. Brazil, for example, has been a net importer of forest products (Rankin, 1985). The recent growth in multi-species and comprehensive tree use, including pulp, wood chips for particle board, and charcoal, has greatly increased the productivity but also the destructiveness of logging operations. Most commercial forestry utilizes only a small number of the most profitable species. Further, logging has often been highly selective and concentrated in areas of easy access, such as seasonally flooded riverine forests and creeks or along newly opened roads. It is reported that most of the Brazilian Amazon wood production is still taken from inundation forests, whereas 95 percent of the tropical forest is in areas of well drained forest (*terra firme*) and relatively unexploited (Myers, 1980). Relatively little effort has been put into regeneration or replanting until the past two decades, although forest conservation practices have been developed over a long period in tropical Asia and in parts of Africa (Gomez-Pompa et al., 1991) (table 14.5). In all, commercial operations are believed to affect around 5 million hectares annually (World Resources Institute, 1985).

Some 20 years ago, Persson (1974) estimated that only 3 percent of the world's tropical forests were being managed on a sustainable yield basis, and it is arguable that the figure has been little improved since. The WRI report (1985) calculated that less than 600 000 ha of tropical timber is planted every year, which nowhere near meets losses (Myers, 1989). The problems of obtaining sustained yield while retaining species diversity are immense, including slow growth rates and therefore long rotation periods, low stocking rates

Table 14.5 Silvicultural systems of the moist tropics

---

1 *Polycyclic*: repeated removal of selected trees in a continuing series of felling cycles, whose length is less than the rotation age of the trees. Forms small gaps. Often relies on natural regeneration

2 *Monocyclic*: removal of all saleable trees in one operation. Cycle length depends on the rotation age of the selected trees. Forms large gaps. Requires replanting
 (a) Shelterwood system: the establishment of young trees under the shelter of older ones before felling. Reduces incidence of gaps. Planting in lines, bands, or blocks
 (b) Malayan uniform system: extracts all marketable trees. Forms large gaps. Depends on natural regeneration from abundant seed years before felling

3 *Combined*: formal planting of selected species as interplantings in a polycyclic management operation. Reduced gaps. Combines natural regeneration and formal planting

4 *Agroforestry*: canopy retention of large trees and/or subcanopy trees. Interplanting of selected trees, shrubs, or short cycle crops. Wide range of combinations. Tries to avoid ground exposure

5 *Conservation silviculture*: Minimum forestry operation to permit natural evolution of the forest and habitats

---

of desired species and therefore commercial volume per unit area, variable log quality, the need for detailed soil, micro-climate, hydrological, and ecological data, skilled labor and infrastructural requirements – all adding up to complex management demands.

One of the difficulties of developing forestry either commercially or sustainably has been the absence of detailed long-term information on species performance or on alternative strategies for management. A review by FAO in 1986 (supported by Poore, 1989) claimed that there were virtually no successful forest management schemes, with the possible exception of the Malayan Uniform System (Schmidt, 1987). On the whole it is believed that the technical methods and ecological understanding are sufficient for sustained yield management but that there is a lack of government finance and priority.

Plantations have often been suggested as the most efficient management solution for using tropical forests. They undoubtedly demonstrate high productivity of desired species, which typically have higher growth rates and shorter growth cycles than native forest species. There are, however, several drawbacks to plantation forestry. They suffer from the same susceptibilities to pests and diseases as monocultural arable crops, they tend to reduce soil fertility and nutrient cycling mechanisms, and they may cause overall environmental degradation through adverse effects on hydrological processes and local climate. Inevitably, too, they reduce the species diversity of the forest. The Jari Scheme established in 1968 in the eastern Amazon is a classic example. The principal aim of the project over a 1.6 million hectare holding was industrial tree plantation for pulp, although there are ancillary lumber, wood fuel, livestock, rice, and mining developments. Despite an immense capital investment, it has been a very mixed commercial success and the long-term viability is uncertain (Fearnside, 1988).

The arguments for conservation

Arguments for conservation range from a philosophical stance to extreme pragmatism. At one end of the spectrum, there are moral arguments emphasizing our responsibilities to protect the diversity of life on the planet, passing on to future generations a full complement of the physical, chemical, and biological legacies that we have inherited. The aesthetic position is similar, underlining the beauty and richness of tropical forests and highlighting an obligation to act as guardians. The notion of responsibility is also inherent in one aspect of the treatment of indigenous forest inhabitants. Quite apart from respecting their right to their own social and cultural systems, there has been a growing fascination in understanding the way in which indigenous societies have adapted to their environment over millennia and the realization that there is much to learn about handling forest resources (Posey and Balee,

1988). Immigrant groups who have lived in the forest for long periods and who have learnt from the older indigenous peoples have a similar position, for example the Brazilian caboclos. Tropical forests have also had a long history of occupation, which is frequently overlooked or disregarded, such as the Maya in Central America, the Niger Empires of West Africa, or the Kampuchean civilization exemplified by the splendours of Angkor Vat.

These arguments are of little avail in hard-nosed economics but fortunately a very good case can be made for conservation on practical grounds. This has been the approach of a number of economists (such as Pearce et al., 1990), although there still seems to be a short-term goal evident in much economic thinking, just as there is in most political decision-making. Biological diversity is unlikely to be conserved by market forces alone (McNeely, 1988). Myers (1985) has given an excellent review of the dependence of life upon tropical products. Food and drink are two of the more obvious consumables: coffee, tea, chocolate, many citrus drinks along with some 250 fruits, nuts, cereals such as corn and rice, pulses such as peanuts and numerous varieties of beans, root crops, and tubers such as yams, cassava, and potato, many of which have yet to appear in world trade. In New Guinea alone, over 250 tree species have fruit consumed by local people but only about 40 have become established as cultivated crops and only 12 have reached the marketplace (Jong, 1979). The humid tropics are not well known for their vegetable products, yet over 1500 plants produce vegetable-like edible materials.

Figure 14.6 World map of endangered communities within tropical moist forest.

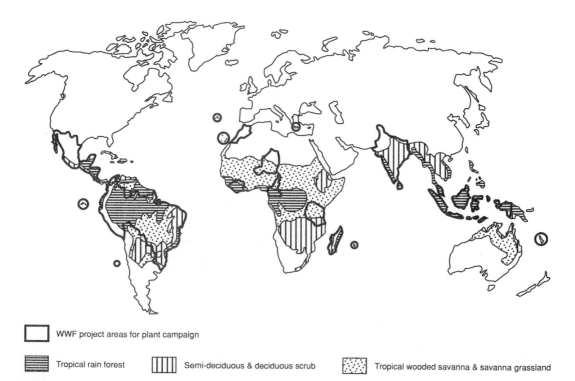

WWF project areas for plant campaign

Tropical rain forest   Semi-deciduous & deciduous scrub   Tropical wooded savanna & savanna grassland

Many of the chemicals used in pesticides and medicines are also derived from tropical species. A number of plants have adapted to insect predation by developing chemical defensive and inhibiting substances. Of the toxic compounds, the rotenoids (from roots and seeds) are found in many plants from Amazonia to southeast Asia. Considerable progress has also been made in biological defense compounds. Herbivorous insects may be used to attack population explosions of unwanted plants and there is potential in introducing

Table 14.6   Criteria for consideration when selecting protected areas

1   *Size*: The conservation value of an area is a function of its size. In principle, the area must be of a size and form sufficient to support entire ecological units or viable populations of flora and fauna.

2   *Richness and diversity*: Richness and diversity of species is usually linked with diversity of habitat. Ecological gradients (catenas, ecotones, altitudinal transition zones) should be represented because of the important transitional communities they support.

3   *Naturalness*: Few places on Earth have not been modified through the influence of man. Areas where this influence is minimal or which have the potential for restoration are particularly valuable.

4   *Rarity*: One of the most important purposes of many national parks and conservation areas is to protect rare or endangered species and communities. Rarity of a species may be related to extremely specialized habitat requirements or to direct or indirect human influences.

5   *Uniqueness*: An area may be unique because the biome it represents is not adequately represented in the national system or because it exhibits particular natural processes.

6   *Typicalness*: In addition to unusual features, it is important to represent typical areas of common habitats and communities of the biogeographic unit.

7   *Fragility*: Fragility is often linked with rarity. Fragile habitats, species and communities have a high sensitivity to environmental changes.

8   *Genetic conservation*: Richness and diversity usually reflect genetic diversity but there may be other genetic considerations that justify protection, such as the occurrence of wild forms of domesticated plants and animals.

9   *Historical records*: Where an area has been studied and monitored for a long period of time, its research value is greater than that of a comparable unmonitored area.

10 *Position in an ecological/geographical unit*: It is better to include within a single large area as many as possible of the important and characteristic formations, communities, and species, rather than having smaller areas representing a single formation or community.

11 *Indispensability*: An area may deserve protection because it represents an entity too valuable to lose, such as a seasonal habitat component of a migratory species.

12 *Potential value*: Areas once known to be of exceptional quality but which have been recently damaged could, with appropriate management and protection, regain the former quality.

13 *Intrinsic appeal*: The area should provide opportunities for recreation, with features having appeal to humans. Conservation is more realistic if it gives more weight to some groups and organizations than others.

14 *Modified landscapes augmenting biological values*: National or cultural sites or particular forms of land use which have a significant influence on the region's biogeography may require protection.

15 *Opportunities for conservation*: The socio-political climate is extremely significant in determining conservation priorities. Lack of support or resentment results in limited conservation success and/or failure, despite the richness or value of the wildland.

Source: after Mackinnon et al., 1986

Alternative strategies shaping future change

Change is continuous in tropical forests. Total preservation of species is impossible as variations in biological and environmental processes occur continuously, including extinctions (Simberloff, 1984; Lugo, 1988). Although understanding of processes at micro, local, regional, and global scales has improved enormously, there are still too many uncertainties to predict how a particular area may change with any confidence. If the argument is accepted that as much of the biologically richest forest should be conserved as possible, then there are several important questions to address.

1   How may these areas be identified? The pace of forest destruction makes the initial selection of protected areas an imperative concern and various criteria have been put forward to help in their determination (tables 14.6 and 14.7). It is already known in very broad terms where the greatest areas of endemism and species diversity are to be found (see, for example, Gentry, 1982). However, until detailed inventories are available together with species ranges and a knowledge of their dynamic change, predictions are largely guesswork. One approach is based on the idea that the greatest diversities occur in the refuges to which tropical forest is supposed to have retreated during glacial maxima (Whitmore and Prance, 1987). A different approach is to explain diversity on the basis of normal evolutionary trends with, for instance, edaphic specialization (Gentry, 1986).

2   What priorities can be given to land outside protected zones but within the naturally forested area? Many scientists argue for a buffer zone around protected areas, shielding them from outside disturbance. The cooperation of local people is considered essential, allowing them to use the forest in return for guarding it (McKinnon et al., 1986). Beyond the protected and buffer zones, alternative strategies have been proposed for using the forest resources with as little disturbance as possible. This may require revaluation of standing forest (see Peters et al., 1989). Economic incentives are considered the best inducement for effective conservation along with full repayment of all social and economic costs. A progressive list of options has been outlined by Goodland (table 14.8). The use of perennial trees and shrubs and agroforestry is considered a more sustainable form of land use. It may also involve a critical examination of the optimum carrying capacity (Fearnside, 1985). At the bottom of most ecologists' lists would appear arable monocultures or grazing on virtually monospecific grasslands.

3   How large an area should be retained? Rational use of tropical forests involves combining conservation with buffer zones and multipurpose land uses with the minimum possible disturbance. To ensure conservation of as great a representative series of protected areas as possible, a number of techniques have been tried. One of the best known methods of assessing the minimum critical area required has been developed in joint Worldwide Fund for

Table 14.8 Preferred land use options for moist tropical forest

1 *Intact forest*
   Scientific repository: gene pool, germ plasm storehouse, phytochemistry

   Environmental protection services; climatic buffer, watershed protection

   Indigenous reservation: natural, legal and moral issues; indigenous knowledge of the plants and animals and their uses

   Collecting, gathering, tapping, game and fish culling

   National parks; tourism, national and international; recreation

2 *Utilization of natural forest*
   Sustained yield management

   Leaf protein, leaf and other chemicals

   Selective felling with careful removal

   Bole removal with roots, stump, bark and branches left on site

   Enrichment planting; reafforestation and regeneration

   Clear cutting small tracts leaving regeneration foci in strips or surrounds

3 *Tree plantation*
   Mixed species polyculture rather than monoculture products

   Mixed species polyculture timber

   Monoculture timber and products

4 *Agrosilviculture*
   Multiple dimension forestry (timber, wood products, understory components)

   Polycropping and intercropping

   Tree plantations and short-cycle crops grown simultaneously

   Trees with pasture (multi-species grazing: legumes forbs and grasses)

   Subsistence rotational gardens (trees, perennial and short-cycle crops with grazing, fishponds and other varied inputs)

5 *Agriculture and pasture*
   Small areas, long fallows, multi-species cropping, bred tolerances for pests and infertile soils, rotations

   Perennial crops preferable to short-cycle crops

   Subsistence crops preferable to export and cash crops

   Oliogotrophic crops for export out of the area: hydrocarbons and carbohydrates preferred to nutrient-demanding crops

   Multi-species pasture for mixed herbivores

   Oligotrophic pasture for monospecific herbivores (the worst case would be non-indigenous cattle for export)

The ranking runs from the better (1) and gradually descends through all five categories to the least desirable (5). Examples at the top of sustained-yield (over 50 years) are preferred on environmental grounds to the non-renewable, degradative examples at the bottom. Within each of the five categories a ranking also has been attempted, so that the environmentally preferred alternatives precede the less desirable. Overlapping occurs both vertically and horizontally.
Source: based on Goodland, 1979

Nature and INPA (Brazilian Amazon Research Institute) research, with replicated patches of forest from 1 to 10 000 ha in area. Monitoring has proceeded for some years but is as yet insufficient to make firm recommendations for anything but the smallest sizes (Lovejoy et al., 1986). Even Barro Colorado Island in Panama, long studied by North American scientists, has been losing species and a number of national protected areas throughout the tropics are really too small to survive. Many scientists have utilized island biogeography theories to assess the size of potentially protected areas, using species : area relationships (Diamond, 1975). A rule of thumb is that a ten-fold increase in area permits a doubling of the number of species (MacArthur and Wilson, 1967; Zimmerman and Bierregaard, 1986). However, the application of island biogeography models depends upon well established examples, which are rare. Much of the emphasis is rightly placed on the larger vertebrates and predators, since they tend (a) to have public appeal, (b) to be determined by plant habitats, and (c) to require a large area and therefore are likely to encompass regional variations. This has been worked out for Indonesian New Guinea by Diamond (1986).

At present, only around 5 percent of tropical moist forests are protected in parks and reserves, 4 percent in Africa, 2 percent in Latin America and 6 percent in Asia (Brown, 1985). Of the 1420 protected areas identified in 1986 by Mackinnon and his colleagues, 40 million hectares (out of a total of 174 million) relate to moist forest. A more useful target might be 10 percent but even this is considered to be too small to protect some habitats, and figures of 20 percent for rain forests and 10 percent for humid savannas have been suggested (Myers, 1979). We depend on fewer than 1 percent of living species for our existence and seem to know remarkably little of the remainder. In nature, such dependence upon a limited resource base would place the future of a species in jeopardy. Unfortunately, the successful management of the moist tropics demands a long-term strategy while planners, like politicians, are faced by short-term pressures with a temptation to forgo the investment needed for the future. In the end, the loss of habitat equates directly with expansion of human population. The chilling conclusion is that without a greater degree of success in protecting tropical moist forest and reducing population pressure, we are dealing a murderous blow both to our environment and to ourselves.

Further reading

Eden, M.J. (1990) *Ecology and Land Management in Amazonia*. London: Belhaven Press.
Cleary, D. (1991) *The Brazilian Rainforest: Politics, Finance, Mining and the Environment*. London: The Economist Intelligence Unit, Special Report 2100.

Furley, P.A., Proctor, J. and Ratter, J.A. (eds) (1992) *The Nature and Dynamics of Forest–Savanna Boundaries*. London: Chapman & Hall.

Goodman, D. and Hall, A. (eds) (1990) *The Future of Amazonia: Destruction or Sustainable Development?* London: Macmillan.

Jordan, C.F. (ed.) (1987) *Amazonian Rain Forests: Ecosystem Disturbance and Recovery*. New York: Springer Verlag.

CHAPTER FIFTEEN

Land Degradation in the Humid Tropics

Ian Douglas

The problem of land degradation in tropical forest regions is clearly stated by the South Commission (1990):

> The soil covered by natural forests is usually fragile, and its use for crop or livestock farming soon degrades it even making it unsuitable for the replanting of trees. The destruction of the natural plant cover quickly results in severe erosion and in water run-offs which damage the natural water-regimes. In addition the clearing of forest on steep watersheds heightens the risk and volume of floods and landslides.

For the seasonally wet and semi-arid tropics of Africa, Sir Harold Tempany (1949) noted:

> It is at least probable that in Africa defective methods of livestock management, including overstocking, overgrazing, uncontrolled burning of pastures and bush, over-concentration of cattle at water sources, particularly during dry periods, and the like, are in the aggregate responsible for greater damage by erosion than defective cultivation, great though the effect of the latter may be.

In addition to diminution of productive capacity due to erosion, land degradation implies a variety of processes, including compaction, salinization, laterization, depletion of soil organic matter and plant nutrient reserves, acidification, and development of specific nutrient deficiencies. Degradation is the combination of effects of these processes and it occurs both during forest removal processes and during subsequent cropping, grazing and harvesting (plate 15.1).

Degradation is also a function of fire frequency, which not only depends on people's actions but may also be part of a natural

Plate 15.1 Rill erosion in northern Sulawesi, on land newly cleared as part of Indonesia's transmigration program. Although erosion such as this is the most visible sign of land degradation, *in situ* changes such as soil nutrient depletion are no less important.

sequence of events in tropical rainforests, particularly close to their climatic limits (see chapter 13, Stott). The association of fire with drought in areas affected by El Niño–Southern Oscillation (ENSO) events has been clearly demonstrated in Borneo (Malingreau et al., 1985). On the other hand, fire in eastern Borneo in 1983, and again in 1987 and 1991, was more prevalent in forest that had already been selectively logged than in undisturbed forest. Whatever the cause, all biomass burning leads to emissions to the atmosphere (Houghton et al., 1987; Robinson, 1989) and release of nutrients to the soil. The risk is that some of the latter will be mobilized by storm runoff and carried out of the area, so leading to degradation of the nutrient stock.

Much variation in the natural vegetation of the tropics stems from contrasts in soils types. Soil is one of four factors, along with climate, soil-water regime, and elevation, which broadly explain differences in tropical rain forest formations (Whitmore, 1990). Many tropical soils are poor in nutrients because they have been subjected to long periods of leaching on old land surfaces, while others are young and thin because they are on actively evolving landforms affected by earthquakes and mass movements (Douglas and Spencer, 1985). These varied environments within the tropics exhibit great variations in natural stability and erosion rates (table 15.1). Specific parent materials, such as detrital sand formations,

Table 15.1   Plate tectonic style, landscapes, and natural rates of catchment erosion in the humid tropics

Tectonic setting	Major channel characteristics	Dominant sediment sources	Estimated erosion rate (range of annual sediment yield) ($t\ km^{-2}\ year^{-1}$)	General depth of soil or regolith on slopes	Example	Information source
Active plate margin rift, or half graben edge	Braided channels, abundant gravel	Mass movement often triggered by earthquake	Up to 10000	Thin	Markham River, Papua-New Guinea	Löffler (1977)
Active volcanic areas of recent lava flows	Braided channels	Volcanic debris, unstable ash deposits	Up to 10000	Skeletal	Toto Amarillo, Costa Rica	Kesel (1985)
Tectonically active mountain areas	Deep gorge sections where river usually occupies whole valley floor in flood. Some braiding, but lateral movement of river restricted	Valley wall failure, land sliding associated with seismic activity	7000–10000	Thin	Upper Fly River, Papua-New Guinea	Pickup (1984)
Late Tertiary tectonic activity and weak mud rocks	Wide channels with gravel bars, frequent undercut banks and grassed flood deposits	Bank erosion in main channel and tributaries. Erosion by saturated overland flow in streamhead hollows	About 1000	Around 1 m	Segama River, Sabah, Borneo	Douglas and Spencer, unpublished
Passive margin with relief of 2000 m in equatorial climate	In upland, boulder strewn channels with mature trees right up to water's edge. Abundant quartzs and between boulders	Comminution of boulders. Surface wash on slopes	50–100	Up to 30 m	Combak, Malaysia	Douglas (1968)
Passive margin with relief of 2000 m in tropical cyclone zone	In upland, boulder strewn or rock-cut channels, with areas of exposed rock or boulders at low flows. Channel capacities much greater than previous case	Disintegration of boulders, surface wash on slopes, bank erosion	50–200	Up to 30 m	Babinda and Behana Creeks, North Queensland	Douglas (1967)
Ancient craton with relief of 2000 m	In upland, channels guided by ancient structural lineaments, resistant angular blocks and some derived gravel. Monsoona¹ climates produce large broad lowland valleys	Little sediment supply except through river action on racks on channel wall	Up to 50	Up to 30 m	Mahawali Ganya system, Sri Lanka	Bremer (1981)
Ancient craton with erosion surface	Wide sandy channels. Little incision. Occasional rock bars which suffer little erosion	Wash from etchplain surfaces	Up to 30	Up to 30 m	Zaire, Africa	Büdel (1965, 1981)
Sedimentary basin on ancient craton	Wide anastomosing channels, often with legacies of past phases of fluvial erosion	Little locally derived sediment, except from bank erosion	Up to 30	Up to 30 m	Amazon basin	Garner (1974), Baker (1978)

limestones, and ultrabasic rocks, weather into soils with special geochemical characteristics, which in turn support distinctive vegetation formations. The sand formations are usually characterized by tropical podzols which are poor in nutrients and whose drainage waters tend to be acid, black waters with a pH as low as 3.

In the seasonally wet tropics, two important types of soil are widespread, the ultisols (latosols) and the vertisols (black or regur soils). The latosols are characteristic of the old land surfaces of the savanna landscapes of eastern South America, Africa, and parts of India, and often give way to areas of ferricrete, iron-rich indurated layers, also known as laterite. Such soils tend to be low in plant nutrients and to lose those that are available quickly when disturbed. The tropical vertisols, long known as black cracking clays, have a variety of parent materials but are often associated with basic igneous rocks. They retain water and swell in the wet season, but dry out and crack in the dry season. Many such soils in southern Africa and India have a low humus content as a result of continuous cultivation and erosion, as increased numbers of people and cattle have led to a change from shifting cultivation to continuous use (Vine, 1966). The wide variety of landscapes, water regimes, and soils in the tropics must be kept constantly in mind when discussing land degradation.

Shifting cultivation and land use

Shifting cultivation is a rotational form of land use. Forests are cleared and crops are grown and harvested for one or more years. The initial benefit of clearance and biomass burning is a build-up of nutrients in the soil, the results of a transfer from the plant stock to the upper parts of the soil profile. When forest in Western Samoa was cut by hand and the trunks and larger branches were piled up and burnt, much debris was left to rot and decompose, adding to the soil organic matter, encouraging weed growth which protected the soil against raindrop splash. Four months after hand clearance, the top 30 cm of the soil had higher nutrient levels than the undisturbed forest (table 15.2). However, these soil fertility gains from biomass burning are gradually reduced with time due to organic matter decomposition, nutrient uptake by plants, and nutrient leaching. Inevitably some nutrients are removed from the system as crops are harvested and others are lost in drainage waters. The rate at which this occurs varies with the soil types and the way it is used. In sandy ultisols this takes about 35 months under traditional cultivation (table 15.3). Build-up of aluminum in the soil poses special soil management problems here.

Following clearance and cultivation, the land is abandoned and left fallow for several years, before the cycle is begun again. The length of the fallow is an important regulator of the degree to which land used by this slash-and-burn, or swidden, type of agriculture becomes decreasingly rich in nutrients and so degraded. On the slopes of Mt Apo, southwest of Davao, Mindanao, Philippines, nitrogen losses under primary forest conditions were only $0.12 \text{ g m}^{-2} \text{ year}^{-1}$. In a hypothetical 17-year shifting cultivation cycle at this location, net annual loss of mineral matter, over and

Table 15.2   Change in soil properties at Togitogiga, southern Upolu, Western Samoa following hand clearance of forest and biomass burning

	Depth (cm)	Organic carbon (%)	N (%)	Cation exchange capacity (m.e. %)	P retention (%)	Ca (m.e.%)	Mg (m.e.%)	K (m.e.%)
1 Uncleared, undisturbed forest	0–7.6	3.5	0.65	24.6	82	2.3	1.22	0.41
	7.6–15.2	1.8	0.34	24.3	85	0.3	0.23	0.13
	15.2–30.5	1.2	0.21	17.8	87	0.7	0.20	0.10
2 Hand-cleared forest	0–7.6	7.6	1.01	39.6	94	6.9	3.32	0.23
	7.6–15.2	5.8	0.81	44.5	94	4.4	2.22	0.13
	15.2–30.5	4.2	0.52	34.1	97	2.1	1.40	0.11

Source: after Reynolds. 1990

Table 15.3 Change in nutrient status of the upper 15 cm of a low-input cropping system following burning on an ultisol at Yurimaguas, Peru

Months after burning	Exchangeable (cmol dm^{-3})				Al saturation (%)	Available P (mg dm^{-3})
	Al	Ca	Mg	K		
3	1.1	0.30	0.09	0.13	68	20
11	1.5	0.92	0.28	0.19	51	13
35	1.7	1.00	0.23	0.10	53	3

Source: after Villachia et al., 1990

above that which would occur in natural forest, would be 2.41 g m^{-2} year^{-1} (Kellman, 1969). Nye and Greenland (1960) suggested that, except for potassium, no serious decrease in available cations is likely to occur during a single cropping period of one to three years, but that deficiencies may well arise with more prolonged cropping, or over successive cropping periods with intermittent short fallows.

Shifting cultivation is widely practiced in tropical forest regions. Some of the land cleared by such activity is subsequently converted to permanent cropland or pasture, but much, often most, is left to revert to forest. The latter, fallowed land is that which is potentially degrading, and may be regarded as the portion affected by true shifting cultivation.

True shifting cultivation accounted for 1.8×10^6 ha, or 32 percent of tropical American deforestation in 1980 (Lanly, 1982, 1985). Between 1500 and 1800, the European contact affected tropical indigenous people adversely and secondary forests regenerated on the lands they previously cultivated (Myers and Tucker, 1987). Not until about 1940 did social conditions begin to promote a rise in shifting cultivators and much more forest clearance at rates of 1.5×10^6 ha year^{-1} for closed forests and 0.3×10^6 ha year^{-1} for open forests (Lanly, 1982). From these observations, the area of degraded, fallow land in Latin America in 1980 was estimated to be between 21 and 74×10^6 ha, according to which assumptions about pre-1940 change are adopted (Houghton et al., 1991).

Between 1980 and 1985, shifting cultivation is estimated to have accounted for 49 percent of the deforestation in tropical Asia, using some 70×10^6 ha of land by 1985 (Houghton, 1991). Calculations of how much forest in tropical Asia is degraded suggest some $15-22 \times 10^6$ ha of tropical rainforest, and a total of $81-118 \times 10^6$ ha of all moist, seasonal and dry forests, about one-third of the total forest area of 326×10^6 ha in 1985. For Africa, less precision is available, but an estimate of 84×10^6 ha of degraded forest is given by Houghton (1990a). In these calculations, degradation is an expression of loss of woody biomass from the forest, or a reduction in carbon stocks. The figures given thus represent the maximum area likely to be affected by soil degradation. However, other nutrients may have been lost from the soils and potential productivity may have been reduced.

In both Latin America and Asia, shifting cultivation is intensifying. In Meghalaya, northeast India, a 119 percent increase in population since 1950 has forced shifting cultivators progressively to shorten the length of the fallow period between cycles of clearing and burning from 30 years to as little as 5 years (Toky and Ramakrishnan, 1983). The short cycles cause exotic herbaceous weeds such as *Eupatorium adenophorum* and *Imperata cylindrica* to replace the forest permanently. In the hundred years after 1880, 56 percent of the forest–woodland vegetation of Meghalaya was lost, at an increasing annual rate, with most of the deforested land being degraded to the grass–shrub complex (Flint and Richards, 1991).

Similar changes were noted in parts of the islands of Tonga where population densities increased by almost 100 percent to around 2 per hectare in the period 1931–66 (Maude, 1970). In the Ha'apai Islands many farmers reduced the length of the fallow period from four or five years to two or three years. The fallowed land was also used for scattered cassava crops among the grasses and other herbaceous plants. The shorter fallow only allowed shrubs such as lantana (*Lantana camara*) and guava (*Psidium guajava*) to grow, instead of the trees which became established after four years of fallow. Fortunately the well-drained, fertile, volcanic soils of Tonga are able to withstand this intensity of use much more than those of many other parts of the tropics.

Not every country has a high density of population, many with shifting cultivators still having relatively low densities (India 2.8 per hectare, Thailand 1.1, Indonesia 1.0, Malaysia 0.5, Brazil 0.17, Papua New Guinea 0.08, for example). Therefore, in some countries the degradation of land as a result of shifting cultivation may be negligible. In Brazil, the degradation is at the margins of the Amazonian forest, rather than in the core of the forest. The same applies in Borneo, although the pressures of in-migration are aggravating the situation in Kalimantan (Potter, 1987). Some areas of slash-and-burn activity in tropical forests thus still conform to the traditional view that shifting cultivation practices leave tropical forests in ecological balance and do not irreversibly degrade the soil (Redclift, 1990). The alarming change in degradation by shifting cultivation is that it is no longer driven solely by self-sufficiency needs of growing rural populations, but is under pressure from local and international markets.

Where highways between major cities and country towns cut through forest areas, people often exploit the easy access by truck to supply urban markets. Along many highways, such as those crossing the main range of the Malay Peninsula east of Kuala Lumpur or the Crocker range in Sabah east of Kota Kinabalu, patches of land on steep slopes have been cleared for bananas and other cash crops. Slash-and-burn techniques are used, but the slopes are so steep, often in excess of 25°, that mass movement and severe erosion may occur. Land degradation through this semi-commercial shifting cultivation is accelerating around many cities.

Elsewhere the vagaries of international market prices exert more subtle pressures on shifting cultivators and their impact on land. In Guatemala, small-scale coffee growers suffered from a fall in international coffee prices and an increase in world oil prices in 1973. Fertilizer costs thus rose as farm incomes declined. Unable to sustain income from coffee at a high enough level to buy food, farmers went uphill into the forest and began clearing patches of land for temporary cultivation of vegetables. They substituted the nutrients available by slash-and-burn agriculture for nutrient inputs from fertilizers. Commodity price fluctuations of this type frequently induce an acceleration of land degradation in this way.

One of the key issues about the biogeochemical sustainability of shifting cultivation is the rate at which the stock of nutrients in secondary regrowth vegetation reaches a level approaching that in the natural forest. Good management of the forest minimizes nutrient loss. Traditional forest societies were long thought to be in such harmony with their environment that their practices returned the nutrient stock rapidly, but that view may be questioned.

Indigenous groups and land degradation

The oft-quoted belief that traditional societies always live in equilibrium with their environment may be wrong. The attitudes of different people to the tropical forest were studied as part of the UN Man and the Biosphere Program (Lugo et al., 1987). A major conclusion was that:

> Indigenous practices can show us better management methods. Isolated groups in Peru, Zaire and Borneo have developed productive systems adapted to conditions which befuddle modern science. Previous analysis of shifting agriculture has been based on faulty assumptions. Long-term agroforestry techniques make the fallow and productive period while protecting and renewing the site. In some circumstances, stable, intensive agriculture can be practiced with high rates of production. Game harvest on a sustained basis provides a continuing source of protein when managed properly. (Lugo et al., 1987, p. vii)

The Lun Dayeh people of interior Borneo have essentially abandoned shifting cultivation in favor of permanent, irrigated wet-rice cultivation. Gaining a surplus of rice, their use of tropical forest land involves little clearance of new forest and seems well adapted to the low density human population conditions. Their success is said to stem from their "profound and intricate knowledge of the highly varied environments they inhabit" (Lugo et al., 1987, p. 11).

In contrast, the Iban people of a neighboring part of Borneo have been said to have "a warrior's view of natural resources as plunder to be exploited" (Geertz, 1969, p. 16). The Iban seriously over-cultivate their land using a single plot for more than three years. Their settlements are large villages and thus movement to new areas is difficult. As a result, the Iban cannot be said to live in harmony with their environment, the way the Lun Dayeh do. The Iban culture has a deeper rooted belief, in the context of the low density of human occupation of Borneo, that "there are always other forests to conquer" (Geertz, 1969, p. 16). In the Iban area, destruction of the forests and land degradation occur despite a lack of rapid population growth.

Rockwell (1991) finds it "romantic to conclude that tribal peoples, if only left alone, would adopt an ecology of survival that leads them towards conservation of the environment." Attitudes and beliefs about nature clearly influence land degradation. Farmers in the rainforests of southeastern Mexico were found to be applying twice the amount of fertilizer they needed to their crops. They had read the instructions on the fertilizer containers but had not believed them. In the past they had always "fed the forest and the ground" with "rituals." Knowing that the fertilizer fed the ground, they believed that the more they fed it with fertilizer, the more the forest and the ground would feed them with crops (Lourdes Arizpe, personal communication, 1991). The problems of fertilizer residuals thus stem from a traditional attitude to the land.

Animals, grazing systems, and land degradation

In undisturbed tropical rain forests, animals affect the pattern of runoff and erosion in many ways. Termites, ants, and other arthropods, as well as beetles and earthworms, turn over the surface soil, incorporate organic matter, and expose bare soil to raindrop impact. Some birds, such as the bower bird of the Australian rain forests, create areas of bare soil for their courtship arenas. Many mammals disturb the ground surface on a larger scale. Occasional bare areas, often with an almost vertical exposed back wall 1–3 m high, found in Bornean rain forests are locally described as rhinoceros wallows, but may owe more to the activities of wild pigs. Forest deer often descend slopes to drink from streams along well-defined tracks which become channels for flowing water and sources of sediment during heavy downpours.

Some landscapes of the seasonally wet tropics are much more dominated by animal activity, especially that of termites. In northern Australia, many parts of the landscape are marked by pillar-like termite mounds, while in southern Zaire and adjacent central African areas, huge round termite mounds create a mammelated terrain. In southern Shaba province, Zaire, the mass of soil disturbed by termites is around $7–20$ t ha^{-1} (Aloni Komanda, 1978).

Table 15.4 Sequence of events in land degradation through cattle grazing in the Upper Nogoa catchment, central Queensland

1   Tall unpalatable grass cover is burned to promote new growth.
2   Regeneration of palatable species occurs with first rains of approaching wet season.
3   Cattle, either domestic or feral, are attracted to the new growth and graze more intensively.
4   Cattle cause trampling, compaction, and destruction of ground cover.
5   Intense, short-duration, early wet season storms cause high soil detachment and splash.
6   Overland flow develops more quickly than before and becomes more concentrated.
7   Rilling occurs.
8   Gullying occurs.
9   The shallow soils are removed and the underlying highly dispersible mudstones are exposed.
10 Gully erosion develops rapidly in the mudstones as channels widen and headwalls recede.

Source: adapted from Blandford, 1981

The disturbance of the ground by animals reaches new proportions when herds and flocks of domestic animals concentrate around water holes and farmyards. The resulting degradation is particularly acute where competition for land arises between agriculture, livestock, wildlife, and settlements. In Kenya, farmers have moved to the foothills of Mt Kilimanjaro and land along perennial streams and swamps, displacing pastoralists. In droughts in the late 1980s, pastoralists were less able to adapt through loss of land to the cultivators and through the gazetting of national parks. The net result is a severe overstocking and overgrazing of Kenya's marginal lands and consequent degradation (Darkoh, 1991).

The way in which pastoralism can convert a relatively stable area into a zone of rapid gully erosion is illustrated by sheep grazing in the upper Nogoa catchment of central Queensland. The area is underlain by highly dispersible mudstones interbedded with thin sandstone layers. The ground surface was originally protected by a thin soils and grass cover. Close grazing by sheep broke the grass cover and allowed intense rains, with 15 minute intensities of up to 80 mm h^{-1}, to erode the soil and attack the dispersible mudstones, creating rills, the most rapidly developing of which enlarged into gullies cutting down to the more resistant sandstone beds. Cattle introduced to the area later continue to initiate rilling on the interfluves (table 15.4) (Blandford, 1981).

Land degradation through commercial forestry operations

Commercial logging includes selective extraction of valuable trees above a certain size, total tree removal for sawlogs, and wood-chip operations to utilize as much of the tree crop as possible. All these operations involve harvest of nutrients in wood and destruction of much of the forest canopy. They also cause compaction of a considerable portion of the soil surface by forest road construction, allowing rain to run downslope along the new routes and log haulage lines and across bare log-loading or high-lead assembly areas. The order of magnitude of sediment loss from an actively logged steepland catchment in the granite hills of the Main Range of peninsular Malaysia is 4400 t km^{-2} year^{-1}, as measured in the 20 km^2 Sungai Batangsi catchment in the calendar year 1988 (Lai, 1991), while an undisturbed catchment, the 4.7 km^2 Sungai Lawing, yielded 53.6 t km^{-2} year^{-1} and the larger, partly cleared and cultivated 68 km^2 Sungai Lui catchment produced 92.6 t km^{-2} year^{-1}. This 40-fold increase in sediment yield during logging activity indicates the degree of land degradation often associated with forestry activity. These Main Range catchments are steeper than most areas where logging occurs in southeast Asia and may represent the highest levels of impact.

The environmental changes that combine to produce such an increase in sediment yield are all linked to the exposure of easily eroded soil and weathered rock to the impact of intense rainfall. Bare soil areas, such as roads, tracks, and log-loading points, are the foci of such degradation. On these compacted sites there is little infiltration and water runs off directly, entraining high quantities of soil particles. A 5 × 1 m plot on an abandoned logging track on the weathered mudstones of the Kuamut formation in Ulu Segama, Sabah, Malaysian Borneo, yielded 152 times more sediment than a nearby undisturbed forest plot (44.6 as opposed to 0.293 kg ha^{-1} year^{-1}) (Sinun, 1991).

Such tracks will become gullied unless they are covered by vegetation quickly. Gullies continue to entrench and widen as long as rain falls directly onto them or overland flow runs rapidly into them (plate 15.2). Where such gullies continue unbroken to the stream channel they effectively increase the drainage density and thus shorten the lag time between rainfall and storm discharge peak. The continued erosion of gully walls and floors yields more sediment to the drainage system and degrades land.

In extreme cases, road and roadside erosion can lead to changes in drainage, where roads run downslope across divides between catchment areas. Spillage of sediment from road drains, collapse of hollow log culverts under roads, and wash of debris from large compacted areas damage adjacent land long after all logging activity has ceased. Thus while logged forest may begin to recover, the tracks and loading areas remain bare much longer and thus continue to be degraded.

Such on-site degradation has its counterpart in downstream deposition of sediment, especially where upland streams debouch on to lowland plains, or in estuarine and deltaic areas. Many such damaged rivers tend to become so encumbered with sediment that they widen their channels and erode their banks, causing further land loss. The constant building up and erosion of mid-channel

Plate 15.2 Gully erosion
is frequently initiated by
the concentration of runoff
along roads, as here in
Sabah, Malaysia.

sand banks, and processes of riparian deposition and bank erosion
in the lower reaches of the Jumna in Bangladesh (Khondaker, 1989)
may well be aggravated by runoff and sediment from degraded
areas further upstream.

Forestry operations remove nutrients from the land not only
through the harvest of forest products, but also in the changed
efficiency of hydrologic and geomorphic processes, as many paired
catchment experiments in the humid tropics have shown (table
15.5). This degradation process is again affected by the rate at
which the vegetation of the logged area recovers (figure 15.1).

Too few enquiries have been pursued for long enough to
establish whether the original level of the natural nutrient circula-

Table 15.5 Sediment yields from forested and cultivated catchments in the
humid tropics

	Forested	Cultivated	Source
Mboya Range, Tanzania	17.3	73.8	Douglas (1981)
Cameron Highlands, Malaysia	52.7	257.8	Douglas (1981)
Cilutung, Java, Indonesia	2250.0	4750.0	Douglas (1981)
Barron, Australia	14.3	34.0	Douglas (1981)
Millstream, Australia	15.5	30.8	Douglas (1981)
Gombak, Malaysia	60.0	1157.5	Douglas (1981)
Northern Range, Trinidad	4.5	40.0	Douglas (1981)
Apiopodoume, Ivory Coast	242.5	4250.0	Douglas (1981)
Apiopodoume, Ivory Coast	5.0	9000.0	Lelong et al. (1984)
Caqueta Basin, Colombia	153.0	400.0	McGregor (1980)

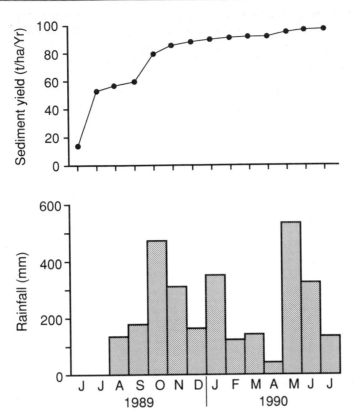

Figure 15.1 Cumulative curve of sediment yield from a 5 m² erosion plot on an abandoned logging track in Ulu Segama, Sabah, Malaysian Borneo, showing decrease in erosion as vegetation became re-established (based on Sinun, 1991).

tion in a tropical forest is again attained as logged-over forest recovers. The scars of the compacted bare areas remain visible for many years, but the degradation caused by commercial logging may be temporary.

Degradation in tree crop and plantation systems

The often high-value plantation crop productions systems are geared toward sustained agricultural production. Soil degradation affects nutrient supplies to their crops and thus plantation managers avoid excessive degradation. Lessons are sometimes learnt in a painful way. When the rubber tree was first brought from Amazonia to what was then Malaya, the trees were planted in bare soil that was kept free of weeds. Much soil erosion occurred and in many places gullies a meter or more deep ran back upslope from the heads of permanent stream channels. Although by 1920 the wisdom of retaining a cover crop on the ground was accepted by all planters, the gullies that had already developed remain visible today and function as an extension to the drainage network, accelerating the rate of stormwater runoff.

Most tree crop plantations are stable agricultural production systems. Loss of soil from them differs little from that under natural forest. Problems arise, however, when the demand for a crop falls and the plantation is no longer properly maintained. On the island of Pinang, Malaysia, old clove plantations, laid out in rows of shrubs on terraced hillsides, have become severely degraded as the terraces are no longer maintained and ravines now dissect the slopes. Degradation in this case is caused by the failure of the economic system to provide the incentive to sustain the soil conservation work.

One of the tropical crops often associated with severe erosion is sugar cane, an annual crop which involves ploughing the soil every year and having several weeks when the bare soil is exposed to the heavy rains. In humid tropical northeast Queensland, where tropical cyclones can bring up to 500 mm of rain in 24 hours and 30 minute intensities of 200 mm h^{-1} are not unknown, soil losses from sugar-cane fields were an extremely severe problem (Pauli, 1978). However, in the 1980s, widespread non-tillage cultivation, in which the residues of the previous crop were left on the ground and the incorporation of organic matter by natural processes was used to fertilize the soil, was adopted and soil losses have been greatly reduced.

Land degradation in lowland agricultural systems

Much humid tropical agriculture relies on the drainage of swampy coastal lowlands. In some places, these lowlands have waterlogged soils through which run small streams of clear, but black looking water. These blackwater rivers are influenced by the waterlogged organic matter covering the grounds and are acidic, with a pH of 3–4. When drained, the reclaimed soils often develop the acid sulfate soil characteristic which makes cultivation difficult. Waterlogging of these coastal soils can only be overcome with expensive remedial works, while liming to overcome the acid sulfate problem may be expensive.

In southern Kalimantan, Borneo, much of the swampy coastal lowlands suffer the acid sulfate problem. At the beginning of the twentieth century, these well-watered alluvial areas were seen as ideal for the expansion of padi rice cultivation and were a target for colonization by landless people from Java under the transmigration program. The effort put into draining the wetlands was sadly wasted as acid sulfate characteristics built up and some settlers abandoned their coastal lands and moved into the lowland rain forest proper, clearing areas for cultivation and so creating two degraded areas through the mis-siting of the original land development.

Many tropical coastal zones also have complex sand ridge barrier systems which are covered with a special version of forest adapted to rapidly draining soils of extremely low nutrient status. Where these forests are on monsoon-affected coasts with a marked drought season, as in Kelantan and Terengganu, peninsular Malaysia, fire is a major hazard, while grazing by domestic animals depletes their foliage in dry periods. Exploitation of these forests by coastal villagers has led to degradation of many sand ridge areas, which now have a surprisingly barren appearance.

An even more widespread problem than acid sulfate soils is that of the saline and saline–alkaline conditions sometimes associated with irrigation developments in the seasonal tropics. The lower Indus valley in Sind, Pakistan, has large-scale salinity problems associated with irrigation, but salt-affected soils occupy some 2 500 000 ha in the Indo-Gangetic plains (Dhruva Narayana and Abrol, 1981), often being interspersed with productive normal soils. About 7 percent of the world's land area (possibly as much as 30 percent of the world's potential arable land) is affected detrimentally by salinity; an estimated 50 percent of all irrigated land has been damaged by secondary salinization, alkalinity, or waterlogging (Brown and Thomas, 1990). The problem is often associated with irrigation, where losses of water by evapotranspiration leave residual salts at the ground surface. Sometimes, overapplication of irrigation water causes rises in the groundwater table which bring water with a high salt content to the ground surface. These problems arise under most water management systems, from Australia and the USA to Pakistan, India, and the African Sahel. In the last region, the construction of new irrigation schemes has barely kept pace with the loss of irrigated land through salinity problems (Brown and Thomas, 1990).

Many land development and irrigation schemes are carefully planned to avoid deterioration of land quality. Tree crop planting usually achieves effective soil protection, providing that the canopy is dense enough to break the fall of raindrops and that there is a ground cover during the early stages of tree growth. Loss of nutrients is minimized in rice cultivation, although fertilizer and pesticide inputs have increased with the adoption of high-yielding varieties. Even in the humid tropics, some irrigation schemes have failed due to waterlogging, while salinity continues to be a severe problem in drier areas. Treatment of saline areas varies according to the particular pedologic and hydrologic situation. In some instances, the movement of salt into the affected area can be reduced by soil conservation works on the surrounding hillslopes. In others, drainage has to be improved so that excess irrigation water is removed, carrying excess salts with it. Restoring the water balance toward a more natural regime also assists in reclamation of salt affected areas.

Where salt is mobilized on hillslopes, through either land clearance for grazing or cereal cultivation, it often moves with subsurface throughflow within the top 50 cm of the soil profile. This throughflow can be held back by building clay-lined contour banks, with the lining descending a meter below ground level. Water so retained has improved land productivity on the hillslopes, while concentrations of salt remain low enough not to affect plants and animals.

Where salts are allowed to migrate to the lowlands and ground-water tables rise, the salts which accumulate at the ground surface can be lethal to all but a few halophytic plants. The problem may be alleviated by improving drainage until the water table is 3 m below the surface (Pereira, 1989). Careful management is then necessary to ensure that the water table remains low and that saline water is evacuated. In the Sind, two drainage channels parallel to the Indus have been built to evacuate saline water to the sea.

Reclamation of alkaline soils in the Indo-Gangetic plain has involved three steps:

1   Land levelling and building small earth ridges around the edges of fields (bunding).
2   A single application of gypsum at around 15 t ha^{-1} depending on the sodium salts in the soil.
3   Rice cultivation in the wet season (June–October) and wheat during the winter (November–April), with appropriate agronomic practices.

Successful operation of this technique has required, in addition to the correct gypsum application, choice of suitable crops and varieties in a rotation, correct time of transplanting and sowing of crops, maintaining adequate plant density, appropriate fertilizer application including micronutrients, proper water management, and adequate plant protection measures (Dhruva Narayana and Abrol, 1981).

Land degradation associated with urban, industrial and mining development

Land clearance for building construction occurs round all tropical cities, often cutting tens of meters into deeply weathered hillsides and temporarily exposing large areas to the erosive power of the rain (plate 15.3). All the processes previously described with reference to forestry operations apply with even more effectiveness in this situation. Around the city, agricultural land may be used for extremely intensive market gardening or may be less well tended and used for cattle grazing while owners wait for land prices to rise to a point when it is propitious for them to sell to developers. Other land on the urban periphery is alienated for waste disposal and landfill operations and is occupied illegally by impoverished squatter settlements or is converted to recreational uses. All these pressures reduce the vegetation cover and accelerate erosion.

Some mining activities involve moving huge quantities of over-burden and working alluvial material. Around the alluvial tin mines of southeast Asia are huge areas of mine waste and piles of sand that has passed through the separators used to sort out the grains of tin. Such land is both degraded and derelict and has special problems of rehabilitation. Other mining operations, such as copper mining, degrade land both through their urban-type infra-

Plate 15.3 Deeply eroded gully on a construction site in Kuala Lumpur. Urban flooding, erosion, and landsliding are common in many Third World cities, where land-use control is often non-existent.

structure problems of roads and yards, and through their need to dispose of tailings. The Bougainville copper mine in Papua-New Guinea had particular problems in its impact on the Kawerong River system (Brown, 1974; Gilles, 1977). Tailings problems are also severe at the Ok Tedi site in Papua and the Muamut copper mine in Sabah. Toxic metals that damage vegetation may be a potent force in plant cover and soil degradation. Although limited in extent, these contaminated areas may have so much toxic material that they are difficult to rehabilitate and pose continuing problems as sources of polluted runoff.

Controlling land degradation in the tropics

Essentially, the control of land degradation is achieved by maintaining crop or other vegetation cover and by avoiding having more or larger bare areas than absolutely necessary. The International Union for Conservation of Nature and Natural Resources (IUCN) adopted the following guidelines for the restoration of degraded lands in 1988:

- Critical sites, especially those liable to erosion or habitats of threatened or endangered species, should be rehabilitated or restored first.

- For the initial revegetation, plants should come from different successional stages to lessen the chances of natural succession being thwarted and to let slow-dispersing species be included.
- Plants that facilitate succession or maintain nutrient and water cycles should be included. Nitrogen fixers and plants whose roots can penetrate beyond upper soil layers are important.
- Social scientists and local people must be involved in rehabilitation programs and community knowledge about soils and plants should be exploited.
- Rehabilitation programs should provide economic benefits to the participants.
- Laws and customs affecting the area to be restored need periodic review. If they lead to environmentally damaging actions, attempts should be made to revise them.
- Tree plantations should be established on degraded land rather than on undisturbed or cultivated areas.
- To preserve the species needed for future land rehabilitation and restoration, protected areas should be set aside to conserve plant and animal wildlife.

Appropriate use of plants and restoration of a good ground cover is the overriding theme of these guidelines. Livestock must also be regulated in order to maintain ground cover. Rehabilitation of tropical grassland ecosystems should involve the following (Brown and Thomas, 1990):

- Destocking of animals should take place on a rotational basis to control numbers and allow for marketing of animals.
- Grassland regeneration should be helped by temporary exclusion of stack, rotational grazing, and replanting and protection of denuded areas.
- Regrowth of vegetation should be monitored and appropriate grazing densities maintained.
- People should be encouraged to continue the traditional practice of keeping mixed herds as different food and water requirements of animals mean that more ecological niches are exploited than when only one type of livestock is grazed.
- Breeding should aim at multi-purpose stock (milk, meat, and use as draught animals).

Even in urban construction projects, the guidelines must emphasize the need to leave as little bare soil exposed for as short a time as possible (table 15.6). The best tropical urban development control systems incorporate terrain evaluation procedures in order to eliminate many of the risks of severe erosion and land degradation. Those developed by the Hong Kong Geotechnical Control Office are a fine example (Styles and Hansen, 1989).

The principles behind these erosion control guidelines should be adopted widely to avoid further land degradation. However, the problem of degraded lands cannot be solved until degradation is

Table 15.6   Guidelines for erosion control during urban construction in the tropics

---

Actions before development begins

1   Evaluate soil, geologic, geomorphic, and hydrologic conditions.
2   Select development compatible with the site.
3   Identify physical and biological features to be retained in the development.
4   Prepare a development plan which minimizes site limitations and includes erosion and sediment control measures.
5   Plan construction in a series of parcels so that only a small part of the project area is disturbed at any one time.

Actions during and after the construction process

1   Keep earth moving to a minimum: retain as many trees as possible.
2   Avoid having areas bare or exposed for long periods of time.
3   Protect areas liable to erosion with mulch or temporary vegetation.
4   Control runoff using diversion tunnels, storm drains, tile drains, or grassed waterways.
5   Construct sediment basins to store sediment and surface runoff during construction.
6   Provide for safe off-site disposal of runoff, including the increased runoff resulting from the development.
7   Establish vegetation on finished areas as soon as possible.

---

recognized as a social and economic issue. The global economy, dominated by the rich nations of the Western world, is a major contributor to land degradation in the tropics, not only through the international market for forest products, but also through the way it drives cash crop production and fertilizer use in agricultural systems. Social systems, attitudes, and work distribution, including gender relations, all play a role in land-use decisions and responses to changes in soil management techniques. Ultimately, the link between land degradation and population increase has to be faced. If no significant progress is made in stemming population growth, all efforts to prevent further degradation and to rehabilitate degraded land may be insufficient to provide enough land to grow the crops to feed the world's people. The technology of restoration is understood. The will to do all that is necessary to achieve adequate agricultural production from the land, to ensure sustainable land use, and to achieve equity in the distribution of the benefits of such measures is not always evident.

Further reading

Blandford, D.C. (1981) Rangelands and soil erosion research: a question of scale. In R.P.C. Morgan (ed.), *Soil Conservation: Problems and Prospects*. Chichester: Wiley, 105–21.

Brown, H.C P. and Thomas, V.G. (1990) Ecological considerations for the future of food security in Africa. In C.A. Edwards, R. Lal, P. Madden, R.H. Miller, and Gar House (eds), *Sustainable Agricultural Systems*. Ankeny: Soil and Water Conservation Society, 353–77.

Darkoh, M.B.K. (1991) Land degradation and resource management in Kenya. *Desertification Control Bulletin*, 19, 61–72.

Douglas, I. and Spencer, T. (1985) *Environmental Change and Tropical Geomorphology*. London: Allen & Unwin.

Pereira, H.C. (1989) *Policy and Practice in the Management of Tropical Watersheds*. Boulder, CO: Westview.

Vine, H. (1966) Tropical soils. In C.C. Webster and P.N. Wilson (eds), *Agriculture in the Tropics*. London: Longman, 28–67.

Whitmore, T.C. (1990) *An Introduction to Tropical Rain Forests*. Oxford: Clarendon Press.

CHAPTER SIXTEEN

Dryland Degradation

Andrew Goudie

Introduction

In the past two decades desertification has been recognized as one of our most important environmental dilemmas, and it is widely reported that the process affects about 65 million hectares of once-productive agricultural land and threatens the livelihoods of 850 million people. However, the belief that degradation is reducing the productive capacity of deserts and their margins, though current and fashionable, is not new. Over a century ago, the concept of progressive desiccation arose (Goudie, 1990). This was based on the assumption, now shown to be substantially incorrect, that wet conditions (pluvials) were a feature of the glacial phases of the Pleistocene and that aridity has increased with the waning of the Pleistocene ice bodies in the Holocene interglacial. There has also been a long-standing interest in the role that humans have played in degrading such areas, and George Perkins Marsh's experience in the drylands of the Mediterranean basin prompted him to write *Man and Nature* (1864), in which he stressed the adverse consequences of deforestation and other human actions.

Dryland degradation as a phenomenon is itself not new. There is an increasing body of evidence which suggests that humans had achieved substantial environmental modifications in prehistoric and historic times. For example, the classic study of Jacobsen and Adams (1958) in Mesopotamia demonstrated that salinization and siltation of irrigation systems was creating problems over 4000 years ago, while Marsh himself (1864, p. 11) attributed the "sterility and physical decrepitude" of the once flourishing countries of the Mediterranean basin and Near East to "the brutal and exhausting despotism which Rome herself exercised over her conquered kingdoms." He pointed to the need (p. 45) "to rescue its wasted hillsides and its deserted plains from solitude or mere nomade occupation, from barrenness, from nakedness and from insalubrity."

Dryland degradation is by no means restricted to Third World areas such as the Sahel of Africa. It is a phenomenon which afflicts large areas of Australia, the former Soviet Union, southern Europe, and the USA. For example, as Sheridan (1981, p. 121) wrote:

> Desertification in the United States is flagrant. Groundwater supplies beneath vast stretches of land are dropping precipitously. Whole river systems have dried up; others are choked with sediment washed from denuded land. Hundreds of thousands of acres of previously irrigated cropland have been abandoned to wind or weeds. Salts are building up steadily in some of the nation's most productive irrigated soils. Several million acres of natural grassland are, as a result of cultivation or overgrazing, eroding at unnaturally high rates. Soils from the Great Plains are ending up in the Atlantic Ocean .... About 225 million acres of land in the United States are undergoing severe desertification – an area roughly the size of the 13 original states.

The term "desertification" was first used, but not formally defined, by Aubréville (1949). It came into common usage in the 1960s and 1970s as a result of two prime stimuli: the environmental movement and the Sahel drought and African crisis. In 1977, in Nairobi, the United Nations convened a Conference on Desertification (UNCOD), the purpose of which was to define the problem, agree on a Plan of Action, and bring the process under control by AD 2000. However, "desertification" and related terms such as "desertization" and "land aridization" have been the subject of a great deal of sterile terminological debate (see Verstraéte, 1986, for a review and discussion).

Definitions

Some definitions stress the importance of human causes (e.g. Dregne, 1986, pp. 6–7):

> Desertification is the impoverishment of terrestrial ecosystems under the impact of man. It is the process of deterioration in these ecosystems that can be measured by reduced productivity of desirable plants, undesirable alterations in the biomass and the diversity of the micro and macro fauna and flora, accelerated soil deterioration, and increased hazards for human occupancy.

Others admit the possible importance of climatic controls but give them a relatively inferior role (e.g. Sabadell et al., 1982, p. 7).

> The sustained decline and/or destruction of the biological productivity of arid and semi arid lands caused by man made

stresses, sometimes in conjunction with natural extreme events. Such stresses, if continued or unchecked, over the long term may lead to ecological degradation and ultimately to desert-like conditions.

Yet others, more sensibly, are more even-handed or open-minded with respect to natural causes (e.g. Warren and Maizels, 1976, p. 1):

A simple and graphic meaning of the word "desertification" is the development of desert like landscapes in areas which were once green. Its practical meaning . . . is a sustained decline in the yield of useful crops from a dry area accompanying certain kinds of environmental change, both natural and induced.

Another controversy surrounds what should and what should not be incorporated within the term. This is brought out starkly when one compares the definitions of Rapp (1987, p. 27) and then Mortimore (1987, pp. 2–3). Rapp defined desertification as:

The spread of desert-like-conditions of low biological productivity to drylands outside the previous desert boundaries. Desertification is severe degradation of drylands, lasting more than one year, and is manifested by the loss of vegetation cover, loss of topsoil by wind or water erosion, reduction in primary productivity through soil exhaustion, salinization, or excessive deposition of sand dunes, sheets or coarse flood sediments.

Mortimore, on the other hand, was more restrictive:

argue

The essence of desertification is "the diminution or destruction of the biological potential of the land" leading to the extension of desert-like conditions of soil and vegetation into areas outside the climatic desert, and the intensification of such conditions, over a period of time. I wish to exclude certain processes of ecological degradation which are commonly, but to my mind confusingly, included within the purview of the term. These are deforestation (which is the normal prelude to agricultural land use and is reversible), salinization of irrigated soils (which is caused by inadequate drainage), and soil erosion by water.

Another restriction that is often placed on the use of the term is that it should comprise some notion of long-term and possibly irreversible or irreparable change (e.g. Wehmeier, 1980, p. 126):

Desertification here will be understood as disadvantageous alterations, not oscillations, of and within ecosystems in arid

Plate 16.1 One view of desertification is that it involves the inexorable frontal movement of desert conditions including the migration of sand dunes into settled areas. In this example from Kolmanskop, near Luderitz in Namibia, this would appear to be the case, but this is a hyper-arid area where dune movement is the norm. Most desertification does not involve sand encroachment of the type illustrated here.

and semi-arid areas. These alterations are usually triggered by man – involuntarily – take place within a relatively short period of time – several years to several decades – and very often cause irreparable or but partly repairable damage.

No less contentious is the spatial character of desertification. Fundamentally, as Mabbutt (1985, p. 2) has explained, "the extension of desert-like conditions tends to be achieved through a process of accretion from without, rather than through expansionary forces acting from within the deserts." Thus, contrary to popular rumor, desertification does not advance over a broad front in the way that a wave overwhelms a beach or volcanic lava flows down a hillside (plate 16.1). Rather it is like a rash which tends to be localized around towns and villages. For that reason it has been likened to dhobi itch – a ticklish problem in difficult places (Goudie, 1981). However, there are few reliable long-term empirical studies of the pattern of change (the monograph of Mortimore (1989) being a notable and praiseworthy exception) and there are even fewer reliable studies of the rate of supposed desert advance. One problem is that arid zones show very substantial fluctuations in biomass production from year to year, so that any attempts to establish a trend by comparing any two years may be fraught with uncertainties (Helldén, 1984; Dregne and Tucker, 1988). Warren and Agnew (1988, p. 4) raise other pertinent issues:

1 How do you set a standard against which to judge the present condition? For example, if the land was denuded of vegetation or severely gullied, was this because of a drought, overgrazing, or had it always been so?
2 How do you assess recoverability or resilience? Resilience is hard to determine and depends not only on physical characteristics of the soil, but the land-use system.

They also point out that a reduction in biomass is not necessarily a good indicator of loss of productivity, arguing that pastures may be more productive when they are grazed low, because moderate grazing encourages growth and because small plants may be less vulnerable to drought. Likewise, the invasion of unpalatable bush into grassland areas that have been overgrazed leads to an increase in biomass and vegetative cover, but a loss in productivity for pastoralists.

For these reasons maps of desertification and figures for the rate and areal extent of the process need to be treated with some caution. As Grainger (1990, p. 157) has pointed out, "The shortage of reliable data on the extent and rate of spread of desertification is a major reason why there are so many differing views on the phenomenon and why there has been little effective action to bring it under control. Unless desertification can be accurately defined and measured, the possibilities for scepticism and misunderstanding abound, and policy makers will be reluctant to allocate funds for programmes to control it."

The 1977 Nairobi UNCOD meeting depicted on its maps four classes of desertification:

(a) Slight
   • little or no deterioration of plant cover or soil.
(b) Moderate
   • significant increase in undesirable forbs and shrubs; or
   • hummocks, small dunes, or small gullies formed by accelerated wind or water erosion; or
   • soil salinity causing reduction in irrigated crop yields of 10–50 percent.
(c) Severe
   • undesirable forbs and shrubs dominate the flora; or
   • sheet erosion by wind and water have largely denuded the land of vegetation, or large gullies are present; or
   • salinity has reduced irrigated crop yields of more than 50 percent.
(d) Very severe
   • large shifting barren sand dunes have formed; or
   • large, deep and numerous gullies are present; or
   • salt crusts have developed on almost impermeable irrigated soils.

Building upon this, Mabbutt (1985) developed a three-fold categorization of desertification in one particular land-use type – rangelands:

(a) Moderate
- significant reduction in cover and deterioration in composition of pastures;
- locally severely eroded;
- would respond to management supported by improvements and conservation measures;
- loss of carrying capacity up to 25 percent of earlier carrying capacity.

(b) Severe
- very significant reduction in perennial vegetation cover and widespread deterioration in composition of pastures;
- widespread severe erosion;
- requiring major improvements;
- loss of carrying capacity 25–50 percent of earlier carrying capacity.

(c) Very severe
- extensively denuded of perennial shrubs and grasses and subject to widespread very severe accelerated erosion;
- large areas irreclaimable economically;
- loss of carrying capacity over 50 percent of earlier carrying capacity.

Using data based on these categories he estimated that on a global basis 4500 million hectares were at risk from desertification and that 945 million hectares were severely or very severely desertified.

UNEP data indicated for different regions the amount of desertification that has occurred in the 3275 million hectares of "productive" drylands. It is estimated that 61 percent showed evidence of desertification, with the greatest percentage being in Sudano-Sahelian Africa (about 88 percent). In all, UNEP has estimated that desertification threatens 35 percent of the Earth's land surface (about 45 million km^2) and 19 percent of its population (some 850 million people) (Stiles, 1984).

A case study: current desertification in the USA

A good way in which to gain an appreciation of the diverse components that make up the phenomenon of desertification is to look at desertification in a country where, as we have already seen, it is a major problem – the USA. Some of the pheneomena concerned are:

- accelerated dust storm incidence;
- dune reactivation;
- arroyo development;
- groundwater depletion;
- ground subsidence;
- salinization and waterlogging;
- overgrazing and bush encroachment;
- lake bed desiccation;

- invasion of alien plants;
- chemical contamination;
- the spread of urban land;
- sedimentation in reservoirs.

As will become evident, technology can be a cause of desertification as much as a panacea.

Possibly the most notorious indicator of desertification in the USA has been the incidence of dust-storms. The 1930s – the so-called Dust Bowl Years – saw a dramatic series of "black blizzards" of dust, which resulted from a combination of dry, hot years and a rapid expansion of cultivation in the High Plains, which was facilitated by the availability of the internal combustion engine. These years have been dramatically described by Worster (1979), but the phenomenon is still a serious problem: some shelterbelts and windbreaks have recently been removed; center-pivot irrigation has enabled the cultivation (and hence baring) of some susceptible soils of aeolian origin (as in New Mexico); and surfaces are being disturbed by such processes as military maneuvers, vehicular recreation, urban expansion, and oilfield development (Wilshire, 1980).

Overgrazing, in terms of area affected, is the most potent desertification force in the country (Sheridan, 1981). Particularly serious deterioration has occurred in the Navajo lands, the Challis Planning Unit (Idaho) and the Rio Puerco basin of New Mexico. This in turn leads to soil erosion and miscellaneous vegetation changes, such as bush encroachment.

One aspect of environmental deterioration which has a long history and uncertain origins is the development of erosional trenches, arroyos, in valley bottoms (Cooke and Reeves, 1976). Past episodes of cut and fill have undoubtedly occurred in response to natural climatic changes or to geomorphic thresholds being exceeded (Schumm et al., 1984) (table 16.1), but such human activities as overgrazing, soil compaction, and the channeling of water from roads and railways, have played a major role in some recent episodes. Arroyo growth leads to loss of land, lowered water tables, increased sediment yield, and sundry other adverse consequences. Moreover, accelerated erosion, as represented by arroyos and other phenomena, has reduced the capacity of reservoirs and the effective life of dams.

Soil structure deterioration has also been highlighted as a significant problem (Sabadell et al., 1982). Compaction, produced by poor tillage methods on susceptible soils, decreases rates of water infiltration, increases runoff and erosion, and reduces the ease of root and water penetration into the soil. A related problem is that of soil crusting caused by exposing the soil to raindrop impact or to sprinkler irrigation.

Salinization has become a severe problem in parts of the Central Valley of California, as a result of raised water tables brought about by inter-basin water transfer and extensive irrigation. The

Table 16.1   Causes of channel incision

Decreased erosional resistance

1   Decreased vegetational cover owing to increased agricultural activity,
    urbanization, timbering, overgrazing, fires, and droughts
2   Surface disturbance causing decreased permeability and cohesion

Increased erosional forces

1   Constriction of flow by dikes, vegetation, bedrock, ditches
2   Concentration of flow by roads, trails, and ditches
3   Steepening of gradient and energy slopes by deposition of sediment
    (valley plug), channelization, and meander cutoffs, base-level lowering by
    decreased lake or reservoir level, and by decrease of discharge or flood
    peaks in main channel
4   Increased discharge and flood peaks
5   Decrease of sediment load

Source: from Schumm et al., 1984, table 2.1

drainage problem has become acute. Some rivers are also becoming increasingly saline, including the Colorado River. A related problem is that in arid areas, with their limited leaching potential, some agricultural chemicals (e.g. fertilizers and biocides) can accumulate to undesirable levels. In all, about 10 million hectares of the western USA (>20 percent of all irrigated land) suffers from salinity-related reductions in yield, and the cost amounts to some hundreds of millions of dollars per year in both the Colorado River basin and the San Joaquin Valley. Many of the issues associated with over-exploitation of water resources are discussed in El-Ashry and Gibbons (1988).

In some parts of the American Southwest, inter-basin water transfers have created lake desiccation problems. As a result of the water needs of the Los Angeles area, the levels of both Owens Lake and Lake Mono have fallen, and the deflation of salts and dust from their drying margins is considered to be an undesirable development (Barone et al., 1981). The conflicts involved in the exploitation of the water resources of the Owens Valley have been graphically described by Kahrl (1982).

Another source of hydrological modification, groundwater "mining," means that water levels have fallen in the High Plains and elsewhere. As a result of the spread of techniques such as center-pivot irrigation (plate 16.2) the water level has declined by 30–50 m in a large area to the north of Lubbock, Texas, and the thickness of the Ogallala aquifer has decreased by more than 50 percent in certain counties. This groundwater exploitation has had another deleterious environmental effect – ground subsidence, which is recorded from such locations as south-central Arizona, Denver (Colorado), the San Joaquin valley (California), Milford (Utah), the Raft River valley (Idaho), and Las Vegas (Nevada). In the central valley of California the subsidence now exceeds 8 m.

The spread of alien plants has also proved to be a manifestation of desertification. In the southwest USA, for example, many stream channels are lined by *Tamarix pentandura*, the salt cedar. It has spread explosively, increasing from around 4000 ha in 1920 to almost 100 times that figure in the early 1960s (Harris, 1966). With roots either in the water table or freely supplied by the capillary fringe, these plants transpire huge quantities of water, so that in the Upper Rio Grande valley in New Mexico they consume around 45 percent of the area's total available water (Hay, 1973). Another vegetational change of great seriousness is bush infestation, whereby noxious and undesirable woody plants spread into rangeland as a result of poor land management practices. Troublesome woody pests include sagebush, cactus, rabbitbrush, mesquite, and acacias.

Plate 16.2 Center-pivot irrigation on the North American High Plains, which has led in some areas to substantial depression of groundwater levels.

The cultural context

The causes of desertification have been the subject of controversy and debate. In particular there has been some discussion about the relative importance of human activities in comparison with climatic deterioration. However, it is generally agreed that the twentieth-century explosion in human population numbers has been of fundamental importance. In the United States, dryland degradation followed the march westwards in the frontier of settlement. In recent decades the southwestern sunbelt states (e.g. New Mexico) have seen some of the most rapid rates of population growth. Many non-western states with extensive desert areas have also shown marked demographic increases. During the twentieth century, between 1900 and 1980 Algeria's population rose by 380 percent, Botswana's by 683 percent, Egypt's by 420 percent, Sudan's by 317 percent, Pakistan's by 509 percent, etc. These increases have led to four pertinent consequences: overcultivation, overgrazing, defor-estation, and salinization of irrigation systems.

Overcultivation has two components: intensification *in situ*, and creep into areas where conditions are fundamentally unsuitable, primarily because of their aridity or because of the fragility and infertility of their soils. Dryland, rain-fed cropping now takes place in areas receiving as little as 250 mm of annual rainfall in the Sahel of West Africa and 150 mm in the Near East and North Africa. Some of these areas have friable soils, developed on late Pleistocene dune fields, and are thus highly prone to water erosion and wind reactivation.

Plate 16.3 Soil erosion and gully development is one of the most serious manifestations of desertification. These examples, locally called "dongas," occur in Zimbabwean bushland, and are near St Michael's Mission.

Related to overcultivation is overgrazing, for in many areas increasing numbers of humans require increasing numbers of domestic stock. These increases in free-ranging stock populations have reduced available pastureland and have intensified pressure on remaining pastures (plate 16.3). The carrying capacity of the land, a difficult parameter to quantify, may be exceeded. There may also be conflicts between pastoralists and cultivators, so that as the front

of cultivation moves into ever more marginal areas, the pastoralists lose more and more of their grazing lands. Nomadic pastoralists, who in many cases had developed various types of sophisticated adaptations which enabled them to maintain the utility of marginal areas, have often seen their traditional systems disintegrate, so that the equilibrium between people and land became disrupted. Seasonal or annual migrations may, for example, have been curtailed by deliberate policies of sedentarization imposed by central governments or by the establishment of national boundaries where none previously existed.

Another cause of overgrazing has been the installation of boreholes and excavated waterholes. These have permitted the rapid multiplication of domesticated stock and thereby promoted excessive grazing. Vegetation in effect replaces water as the main limiting factor on stock numbers.

A third cultural cause of desertification is deforestation and removal of woody material. Many non-Western people depend on wood for domestic uses (cooking, heating, brick manufacture, etc.), and the collection of wood for charcoal and firewood is an especially serious problem in the vicinity of urban centers.

The fourth prime cause of desertification is salinization. This often results from the spread of irrigation with inadequate provision for drainage, and kills plants, destroys the soil structure, and reduces plant growth. It can, however, also be an unwanted consequence of vegetation clearance. This reduces the amount of moisture lost to the soil as a result of interception and evapotranspiration, so that groundwater levels rise, creating seepage of saline water in low-lying areas like valley bottoms. This is a serious cause of salinization both in the prairies of North America and in the wheat belt of Western Australia (Peck, 1978).

In addition to these four fundamental cultural causes of degradation there are others that may be locally significant: soil compaction and crusting caused by trampling, heavy farm machinery, or raindrop impact on bared surfaces; soil fertility reduction as a result of use of dung for fuel instead of manure; reduction in groundwater levels as a result of mining of finite groundwater reserves; the desiccation of lakes as a result of inter-basin water transfers and irrigation losses; the disruption of fragile surfaces and their vegetation by accelerated vehicular access; pollution and disturbance by mining; the abandonment of traditional conservation and husbandry methods; vegetation change as a result of diminution of wildlife; and loss of land to burgeoning urbanization.

The context of culturally changed climates

A further possible, but contentious, cultural role in promoting desertification is in creating climatic deterioration. Most hypotheses are still primarily speculative. Bryson and Barreis (1967) have proposed that if wind deflation from bared agricultural surfaces increased over natural levels, then this would serve to modify the scattering and absorption of solar radiation in the atmosphere, which in turn might reduce convective activity and associated rainfall. Walker and Rowntree (1977) suggested that once soil moisture levels are reduced by such a process, positive feedback will set in to accentuate desert conditions and allow them to persist, in that rain-giving depressions will no longer be nourished by the evaporation of soil moisture. Another mechanism has been proposed by Charney et al. (1977), who believe that an albedo increase brought about by deforestation and overgrazing would mean that the more reflective ground would absorb less solar radiation. This would mean that there would be less sensible and latent heat transfer to the atmosphere, which in turn would produce less rain-giving convective clouds.

The climatic context

As we saw in the discussion of the definition of desertification it is sometimes proposed that natural climatic deterioration may contribute to dryland degradation and the spread of desert-like conditions. Among the types of climatic deterioration that have been postulated are post-glacial desiccation (now a largely discredited concept), long-term cyclic deteriorations with a duration of several centuries (e.g. that proposed by Winstanley, 1973), and a long-continued series of drought years of the type that afflicted the Sudan–Sahel zone of Africa from the late 1960s.

Examination of rainfall data for recent decades reveals that some arid areas show relatively clear evidence for a downward trend in values, while others show either no trend or an upward trend. A downward trend has been established for the Sudan and Sahel zones of Africa (Walsh et al., 1988), and this has had a suite of consequences, including a substantial rise in dust-storm activity (Middleton, 1985) and a severe diminution in the area and volume of Lake Chad (Rasmusson, 1987). By contrast, the latest analyses of monsoonal summer rainfall for the Rajasthan Desert in India indicate that there has been a modest upward trend in precipitation levels between 1901 and 1982 (Pant and Hingane, 1988). Data for northeast Brazil (Hastenrath et al., 1984), much of Australia (Hobbs, 1988), and California and Arizona in the USA (Grainger, 1979) show no clear trend in either direction.

Dryland rehabilitation

There is abundant technology available that can be used to reduce desert degradation or to rehabilitate degraded areas. A major priority is to protect and stabilize soils from wind and water attack. Watson (1990) has outlined the methods that are available to control blowing sand and mobile desert dunes:

1 *Drifting sand*
  - enhancement of deposition of sand by the creation of large ditches, vegetation belts, and barriers and fences;
  - enhancement of transport of sand by aerodynamic stream-lining of the surface, change of surface materials, or panelling to direct flow;
  - reduction of sand supply, by surface treatment, improved vegetation cover, or erection of fences;
  - deflection of moving sand, by fences, barriers, tree belts, etc.

2 *Moving dunes*
  - removal or mechanical excavation;
  - destruction by reshaping, trenching through dune axis, or surface stabilization of barchan arms;
  - immobilization, by trimming, surface treatment, and fences.

However, as he points out, it is preferable to avoid disruption of naturally stable aeolian landscapes in the first place, for the destruction of vegetation, the breaching of stone pavement, and the disintegration of surface crusts can lead to rapid scouring of unconsolidated materials. Comparable solutions are required for the control of dust storms (Middleton, 1990). Middleton identifies three main approaches: agronomic measures, soil management, and mechanical methods. Agronomic measures utilize living vegetation or the residues from harvested crops to protect the soils (e.g. "stubble mulching"). Likewise, strip cultivation, though not easily compatible with highly mechanized agriculture, reduces wind velocity across a fallow strip, reduces the distance the wind travels over exposed soil, and localizes any soil drifting. Soil management techniques use different methods of tillage to protect the soil surface (table 16.2). Mechanical methods to control wind erosion include the creation of barriers to wind flow, such as fences,

Table 16.2 Tillage practices used for soil conservation

Practice	Description
Conventional	Standard practice of ploughing with disc or mouldboard plough, one or more disc harrowings, a spike-tooth harrowing, and surface planting
Strip or zone tillage	Preparation of seed bed by conditioning the soil along narrow strips in and adjacent to the seed rows, leaving the intervening soil areas untilled; e.g. plough-plant, wheel-track planting, listing
Mulch tillage	Practice that leaves a large percentage of residual material (leaves, stalks, crowns, roots) on or near the surface as a protective mulch
Minimum tillage	Preparation of seed bed with minimal disturbance; use of chemicals to kill existing vegetation, followed by tillage to open only a narrow seedband to receive the seed; weed control by herbicides

Plate 16.4 The scale of
soil erosion demonstrated
in this example from the
semi-arid Baringo area of
Kenya is such that action
needs to be taken to
restore the vegetation
cover. The degree of root
exposure shows the
amount of top soil that
has been lost in just a few
decades.

windbreaks, and shelter belts, and altering of surface topography as
by ploughing furrows.

Water erosion of soils (plate 16.4) can be reduced by either
preventive or control measures (Lal, 1990). Erosion-prevention
measures involve cultural practices that serve to minimize raindrop
impact, increase or enhance structural stability of the soil, and
improve the infiltration rate. Erosion control measures involve
management of surplus water or overland flow for its safe disposal
at low velocities. Erosion-prevention measures are preferable to
control measures, as the latter are expensive, usually less effective,
and often too late to be useful.

Salinity problems also require a variety of management techniques, though once again prevention (e.g. by judicious use of
irrigation water by provision of adequate drainage or by avoiding
vegetation removal in susceptible areas) is preferable to controls
(Rhoades, 1990). The following list gives an indication of the range
of practices that are available:

- growing suitably tolerant crops;
- managing seedbeds and fields to reduce local salinity accumulation;
- irrigating to maintain high soil water potential and periodically to leach salts;
- managing soils (e.g. by organic additions) to reduce puddling
  and crusting;

- accurate measurement of flow and water application to reduce over-watering in irrigation systems;
- seepage control from irrigation canals;
- efficient irrigation by proper scheduling and the use of water conservation methods (e.g. drip and trickle irrigation);
- efficient soil drainage;
- separation of saline discharges so that they do not re-enter rivers used downstream;
- provision of adequate drainage to remove surplus irrigation water.

It cannot be stressed too highly that irrigation systems, unless they are managed with care and sensitivity, can have a whole suite of adverse environmental consequences, of which salinity is but one (Adams and Hughes, 1990). In many cases it is preferable not to construct new irrigation schemes but to reorganize, rehabilitate, and improve existing irrigated areas or to take steps to improve rain-fed agriculture. It is probable that various types of small-scale water management systems (e.g. check dams and tanks) offer greater potential for increasing the bioproductivity of dryland areas (Biswas, 1990), and that there is substantial scope for water conservation through re-use of wastewater.

Crucial to many aspects of land degradation in arid and semi-arid lands is the status of the vegetation cover. Trees and shrubs offer a range of benefits which include productive ones (e.g. fuel, building material, fodder, food, and fiber), environmental ones (e.g. reduction of soil erosion, reduction of flood damage, improvement of soil structure, and improvement of microclimate), and social ones (diversity of diet or income, counter-seasonality, and mainte-nance of standing capital). However, it is difficult to encourage the planting and maintenance of trees in drylands, particularly where the dominant form of land use is pastoralism (Burley, 1990). One way forward is to develop agroforestry systems in which trees and shrubs are deliberately used on the same land management unit as agricultural crops and/or domesticated animals. Social forestry, in which local people plant trees outside regular forest areas, seems to be a viable and valuable long-term strategy to meet afforestation targets. Crucial to the success of such schemes will be the planting of multipurpose trees, including species of *Acacia*, *Leucaena* and *Prosopis*, which can tolerate poor soils and yield a range of useful products besides providing general environmental benefits.

The cultural basis for change

Although the technological mechanisms of rehabilitation that we have just outlined offer a considerable scope for improvement of arid realms, they do not get to the root cause of degradation. To do this, social, economic, and political structures need to be changed.

As Warren and Agnew (1988, p. 8) remarked, "Moving dunes often take the blame for degradation, where the serious problems lie elsewhere in the structure of the rural economy." They went on to note (p. 19) that "Blaming technique and technology, however, is almost as pernicious as blaming the environment itself: it erodes the issue. Why are people driven to adopt damaging methods of using their environment? This is a much more complicated and sensitive question, and one that many would rather avoid."

The severity of the problem of getting to root causes has been dramatically encapsulated in the words of Cloudsley-Thompson (1978):

> Desertification results from Man's overexploitation of arid lands, and can be stopped almost as easily as the world population explosion! For, after all, a smoker only has to stop buying cigarettes, an alcoholic can lay off the bottle, and a junkie can give up heroin. It would perhaps be a little more difficult for a desert dweller to forfeit his herds because, if he were to do so, he would have to abstain from eating.

Grainger (1990) has highlighted some of the underlying cultural causes of landscape deterioration. Among these he recognizes "That traditional rainfed cropping systems are now breaking down as rising demand for food to feed growing populations forces farmers to increase production, either by decreasing fallow periods or by increasing the area cultivated." In some areas overcrowding is exacerbated by land-tenure restrictions or political considerations which restrict people to certain limited areas, often of low fertility (e.g. in South Africa). Related to this is the resettling (sedentarization) of nomads into villages. The expansion of cash cropping, to produce foreign exchange earnings, means that there is an expansion in the area devoted to the cultivation of what are often very demanding crops. Moreover, these cash crops may displace subsistence croppers and pastoralists to more marginal and degradable lands. In many countries the size of herds has greatly expanded in recent decades, and there are a variety of reasons for this. First, as other sources of income (e.g. from trans-Saharan trade, the local manufacture of goods) have declined as a result of changes in transport technology and industrial methods, nomadic tribes have become more dependent on their herds. Second, in countries like Botswana, increased beef prices from European importers have encouraged expansion of herds. Third, improved veterinary services have considerably reduced mortality rates among stock, while new boreholes have enabled higher concentrations of animals. Increasing human populations also require that more and more fuelwood is obtained from a decreasing area. The poor frequently have no effective access to any other fuel source.

The inequitable distribution of resources can also be postulated as a cause of land degradation, for people who have lost control over the land they work can have little interest in the long-term resilience of that land, and little power to influence it (Blaikie, 1985). Likewise, poverty is a likely cause of degradation, because if a community is too poor to raise the necessary capital for land restoration, then degradation is likely to continue and accelerate. The poor cannot easily postpone immediate consumption in the hope of future returns, not least because the risks of failure are particularly great in these harsh, marginal environments.

In addition to the fundamental demographic problem, desertification is often being promoted by political decisions: the encouragement of a resource-hungry urban elite causes degradation in the urban hinterland; the need for foreign currency and tax revenue to maintain central government bureaucracy promotes cash cropping; decision-making is done by urban elites who are ignorant of the problems of marginal areas; and political instability and conflict may devastate the land and traditional social structures. It can be argued that the power of central governments can be deleterious, and that people will be more likely to react positively to development schemes if they are given more control over land, water, and marketing, and if power is transferred from the city to the countryside.

Political viewpoints and propaganda also bedevil discussion about desertification. Consider, for example, the following contribution from China to the Nairobi UNCOD in 1977 (in Biswas and Biswas, 1980, p. 111).

> Before liberation, the people of China had suffered from oppression and exploitation under imperialism, feudalism and bureaucratic capitalism . . . the desertification of the land before liberation was the result of destruction and plunder of natural resources by the feudalistic ruling class, the imperialists and the Kuomintang reactionaries . . . . Since the founding of the People's Republic of China, people of all nationalities in the desert areas got organised under the leadership of Chairman Mao . . . they launched an all-out war on the deserts . . . and finally won their great victory.

A rather different viewpoint is presented by Smil (1984), who believes that desertification processes in China have accelerated since 1949 and (p. 199) "that many of the developmental policies of the past three decades have led to unprecedented destruction and degradation of the country's environment."

Another fundamental problem is that desertification requires long-term planning and investment of resources. As Tolba (1987, p. 112) has remarked,

> Investment in antidesertification, by its very nature, can only produce modest, though largely secure returns on a new preparedness of governments to change their time horizons for planning. With so many apparently more immediate and pressing problems, governments may feel they cannot afford

to concentrate on long-term development projects like range management and agro-forestry which produce so few exportable surpluses.

Conclusion

Desertification, or land degradation in drylands, is a long continued process (or processes) that afflicts developed and developing countries alike. It is caused by a blend of human misuse and climatic stress, but there are a wide range of technological solutions that can help to alleviate adverse trends. However, ultimately, as with other types of environmental decay, such technological changes will only be fully successful when underlying economic, social, and political structures are favorable. Moreover, as is evident from the experience of people in the American west or in Central Asia, large-scale, high-technology activities associated with manipulation of water resources (e.g. by center-pivot schemes or inter-basin water transfers; see chapter 12, Petts) can cause disastrous degradation. Many desert landscapes are highly sensitive and need to be treated with circumspection.

Further reading

A good general guide to arid lands and the use that is made of them by humans is Heathcote, R.L. (1983) *The Arid Lands: Their Use and Abuse* Harlow: Longman. The best treatment of desertification is Grainger, A (1990) *The Threatening Desert: Controlling Desertification*. London: Earthscan.

A useful source of general information on dryland degradation is a journal called *The Desertification Control Bulletin*, published by the United Nations Environment Program.

A discussion of methods for ameliorating conditions within deserts and on desert margins is provided by Goudie, A. (1990) *Techniques for Desert Reclamation*. New York: Wiley. Binns, T. (1990) Is desertification a myth? *Geography*, 75 (2), 106–13 probes the whole meaning of the desertification concept.

PART VI

Case Studies of Human Impact

CHAPTER SEVENTEEN

Case 1: Changing Use of the Sahara Desert

Erhard Schulz

Introduction

It is a time of "global change" and "planetary engineering" (Heuse, 1987), in which the "spreading of the desert" represents one of "humanity's greatest challenges" (Cloudsley-Thompson, 1977). To combat this, continent-wide programs of afforestation have been proposed (Agnew, 1990a, b; Faure and Faure-Desnard, 1990), while large-scale exploitation projects such as Libya's "great man-made river" aim to use the remaining resources of the desert (Joussiffe, 1990). But plans to use the resources of a particular region should be guided by our knowledge of that region. This chapter concerns the landscape of the central part of the Sahara along with adjacent lands (figure 17.1), an area which the author knows at first hand from fieldwork.

In dealing with arid zone resources and their exploitation, one has to consider first of all the nature and limits to the desert. But in trying to describe or define the desert one enters automatically into a chaos of different definitions, which attempt to cover up the fact that desert is in many respects a subjective and emotional term meaning "no man's land" or "dangerous land." Curiously, the search for some solid limits very often results in having to adapt the limits defined by annual precipitation or evaporation coefficients (Dubief, 1959–63; Mensching, 1981; Schmida, 1985). If one looks at the Sahara, the most characteristic climatic feature is the interaction of the summer monsoonal rainfall regime with that of the Atlantic or Mediterranean depressions. It is not so much the low amount of rainfall but the irregularity of precipitation which probibits living organisms normally adapted to a regular quantity of water. Moreover, the definition of desert limits by the amount of annual rainfall can easily lead to the idea that the desert is spreading or retreating when this is assessed over a short time period (Bernus, 1983; Reichelt, 1989). The restricted number of

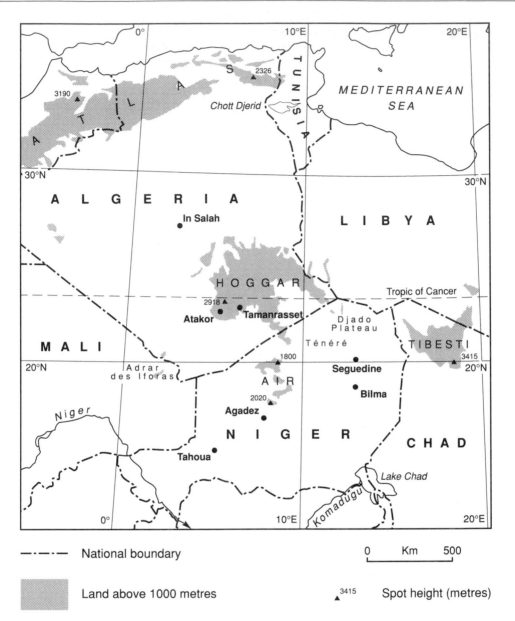

National boundary

Land above 1000 metres

0   Km   500

▲3415   Spot height (metres)

Figure 17.1 Location map of the central Sahara.

meteorological stations in the Sahara and the often short period of measurements further limit the value of purely climatic definitions.

If one tries to find some stable indicators, the definition of the Sahara becomes very complicated. Geomorphic features are different at the northern and southern margins, and the value of endorheism (Schiffers, 1951) is counteracted by large rivers, such as the Nile and Niger, coming from other regions. For soils there is discussion over whether any pedogenesis is possible in the desert and if yermosoils or the presence of calcic or argillic horizons might define a desert environment (Aguilar, 1987; Blume et al., 1985;

Skowronek, 1987; Valentin, 1981; Völkel, 1989). One also has to consider relict paleosols as strata in desert soil profiles (Vogg, 1987). Perhaps two phenomena are most characteristic in desert soils. These are, first, the formation of vesicular soils exposing superficial layers rich in enclosed air bubbles within fine grained sediments, which impede the germination of plants (Volk and Geiger, 1970) or superficial pellicular organization (Valentin, 1981), and, second, the spatially isolated distribution of true soils dependent on locally favorable conditions (Schulz and Pomel, 1993).

In the landscape there is one other clearly visible phenomenon which could be considered to summarize the effects of a range of different environmental parameters, which are themselves hardly measurable. This is the vegetation and its distribution. In the Sahara there are two regions where the vegetation distribution changes remarkably over a distance of a few kilometers. This is the transition from a diffuse permanent vegetation dependent on sufficient annual rainfall to one of a spatially contracted permanent vegetation, usually linearly restricted to wadis and depressions. In this latter zone the annual precipitation is no longer sufficient to allow a diffuse cover of trees and bushes, and they can only survive in places where runoff and groundwater are available as compensation (Monod, 1954; Schulz, 1988; Walter, 1964).

The vegetation of the Sahara and the adjacent regions

Two zones of change from a diffuse to a linear distribution of permanent vegetation are visible in the Sahara, one at about 30°N latitude, the other at 16–17°N (figure 17.2). These transitions are very clear and sharp and they can be used to limit and define the nature of the desert itself. The boundary at about 30°N, marking the northerly transition from semi-desert to desert, sees the sparse but diffuse cover of bushes and tussock grasses change to a linear distribution of trees and tussock grasses in wadis and depressions. This coincides with a shift in floristic composition too, going from the *Artemisia–Ephedra–Chenopodiaceae* communities of the Mediterranean influenced semi-desert into the *Acacia–Panicum* or *Tamarix–Stipagrostis* units of the desert (plates 17.1 and 17.2), which are mainly composed of Sahara–Sindian elements (Barry et al., 1976; Bornkamm and Kehl, 1985). This northern vegetation transition also runs parallel to an annual precipitation of about 50 mm in the Mediterranean winter rainfall regime.

Contrary to the almost treeless semi-desert where the stress of frost and aridity restricts the presence of tree lifeforms (Stocker, 1962), the desert vegetation can be described as a permanent tree, bush, and tussock grass vegetation along wadis and in depressions. As the main ecological factor for vegetation is water amount and regularity, all the minor factors, such as substratum type, relief,

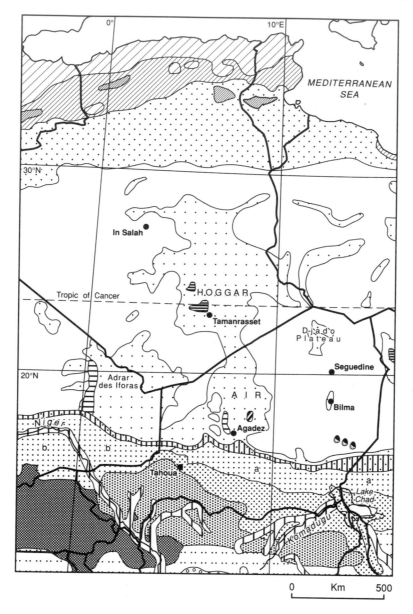

Figure 17.2 Schematic vegetation map of the central Sahara and the surrounding areas (modified after Schulz, 1987).

aspect, and salt content, become important for the establishment and for the survival of plants.

In principle there are two strategies for vegetation to survive in harsh arid environments. The first is to equip a small number of organisms with a range of protective adaptations against desiccation. These include the C_4 photosynthesic pathway which is typical of many tropical plants and which allows the uptake and storage of

Vegetation Legend

Mediterranean shrub & forest *(Quercus, Pistacia, Pinus, Cedrus)*

Mediterranean steppe *(Stipa, Lygeum)*

Chenopodiaceae vegetation of the coasts & the chotts (playas)

Semi-desert *(Artemisia, Ephdra, Chenopodiaceae, Stipagrostis)*

Contracted desert vegetation *(Acacia-Panicum, Tamarix-Stipagrostis)*

Ephemeral achab vegetation

Diffuse *Artemisia-Ephedra* vegetation of the high mountains of the central Sahara (semi-desert)

Acacia-Commiphora-Rhus savanna of the high plateaux of the Air mountains

Maerua-Acacia savanna of the high plateau of south-east Niger

Semi-diffuse *Acacia-Panicum* vegetation (enlarged Wadi vegetation)

Acacia-Panicum savanna

Commiphora-Acacia savanna

Acacia-Leptadenia-Commiphora savanna

Piliostigma-Bauhinia-Acacia savanna

Acacia thorn bush around Lake Chad

Combretaceae savanna

Parkia-Butyrospermum-Terminalia savanna

Isoberlina-Daniellia Pterocarpus savanna

Isoberlina-Carissa-Ficus savanna of the Jos Plateau

Afzelia-Lophira savannas & open forests

Riparian woodland

Figure 17.2 *(cont'd)*

CO_2 during the night and photosynthesis during daytime by closed stomata (Burkis and Black, 1976). The establishment of trees and bushes depends on the possibility of reaching the groundwater table, and the first stages of growth are underground. Once the roots have arrived at the groundwater level trees will follow their internal life cycles involving transpiration and flowering more or less independently of the annual rainfall (Ullmann, 1985). Reproduction takes place vegetatively by way of shoots from the superficial roots as well as by seed germination. This results in the typical grouped structure of trees in wadis. Many of the tree species have a bipartite root system in which the lateral superficial roots can exploit rainfall and runoff and the tap roots reach to the groundwater, often more than 30 m deep.

Plate 17.1 Contracted *Acacia–Panicum* vegetation in front of the Messak Mellet escarpment, Fezzan, southwest Libya.

Plate 17.2 Contracted *Tamarix–Stipagrostis* vegetation on the Djado plateau, Enneri Achelouma, northeast Niger.

Plate 17.3 Achab flora of *Astragalus* and *Fagonia* between Djebel Acacus and Tassili-n-Ajjer, Algeria.

The second strategy is that of the mass response. The infrequent rainfall events are answered and exploited by the achab floras, short-lived plots of annual grasses and herbs which have little or no protection against desiccation and which have to fulfill their life cycle within a few weeks (plate 17.3). The dormancy of the seeds may last several years before the next rainfall. These achab floras are very common on sandy plains as the substrate can store humidity more effectively, and the germination is not impeded by the superficial air layer in the clayey or silty substrata. Substrate types are also important for the distribution of the permanent vegetation. On coarser sands the *Acacia–Panicum* communities are common in the wadis, whereas *Tamarix* and *Stipagrostis* units colonize the fine-grained material. Along wadis one can often see these changes in the vegetation types depending on the sediment type. In gravel and boulder areas one will often find *Maerua* and *Aerva* communities together with several *Fagonia* species.

In the mountains of the central Sahara the vegetation shows a characteristic altitudinal zonation. The Hoggar mountains and, comparable to them, the Tibesti massif have, at about 2000 m altitude, a transition from the linear *Acacia–Panicum* vegetation to a sparse but diffuse bush cover of *Artemisia* and *Ephedra* (plate 17.4). It represents the change to a semi-desert flora dependent on winter rains, whereas at Tamanrasset at about 1400 m altitude the maximum rainfall occurs during the summer.

Plate 17.4 *Artemisia–Ephedra–Moricandia* semi-desert in the Atakor massif, Hoggar mountains, southern Algeria.

The altitudinal zonation of vegetation in the Air mountains (northern Niger) stands in contrast to that in the Hoggar and Tibesti mountains. In the Bazgan and Egalak ranges of the Air massif, for example, the transition at around 1600 m altitude to a diffuse tree formation of *Acacia–Commiphora–Rhus*, representing a mountain savanna, is restricted to the area underlain by solid granite. This forms large blocks between which natural cisterns have been formed. Moisture and fine-grained sediment can collect in these hollows, and they provide food niches for the establishment of bushes and trees (Schulz, 1988; Schulz and Adamou, 1988). The two different types of altitudinal zonation are shown in block diagram form in figure 17.3, which also explains the typical arrangement of the different tree species along the wadis.

Both of the principal vegetation types of the high mountains contain a certain number of isolated floristic elements, such as *Erica arborea* for the Tibesti, *Olea laperrini* for the Hoggar and the Air mountains, as well as *Dichrostachys glomerata* and *Boscia salicifolia* in the Air mountains. Other plant species showing a disjunct distribution are *Moricandia arvensis* (Cruciferae), *Rhus tripartita* (Anacardiaceae) and *Ziziphus lotus* (Rhamnaceae), these being present in the Atlas mountains, the pre-Saharan semi-desert, and the Hoggar mountains. Some groups of *Olea laperrini* and *Myrtus nivellii* are still present in the Atakor (Quezel, 1965). Isolated elements are often interpreted as relicts of former more favorable climates (Quezel, 1958, 1965), but they may also represent taxa dispersed over long distances by animals or birds.

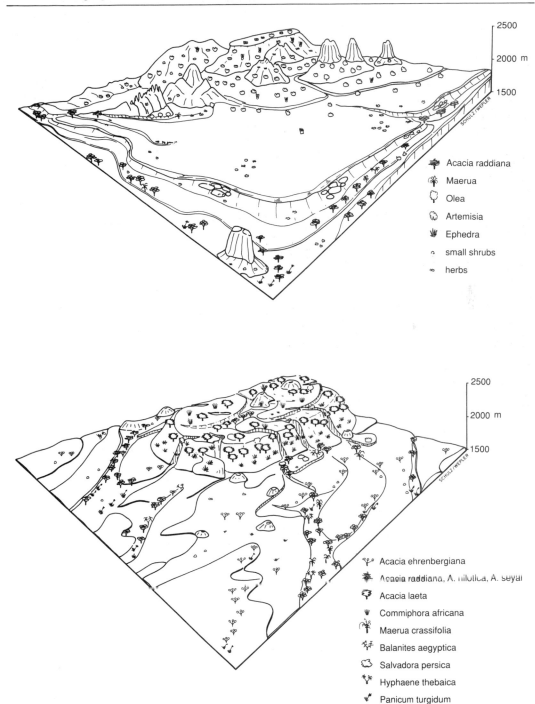

Acacia raddiana
Maerua
Olea
Artemisia
Ephedra
small shrubs
herbs

Acacia ehrenbergiana
Acacia raddiana, A. nilotica, A. seyal
Acacia laeta
Commiphora africana
Maerua crassifolia
Balanites aegyptica
Salvadora persica
Hyphaene thebaica
Panicum turgidum
Cymbopogon sp.

Figure 17.3 Block diagrams showing the altitudinal zonation of vegetation in the Atakor/Hoggar mountains of Algeria from the desert to the semi-desert (top) and in the Bagzan/Air mountains of Niger from desert to savanna (bottom).

The vegetation transition marking the southern edge of the desert occurs at about 16–17°N. Here the transition is made within the *Acacia–Panicum* complex, and represents a change to a diffuse tree and grass vegetation under a tropical summer rainfall regime. This is the transition from desert to savanna, and it coincides with an effective annual precipitation of about 150–200 mm and a three-month rainy season. The change takes place on a range of different substrata and relief conditions. It is climatically controlled, and the change from a linear to a diffuse distribution of the permanent vegetation represents the only climatically determined boundary in West African vegetation.

For the savanna regions south of the Sahara there is still considerable debate as to whether the Sahel is only a transition to the Sudanian savannas (Barry, 1982; Barry et al., 1986) or whether the Sahel represents a savanna zone in itself with defined limits (Le Houerou, 1989; Monod, 1986; Schulz, 1988). The phenomenon of the diffuse versus contracted pattern of permanent vegetation is apparently now accepted as the limiting criterion for marking the definite changes in the landscape system of the southern Sahara. This vegetation change is often taken only to be representative of a certain isohyet for annual precipitation (e.g. 150 mm; Le Houerou, 1989). Because the annual precipitation has changed remarkably during the past few years – with good rainy seasons in 1986–8 (Agnew, 1990b; Malo and Nicholson, 1990) – it is evident that isohyets do not truly reflect landscape boundaries.

By looking at the landscape and the vegetation south of this limit, one can recognize that after 10–20 km there is a second change in the vegetation. The sparse tree cover is still very low but is now characterized by *Commiphora africana* together with several *Acacia* species, *Maerua* and *Ziziphus*. The herb layer is exclusively composed of annual grasses and herbs which become denser and richer in their floristic composition toward the south. This represents the transition to the landscape and the savanna system of the Sahel, which also incorporates *Acacia–Piliostigma–Bauhinia* or *Combretaceae* savannas in the south. Rainfall increases gradually to about 600–700 mm per annum and the rainy season lengthens to 4–6 months.

The Sahelian savannas are for the most part semi-natural ecosystems, resulting from former and current agricultural systems, such as that for gum production. A related impact has derived from the nomads from the Sahara and the Sahel and the sedentary cultivators, whose use of fire and whose grazing animals have significantly altered the character of pasture lands (Frederiksen et al., 1990; Trollope, 1982) (plate 17.5). In consequence, it could be argued that the Sahelian savannas cannot serve as climatically determined vegetation units (Obeid and Seif, 1970; Seignobos, 1982).

Historical descriptions of the Sahara and its vegetation from the nineteenth century very often report denser and more varied vegetation types, but in principle they closely resemble the present

Plate 17.5 Degraded *Acacia–Balanites* savanna north of Tabalak, Niger; monotypic *Cenchrus* grass cover; *Calotropis* indicates overgrazing.

ones. The most striking fact is that all reports agree about the location of the southern margin of the Sahara, namely as the change from linear to diffuse vegetation, the latter under names such as desert-steppe or steppe forest (Denham, 1822–5; Barth, 1857–8; Rohlfs, 1874–5; Nachtigal, 1879). This transition is reported along the old Borno road from Lake Chad to Tripoli, where it could be observed at the same locality in 1984 in a poor state but still with several generations of living trees (Schulz, 1988). This implies that the desert in the Sahara as a system has remained relatively stable in its limits on a secular timescale and that the term of desertification has nothing to do with a spreading of the desert itself. Some recent rumors about a shrinking of the Sahara (Diegne, referred to by Booth, 1991) are apparently based on a misunderstanding of the rapid reaction of the ephemeral achab vegetation to changing rainfall conditions.

In summary, the present distribution of the vegetation in the Sahara and the Sahel represents a zonation which is, in a certain manner, dependent on climate, especially on the effective amount of precipitation, but only the boundaries of semi-desert to desert and desert to savanna are really climatically controlled. The type and density as well as the floristic composition of the present savannas in the Sahel are more or less anthropogenic, representing a large variety of ancient and modern agricultural and pastoral systems; in other words, they form a cultural landscape.

Table 17.1   Vegetation zonation for the central-southern Sahara and the Sahel

Sahara	Semi-desert Diffuse permanent vegetation, *Artemisia–Ephedra–Retama–* *Stipagrostis, Chenopodiaceae.* Shrub and tussock grass communities, achab floras.
	Desert Contracted permanent vegetation, *Acacia–Panicum/Tamarix–* *Stipagrostis.* Tree, bush and tussock grass communities. Achab floras.
	Savanna Diffuse permanent vegetation, thorn-tussock grass savanna. *Acacia–Panicum* savanna.
Sahel	*Commiphora–Acacia* savanna, thorn savanna. *Acacia–Bauhinia–Philiostigma* savanna. *Combretaceae* savanna. Dry savanna.

A simple scheme (table 17.1) shows that the Sahara consists of semi-desert, desert, and savanna and that the schemes of Jaeger (1956) and Troll (1952) for the savanna zonation are still valid.

The different forms of human use of vegetation

The traditional use of vegetation in the Sahara and the adjacent drylands includes pastoral nomadism as well as fodder supply for animals kept near the house, human consumption, medicine, supply for construction, and as fuel for energy. Of these, the human impact is greatest in the use of the plant cover to feed domesticated animals. It varies according not only to the number of animals but also to their type. The different behavior of grazers (sheep and cattle) or browsers (camels and goats) is clearly visible in the state of vegetation. Sheep and cattle graze systematically, cutting the grass and herb cover down to a sort of lawn. Cattle rarely attack trees in the Sahara, with the exception of *Acacia albida*, whereas in the Sudanian savannas they are kept in the forests or thickets to feed on the leaves (Boutrais, 1980). As browsers, camels and goats take parts of the grasses, herbs, bushes, or trees and tear them off, while donkeys very often attack the bark of the trees. On the other hand, goats and camels touch a greater number of plants instead of over-exploiting a few at a time.

   The second differentiation is the period of time over which these animals can exist without water. Cattle, goats, and sheep have to be watered at least every second day whereas camels can survive longer than two weeks without drinking, depending on the water content of the pasture (Gauthier-Pilters and Dagg, 1981). These facts limit the pasture area of the different animals. Goats and

Plate 17.6   Lake Elki
Guernana, Kawar,
northeast Niger.
Groundwater fed lake in
front of the Bilma
escarpment. Oasis
vegetation. Palm gardens.
Salt evaporation in the
foreground between
Cyperus swamp.

sheep use only restricted areas around water-points or along a wadi (plate 17.6). These conditions will often provoke heavy damage to the vegetation as the herds are kept in limited areas in the neighborhood of water-points or in the villages. This is one of the most important causes of desertification in the savanna regions.

In contrast, camels can exploit pastures in the very remote areas and they can use the different achab floras without regard to water-points or wells. Camel nomads very often have a highly evolved knowledge of where achabs will possibly develop in a given meteorological situation, and they will guide their herds according to their experience. In several nomadic tribes of the Sahara there are traditional circulation patterns during the year (Toupet, 1963).

An additional effect lies in the sociological organization of the tribes. For example, in the Tuareg society of the Air mountains in Niger, the goats belong to the women who are free to herd them as they wish, while cattle or camels normally belong to the men.

There is also a difference in the reproduction rate. A herd of goats may be re-established after drought periods within a short time, whereas a herd of camels requires several years. So when a short-term livestock strategy predominates goats are preferred, although the risk of overgrazing on already damaged vegetation is then high. Indeed, a goat herd may be re-established after a drought period more rapidly than the vegetation, further increasing the risk of degradation. These different reactions of the animals are some-times used in a kind of strategy in responding to drought periods or other difficulties. The livestock is diversified to give the chance to use the different animals in various ways and also to have some of them left over to sell in times of need (Klute, 1990; Swift, 1973).

The effects of pastoralism are certainly not simply negative; rather, they create a certain type of vegetation. They change the vegetation either by eliminating palatable plants or by reducing their numbers and replacing them by others which are less browsed and which act as weeds under these conditions. The development of scrub formations as a result of overgrazing is reported from many arid regions (Walter, 1964), and in the Sahara and the Sahel large stands of *Calotropis procera* (Asclepiadaceae) develop on overgrazed areas (Le Houerou, 1980).

Putting animals out to pasture can also directly affect the diffusion of plant species. The seeds of species such as *Acacia laeta* and *Acacia raddiana* may be transported in animal intestines and thus be prepared for better germination in the ground. The spread of *Cenchrus biflorus* – the famous "cram cram" grass – which represents the pasture in the savannas of the northern Sahel, is mainly a result of its "sticking" fruits. Direct transport, perhaps involuntarily, will have been responsible for the distribution of elements like *Acacia albida* and *A. laeta* and equally many Sahelian and Sudanian trees found in the Saharan mountains (e.g. *Anogeissus leiocarpus*, *Boscia salicifolia*, *Dichrostachys glomerata*). Their presence is often interpreted as relict of older, more favorable climatic periods (Bruneau de Mire and Gillet, 1956), but it is equally possible that they represent a certain type of "caravan flora," having been transported with the camels or their loads, and consequently being present around the mountain pasture areas. Similar observations were made in the Jebel Marra in the Sudan republic (Miehe, 1988; Wickens, 1976). Very often trees here do not have the chance for natural regeneration and so they reflect instead the repeated colonization of mountain areas by plants carried by the camels or their drivers (Schulz et al., 1990).

Contrary to the savanna regions, fire is not used as an instrument for clearing and preparation of pasture in the desert proper.

Human alimentation

Research on the food procurement of the Tuareg societies of the Hoggar and Air mountains has been conducted by Bernus (1981), Gast (1968), Nicolaisen (1963), Spittler (1989), and Schulz and Adamou (1988). Klute (1990), Smith (1980), and Swift (1973) investigated the economy and culture of the nomadic tribes of the Malian southern Sahara. All these investigations have highlighted the nomads' broad and profound knowledge of the vegetal resources as well as the organization systems which protected certain areas against grazing when the wild millet was ripe. This was still

practiced in the 1950s in the Hoggar mountains (Gast, 1968; Nicolaisen, 1963). In the southern Sahara and northern Sahel there are some effective harvesting methods still in use to collect the wild millets. These are the use of baskets or a simple sweeping of grains with a broom (plate 17.7). Sometimes ants' nests are also exploited for their content of small grains or seeds.

Plate 17.7 Harvesting of wild grasses for human consumption by using a swinging basket; Tilemsi valley, west of Adrar des Iforas, northeast Mali.

These various techniques have been the subject of much discussion in terms of their effects on the origins of agriculture and domestication of plants; more specifically, harvesting with a knife or sickle selects for a tough rachis in the grain head, while beating into baskets preserves the naturally brittle rachis of grasses (Harlan, 1989a, b; Smith, 1980). In contemporary Saharan cultures, harvesting of wild cereals is often considered to be restricted to famine periods when other food is no longer available. Spittler (1989), for example, points out that this is typical of the Kel Ewey Tuareg of the Air mountains, Niger. They are very proud of their system of food procurement, which is exclusively based on imported millet from the Sahel and the Sudan. They take it as a sign of civilization and the use of food collected from wild plants is regarded as a fall-back to a more primitive cultural state.

The kind of alimentation varies from tribe to tribe and it also depends on the type of economy. The traditional economy of the Sahara includes full nomadism, semi-nomadism with some garden culture or tree exploitation, and fully sedentary oasis garden culture. But one has to consider that it is very rare to find only one

Table 17.2 Commonly used plants for human consumption in the Sahara and Sahel

Grains, seeds

+++Acacia albida	+++Boerhavia repens	+++Dactyloctenium aegyptium
Acacia raddiana	Cassia tora	Digitaria ciliaris
++Balanites aegyptiaca	Cirullus lanatus	+Echinochloa colona
		+Eragrostis cilianensis
+Celtis integrifolia		+Eragrostis tenella
	++Colocynthis vulgaris	++Panicum laetum
Dichrostachys glomerata	——	+Panicum turgidum
Parkia biglobosa		++Sorghum aetiopicum
Tamarindus indica	Eruca sativa	Stipagrostis adscensionis
	Farsetia ramosissima	
——	Limeum indicum	+Stipagrostis plumosa
		++Stipagrostis pungens
++Boscia senegalensis	Portulaca oleracea	
	Sesbanisa pachycarpaea	
——	++Tribulus terrestris	
Aizoon canariensis	Cenchrus biflorus	

Fruits

+Acacia albida	++Maerua crassifolia	++Grewia tenax
+Acacia raddiana	Parinari crassifolia	Grewia villosa
Adansonia digitata	++Rhus tripartita	Leptadenia pyrotechnica
		Myrtus nivellii
++Balanites aegyptiaca	Sclerocarya birrea	++Salvadora persica
+Borassus aethiopum	++Tamarindus indica	Strychnos spinosa
++Butyrospermum parkii	Ximenia americana	
++Celtis integrifolia	++Ziziphus mauretania	++Tapinanthus globiferus
	++Ziziphus spinachristi	
	——	
++Cordia sinensis	Boscia angustifolia	++Citrullus lanatus
	++Boscia senegalensis	
Diospyros mespiliformis	Capparis spinosa	Glossonema boveanum
Ficus gnaphalocarpa		+Leptadenia hastata
		Nymphaea lotus
Ficus salicifolia		
++Hyphaene thebaica		
Lannaea acida		

Leaves

+Acacia raddiana	Myrtus nivellii	Moricandia arvensis
		Polygonum salicifolium
++Adansonia digitata	Pithuranthus scoparius	Portulaca oleracea
	+Salvadora persica	Schouwia purpurea
+Celtis integrifolia	——	Tribulus terrestris
Combretum paniculatum	Achryanthes aspera	
Lannaea acida	++Amaranthus angustifolius	Veronica anagallis-aquatica
++Maerua crassifolia	++Amaranthus graecians	——
Moringa oleifera	Boerhavia elegans	Panicum turgidum
	++Cassia tora	
++Rhus tripartita	++Corchorus tridens	
+Tamarindus indica	Eclipta prostrata	
——	Eruca vesicaria	

Table 17.2   *(cont'd)*

Boscia salicifolia	*Glossoneam boveanum*	
++Boscia senegalensis	*Inigofera astragalina*	
Cadaba farinosa	*Launaea nudicaulis*	
Crataeva religiosa		
++Leptadenia hastata		

Rhizomes, roots, twigs, gum etc.

+Acacia raddiana	*+Cistanche phelypea*	*Panicum turgidum*
	Cynomorium coccineum	
Acacia seyal	*Nyphaea lotus*	*Phragmites communis*
	———	*Scirpus holoschoenus*
+Borassus aetiopum	*Cymbopogon*	*Typha australis*
	schoenanthus	
++Tamarix aphylla	*Juncus maritimus*	
———		
Calligonum comosum		

+++ very common; ++ frequent; + occasional.
Source: after Busson, 1965; Bernus, 1981; Schulz and Adamou, 1988

type of economy. For the most part there is a combination of long-distance pastoralism with camels or caravaning, middle-distance pastoralism of sheep and goats, along with oasis activities. This is either maintained in the same clan or organized hierarchially, with the aristocracy taking care of the long-distance nomadism, and ordinary tribe members or captives and their descendants caring for the garden culture in the oasis. In the Air mountains this organization still operates in some oases and creates great problems between the cultivators and the nomads, as the cultivators will no longer accept the attitude of dominance by the old aristocracy. Other sedentary work, such as the exploitation of salines or salt-mines, is included in the same way. This mixed economy is also represented in the organization of the traditional caravan cycles between Sahara, Sahel, and Sudan.

Animal-keeping assures a basic supply of animal protein, not so much from meat, but from milk and milk products. The proportion of milk and milk products in the diet of the Tuareg societies varies between 50 and 80 percent, also largely fulfilling calorific needs (Klute, 1990; Felix-Katz, 1980). Felix-Katz (1980) reports from cattle-keeping Tuareg in the Azawak region, Niger, a mean calorific intake of about 2700 kcal per person per day based on milk, imported millet, and very little meat. The consumption of meat is very rare and mostly restricted to festivities. Beside the meat of goat and sheep that of wild game is highly appreciated, even if it is very rare now. The meat of gazelles is most preferred. Animal protein comes from other sources too, including turtles, ravens, lizards, and insects. Locusts are especially appreciated, while in the lakes of Kawar of northeast Niger insect larvae (Syrphidae) are systematically collected for protein supply.

If one asks about the quantity of collected plants, fruits, and grains in the southern Sahara, the effective yield of wild millets and grasses varies between 40 and 150 kg ha^{-1} according to Schulz and Adamou (1990), or up to 600 kg ha^{-1} reported by Harlan (1989a). This indicates that in normal rainy seasons a basic food supply from the exploitation of the wild plants is assured, at least for a restricted number of people. The cattle-keeping Tuareg in the Azawak region (Niger) in 1988 reported that the yield of wild millet (*Sorghum aethiopicum*) was sufficient for the next winter in combination with milk and cheese. These wild millet fields re-appeared after five years of drought and the traditional behavior of collecting was re-activated. The importance of wild millets may also be seen in the fact that grains of *Cencrus biflorus* were a normal part of caravan trade goods at the beginning of this century (Fontferrier, 1923).

One has to remember that trees and bushes are equally exploited. Table 17.2 shows the large variety of trees, bushes, herbs, and grasses which are normally used for human consumption in the Sahara, the Sahel and the northern Sudan. Trees and bushes serve in two kinds of way. For the first, they can assure a basic supply of certain dietary elements, such as grease (e.g. from *Balanites* and *Hyphaene*) when milk is not available. They are also collected for sweet fruits (*Hyphaene*, *Grewia*, *Ziziphus*, *Salvadora*, or the various *Acacia* gums) and for the vitamin-rich leaves of *Maerua* or *Rhus* (van Maydell, 1983) (plate 17.8). The second way is the assurance of a basic level of sustenance in famine periods. This is the case for the fruits of *Acacia* or the leaves of *Maerua* and *Leptadenia*. *Maerua crassifolia* assured the survival of several

Plate 17.8 Collecting of *Rhus* leaves by Tuareg woman, in between granite blocks, northern Bagzan, Air mountains, Niger.

Tuareg in the Air mountains during the drought period of 1984/5. These leaves were mixed with millet, extremely rare at that period, or they replaced the millet for a certain time (Spittler, 1989). Very often the fruits of *Hyphaene* are collected for their starch and sugar content in the exocarp.

The collection of food from trees and bushes employs a range of different methods, as for the grasses and herbs. As trees are inhabited by ghosts or devils according to local cultural belief, collection must not exceed three handfuls of leaves or fruits from each tree. In this way a direct mass exploitation of biological resources is avoided in the desert. Contrary to the grasses and herbs, a quantitative evaluation of potential collection yield is not possible, because of the ritual restrictions as well as other taboos, or simply because of ownership, which often plays a role for important tree species.

In summary, traditional human alimentation in the Saharan desert has often been assured from wild plants in combination with milk. The former are incorporated in the combination of different types of economy and should not be regarded as a primitive stage of human culture or only as a food surrogate in times of famine. Moreover, this use plays an important role for the people travelling in remote areas, either following their herds or avoiding contacts with sedentary people, which is the case for the border region of Niger–Chad–Libya, where the Tubu from the Tibesti region are herding their camels or where there is a lot of clandestine traffic (plate 17.9). In the caves of the Djado region, for example,

Plate 17.9 Tuareg with his camel going to Tarheranet, central Hoggar mountains, southern Algeria.

enormous numbers of shells of *Colocynthis* seeds are quite common, together with the grinding stones. These seeds give a sort of flour after complicated processing to eliminate the bitterness. Another example is the Kel Adrar, the Tuareg tribes of the Adrar des Iforas in northern Mali. They still follow an economy of animal-keeping, full nomadism, and the exploitation of the extended fields of the wild grasses for their alimentation. It gives them a certain economic independence, which was important in the struggle for political autonomy within the state of Mali (Klute, 1990; Swift 1973).

Medicinal use

A profound knowledge of the medicinal sources in the vegetation of the Sahara, Sahel, and Sudan is documented by Adam et al. (1972), Bernus (1981), Boulos (1987), Dupire (1962), Gast (1968), and Schulz and Adamou (1988). Recent investigations in the Air mountains of Niger by the author have shown that there is still a wide use of plants for medicinal purposes. This is especially the case for the Tuareg women, who need this knowledge for their own treatments of diseases, wounds, or inconveniences when they are isolated with their herds in the bush for months. Some of these women are specialized in healing particular diseases and they are consulted from far distances for their healing capacities. There is also a high degree of concurrence between different kinds of medicine, not only between modern European medicine and traditional herbal medicine but also between herbal medicine and the practice of the Islamic mullahs. This is based on the one hand on religion, as the herbalists are suspiciously regarded for their connection with ghosts and devils living in the trees, and on the other hand on the cost of treatment. Successful treatment of disease will cost about the same amount of money whether for herbal treatment, for treatment by the ablution of written Koranic verses, or by modern medicine. The Tuareg themselves use all the three medicines equally until one works, if they can afford the money.

Herbal medicine as well as the collection of herbs and leaves for spices or vitaminization of food, and for infusions, assure a good income for the women. For example, the herb and leaf mixture *Ilatan* from Mount Bagzan in the Air mountains is sold in proportion 1:100 for millet or directly for money. It is much valued and commercialized and widely distributed (Schulz and Adamou, 1988). The general problem of the herbal medicine today is not its acceptance in the Sahara and the Sahel, but the risk of loss of knowledge about the medicinal uses of plants. The older generation transmit this information only in confidence to a successor or they will take it with them to the grave. Because of this confidential barrier there is also restricted opportunity for investigation by foreigners. These traditions vary from tribe to tribe. The treatment of diseases is almost always done with mixtures of different plants and only very rarely by single plants.

On the open markets of the Sahara and the Sahel, between In Salah in Algeria and Tahoua in Niger, there is a certain fondness for herbal drugs. These herbal compositions are very much sought after and are relatively expensive to buy. They mostly consist of *Artemisia herba-alba*, *A. judaica*, *Pituranthus scoparius*, *Ammodaucus leucotrichus*, and especially *Myrtus nivellii*. They are used against stomach or intestinal problems, colds or bronchial ailments. *Artemisia* was in former times exported from the Hoggar to the south in large quantities and even now there is a tradition of travellers from the Air mountains coming back from Algeria to buy these medicines in large quantities in Algeria and sell them for good prices in the Air (M. Araghli, Timia, personal communication). A similar case is the collection of *Solenostemma oleifolium* as a surrogate for tea and as a medicine for bronchial diseases. This once represented an export good from the Tibesti and northern Niger to Libya and Egypt.

Exploitation of plants for construction, instruments, chemical processes, and energy supply

For these purposes it is mostly trees and bushes that are exploited. Traditional construction is based on wood for beams in house or hut roofs and grass mats to cover them. As solid hardwood is very rare and in consequence now expensive, the quartered palm trunks and branches from *Acacia* or *Tamarix* have been used for several generations of houses or huts. The trees mostly exploited for building are *Acacia raddiana*, *Hyphaene thebaica*, and *Phoenix dactylifera*. Smaller dwellings are often built on the stems and branches of *Caloptopis procera* or the leaf rachis of *Phoenix dactylifera*. Stems of *Phragmites* and *Typha* often serve for the walls and fences of huts or groups of huts. The thorny branches and twigs of *Acacia ehrenbergiana* or *Balanites aegyptiaca* are gathered in large quantities to fence the goats and sheep in, near the house or during the night in the bush.

The depilation and tanning of goat skin requires the latex and fruits of several plants. Plants from the *Asclepiadaceae* family serve for the latex in the depilation process. *Calotropis procera* and *Pergularia tomentosa* leaves and branches are collected and crushed, then the skins are covered with them and rolled and buried for a while to lose the hairs. The tanning itself is made with a maceration of *Acacia nilotica* fruits which are collected for these purposes in the whole area between the Sahara and the Sudan.

Mats are woven either of *Panicum* stems, which is the favored grass for these purposes, or from the leaves of the dum-palm (*Hyphaene thebaica*) in the southern Sahara and northern Sahel. The leaves of *Hyphaene* are cut in large quantities for their upper parts by the women of the blacksmith caste. They are roasted (fermented) in water for a certain period before they are pliable enough for the weaving of mats and baskets. Mats and baskets of different shapes are the basic equipment for caravan transport. It is the traditional work of the women in the mountains as well as in the towns, and it gives a certain direct income for them.

Hardwood of high quality is required for traditional equipment such as mortars and their pounding sticks, and sticks for working or crushing salt crusts. For this mostly *Acacia nilotica* or *A. raddiana* are exploited. Bowls, spoons, and beds are made from *Balanites aegyptiaca* wood, with the blacksmith caste being responsible for their manufacture. Because of these various activities, several tree species are endangered in the Sahara, such as *Acacia nilotica* and *Hyphaene thebaica*.

The preparation of salt in the salines, where saline groundwater is enriched by evaporation and then refined by boiling, is another wood-consuming process. Moreover, the direct leaching of salt from the ash of twigs and leaves of some bushes and grasses is common in the area surrounding Lake Chad, where *Salvadora* has been exploited for long periods (Barth, 1857).

The most important use of wood is as an energy supply for heating and cooking. Just about all trees and bushes in the Sahara and in the Sahel are heavily exploited and stripped from top to bottom. Even if there is a certain respect in the belief that the trees are inhabited by ghosts and devils and they are only deprived of one branch at a time, the result is the same. Moreover, not all the wood is exploited for energy and cooking. People always prefer freshly gathered *Acacia* wood or that of *Balanites*. *Tamarix* wood is less used because of the smoke, while that of *Commiphora* is refused. Dead wood is regarded as second class even though it is now forbidden by governments to cut fresh wood. But in the present situation all kinds of wood are collected, and over long distances. The steadily growing demand for energy forces people to go to remote areas to seek for wood, and wood becomes more and more expensive, so that on the Agadez market the price of wood for cooking exceeds the price for the meal itself.

The collection of wood is now a certain job for unemployed people, and along the main tracks and roadsides there is a large number of stores of wood sellers. The same can be seen on the main track to the Kawar in East Niger, which crosses the Air and the Ténéré. The traditional caravans and now also the lorry drivers have a lucrative job in buying wood and charcoal in the southeastern Air and selling it to the Kawar people. The production of charcoal, which is a reserved job for the blacksmiths, demands wood of even higher calorific value. Finally, the biggest threat to the wood reserve lies in the mode of cooking itself. The traditional cooking on open fire with branches and three stones for the pot wastes an enormous amount of energy.

In the Sahara there are only two ways to reduce this waste of energy. The Algerian government decided to subsidize the price of natural gas as cooking energy. Nowadays gas cookers are common in the remotest oases and the supply of gas is assured. There is no chance of similar measures being adopted in the Sahel countries because of the lack of their own fossil fuels. The only solution which seems to be successful is the propagation of iron stoves made by the local blacksmiths for reasonable prices. This can at least reduce the consumption of wood for cooking and seems to be widely accepted. The recent rise in petroleum prices has practically finished all discussion of a possible substitute for the use of wood for energy in these countries for the time being.

The government policy in these countries is clear: trees are protected, in some cases entirely. Forestry administration is under military government and there is a strict ban in large areas on touching trees. However, reality is different. Even though the protection of tree resources is widely accepted by the people in principle, in practice this is not respected, for two reasons. The first reason is desperation. People recognize the value and the importance of the trees but if other people come and exploit them, often raiding them in a kind of attack, they decide to exploit the few remaining resources for themselves. For the second reason, one can use the blacksmiths as an illustration. They have a certain reputation as outlaws in the Tuareg society, but on the other hand they are respected because of their contact with fire and "magic" processes. They will do a lot of work that other people would not dare to attempt.

There is also the question of whether a complete ban on all exploitation of the trees is logical. An intelligent use might actually protect and propagate them. As with hunting, the strict ban by the administration is only respected in public: in fact everybody does it, in secret, and the military are first among them.

Finally, the long-lasting tradition of metal smelting (gold, silver, copper, and iron) which dates back to ca.2000 BP (Grebenard, 1985) and has continued to modern times in the entire Sahel and southern Sahara has changed the landscape system. The ratio between wood and metal for the smelting process is comparable in different regions: 100 kg of high calorie wood are necessary to produce just 5 kg of metal (Devisse, 1990; Kadomura, 1989; Otto, personal communication). These proportions highlight the anthropogenic nature of the Sahelian and Sudanian savannas.

Conclusion

The Sahara desert is a clearly defined ecosystem manifested by a contracted, or linear, distribution of the permanent vegetation. The boundaries of the desert and the savannas are stable on a secular timescale and have changed only in geological times (e.g. early Holocene; see chapter 3, Street-Perrott and Roberts). This underlines the fact that the degradation both in the savanna and in the desert regions represents *in situ* modifications of the different ecosystems, and that it does not indicate a spatial shift in one landscape system at the expense of another. The desert and the savanna vegetation respond in different ways to changes in climate. The resilience of the desert and the savanna ecosystems seems to be relatively high as there are two ways in which they respond to climatic or ecological changes, first by ephemeral vegetation and only subsequently by permanent vegetation.

The dualism of restricted permanent vegetation and of ephemeral achabs which are variable in space and time forces a flexibility in the human utilization of biotic resources and therefore also in socio-economic structures. This was for a long time assured by the profound knowledge of the nature of these resources as well as by the traditional combination of long-distance pasture or caravan organization with short-distance pasture and oasis culture. The variability of the resources forced an adaptive exploitation, some-times responsive, sometimes planned. It also incorporated periods of hunger and misery when people could only survive if they left the region for some years. In this connection it is remarkable that former periods of drought and hunger were for the most part periods of vegetation recovery, because the population left the area. The Air mountains in Niger serve as an example. The large *Acacia* woods were all established after the Kaoucen revolt in 1917. However, during the last two severe drought periods the people could not move to other regions to the south, largely because of problems of political administration. For the Air mountains it is the first time that the people have had to earn their living entirely within the region itself (Spittler, 1989). Finally, the refugee problem created by the most recent drought periods has been one of the major causes of conflict and bloodshed between the governments of Niger and Mali and the Tuareg populations of these countries.

Demographic increase, the establishment of political boundaries which cut across the traditional economic regions of different ethnic groups, and the declining economy will be a threat to the scarce resource base for a long time. Protection plans against further degradation are numerous. They range from forest belts against the "spreading of the desert" to afforestation in the desert wadis, as well as stabilization projects for all kind of dunes in the desert and the Sahel. Well-meaning activities to rebuild the reduced lifestock, herds of the type "a goat for the Sahel," will add a further threat to the endangered regions.

As a result of the severe droughts since the late 1960s there is a virtual absence of shallow groundwater which will need a long time to be recharged, if it was not all fossil, that is non-renewable. Like the vegetation, groundwater is a very fragile resource for human survival in arid lands. If groundwater is not fossil it is only occasionally recharged by runoff in the wadis and the consequent infiltration into fine-grained wadi sediments. Artificial changes in the wadi systems have included the protection of garden areas against bank erosion by the building of revetment walls. These development aid projects have succeeded in protecting the riverbanks but have changed the hydrological system permanently in impeding the slow infiltration of water and consequently the recharge of local wells. Comparable careless exploitation of fossil groundwater often creates damage such as salinization or compaction of soils following irrigation without the necessary drainage.

Concerning forest activities, one can plant as many trees as one wants, but these trees will begin to consume water when they reach the age of three or four years, and it is not sure that there will be enough water for them. It is easily possible to destroy or waste the last remaining resources by introducing well-meaning but inappropriate large-scale projects. This is not to mention the introduction of non-native trees like *Eucalyptus* and *Azadirachta*, which consume large quantities of water to support their rapid growth, and do not have the range of uses of indigenous species.

Knowledge about the traditional use of desert vegetation resources runs the risk of dying out. Over-exploitation and overgrazing could cause the disappearance of many species which nowadays are desperately sought after as the last genetic resources for the heavily exploited and ecologically weak cultivated plants of European-type agriculture.

Preservation of these limited resources requires knowledge about their nature and origin and an adaptive small-scale form of management. It must be based on the diversity and the variability of desert ecosystems on the one hand and on a recognition of the indigenous population on the other. A recovery of the now heavily damaged vegetation will take a long time and will follow a succession which is poorly understood. So far the recovery potential of the different savanna types and the desert vegetation does not seem to be totally exhausted. This succession will be the only way for the establishment of the plant cover which can be adapted to the changed climatic and pedological conditions. It will then tolerate a modest and intelligent exploitation. For instance, the re-establishment of lifestock by foreign development projects will certainly be more rapid than the recovery of the vegetation itself, and so will lead to further degradation.

Unfortunately, the present combination of demographic increase, political instability, and rising prices for energy does not provide great hope for the preservation of the knowledge about a wise use of the desert's resources and the survival of the variety of cultural traditions.

Further reading

Le Houerou, H.N. (1989) *The Grazing Land Ecosystems of the African Sahel.* Ecological Studies, 75. Berlin: Springer.

Nicolaisen, J. (1963) *Ecology and Culture of the Pastoral Tuareg.* Copenhagen: Nationalmuseets Skrifter.

Quezel, P. (1965) *La Végétation du Sahara du Tchad à Mauretanie.* Stuttgart: Geobotanica Selecta.

Spittler, G. (1989) *Dürren, Krieg und Hungerkrisen bei den Kel Ewey (1900–1985).* Wiesbaden:

Williams, M.A. and Faure, H. (eds) (1980) *The Sahara and the Nile.* Rotterdam: A.A. Balkema.

CHAPTER EIGHTEEN

Case 2: The Chesapeake Bay Estuarine System

Grace S. Brush

Introduction

More than almost any other area, estuaries, representing much of the Earth's land–ocean interface, have been exploited by humans. Their rich food resources and open transport routes have made them among the most densely populated regions on Earth. Buffering the ocean from the land, they suffer the effects of human activity on land as well as water. The degree of exploitation differs for different parts of the Earth, but in general has intensified with the growth of large-scale agriculture and urban development.

This chapter is about the effect of human activity on an estuarine system. The story revolves around Chesapeake Bay, the largest estuary in North America, and the fourth largest in the world. Due to a scarcity of written documents, this historical account relies heavily on data archived in sediments deposited in the estuary from its inception. The reasons for focusing on the Chesapeake Bay are two-fold. The major human impact has occurred within the past two to three centuries following European colonization, as the human population in the region drained by the Chesapeake soared from 2 million to 14 million. Prior to that time, small populations of native people, whose activities were mainly hunting and fishing, occupied the area. Thus the impact of human activity on the estuary can be evaluated by analyzing environmental indicators preserved in sediments deposited before and after European settlement. The second reason is that the record preserved in the sediments has been used extensively to reconstruct a long-term history of the Chesapeake system as a guide for managing the estuary. This large geographic database, spanning several millennia, provides insight on the effect of human activity on estuarine systems within the longer context of natural variations in climate.

What is an estuary?

Of the numerous definitions of estuaries put forth in the literature, the one proposed by Cameron and Pritchard (1963), which defines an estuary as a "semi-enclosed coastal body of water having free connection with the open sea and dilution of sea water by land drainage," is the most inclusive. Estuaries, so defined, include highly stratified fjords typical of the Alaskan and Scandinavian coasts (figure 18.1a), salt wedge or bar-built estuaries such as Pamlico Sound off the North Carolina coast in the USA (figure 18.1b), and partially mixed estuaries in the drowned river valleys of the coastal plains throughout much of the world, including Chesapeake Bay and its tributaries (figure 18.1c).

Estuaries form along oceanic edges of continents during warm interglacial periods, as rapid increases in sea level follow the retreat of glaciers. Geologically, they are momentary features on the landscape, with a maximum lifetime of 10 000–15 000 years. Coastal plain estuaries, the focus of this chapter, were formed by the encroachment of seas into previously formed river valleys. Typically shallow, but often containing deep holes or trenches

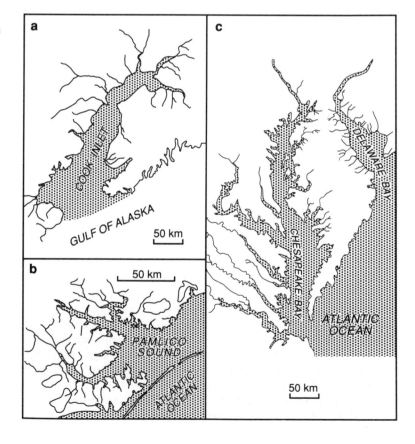

Figure 18.1 Drawings of representative estuaries in North America. (a) Cook Inlet, Alaska, an example of a fjord estuary; (b) Pamlico Sound, North Carolina, an example of a bar-built salt wedge estuary; (c) Chesapeake and Delaware Bays along the mid-Atlantic coast of the USA, two examples of partially mixed estuaries. The Chesapeake is characterized by a much more extensive land–water interface.

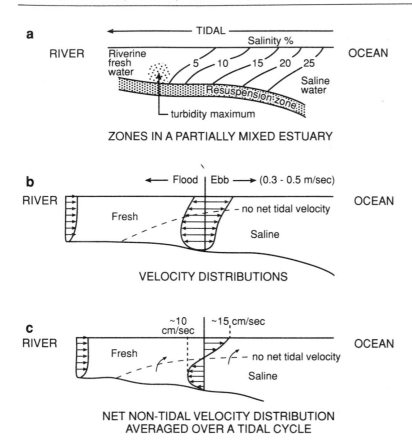

Figure 18.2 Diagram of some characteristics of an idealized partially mixed estuary (in part after Pritchard, 1967). (a) Zones of freshwater, salt water, resuspension, and the turbidity maximum; (b) velocity distributions during maximum ebb and flood tides; (c) net non-tidal velocity distribution averaged over a tidal cycle.

reflecting older river channels, these estuaries depend on tides, freshwater inflows, and sometimes wind to establish the salinity fields and degree of mixing which are their characteristic features. They act as conduits between land and ocean and serve as sinks for sediment and other substances transported into them. While these overall processes are occurring, the semi-diurnal tidal ebbs and flows cause water to move back and forth several kilometers in a tidal excursion. Each partially mixed estuary has its own particular set of characteristics, but the general processes are similar. Figure 18.2 is a schematic of the dynamics of Chesapeake Bay.

The Chesapeake Bay estuary

Geography, geology and biology

Chesapeake Bay, located in the mid-Atlantic region of the USA between 37° and 39°40′ North latitudes, is about 290 km long with at least 25 major tributaries and many smaller ones, draining a total area of about 166 000 km² (figure 18.3).

The present Chesapeake Bay is the latest of several estuaries that formed along the Coastal Plain of Maryland and Virginia, USA,

Figure 18.3 Map of
Chesapeake Bay.

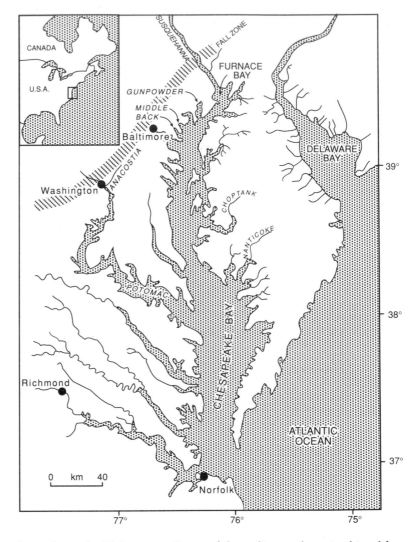

Figure 18.3 Map of Chesapeake Bay.

throughout the Pleistocene. Some of the sediment deposited in older estuaries was eroded during the intervening longer glacial periods, when sea level was much lower. Evidence of at least four ancient estuaries is seen in patchy occurrences of paleochannels filled with estuarine sediments and terraces of different elevations throughout the mid-Atlantic Coastal Plain (Colman et al., 1990). Whereas all of the Chesapeake estuaries appeared, developed, and disappeared as climate and sea level changed with the advance and retreat of glaciers throughout the Pleistocene, the rapid expansion of the human population within the past few hundred years distinguishes today's Chesapeake from past estuaries.

Except for the deep trough and holes representing the channel of the ancient drowned river, where water depths average 30 m, the remainder of the Chesapeake is shallow, with mean water depths of 9–10 m. The tributaries are likewise shallow, averaging 8–9 m in depth. Hence, estuaries, unlike oceans and deep lakes, have little dilution capacity and are particularly sensitive to influxes of materials, especially if rates of input are rapid and exceed the rate of rise of sea level.

The landscape drained by Chesapeake Bay is underlain by widely diverse geologic substrates. These different rock and soil types influence inputs into the estuary. Of the total area, 36 000 km^2 is in the Coastal Plain, where the substrate consists mostly of unconsolidated sediments including sand and gravel. The remaining 130 000 km^2 extends from the Piedmont into the Appalachian provinces, where thick saprolites weathered from crystalline and metamorphic bedrock alternate with shallower, coarser soils derived from quartzite and sandstone. There are interspersed patches of bare serpentine rock.

Most of the Coastal Plain is between 0 and 24 m above sea level. As pre-Pleistocene incised valleys became inundated with the rise of sea level during the Holocene, the landscape was criss-crossed by a network of tidal tributaries draining into the Chesapeake estuary. The result is a mosaic of wetlands, marshes, and high ground underlain by many soil types with different water retention capacities including sand, gravel, fragipan, silt, silt loam, and clay. In the north, a mix of well-drained areas is interspersed with swamps, while in the south where the middle and lower reaches of the rivers are tidal, extensive salt marshes occur along Chesapeake Bay, and barrier island–salt marsh complexes border the Atlantic Ocean. The area is a vast array of habitats supporting a richly diverse fauna and flora, including marsh plants, submerged macrophytes, shellfish, finfish, and waterfowl.

The contact between Coastal Plain unconsolidated sediment and Piedmont crystalline and metamorphic rocks to the west constitutes the Fall Zone (figure 18.3), and represents the farthest inland excursion of salt water, where fresh tidal wetlands form a buffer between the watershed and the estuary. These natural waterways, providing easy access to and from inland areas, fostered the early concentration of human populations along the Chesapeake and its tributaries, and are the location of major ports and cities.

Although the navigable channels of Chesapeake Bay were an important economic asset providing cheap transportation of farm products, lumber, and other commodities, the major economic resource of Chesapeake Bay has always been its biology. Historically, oysters, striped bass, shad, herring, and crab supported by fresh tidal wetlands and salt marshes surrounding the Bay and submerged aquatic vegetation in the tributaries, have been the mainstay of the coastal economy (plate 18.1).

History

In addition to climate change which forged the present Chesapeake Bay over several thousand years, European colonization etched its own signature through rapid deforestation, agriculture, and urbani-

Plate 18.1 Traditional oyster fishing in Chesapeake Bay. Even by the late nineteenth century, when this photograph was taken, human impact on the landscape had begun to modify the ecology of the estuary.

zation. All three affected navigability, water quality, and the biological resource of the Chesapeake through increased sedimentation, eutrophication, toxicity, and habitat modification (figure 18.4).

Prior to European settlement, the entire region was forested except for tidal wetlands and small Indian clearings (Bruce, 1896; Stetson, 1956). After colonization, land use can be divided into four major periods. The periods are not synchronous, as Europeans settled at different times throughout the region, and practiced different kinds of agriculture. From the late seventeenth to mid-eighteenth century 20–30 percent of land was cleared, much of it in the southern region for tobacco farming. From the late eighteenth to mid-nineteenth century, the number of small farms engaged in both tobacco and grain farming increased, resulting in 40 percent deforestation. By the late nineteenth century, and continuing into the 1930s, many small farms were combined into large commercial operations, and the amount of land under cultivation ranged from 60 to 80 percent. In addition, the use of heavy machinery resulted in a deep plough zone. From the 1920s through the 1940s extensive areas of wetland, particularly those bordering tributaries on the eastern shore of Chesapeake Bay, were drained for arable land.

PATHWAYS TO DETERIORATION OF ESTUARINE HABITAT THROUGH DEFORESTATION, AGRICULTURE AND URBANIZATION

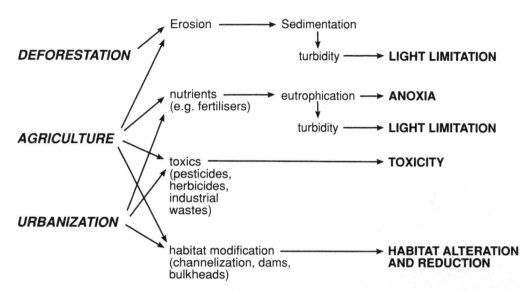

Figure 18.4 Diagram showing interrelationships between effects of deforestation, agriculture, and urbanization on estuarine ecology.

After 1930, depressed economic conditions led to farm abandonment in many areas. Since then, approximately 40 percent of the land in the Chesapeake Bay region has been forested, mainly from afforestation.

Sedimentation into the tributaries, the larger of which could accommodate ocean-going vessels, increased dramatically with European colonization (Gottschalk, 1945). Within 50 years, open water ports became mudflats. Joppatown, a port established in 1707 at the head of Gunpowder River, was abandoned in 1768 and port commerce moved to Baltimore on Patapsco River, because ships could no longer navigate the sediment-laden channel of Gunpowder River. Today, Baltimore Harbor remains a viable port only because of periodic dredging. Along with making shipping channels unnavigable, sediment runoff increased turbidity and decreased light in the water column, thereby stressing biological populations, particularly the benthic flora and fauna.

European settlement also contributed large amounts of nutrients to the Chesapeake waters from fertilizers and sewage. Heavy applications of fertilizers, first guano and then chemical fertilizers, accompanied agriculture. The forested landscape, which had been converted to agriculture during the eighteenth century, was in many places transformed into an urban landscape during the twentieth century, as the human population expanded from 2 million in 1600 to 6.5 million in 1940 and 14 million in 1990 (Brugger, 1988; US Census data). This expansion led to further eutrophication of estuarine waters with the discharge of nutrients from sewage treatment plants. In 1911, the Baltimore Waste Water Treatment Plant was the first facility to discharge secondarily treated sewage into Back River, a tributary of Chesapeake Bay. Flow from the Baltimore Waste Water Treatment Plant increased from 1.7×10^5 m^3 per day in 1917 to 6.9×10^5 m^3 per day in 1990. At present, a total of 2.4×10^6 m^3 of waste water is discharged daily into Chesapeake Bay and its tributaries from 280 municipal plants with a capacity of > 3.75 m^3 per day and from 21 large industrial plants. In addition, an unknown volume is contributed by many smaller facilities (Maryland Department of the Environment, personal communication).

The resulting algal growth from increased nutrients has increased turbidity and decreased light, thus stressing further the benthic populations. Submerged aquatic vegetation, which provides food and habitat for many estuarine species, was particularly affected. Aerial photographs and historical records show that widespread reductions of all species in most parts of the Bay since 1965 became significantly greater from 1970 to 1975 (Orth and Moore, 1984).

Increased eutrophication from nutrient inputs can also have the added deleterious effect of depleting oxygen in the water column. Seasonal anoxia occurs in the region of Chesapeake Bay where a wedge of freshwater overlying tidally driven saline water creates stratification and prevents vertical mixing. The strength of the stratification is related to the amount of freshwater discharged into the estuary. Fifty percent of freshwater enters the Chesapeake from Susquehanna River and 80 percent comes with spring runoff. Consequently, estuarine water is strongly stratified throughout spring and summer for about half its length. Along with the demand for oxygen by benthic organisms, oxygen in the water column is also utilized by an increasing amount of decaying biomass from excess algal growth. Under stratified conditions, the bottom saline water can become depleted of oxygen and toxic to bottom-dwelling organisms.

Until very recently during the twentieth century, pesticides and herbicides were used extensively in agriculture. These, along with discharges from industrial plants, have contributed to estuarine toxicity. Urbanization has also been accompanied by physical modifications of rivers, shorelines, and marshes by dams and bulkheads, which serve residential and industrial needs. These features result in reduction and alteration of habitats for many estuarine species.

Reading the history from the sediments

In order to document the effect of deforestation, agriculture, and urbanization on the Chesapeake Bay estuary, and to compare effects of human activity with effects of climate change throughout the Holocene, sediment cores were collected from the estuary and tributaries. Organisms or parts of organisms, charcoal, chemicals, and other substances are deposited with the sediment, buried, and fossilized. These fossilized components of the sediment are used to reconstruct biological populations and estuarine water quality at the time of sediment deposition. Sediment cores collected range in length from < 1 to > 3 m and, depending on the sedimentation rate, span time periods from 200 to 11 000 years. Sediment cores used for these studies were restricted to those showing no sign of physical or biological mixing.

In order to construct a times series of events from the stratigraphic record, it is necessary to date the sediment. Chesapeake Bay cores were dated using carbon-14 analysis, and identifying historically dateable pollen horizons (table 18.1). In eastern and midwestern North America, ragweed (*Ambrosia* sp.) is a colonizer of disturbed soil and a prolific pollen producer. Consequently, the agricultural horizon is recognized by an increase in ragweed pollen. The date assigned to this horizon differs for cores from different locations because, as mentioned above, agricultural activity was not synchronous throughout the area. The demise of American chestnut in eastern North America in the late 1920s and early 1930s through disease is also identified in many cores by the disappearance of chestnut pollen at a particular horizon. Sedimentation rates are then averaged between dated horizons.

Because sedimentation in the estuary is highly variable and changes related to human activity have occurred in relatively short time periods, a more precise resolution than that provided by average sedimentation rates is essential for interpreting stratigraphic changes. Therefore, sedimentation rates for individual increments of 1–2 cm between dated horizons were estimated by adjusting the average sedimentation rate according to the pollen concentration in the sediment (Brush, 1989). Although there are laminations in some cores, a lack of understanding of the mechanisms for the layering prohibits using these features for dating in the manner that varved sediments provide a chronological tool in some lakes.

Having established a sedimentation rate for each sub-sample or increment of the sediment core, the number of years required for the deposition of the sub-sample can be calculated, and actual calendar years assigned to each sub-sample. The vertical (depth)

Table 18.1 Pollen horizons used for dating sediments

1930 (1923–1937)	decline of American chestnut	decrease in chestnut pollen to < 1%
1840 (1820–1860) 1780 (1760–1800)	40–50% of land cleared for agriculture	% ragweed pollen < 10; oak to ragweed ratio < 5 (generally < 1)
1730 (1720–1740) 1650 (1640–1660)	< 20% (generally < 5%) of land cleared for initial and tobacco farming	% ragweed pollen < 1 to < 10; oak to ragweed ratio < 5 (generally < 10)

axis of the core is thus replaced by a chronological or time axis. Influxes of different indicator components, obtained by multiplying the concentration of the component in the sub-sample by the sedimentation rate for that sub-sample, can then be plotted against the time axis, and changes in indicators related to historically documented human activity. Similarly, changes in indicators deposited in prehistoric time can be assigned to real time, and comparisons of conditions in Chesapeake Bay prior to human influence compared with conditions after human occupation.

Effects of climate change and human activity on the estuary were reconstructed from the stratigraphic record using the fossil indicators described above. Presence, absence, and changes in abundance and ratios of the stratigraphic indicators are used to identify and interpret change.

Chesapeake Bay before European settlement

Pollen contained in radiocarbon-dated sediments deposited at the mouth of the present Chesapeake Bay 15 000 years ago include abundant spruce, pine, and fir with some birch and alder (Harrison et al., 1965). The climate is estimated to have been 3–8°C colder than at present. The same sediments show oak pollen becoming dominant 10 000 years ago, as climate became warmer. This was followed by increases in hickory and hemlock. Pollen in cores from a floodplain of Anacostia River, a tributary of Potomac River close to Washington, DC, show an abundance of pine, spruce, and fir 11 000 years ago, with spruce and fir disappearing about 9000 years ago. Oak became plentiful 5–6000 years ago, followed by hickory 4000 years ago. About 3000 years ago pollen of sweet gum and black gum became abundant (Brush and Yuan, in preparation).

The history of the past few thousand years has been reconstructed from roughly 100 sediment cores collected throughout Chesapeake Bay and its tributaries. These cores span European occupation of the area, the Little Ice Age which extended from the thirteenth into the eighteenth century, and the Medieval Warm Period from about AD 1000 to 1200. These two climatic intervals, which preceded European settlement in North America, are well documented by historical, stratigraphic, and tree ring data in both Europe and the United States.

Temperatures worldwide were anomalously high for approximately two to five centuries, from about AD 800/900 to 1200/1300, and agriculture expanded into northern latitudes in Europe and Asia (Lamb, 1977). Crops including wheat and corn (maize) were grown throughout Greenland and Iceland, and a rich fishing economy resulted from changing fish migrations as waters became warmer in northern Europe. The warm period, which extended for different lengths of time in different places, was followed by a sharp drop in temperature of around 2°C (Grove, 1988), resulting in cold winters and cool wet summers, leading to widespread famine all over Europe. The North Sea became extremely stormy, and within < 50 years, Greenland, which had an active agricultural and fish trade, was isolated by ice. In southern Europe, rivers like the Tiber froze over repeatedly (Lamb, 1977).

The Medieval Warm Period and Little Ice Age are recorded in sediments deposited in Chesapeake Bay and the surrounding tributaries and marshes by changes in pollen and seeds of terrestrial and aquatic plants, and changes in influxes of charcoal, sediment, metals, and nutrients (figure 18.5). High sedimentation rates, charcoal and metal influxes, along with a shift from wet to dry plants, indicate that this was a dry period characterized by numerous fires. Increased total organic carbon, nitrogen, and sulfur in cores from an area of Chesapeake Bay that has been anoxic much of the time since European settlement indicate a change in environmental conditions during this prehistoric warm interval (Cooper and Brush, unpublished data). The stratigraphy suggests high river discharge of freshwater and intensified estuarine stratification. Increased river discharge could result from deforestation caused by burning, high rainfall following fires, or both. In either case, a higher river discharge would increase runoff of sediment, metals, and nutrients from the burned landscape, and also increase the freshwater supply to the estuary, thus increasing eutrophication and intensifying stratification. Cores from marshes show large amounts of charcoal during this interval as well as a shift in pollen and seeds from marsh plants to submerged macrophytes, signifying open water in parts of the marsh, possibly reflecting a local rise in sea level.

Figure 18.5 Sedimentation rates, charcoal influxes, and amounts of zinc, manganese, cadmium, copper, and lead relative to iron plotted against a vertical time horizon from a sediment core collected in the Nanticoke River, a tributary on the eastern shore of the Chesapeake Bay. Influxes of metals were measured as mg cm^{-2} year^{-1}.

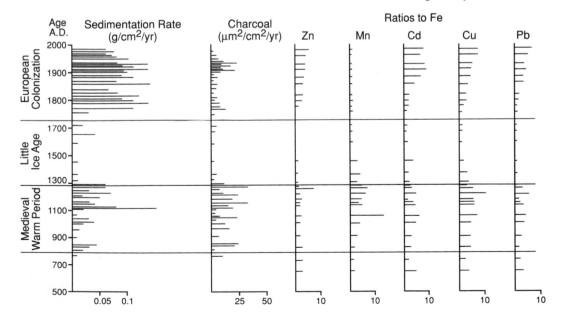

In contrast, the Little Ice Age is characterized by lower sedimentation rates and lower charcoal, pollen, and metal influxes (figure 18.5). Seeds of submerged macrophytes are replaced by seeds first of low marsh plants and later of high marsh plants. Pollen is dominated by pollen of plants that grow in wet rather than dry habitats. The stratigraphic record of the Little Ice Age suggests a wetter climate, where fires were virtually non-existent and freshwater input decreased because the area was largely forested. Wetlands were predominantly low and high marsh. Seeds of submerged macrophytes, signifying open water, are abundant in estuarine sediments, but are no longer found in the marsh, possibly indicating lowering of sea level locally.

Chesapeake Bay after European settlement

The effect of human activity on the Chesapeake estuary is illustrated by changes in sedimentation rates, chemistry, submerged aquatic vegetation, and diatoms in sediment deposited over the past few centuries.

The average pre-European sedimentation rate in the shallow main stem of Chesapeake Bay is 0.06 cm year^{-1} based on carbon-14 dates of 14 cores. The average post-European rate is 0.17 cm year^{-1}, based on identification of the agricultural horizon in 19 cores, including the 14 for which a pre-European rate was obtained (G. S. Brush, unpublished data). The rates are highly variable throughout the Bay, but in general are higher in the northern section. In some areas, sedimentation has increased 10-fold since European settlement, while elsewhere there has been little change. Sediment accumulation rates are likewise variable in the tributaries, but are consistently higher in the upper and middle stretches than in the lower regions close to the mouth. Average sedimentation rates for all tributaries studied are 0.14 cm year^{-1} prior to European settlement and 0.30 cm year^{-1} after European settlement. However, in post-European time, average upstream and midstream rates are 0.32–0.35 cm year^{-1}, compared with an average rate of 0.17 cm year^{-1} in the lower stretches, indicating that the major effect of land clearance was in the upper and middle sections of the tributaries (Brush, 1984a). Detailed analyses of sedimentation rates and history of land use in several tributaries show a clear relationship between the two. For example, sedimentation rates in Furnace Bay, a small embayment at the head of Chesapeake Bay, reflect regional deforestation and agriculture, local charcoal and quarrying activities, and major storms (Brush, 1989) (figure 18.6). Sedimentation rates show a gradual increase as land clearance increased from < 20 to 40 to 60 percent; a more rapid increase as deforestation reached 80 percent. Farm abandonment and possibly soil conservation practices which were initiated in the 1930s are reflected in a lessening of sedimentation after 1930.

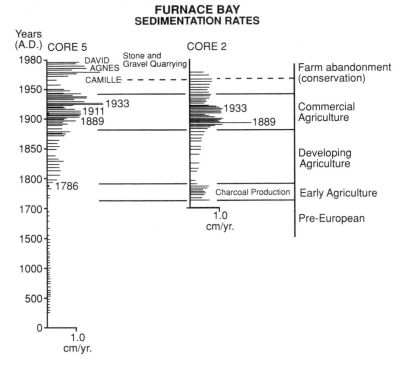

**FURNACE BAY
SEDIMENTATION RATES**

Figure 18.6 Sedimentation rates in two cores from Furnace Bay, an embayment in the upper Chesapeake Bay. Sedimentation rates reflect intensity of agriculture, local stone and gravel mining, and charcoal production as well as some major storms (from Brush, 1989).

Effects of dam construction cannot be identified in sedimentation rates in either the tributaries or the Bay proper. This may be because there are few dams in the Chesapeake that are not several kilometers upstream in the tributaries. The dams that do exist do not appear to maintain a high retention volume of sediment during large storms.

Mineral magnetic studies of sediments deposited in two tributaries of Chesapeake Bay over the past 100–150 years show that the source of the sediment is soil eroded from the landscape (Thompson and Oldfield, 1986; Oldfield and Maher, 1984). Suspended sediment has a similar magnetic signal, thus confirming the contribution of human land use to recent estuarine sedimentation (Thompson and Oldfield, 1986).

Superimposed on this general picture is the effect of local activity such as charcoal manufacture in the early eighteenth century and stone and gravel mining in the mid-twentieth century. Many large storms and hurricanes which transected this part of Chesapeake Bay also left their mark in anomalously high sedimentation rates.

Wolman and Schick (1967) estimated that intensive farming in the early twentieth century caused sediment yields of 400 t km^{-2} year^{-1} in Maryland. By comparison, yields from urban construction ranged from several thousand to 55 000 t km^{-2} year^{-1}. Sediment concentrations in rivers from urban construction areas ranged from 3000 to > 150 000 p.p.m. and 2000 p.p.m. from agricultural watersheds.

Increased eutrophication resulting from nutrients introduced in estuaries is recognized in Chesapeake Bay sediments by increased sedimentary chlorophyll and total organic carbon, a shift in the ratio of pennate to centric diatoms, and increased concentrations of diatom species that prefer water high in nutrients, such as certain species of *Cyclotella*.

All cores show gradual increases in total organic carbon above what would be expected from natural degradation of carbon, since European settlement (figure 18.7). Cores which were analyzed for diatoms show a replacement of pennate diatoms by centric diatoms. The majority of pennate species in Chesapeake Bay are benthic; many are epiphytic, living on submerged macrophytes and macroalgae. Centric diatoms in the Bay are predominantly planktonic forms. The shift from pennate to centric forms during the past few centuries reflects in large part increases in the centric *Cyclotella* spp. Chlorophyll influxes in cores from Back River mirror the volumes of sewage discharged into the river since 1911 (figure 18.8). In other tributaries where waste water is being discharged, *Cyclotella* spp. increased by an order of magnitude.

A combination of high sedimentation rates, increasing eutrophication as measured by total organic carbon and nitrogen, and the

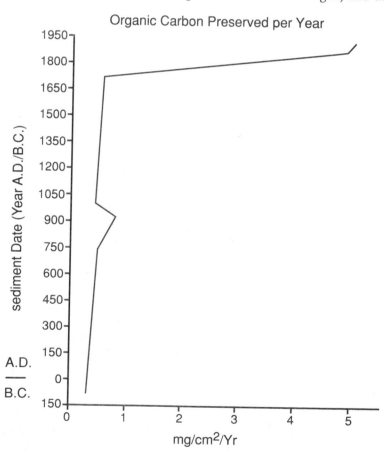

Figure 18.7   A composite profile of organic carbon in sediment cores from the main stem of the Chesapeake Bay in the area of the Choptank River. The large increase in organic carbon from the time of European settlement is characteristic of many cores collected in Chesapeake Bay tributaries. Note also the increase in carbon during the Medieval Warm Period from about AD 800 to 1000.

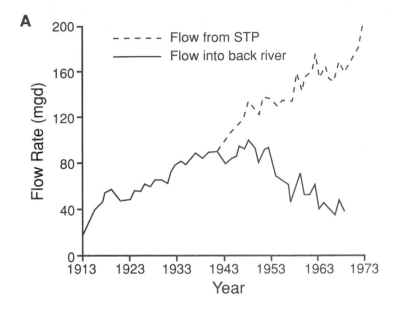

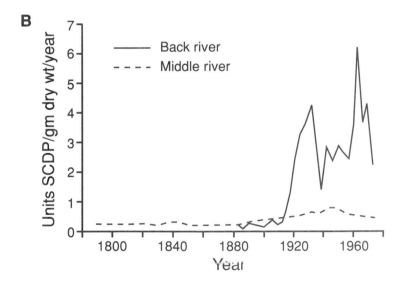

Figure 18.8 (A) Diagram showing the flow of waste water from the Baltimore Waste Water Plant (STP) since 1913. All of the flow was discharged into Back River until 1940 when some was diverted to Patapsco River, where it was used as cooling water in a steel-making plant. (B) Amount of sedimentary chlorophyll degradation products (SCDP) preserved in sediment of Back and Middle Rivers. Middle River is a tributary adjacent to Back River, but does not receive sewage effluent. SCDP decreases in Back River about 1940, when some of the waste water was diverted to Patapsco River. A peak of SCDP around 1958 records a short period of time when the steel plant was shut down and all of the waste water flow was discharged into Back River.

production of hydrogen sulfide as measured by the amount of sulfur and iron in sediments deposited after European settlement, suggests that the present condition of anoxia, although enhanced by freshwater inflow and intensified stratification, has become more persistent with the introduction of increased nutrients into the water (Cooper and Brush, 1991).

The effect of sedimentation, eutrophication, and anoxia on benthic populations is documented in the sediment cores by analyzing seeds of submerged macrophytes, the habitat for many estuarine species. Seed influxes of submerged macrophytes show random fluctuations in populations both before and after European settlement. However, increased numbers of seeds of all species are found in sediments deposited immediately following colonization, the time of initial enrichment of Chesapeake waters with nutrients from the watersheds. As the waters became more turbid and light more limiting with increased sediment and planktonic algal growth, the submerged macrophytes were seriously stressed. The severe declines and local extinctions, documented by aerial photography and field surveys since the 1970s (Orth and Moore, 1984), are registered in the stratigraphy by a reduction or absence of seeds of submerged macrophytes in sediments deposited since 1970.

One effect feeds on another. Water in the upper parts of the tributaries and shallow embayments became more turbid with increased sedimentation and eutrophication. As a result, there was insufficient light for the aquatic grasses, which also housed many of the epiphytic diatoms. Benthic diatoms were replaced by planktonic species, which could live where there was still sufficient light in the upper water column (figure 18.9). Schelske (1975) and Schelske et

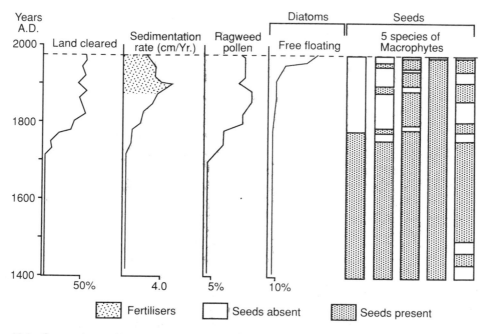

Figure 18.9 Composite profiles of the amount of land cleared, sedimentation rates, ragweed profiles (used for dating sediments), planktonic diatoms, and presence or absence of seeds of submerged macrophytes in the upper Chesapeake Bay. Data for percent land cleared are from US census data. All other data are from sediment cores. The stratigraphic data show that the dominant macrophytes were present for a few centuries prior to European settlement. After European settlement, some species were absent some of the time, and one disappeared completely. Diatom assemblages changed from predominantly benthic to predominantly planktonic. This transformation is believed to be the result of stress on benthic populations due to light limitation in the water column as the waters became more turbid with increased sediment and algal growth. (Redrawn from Brush, 1986.)

al. (1986) have shown that silica became limiting in the Great Lakes where there was high nutrient enrichment, resulting in replacement of diatoms by other algae. Historical records of nutrient inputs and water quality measurements in the Mississippi River show that the amount of silica in the water column decreases with increased phosphorus and nitrogen inputs (Turner and Rabalais, 1991). Diatom abundance increases dramatically and diversity is greatly reduced in the upper Chesapeake Bay in sediments deposited immediately following the discharge of effluent from a sewage treatment plant. Ten to twenty years later, diatoms are rare in the sediments, suggesting that increased eutrophication might also result in silica limitation in partially mixed estuaries.

Conclusions and discussion

The stratigraphic record shows that at least the northern half of Chesapeake Bay was transformed from a benthic to planktonic system since European settlement. This transformation followed the introduction of sediment and nutrients from increased deforestation and agriculture throughout the region. Urbanization was accompanied by the discharge of toxic materials in addition to nutrients. The end result was a near crash of the system in the 1970s, when the submerged macrophytes registered severe declines. The stratigraphic record also shows increasing persistence of anoxia throughout parts of the Chesapeake Bay since European settlement. The gradual deterioration of water quality, accompanied by overfishing, has resulted in decreased populations of economically important shellfish and finfish populations.

   Stratigraphic records of agriculture and urbanization in other systems throughout the world show similar patterns, and not only in recent times. For example, sediment cores from lakes in the central Peten region of northern Guatemala show large increases in phosphorus at the time of extensive forest clearance by the Mayans. Deposition of colluvium in the lakes increased sharply with urban construction (Deevey et al., 1979).

   Large numbers of people settled in the region of the Great Lakes in central North America during the nineteenth and twentieth centuries. Intensive agriculture was followed by industrialization, particularly in the southern region. The shallower Lakes Erie and Ontario became highly eutrophic. Deteriorating water quality, including summer anoxia, resulted in extensive fish kills in the 1960s. Similar conditions occurred in the shallow parts of Lakes Michigan and Huron. However, Lake Superior, the deepest of the Great Lakes, where human populations are less concentrated and there is little industrialization, has remained oligotrophic (Schelske, 1975). Thus the combined effect of shallow water depths and high concentrations of people has had similar effects on the water quality of a lacustrine and estuarine system, even though the systems are physically quite different.

A synthesis of heavy metals in sediment cores collected through-out the world shows a gradual increase coincident with the growth of industrialization (N. J. Vallette-Silver, personal communication). When these data are converted to regional data, they reflect clear relationships with patterns of regional urbanization.

Declines in fish and shellfish populations have been attributed to over-harvesting, more efficient methods of harvesting, and unregu-lated offshore fishing. Efforts to improve the striped bass fishery in the Chesapeake by imposing bans on all fishing of the species for a few years is resulting in rather dramatic increases in the population (Maryland Department of Natural Resources, personal communi-cation). But the stratigraphic record shows clearly that deteriorat-ing water quality began with increased sediment and nutrients entering the estuary as land was cleared for agriculture, and escalated with nutrient and toxic inputs associated with urbaniza-tion. The areas most affected are the upper stretches of the tributaries, closest to the land – the spawning grounds for ana-dromous fish and nursery habitats for fish and shellfish. Regulating fishing, though it may be beneficial in the short term, does not restore habitats. Disease, which has severely affected the oyster population, could also be related to conditions of habitat and water quality. It is difficult to estimate the potential of the estuarine harvest under optimal habitat conditions.

Deteriorating water quality has been the legacy of human activity in many of the world's estuaries. But estuaries respond quickly to change and one would therefore expect rapid response to policies designed to improve water quality. Indeed there are dramatic success stories, such as the restoration of many species in the Thames estuary in England as well as the Potomac and Delaware estuaries in the USA. In all cases where there has been a significant decrease in eutrophication and anoxia, the major pollutants affect-ing the systems were nutrients from sewage treatment plants and toxics from industrial sources associated with large urban areas, referred to as point-source pollution. Other systems like Chesapea-ke Bay and the lower Mississippi River differ in that a major source of pollution is from agriculture. Furthermore, the configuration of the Chesapeake system, with its many long tributaries interdigitat-ing throughout the landscape, provides an extensive land–water interface through which sediment, nutrients, and toxics can enter from areas of intensive agriculture as well as urban development.

The stratigraphic record shows that the Chesapeake maintained a diverse estuarine flora throughout periods of disturbance asso-ciated with variations in climate. On the other hand, anthropogenic disturbance has resulted in reduced diversity and local extinctions of whole communities. There is a fundamental difference between the two types of disturbance. Natural disturbance is episodic in nature, but post-European anthropogenic disturbance is non-episodic and has intensified as the human population continues to increase. Pollution from urbanization on the one hand and from agriculture on the other hand can be considered a spatial analogue of episodic natural disturbance and unrelenting human disturbance. The stratigraphic record suggests that estuarine recoverability is related to the episodic nature of disturbance, and the historic record of relatively rapid recovery where pollution sources are spatially discrete suggests a similar response. Efforts are underway to combat non-point-source pollution by regulating the amount of phosphorus and nitrogen applied in fertilizers, using no-till farming, and creating buffer zones of vegetation along the land–water interface. The success of these policies in restoring diversity within the system may well depend upon the degree to which the effect of the policies mimics the natural pattern of disturbance.

Recognition of early signs of deteriorating water quality is needed to maintain habitats in estuaries as well as lakes and other areas where human populations are concentrated. This can be accomplished by long-term monitoring of water quality parameters, including the use of remote imagery (see chapter 2, Haines-Young). Algal blooms, sediment inputs, and toxic discharges are identifiable from aerial and satellite photographs, allowing quick detection of potential problems. A long-term monitoring program has been established for Chesapeake Bay, where water quality measurements are made routinely at prescribed stations and times of the year. Similar programs are being set up in several aquatic systems. However, long-term monitoring data are not a substitute for information retrieved from stratigraphy. The latter provides a record of the past that includes conditions prior to and during human occupation and also during different climatic regimes. Long-term monitoring extends the stratigraphic record, and places present events within the context of the past record. The record from the Chesapeake indicates that climate change has been an important factor, not only several thousand years ago when the estuary began to form, but as recently as 1000 years ago. The historical Medieval Warm Period and Little Ice Age are recorded in the sediments as very different climate regimes than have occurred since European settlement. Thus habitats have changed as climate changed. What would Chesapeake Bay be like today, if European settlement had occurred a few hundred years earlier? Historical events dictated the time of human occupation along this estuary, but climate as well as history have influenced human occupation of much of the world (e.g. Lamb, 1977). What will Chesapeake Bay be like in the event of future global warming – or cooling? The stratigraphic record provides information regarding conditions in the coastal region during different climates and human occupation, and thus offers insight on future conditions. This knowledge, in concert with long-term monitoring, is essential for effectively protecting the rich but limited estuarine resource.

Further reading

Brush, G.S. (1989) Rates and patterns of estuarine sediment accumulation. *Limnology and Oceanography*, 34, 1235–46.

Cameron, W.M. and Pritchard, D.W. (1963) Estuaries. In M.N. Hill (ed.), *The Sea* vol. 2. New York: Wiley, 306–424.

Doxat, J. (1977) *The Living Thames: the Restoration of a Great Tidal River*. London: Hutchinson Benham.

Nelson, B.W. (ed.) (1972) Environmental framework of Coastal Plain estuaries. *Geological Society of America Memoir*, 133.

Nichols, M.M. and Biggs, R.B. (1985) *Estuaries*. Berlin: Springer-Verlag.

Thompson, R. and Oldfield, F. (1986) The Rhode River, Chesapeake Bay, an integrated catchment study. In R. Thompson and F. Oldfield (eds), *Environmental Magnetism*. London: Allen and Unwin, 185–97.

Wolman, M.G. and Schick, A.P. (1967) Effects of construction on fluvial sediment, urban and surburban areas of Maryland. *Water Resources Research*, 3, 451–64.

CHAPTER NINETEEN

Case 3: China's Yellow River Basin

Edward Derbyshire and
Jingtai Wang

Introduction: Yellow River and yellow soil

The particular combination in northern China of a great and
seasonally violent river with a huge source of soft, readily erodible,
young sediments has given rise to one of the world's greatest single
catchment sedimentation problems. The Yellow River (Huang He)
is 5464 km long and flows through the seven provinces of Qinghai,
Szechuan, Gansu, Shaanxi, Shanxi, Henan and Shandong as well as
the two autonomous regions of Ningxia and Inner Mongolia. Its
drainage area (figure 19.1) covers 750 000 km², in which lives a
population of 84 million, tilling some 13 million hectares of land.
Given the historical record of wide shifts in the lower course of the
river, however (see below), and the potentially threatened catchments of the Hai He and Wei He rivers (Hebei province), the
population at risk rises to 120 million on over 20 million hectares
of land.

The Yellow River displays remarkable variety as it flows through
highly contrasting landscapes. It rises at an altitude close to 5000 m
in the Guzunglie Basin of Qinghai Province, a periglacial plateau
area of many small lakes and poorly integrated drainage on the
northern slopes of the Bayan Har Shan, which separates it from the
drainage of the Yangtze River (Chang Jiang). At Madoi, 50 km
downstream of its outfall from Gyaring Lake, the channel is
shallow, clear, and only 40 m wide in a valley some 20 km across.
The upper reaches of the river between Madoi (4000 m) and the
gorge of Qingtong Xia (in Ningxia) at 1200 m above sea level are
very sinuous, the channel being cut deeply into rock with many
gorge sections alternating with structural basins. Longitudinal
gradients are steep; for example, in the Liu Jia Xia reach, the river
falls 580 m in 216 km on the edge of the Tibetan (Qinghai–Xizang)

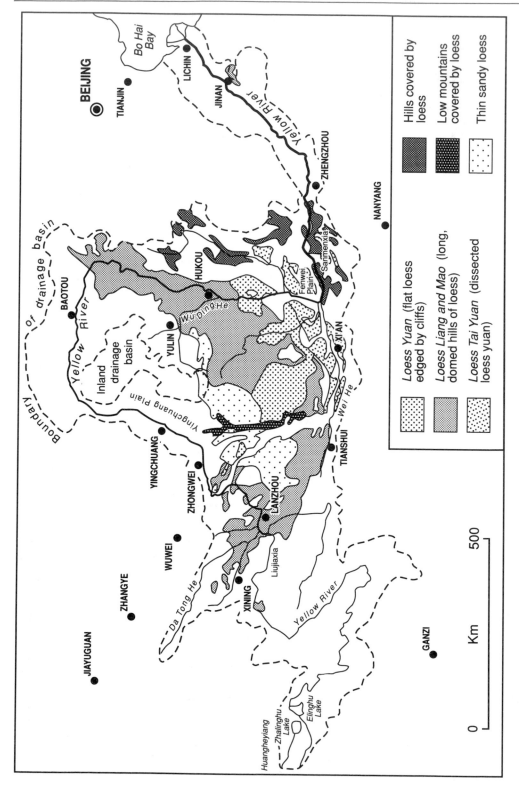

Figure 19.1 Map of Yellow River catchment and distribution of different types of loess terrain.

Plateau (figure 19.2). At least 14 tributaries join the Yellow River in these upper reaches, notably the Daxie He (near Linxia), the Tau He (near Yongling), and the Huang Shui about 40 km upstream of Lanzhou (plate 19.1). As it flows eastwards between the Longyangxia gorge and Lanzhou, the Yellow River is increasingly influenced by a steep hill and mountain landscape mantled by the wind-blown silt known as loess and, in China, as *huang tu* (yellow soil: plate 19.2). Loess reaches thicknesses in excess of 330 m around the Lanzhou basin, the thickest known pile of wind-blown dust on Earth (plate 19.3).

It is in the middle reaches of the river, between the Qingtong Xia gorge and Tan Hua Yu (20 km upstream of the city of Zhengzhou), that the interaction of Yellow River (and its many major tributaries, notably the Wei He) and yellow soil gives rise to a catchment sedimentation problem which is unique in its extent and the severity of its impact on an enormous human population. Before it reaches the highly erodible thick loess of the Loess Plateau in the provinces of Shaanxi and Shanxi, however, the river traverses the

Figure 19.2 Longitudinal profile of the Yellow River, and water discharge and sediment changes downstream, 1950–79 (modified after Ren and Shi, 1986). Note the sharp increase in sediment concentration after the river flows through the main part of the Loess Plateau.

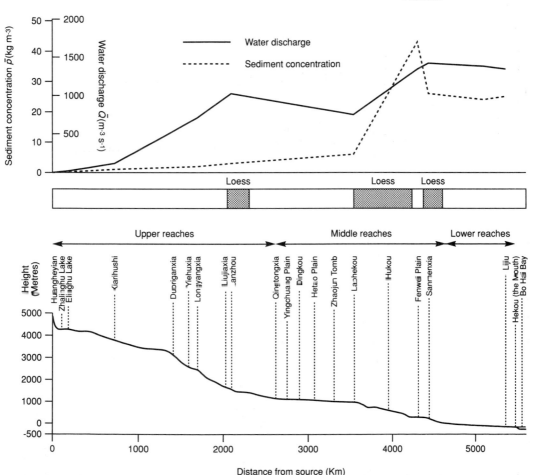

Plate 19.1 Where the Yellow River becomes yellow: the confluence of the Huang Shui (foreground) with the Yellow River.

Plate 19.2 Mountainous bedrock landscape mantled by thick loess near Lanzhou, showing landslide and loess flows (left center) on the steep slopes.

desert region of Ningxia and Inner Mongolia, where annual rainful is everywhere below 300 mm and as low as 50 mm on the northernmost reach of the "big bend" of the Yellow River. Annual surface runoff is below 10 mm over extensive areas and, in contrast with the thick loess region around Lanzhou where the input of suspended sediment to the river is over 1000 t km^{-2} year^{-1}, both river discharge and suspended sediment input rates decrease downstream. The channel is a series of wide braids, the regime of both the main river and its tributaries being highly seasonal. South of Hekou the river enters a region of heavily dissected hills of loess and the climate changes from semi-arid to sub-humid (over 400 mm annual rainfall). It is in this semi-arid part of the catchment, between the Yellow River and the town of Yulin, on the Great Wall in northern Shaanxi, that the remarkably high soil erosion rate of over 25 000 t km^{-2} year^{-1} was recorded, with rates from individual catchments exceeding 50 000 t km^{-2} year^{-1}. South of the Longmen Xia gorge and Yumenkou, as far as the confluence with the Wei He at Fenglingdu, annual sediment yields decline to about one-third of the values for northern Shaanxi in the generally more humid conditions. However, mean suspended sediment loads in the Yellow River continue to rise as the huge catchment area of the Wei He river makes its contribution, giving rise to the enormous sedimentation problems of the lower reaches (Douglas, 1989). Here, between Tan Hua Yu and the Bohai Gulf of the Yellow Sea, is the huge alluvial fan of the Yellow River on which are found some of the highest rural population densities in east Asia (about 500 per

Plate 19.3 Thick loess resting on the sixth terrace of the Yellow River cut across gently inclined Mesozoic and younger red beds (left middle distance).

km^2). On gradients as low as 0.1–0.2 percent, the river flows between levées up to 10 m above the plain, its rapidly shifting channel sometimes flooding as much as 250 000 km^2. The situation is made worse by the addition of over 70 tributaries with catchment areas of more than 1000 km^2. It is in this lower course, with its suspended sediment concentrations in excess of 30 kg/cm^{-3}, that its violent shifts have earned this river the name "China's Sorrow."

Loess: a fragile resource

Loess has been accumulating in north China since the beginning of the Quaternary, about 2.5 million years ago (Heller and Liu, 1982). Resting on Pliocene red beds containing a fauna including the ancestral horse *Hipparion*, the transition to colder climates is marked by mammal remains in the lower (Wucheng) loess of true horse (*Equus* sp.), bison, boar, dog, gazelle, and lower Paleolithic "Lantian man" with plant remains suggesting a forest-steppe environment. The middle (Lishih) loess appears to have accumulated in a continental wooded-steppe environment. There is a rich mammal fauna and, near the top, remains of the Middle Paleolithic "Dali man" have been found. The base of the upper (Malan) loess marks a major environmental shift toward desiccation and cold. There is little evidence of trees, and mammalian remains are typical of dry steppe. This general sequence of wind-blown silts deposited in a cold and relatively dry climate is, however, interrupted by over 20 buried soils (paleosols) and weathered zones developed during milder and moister climatic episodes (Liu and Chang, 1964; Liu, 1985). Taken overall, it represents a remarkable record of terrestrial environmental change through the Quaternary, the number of major climatic changes bearing close comparison with the record of global climatic change derived from the oxygen-isotope record from the equatorial Pacific Ocean (Heller and Liu, 1984; Liu et al., 1985; figure 19.3). There is no doubt, however, that the vulnerability of the loess to erosion has produced many local unconformities and gaps in the sedimentary record: erosion of the loess, although enhanced by human activity, is not a modern phenomenon.

Loess covers an area of 631 000 km^2 or about 6.6 percent of the land area of China, about 317 000 km^2 of this being in the Loess Plateau of the middle Yellow River (Liu, 1964). The geomorphology and susceptibility to erosion of the Chinese loess is a product of its distinctive physical and behavioral properties. Loess consists of coarse to medium silt (0.01–0.06 mm) made up mainly of quartz particles. It is generally poor in clay, but clay size material may exceed 30 percent, especially in the buried paleosols. The airfall deposition of most loess results in very high porosities (over 40 percent in the youngest loess) and low bulk density values (about 1400 kg m^{-3}: Derbyshire and Mellors, 1988). The combination of high porosities and low average moisture contents in the prevailing dry conditions gives rise to high porewater tension which, together

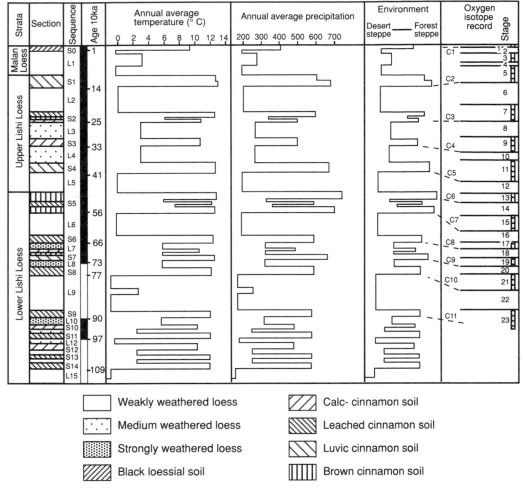

| | | | | Weakly weathered loess | | | | Calc- cinnamon soil |

Weakly weathered loess   Calc- cinnamon soil

Medium weathered loess   Leached cinnamon soil

Strongly weathered loess   Luvic cinnamon soil

Black loessial soil   Brown cinnamon soil

Figure 19.3 The loess/paleosol record in the upper part of the Luichuan sequence covering approximately the last 1 million years. Paleoenvironmental indicators are based on comparison with present-day soils.

with the scattered calcium carbonate cements between the quartz grains, accounts for the nearly vertical slopes so common in loess. It is this property which was taken advantage of by the Chinese to dig "cave houses" beneath the plateau surfaces to provide dry living accommodation which was cool in summer and warm in winter (plate 19.4). The notable cohesive strength of loess arises from its perennially under-saturated condition over huge extents of north China. However, during prolonged rainfall events parts of the loess cover may reach saturation: it then disaggregates instantaneously under its own weight. This process of *hydroconsolidation* causes slope failure and may even result in large volumes of loess flowing as a huge slurry to engulf farmlands. The higher porosities and less frequent overgrowths and cements within the sediment fabric render the younger (Malan) loess much more susceptible to rain-induced erosion than the older loesses, gullies in Malan loess during severe summer rainfalls producing sediment concentrations of as

Plate 19.4 Cave houses cut into thick loess with the farm fields above. Near Luochuan, Shaanxi Province.

much as 1000 kg m^{-3} (over 61 percent by weight: Gong and Jiang, 1979). Gullying in the Chinese loess, producing drainage densities of up to 6.8 km km^{-2}, is conducive to the development of sub-vertical unloading joints and the relaxing of the prismatic fissure systems typical of loess. This sequence of processes produces natural bridges and stacks which ultimately fail. Unloading joints lying roughly parallel to the surface slope are also common, especially in the older loesses, slickensiding indicating relative movement of the adjacent blocks (Derbyshire and Mellors, 1986). Jointing is clearly important in the conduction of water within the Malan loess. Joints become enlarged into natural piping systems, tubes 1–2 m in diameter running parallel to the ground surface. Local saturation gives rise to collapse of some sections of these pipe systems, the resulting "loess karst" being characteristic of, if not entirely restricted to, the Malan loess (plate 19.5). Failure of slabs and columns of loess along sub-vertical joints, followed by their disaggregation on a basal shear plane, produces a characteristic set of slopes similar to those described from the Mississippi loess by Lohnes and Handy (1968). Such slope elements are typical of actively developing gullies in the plateau (*yuan*) and ridge (*liang*) landscapes of the Loess Plateau (figure 19.1). The slopes on stable *liang* and *mao* (rounded hill) forms, however, lie in the range 35–39° below upper convexities. Rainbeat, splash, rainwash, and rill processes turn a thin upper layer of loess into thin slurries on bare ground in the Loess Plateau. In contrast, the combination of highly porous, coarse-grained Malan loess and rather low rainfall

Plate 19.5 Large sinkholes leading into pipe systems up to 2 m in diameter on the loess-covered slopes of the Tse Er valley, south west of Lanzhou.

totals, even in summer showers, in the Lanzhou region leads to surface crusting but a relatively low proportion of overland flow drainage, rills being rare below the historical-maximum rainfall rate of 87 mm h^{-1} at which the runoff coefficient reaches 97 percent.

The practical implications of the distinctive behavioral properties of loess have been recognized in China for many years. Studies of the microscopic fabric (structure) of loess have formed the basis of an engineering zonation (Lin and Liang, 1982) used in regional planning, site selection, and the establishment of construction guidelines. Five major zones have been recognized (figure 19.4). West of the Luipan Shan (zones VI and V) all loesses tend to collapse under their own weight on wetting. Foundation collapse and cracked walls are common soon after construction in the Gansu and Xizang loess regions. Measures such as consolidation, tamping, and structural reinforcement are rarely successful here, only piling into non-collapsible layers or bedrock being effective. North and south of the Wei River (zone III) only the upper terrace loesses are self-subsident on wetting, although significant hydro-consolidation does occur in large test pits. Tamping, consolidation, structural reinforcement, and leakage detection are quite successful here. East of the Luliang Shan only the upper loess on the terraces of the Fen River is self-subsident on wetting. Otherwise the degree of collapsibility is lower than further north and west, in part a reflection of the fact that extensive areas of the surface loess have

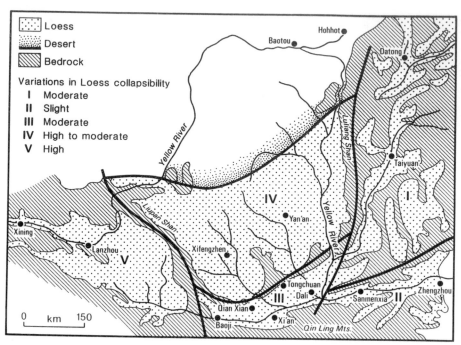

Figure 19.4 Distribution
of zones of different
degrees of loess
collapsibility in the middle
reaches of the Yellow
River.

been remolded during re-deposition (loessic colluvium and alluvium). Generally speaking, the south and east margins of the Loess Plateau (zone II) show low collapsibility with only minor amounts of self-subsidence on wetting. Loess is thin and discontinuous on the eastern margins (northern Henan and southern Hebei). Much of it is loessic alluvium, and susceptibility to collapse is therefore slight. The transition zone from loess to sandy desert (north of zone VI) is characterized by thin, sandy loess mantles with rather low collapsibility.

The collapsibility of loess has resulted in quite severe effects on flat ground well away from the hillslopes. Ground fissures over 1 km long and up to 14 m wide, some probably caused by movement along active faults but others coinciding with severe rainfall, have been described from the Wei He basin by Sun (1988). A series of seven fissure zones cuts across the gentle slope on which the ancient Chinese capital, Xian, is built. Some individual belts can be traced for 8 km. Fissuring and subsidence have certainly been enhanced by abstraction of groundwater to supply urban populations in cities such as Xian and Taiyuan. Total subsidence near Xian in the period 1962–83 was 777 mm (30–70 mm year^{-1}) and extreme rates of up to 123 mm year^{-1} are known, the pattern of subsidence closely following the pattern of watertable fall (figure 19.5). At Taiyuan, the total subsidence between 1980 and 1986 was close to 700 mm.

A major engineering problem in the loess lands is the occurrence of very large slope failures, mainly of the flowslide type. These, and the associated slurry flows, destroy communities and devastate

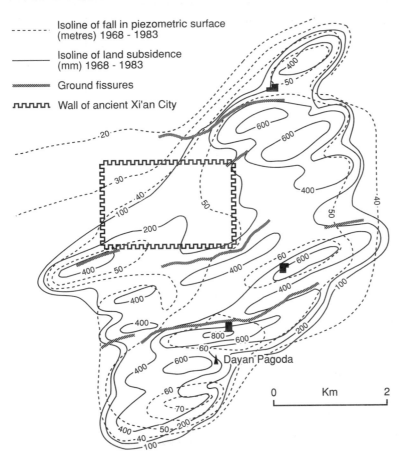

Isoline of fall in piezometric surface
(metres) 1968 - 1983

Isoline of land subsidence
(mm) 1968 - 1983

Ground fissures

Wall of ancient Xi'an City

Figure 19.5 Measured land and groundwater table subsidence in Xian city (after G. Q. Yang, 1986, adapted by Sun, 1988).

adjacent cultivated areas, delivering concentrated sediment plumes into the Yellow River and its tributaries. Although the dry loess flows triggered by the 1920 Gansu earthquake are well known and widely quoted in the scientific literature (see below), this is just one of a great number of such events, none of which has been closely documented, analyzed, and modelled. Historical records and local knowledge, however, indicate that major loess landslides are a widely distributed geomorphological phenomenon (figure 19.6). The current view, based on documentation of 200 gullies in loess in Gansu Province, is that minimum rainfall intensities for the initiation of loess flows and flowslides is 1.6–2.4 mm h^{-1} and that widespread effects are felt with rates of 10–30 mm h^{-1}. Events are concentrated in the period June to August, events occurring particularly on August afternoons or evenings, making up perhaps 70 percent of the total.

The world-renowned Haiyuan earthquake (magnitude 8.5) occurred in 1920 on the Gansu–Ningxia border. It caused the death of almost a quarter of a million people, villages close to the epicenter losing entire populations. Damage extended across more than 60 counties in Gansu, Ningxia, and Shaanxi. Some 647

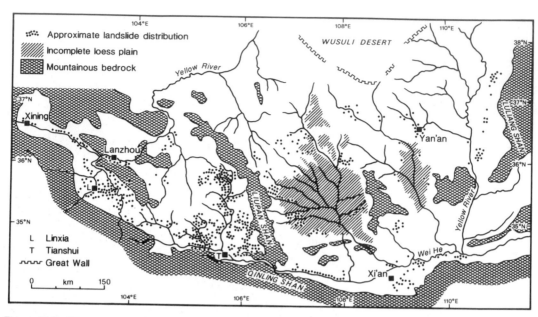

Figure 19.6 Distribution of landslides in the middle reaches of the Yellow River, northwestern China (after Derbyshire et al., 1991).

landslides were triggered by this earthquake and more than 40 landslide-dammed lakes were impounded. Many subsequently burst, sending pulses of muddy water down tributaries and causing further death and destruction. Others, however, including the Dongjiacia Lake with an area of 3.5 km^2 and a depth of 15 m, persisted to the present day, constituting a continuing hazard.

It is estimated that about 30 percent of the total area of Gansu Province is subject to landslides and flows in thick loess, over 2000 slides and 1000 loess flows having been identified. The principal trigger mechanisms are torrential monsoonal rains and earthquake shock, the latter accounting for many of the largest failures in this seismically active region close to the northeastern margins of the Qinghai–Xizang (Tibet) Plateau. The problem is exacerbated by discontinuous vegetal cover and the decimation of plants in dry years, such that they provide little protection against sheet erosion, suffosion, and large-scale piping, gully erosion being a process of less importance here in the drier loess region of eastern Gansu. Moreover, it is a region characterized by steep-sided, flat-floored valleys. The population is locally high, the villages and their intensively cultivated fields tending to be concentrated on the valley floors, on breaks of slope and terraces, where they are most vulnerable to landslide and debris flow hazards. In addition to destroying the small but vital rural reservoirs, these hazards also involve serious soil erosion loss.

Between 1965 and 1979, over 1000 large landslide events took place in the loess highlands of eastern Gansu, more than 2000 people being killed. In 1975, a landslide caused by the wave from a breached reservoir in Zhuang-Long county destroyed over 3000 houses and 20 000 ha of croplands. Five hundred people were

Plate 19.6 Huge landslide (left distance) at Sale Shan. The slide, which buried four villages in less than one minute, occurred in 1983, pushing the Ba Xie River half a kilometer across the valley and destroying two reservoirs.

killed. The Sala Shan landslide, which happened in 1983, was a loess landslide of the rapid type which destroyed four villages, 227 people being killed and 2000 ha of farmland becoming a sea of loess instantaneously (plate 19.6). During the heavy storms in August 1979 in Yun-Jing county, many landslides and debris flows occurred, some 799 people being killed or injured and over 200 ha of farmlands being damaged. In the 1984 rainy season 2569 landslides occurred in Li county alone (Tian Shui prefecture in southeastern Gansu) and losses were very heavy. Between 1964 and 1978, just seven of the largest loess flows destroyed no fewer than 17 544 houses, inundating 22 500 ha of farmland and killing 1142 people.

The combined effect of sheetwash, gullying, and landsliding in the middle reaches of the Yellow River (eastern Loess Plateau) contributes at least 75 percent of the suspended sediment load of the lower Yellow River: mean sediment yields from this region range from 4000 to 25 000 t km^{-2} year^{-1}. Suspended sediment concentrations in some middle Yellow River tributaries have long-term averages over 717 kg m^{-3} (Beiluo He) with extreme values of more than 1300 kg m^{-3} (the hyperconcentrated condition which may temporarily block river flow, as on the Yellow River at Tongguan in 1977, with increased threat to the levées). Such extreme erosion is a reflection of land-use practices in the Loess Plateau. It has been calculated that the loess removed from the plateau in a single year expressed as thickness is equivalent to the thickness of loess accumulated over a period of 100–300 years: this vital resource is thus diminishing rapidly in the present land-use regime (Dai, 1987). Many aspects of the land use are ancient and traditional and are increasingly recognized as damaging. Migration of peasants from the intensely cultivated plainlands of the lower Yellow River to the dry, mountainous loess lands of the upper reaches has resulted in a little-modified transfer of techniques and land use involving, over wide areas, cultivation on slopes as steep as 39° (plate 19.2), sometimes with irrigation pumped up from river sources 200 m or more below. The problem is not new, although there is abundant evidence that human activity has long been an important ancillary cause. In the latter part of the Tang Dynasty (AD 618–907), for example, the loess plateau known as Dungzhi Yuan was 32 km wide but is now only 1 km wide in places, an areal reduction from 1344 to 756 km² in little more than 1000 years. Soil nutrient losses are high. For example, in Shanxi Province loss of nitrates, phosphates, and potash total 10.4 million t year^{-1}, which is over 30 times more than is replaced by chemical fertilizers (Li, 1985). Such depletion has resulted in reduced grain yields and reduced carrying capacities in grasslands (e.g. sheep per hectare in the Yanan district in northern Shaanxi fell from 3 in 1950 to 1.4 in the 1970s: Fu, 1983). Wider scavenging for firewood and herbs has contributed to increases in runoff and consequent soil loss. The historical trend involving higher human populations with consequent increases in food needs, notably grain, led to severe reduction in woodland and grassland cover. During the famine in the years 1959–61, over 660 000 ha of grassland in Shanxi, Shaanxi, and Gansu provinces were converted into cropland. Around Zhidan in Shaanxi province, woodlands covered about 56 percent of the area in 1949 but by 1980 this had fallen to only 18 percent. There is little doubt that a combination of population pressure and inappropriate land-use policies is a major cause of enhanced soil erosion.

Given the relatively unprotected surface of much of the Loess Plateau, owing either to an incomplete natural shrub cover (in the drier north and west) or to the intensive cropping patterns, loess is everywhere subject to erosion by water. Splash erosion by raindrop

Table 19.1 Variation in the effect of cultivation on the permeability of grey-drab forest soils

Soil type	Utilization	3.5–minute permeability rate (mm min^{-1})	Final permeability rate (mm min^{-1})	2-hour total permeability rate (mm min^{-1})
Grey-drab forest soil	Forest land	12.66	3.10	539.7
Cultivated grey-drab forest soil	Farm land	2.86	1.40	200.0

impact affects both dry and agriculturally moistened slopes, contributing to mechanical illuviation (washing in) and sheet erosion. The combination of rainbeat and overland flow (predominantly of the surface saturation type) breaks up the weak soil aggregates and both dissolved and suspended solids are transported downslope. This tends to render the surface layer more sandy and therefore less valuable agriculturally. For example, studies of grey forest soils developed on the loess near Lanzhou have shown that water-stable aggregates constitute 57.5–68.5 percent of the forest soil but only 21.7–32.6 percent of the same soil cultivated for many generations, and that the two-hour infiltration capacity of the cultivated soil is only 37.2 percent of the value for the uncultivated forest soil (table 19.1). Rill and gully erosion tends to increase with slope gradient up to 35° but, because of variations in infiltration losses, the relationship is not simple (table 19.2). Quite small increases in the percentage of clay in different loess materials can greatly reduce its susceptibility to erosion (table 19.3). The effect of raindrop impact and overland flow is materially attenuated by a vegetation canopy, a mature forest frequently intercepting 75 percent or more of the

Table 19.2 Test data on different slopes under 46.7 mm rainfall over 36 minutes, with intensity of 1.3 mm m^{-2}

Degree of slope	Amount of runoff $(cm^3\, m^{-2})$	Amount of erosion			
		Total amount $(g\, m^{-2})$	Rill erosion $(g\, m^{-2})$	Rill erosion as percentage of total erosion
5	36 700	1210	822	68
10	36 600	2370	1650	70
15	36 200	5720	3840	67
20	29 100	6390	5320	84

Table 19.3 Erodibility of different soils at Ziwuling Ridge, Gansu Province

Soil type and utilization	Sampling depth (cm)	Organic matter content (%)	Colloidal content (< 0.005 mm) (%) (A)	Moisture equivalent (%) (B)	Specific value (A/B)	Dispersion ratio (%)	Erosion ratio (%)
Bush: black fertile soil	0–9	3.05	30.8	23.3	1.32	35.8	25.6
	9–19	1.80	29.8	20.9	1.43	31.4	22.0
High forest: black fertile soil	4–19	2.31	25.6	25.1	1.02	37.1	36.0
	19–31	1.08	25.7	22.5	1.14	43.7	38.3
Sward: raw loessic soil	0–13	2.47	25.4	25.0	1.02	34.5	33.8
	13–26	1.16	24.4	22.5	1.08	37.2	34.4
Farmland: loessic soil	0–20	1.73	21.0	24.5	0.86	45.8	53.3
	20–39	0.41	19.8	21.4	0.96	56.4	61.3
Farmland: yellow common soil	0–17	1.67	24.6	23.7	1.04	50.4	48.5
	17–32	0.68	25.1	23.6	1.06	56.0	52.0
Farmland: red clay	0–14	2.12	44.6	24.3	1.81	34.3	19.0
	14–29	0.29	56.7	30.3	1.87	49.0	26.2

Table 19.4 Comparison of young black locust forest with farmland in terms of soil conservation efficiency

Observation year	Land utilization	Age of trees (years)	Degree of slope	Vegetation (%)*	Total rainfall in rainy seasons (mm)	Gross soil loss (kg mu^{-1})	Total runoff amount (m^3 mu^{-1})[a]
1955	Young black locust forest	2	29	–	292.9	0	0
	Sorghum and string bean		22.5	–	292.9	193.1	886.3
1956	Young black locust forest	3	29	–	585.8	5520.0	27 375.2
	Farm land		28.3	40	585.8	17 046.0	7933.3
1957	Young black locust forest	4	29	–	202.4	0	0
	Sorghum and string bean		21.3	–	202.4	424	1410.0
1958	Young black locust forest	5	29	93	398.33	795	2587.8
	Farm land		21	63	366.4	3910	19 325.1

[a]15 mu = 1 hectare.

rainfall, although the amount intercepted decreases rapidly above certain thresholds. For example, interception rates of 25–42 percent with rainfall amounts of 0.1–5.5 mm, have been shown to fall to 6–22 percent with rainfall totals over 40 mm. Thick grass and bush cover both scatters the rainfall impact and impedes soil movement by surface runoff, as does forest litter. Cultivated crops are much less effective, however (table 19.4). Plant roots break up compacted loess soil layers into aggregates and improve the infiltration capacity: such aggregates are highest in loess forest soils (about 90 percent) and lowest on the cultivated loess (about 10 percent).

River water: the threatening resource

The huge alluvial fan (about 250 000 km^2) of the Yellow River below Zhengzhou provides a classic example of the ways in which conditions in the upstream reaches influence events, often catastrophic, in the lower reaches of a drainage system. The Yellow River's alluvial fan is a complex feature, detailed shifts across it of the river channel having been recorded for at least the past 4000 years. The reach of the Yellow River from Taohuayu to the shores of the Bohai Gulf is 768 km long and the catchment 2200 km^2, on which over 100 million people live. It is not only the breadbasket of China: it carries dense communication systems and many towns forming the industrial base of north China.

The rapid reduction in regional slope in these lower reaches is expressed in deposition in the lower reaches of over one-quarter of the sediment load carried past Zhengzhou. Very high rates of aggradation with natural levées and frequent widespread flooding have occurred throughout the history of human settlement of the plain. About 1600 breaches of the levées have been recorded in the past 1000 years, 26 of which produced extensive flooding and marked changes in the course of the river. About 50 percent of the sediment load recorded at Zhengzhou is deposited in the estuary, causing a seaward growth of the delta at this point of about 3 km year^{-1}. Over the past millennia, the Yellow River has shifted several times to reach the sea on either side of the Shandong peninsula (figure 19.7). About 800 years ago (Ming and Qing dynasties) the main course of the river ran from Dongbatou to Yuntiguan, persisting for 661 years (AD 1194–1855). The Haihe and Sishu rivers were without levées and high suspension loads before the sudden southward shift of the Yellow River in AD 1194. There is a clear relationship in the historical record between flooding and neglect of the levées, as occurred, for example, in time of war (figure 19.8), and the maintenance of the levées and the seaward progradation of the river mouth with consequent adjustment of the longitudinal profile.

Major breaching of the levées occurred in June 1855 at Tongwaxiang, resulting in a complete shift in the lower course from the south to the north side of the Shandong Peninsula (figure 19.7). Within 24 hours the old channel near Tongwaxiang carried no flowing water, a huge alluvial fan being produced and reaching the Bohai Gulf in the north and the Yellow Sea coast in the south. Further riverbank bursts occurred in 1863, 1871, 1873, and 1880, with drainage flowing along both branches of the new fan. Further serious breaches occurred in 1933, 1934, and 1958 (figure 19.9).

Since 1949, increasingly ambitious and integrated plans and major works have been applied to the problems posed by the Yellow River basin. They have involved many types of scheme, including dam construction, levée building, and changes in land use, all designed to reduce the sediment influx into the lower Yellow River plain as a means of controlling the very high rates of aggradation. It is an enormous undertaking because it involves environmental management over the entire basin from Tibet to the ocean. Expenditure of the equivalent of more than one billion US dollars has been incurred in the past three decades in schemes designed to increase control over this treacherous river system, and a comprehensive watershed plan now exists for the control of flooding, extension of irrigation, and optimization of the hydro-electric potential of the Yellow River basin.

In a river system with a total fall of 3840 m extending over several climatic provinces, with many reaches showing strong control by the geological structure, the hydro-power potential exceeds 28 000 MW; almost 13 000 MW of this total is in the upper reaches, as defined above, which have the added advantage of low suspended sediment loads. By 1985, more than 150 reservoirs had been constructed, giving a total storage capacity of over 53×10^6 m^3, and power stations had been built on 80 of these dams, giving a total installed capacity of more than 2500 MW.

Features of the present delta

Region outsideYellow
River delta

Yellow River delta below Mengjin

Alluvial fan formed after
breaching at Tongwaxiang

Present river channel

Ancient river channels:

1   in the Yu dynasty
    (2140 -2095 BC)

2   in the Western Han dynasty
    (206 BC - 24 AD)

3   during or between the Eastern
    Hanand & Tang dynasties
    (25 - 907 AD)

4   in the Song dynasty
    (960 - 1279 AD)

5   in the Ming & Qing dynasties
    (1368 - 1911 AD)

6   over-flowing on the Plain after
    being breached at
    Huayuankou in 1938

0   Km   200

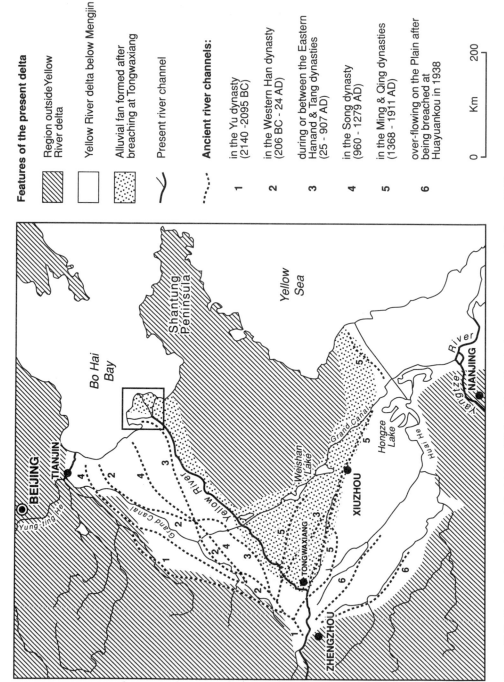

Figure 19.7   Former channels and alluvial fan of Yellow River (inset shows area covered by figure 19.9).

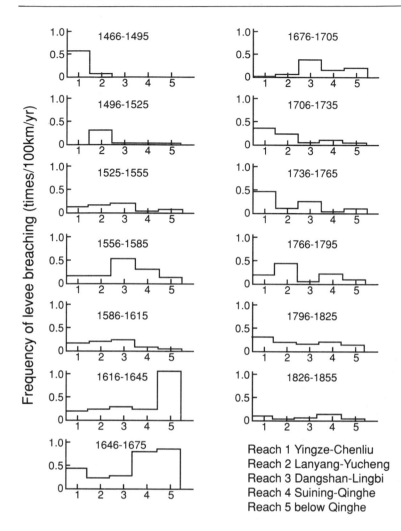

Figure 19.8 Frequency of levée breaching 1466–1855 (from historical records: after Qian, 1990).

Reach 1 Yingze-Chenliu
Reach 2 Lanyang-Yucheng
Reach 3 Dangshan-Lingbi
Reach 4 Suining-Qinghe
Reach 5 below Qinghe

Notable schemes include Yianguoxia (352 MW), Bapanxia (180 MW), Qingtonxia (275 MA), and Liujiaxia. The scheme at Liujiaxia near Lanzhou city was designed as a multipurpose development, providing flood control, irrigation, control of the floating ice hazard, improved up-river navigation, and important freshwater fisheries in addition to its huge hydro-power potential. With a height of 147 m and a storage capacity of 5700×10^6 m³, Liujiaxia's total installed power capacity of 1225 MW and its annual energy output of 5330×10^6 kWh satifies the power needs of the huge area represented by Gansu Province and the Ningxia and Inner Mongolia autonomous regions. Upstream in Qinghai Province there is nearing completion the Longyangxia project. As at Liujiaxia, this is a concrete structure sited at a constriction in the river but, in this case, within a granite gorge. This is the largest hydro-electric reservoir scheme yet built in China (plate 19.7), making even Liujiaxia appear modest. Although the dam is only

Figure 19.9 Migrating channels of Yellow River delta in modern times.

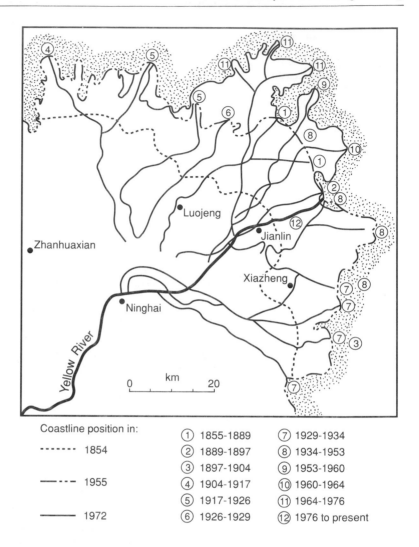

Coastline position in:

- - - - - - - 1854

— - - - 1955

——— 1972

① 1855-1889	⑦ 1929-1934
② 1889-1897	⑧ 1934-1953
③ 1897-1904	⑨ 1953-1960
④ 1904-1917	⑩ 1960-1964
⑤ 1917-1926	⑪ 1964-1976
⑥ 1926-1929	⑫ 1976 to present

slightly higher (178 m), its storage capacity (24.7×10^9 m^3), power capacity (1270 MW), and annual power generation (6×10^9 kWh) are much greater. However, this project, together with another five of similar scale planned for sites upstream of Longyangxia, is a matter of concern in terms of long-term environmental hazards. The shoreline of the newly developed lake is developing toward the base of a succession of over 1000 m of unlithified Quaternary lacustrine sediments. Plans include the further raising of the water level some 9 m higher than shown in plate 19.7, thus extending the wave-energy undercutting of the steep slopes, which already show signs of large landslide failures. Moreover, it must be a matter of concern that Longyangxia is a concrete structure in a region which is tectonically very active. The April 1990 earthquake (magnitude 6.9), which was felt violently in Lanzhou city some 320 km away to the east, had its epicenter at Gonghe town only 30 km from the

Plate 19.7 The Longyangxia reservoir in the upper reaches of the Yellow River, the biggest hydro-electric scheme in China. The cliffs are of soft Pleistocene lake sediments now being undercut by wave action. Note (center) landslide into lake, April 1990.

dam. Hydro-power generation schemes are not restricted to the upper reaches of the river, however. There exist eight hydro-electric stations in the middle reaches, with a total capacity of 4400 MW and an annual power generation potential of 17×10^9 kWh.

Schemes involving diversion of the waters of the Yellow River to provide irrigation water have a long history, notably in the drier upper and middle reaches in Shaanxi, Gansu, Inner Mongolia, Ningxia, and Qinghai. Most of these were small scale and many have been renovated and enlarged in the past four decades, the total irrigated area upstream of Huayuankou exceeding 3.6 million hectares, for example.

The enormous sediment load of the river and its tributaries is also viewed as a resource as well as a hazard. The process of diverting river waters so as to build up fresh sediment surfaces below the level of the river's levées, or warping, is now well developed and a common practice in the lower reaches. Plans include the development of widespread warping in 90 irrigation districts with a total extent of almost 2 million hectares. A complementary process to warping is soil conservation. The susceptibility to subsidence, landsliding and gullying of the Loess Plateau region referred to above has led to a series of schemes and strategies designed to reduce the enormous soil losses and to prolong the life of farmland *yuan* and *liang*. A variety of techniques is now in use. These include construction of ponds and cisterns to conserve and store water, contour ploughing, and slope management. Overland flow of water is impeded by construction of 0.5 m high banks along the edges of hand-dug terraces and scallop-shaped benches known as "fish-scale pits." There has been a gigantic and concerted planting of trees and herbs on terrace slopes since the early 1960s, essentially as a soil erosion and water conservation measure, and orchard land has increased, in part as a response to the soil erosion threat. A staircase system of small dams is commonly built in small (first and second order) gullies as both a runoff inhibitor and a warping device. Over half of the hillslopes and small gullies in the southern half of the Loess Plateau have been subjected to these measures in the past 30 years, with a consequent tripling of average annual farm income in many areas. Nevertheless, there is a considerable heritage in the loess lands of the middle reaches of the Yellow River of degradation arising from both historical times and from unwise policies over the past 40 years. For example, the early 1960s policy of "grain first" required every county to attain food self-sufficiency even in disastrous years and to make a contribution to the nation's sufficiency in good years. Implementation of this quite unrealistic policy caused unprecedented land degradation because land quality was ignored in deciding land use. The economic imbalance so created is still evident in many townships because of the time lag involved in returning land, unwisely put to the plough, to woodland and grass.

In the lower reaches of the Yellow River, the premier management concern is with the flood hazard. Measures have included a systematic raising and broadening of the levées, berm construction by diversion and warping, river-bed riffles and bank-edge dikes, spillway and detention basin construction, and the building of both large and small dams. Levées are now some 10 m above the plain over long stretches of the alluvial fan reach of the river, spillways and warping ponds having been created along more than 600 km of the river's length. The larger dams, such as Sanmenxia, have played a key role in flood mitigation. The exceptional peak flood at Huayuankou in 1958 of 2300 $m^3 s^{-1}$, which destroyed two spans of the Yellow River bridge at Zhengzhou, was effectively controlled by the combined use of the larger dam sites and emergency work on levées. In 1958 peak discharge at Huayuankou was reduced by 20 percent by timely use of the dam at Luhun on a tributary of the Yellow River at the same time as 400×10^6 m^3 of water was diverted into the Tongping Lake basin. Between 1965 and 1985 nine floods with peak discharges in excess of 10 000 $m^3 s^{-1}$ have been safely managed using such strategies.

Conclusion

The Yellow River and the loess lands of its middle and lower reaches present a complex of problems and natural hazards which are, to a notable extent, products of the region's distinctive geological and climatic evolution, particularly over the past three million years. The juxtaposition of highly contrasting climatic, geomorphic, and tectonic provinces along the course of this great river has produced the essential ingredients of one of the world's great problem regions: abundant soft rock (loess) in the middle reaches and the considerable geodynamic energy of the system in the form of sharp climatic and meterological gradients and a high frequency of earthquake activity. A burgeoning human population, quick to take advantage of the inherent fertility of the loess, greatly exacerbated the problem, the enormity of which has been fully recognized only in the past few decades. The ancient Chinese saying, "Knife and fire make new land, the land exhausts itself, and the knife and fire make new land again," graphically encapsulates the entrenched attitudes which must be faced by authorities seeking to alleviate the worst effects of the river and its fragile catchment. It is self-evident that, among all the remedial measures undertaken, the key is the minimizing of soil loss from the slopes. Yet this is a gargantuan task, for the utilization and conservation of most slopes in the loess country requires special and continued treatment, which is not readily mechanized. Some impressive results have been achieved in the past 40 years but the geological, ecological, and land-use inheritance is such in the Yellow River basin that, even given stable population numbers, the scale of the challenge posed by environmental management will remain a severe drain on the gross national product well into the next century.

Further reading

Derbyshire, E. and Mellors, T.W. (1988) Geological and geotechnical characteristics of some loess and loessic soils from China and Britain: a comparison. *Engineering Geology*, 25, 135–75.

Douglas, I. (1989) Land degradation, soil conservation and the sediment load of the Yellow River, China: review and assessment. *Land Degradation and Rehabilitation*, 1, 141–51.

Lin, Z.G. and Liang, W.M. (1982) Engineering properties and zoning of loess and loess-like soils in China. *Canadian Geotechnical Journal*, 19, 76–91.

Liu, T.S., An, Z.S., Yuan, B.Y. and Han, J.M. (1985) The loess-palaeosol sequence in China and climatic history. *Episodes*, 8, 21–8.

Milliman, J.D. et al. (1987) Man's influence on the erosion and transport of sediment by Asian rivers: the Yellow River example. *Journal of Geology*, 95, 751–62.

Qian, N. (1990) Fluvial processes in the lower Yellow River after levée breaching at Tongwaxiang in 1855. *International Journal of Sediment Research*, 5, 1–13.

Sun, J. Z. (1988) Environmental geology in loess areas of China. *Environmental Geology and Water Science*, 12, 49–61.

CHAPTER TWENTY

Case 4: Deforestation in the Himalaya

Martin J. Haigh

Introduction

For many travellers, first contact with the Himalaya is the trip from
Dehra Dun to the hill station of Mussoorie in North India (figure
20.1). The motor road climbs through scenes of immense grandeur
and spectacular environmental devastation: lands degraded by
deforestation and overgrazing, scarred by landslides, cut by quar-
ries and streams choked with debris (see Ehrich, 1980, p. 196). My
early research in the Himalaya during the late 1970s involved
studying environmental change due to road construction near
Mussoorie, measuring landslides scars and outfalls (Haigh, 1979a,
1984a). During 1978 northern India suffered its worst flood
disasters on record (Haigh, 1989a). At Varanasi, parts of the city
flooded which had never been submerged in that city's long history.
Flood levels, recorded on the city's ghats, reached new heights. In
the old colonial summer capital, Simla, I stood outside Clarks
Hotel, and watched a tiny stream swell to overwhelm its culvert,
the road, and bury an *ad hoc* car park axle deep in sediment (plate
20.1). The whole process took less than one hour.

The Himalaya tend to make very deep first impressions on
foreign visitors but after a while one becomes relatively insensitive
to everyday sights. This commonplace includes the zone of deforested land around each hill village, the columns of villagers
carrying firewood to the towns, the timber lorries that only seem to
run at night, the ubiquitous grazing animals and the criss-cross
tracks they weave across the hillsides, and landslides, in some cases
more than 20 per km of roadway. Subsequent travels have taken
me to Kumaun, where the situation is not quite so bad as at
Mussoorie, and the NWFP (Swat Valley) of Pakistan, where things
are rather worse. Together, these experiences had convinced me

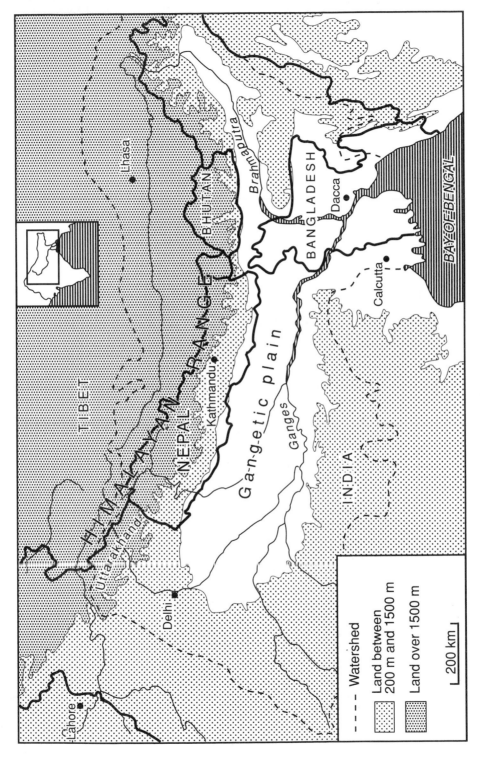

Figure 20.1 Map of Ganges–Brahmaputra river catchment in South Asia.

Plate 20.1 Car park at Simla, north India, buried by debris as a result of a monsoon storm.

that the situation was much the same the mountains over. Stories of frightened or assassinated forest guards, of springs that have dried up, rivers where the flow has declined, fields lost to flood and landslide, are everyday "tea-shop" gossip and persuasive indicators of dramatic environmental change. So recent revelations, beginning in the writings of Western academics based in Nepal, have come as a considerable surprise.

Current wisdom in the West declares that the environmental crisis in the Himalaya is an illusion. There is no spiral of environmental decline and devastation, no evidence for widespread deforestation, no accelerated erosion, no significant hydrological change, no crisis at all, just some unfortunate trends which need careful, scientific examination (Ives, 1989; Ives and Messerli, 1989). Today, the scientific pronouncements from this region are being closely examined for signs of miscalculation, exaggeration, or overgeneralization and the controversy has shrouded the whole issue in a veil of "uncertainty on a Himalayan scale" (see Thompson and Warburton, 1985).

This chapter will try to examine this controversy from three perspectives. First, it will outline the conventional theory of environmental degradation in the Himalaya. Second, it will review its antithesis in the more recent notions of Western scientists in Nepal. Third, it will summarize results produced independently by the Himalaya's own scientists.

The environmentalist model of the crisis in the Himalaya

Once, the Himalayan mountains were mantled in trees. Official statistics for the Lesser Himalayan districts still classify up to 66 percent of the land as forest, even though most is completely devoid of trees. Originally, the dense tree canopy, its undergrowth, the leaf litter layer covering the forest floor, and the dense network of roots below, intercepted a large proportion of the rainfall and returned much, perhaps most, back to the atmosphere as evapotranspiration. The remainder percolated through deep, open-textured, soils beneath the forest and/or down through the rocks below as groundwaters. Later, these waters found their way back to the surface as the springs, seeps, and marshes that contribute the day-to-day base-flow of perennial streams. Additions of rainfall at the ground surface helped to "push through" waters moving through the ground downslope, so springs responded fairly quickly to rainfall provided there was a high level of antecedent soil moisture and part of the land close to saturated overflow (Bruijn-zeel, 1990, p. 53). In Nepal, Chyurlia (1984) has correlated monthly runoff coefficients with rainfall in the preceding month. The important point is that streams in forested headwaters keep flowing long after the last rainfall.

Hydrological records from forest headwaters in the Himalaya are few and far between. However, records from the Oxford Brookes University/ Kumaun University instrumented catchment at Animal Park, Almora (which drains a 1 km^2 headwater of undisturbed forests on $> 25°$ slopes over schist/quartzite rocks) show that, in 1987, 4.75 mm of rainfall fell on June 18, near the end of the dry season. This raised the flow in the main channel from 3.97 to 5.11 $m^3 h^{-1} km^{-1}$. It took 4 days for the flow to return to pre-event levels and 3 weeks later the stream still flowed at 2.27 $m^3 h^{-1} km^{-1}$.

History and consequences of deforestation

The later colonial period in India saw the onset of commercial forest exploitation (Pant, 1922; Rawat, 1983). Forests were harvested for railway sleepers, then to make way for roads, and then because new roads make forests more accessible (see Bahuguna et al., 1981/3; Haigh, 1982b). The growth of new urban centers demanded timbers and fuelwood. The suppression of malaria in the "Terai" zone, the swampland between the Gangetic plains and the foot of the Himalaya, allowed in massive forest conversion to agriculture. Finally, the general expansion of economic activity on the plains and in the larger towns of the Himalayan fringe created an increased demand for paper, timber, and foreign exchange. Recently, the pace of change has accelerated. Demand is increasing and the forests are dwindling. Some of the increased pressure may be due to the expansion of Himalayan village communities (Ashish, 1979) but most comes from outside. Today, the value of timber is such that forest guards carry arms against thieves and those who oppose tree-theft stand in fear of their lives. Meanwhile, serious overgrazing and burning to improve pasture hold back forest regeneration.

The reduction in the forest cover of the front ranges, foothills, and Terai zone of the Himalaya has had serious environmental consequences (figure 20.2). The degraded forests trap less of the rain than before, there is a much reduced litter layer on the forest floor, and the percentage of bare earth or rock exposed at the ground surface is increased. More rainfall strikes the ground directly and these rainsplash impacts pack down the ground surface, sometimes armoring it with a soil crust. As forests are thinned, the volume of undergrowth increases and they become a better environment for grazing animals. Grazing animals nip off the shoots of plants and their activities tend to favor species which grow from the base, like grass, rather than those whose growth comes from a lead bud – like trees and many forest herbs – so grazing prevents forest regeneration. Grazing animals also trample down and compact the soil. Soils under grazing have higher bulk densities and lower infiltration capacities than those which are not grazed (Hewlett, 1982a, p. 52).

Forest thinning increases the rate at which water reaches the soil as less is delayed by interception on the forest canopy, and it

Figure 20.2 Schematic illustration of hydrological processes on a forested and on a degraded Himalayan valley side.

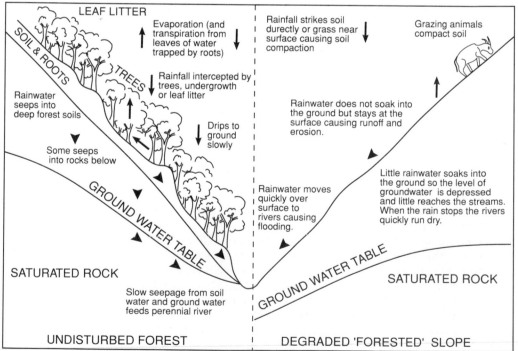

encourages more undergrowth and increased grazing. The result is increased soil compaction and a reduction in the infiltration rate, so more water remains at the soil surface where it may generate surface runoff. Surface runoff causes accelerated soil erosion and decreased soil depth. In general, thinner soils store and transport less moisture. Thus, forest thinning allows increased surface runoff and sediment discharge from headwater hillslopes (Pereira, 1989, p. 58 et seq.). In undisturbed forest, there is little undergrowth or bare ground because the soil is protected by leaf litter. The relative absence of rainsplash impact and trampling allows the soil to retain a looser and more open structure. This permits a more rapid infiltration of water, so more rainfall enters the soil. Forests, however, have a very large biomass relative to grass or scrubland and use a much greater volume of water. This is trapped by their roots and returned to the air by evapotranspiration. In effect, forests create a strong suction pressure in the soil, which helps to pull surface waters into the soil and soil-waters into their root systems. As much as 50 percent of the water that falls on a forest may be trapped in this fashion, but as forest biomass is reduced this declines, so more water is left in the soil and at the soil surface.

Sometimes, forest thinning is accompanied by the decline in spring discharge. Springs are fed by groundwaters which have percolated through the soil and into the rocks below. Groundwater recharge requires that there is a steady supply of water to the upper surface and/or structural discontinuities (faults, joints, bedding plants, schistosity, etc.) in the bedrock. The thin, mineral soils of degraded forest lands are more exposed to the sun and tend to dry out and drain quickly. By contrast, organic-rich forest soils are deep and shaded, and, despite evapotranspiration, may provide more water to groundwater recharge. It all depends upon whether more water is lost by the accelerated surface runoff that may follow forest degradation than was consumed by evapotranspiration in undisturbed forest. Clearly, this balance can be affected by many local conditions: soil, topography, climate, and geology. Even so, environmentalists argue, the net result on many steep Himalayan hillsides is a reduction in groundwater recharge and in the discharge of springs and streams. This has led some scientists to call for the preservation of forests on the higher slopes as "spring sanctuaries" (Bartarya, 1991).

In any case, forest thinning is usually associated with reduced infiltration rates and/or soil water storage capacity, and increased surface runoff, soil erosion, and total runoff (Goudie, 1990, pp. 140–3). If the forest thinning affects the upper hillslope, much of the additional surface runoff may infiltrate in less affected forest, agricultural, or natural stream terrace lands downslope (Meijerink, 1974). When this occurs, mobilized sediment will be deposited and the net impact on any stream channel below the hillslope will be small. Elsewhere, forest thinning may result in increased storm flow, increased sediment load, and dramatic changes in stream-channel morphology. Increases in the mean annual flood encourage

Plate 20.2 The
Mussoorie–Tehri road
cutting across a
deforested hillside at
Masrani, Garwal.
Deforestation and road
construction are two of
the principal causes of
accelerated erosion and
landsliding in the
Himalaya.

a stream to entrain sediment, and enlarge and incise its channel. Increases in sediment load encourage a stream to aggrade its bed, become shallower, wider, steeper, and less sinuous, and to transmit a higher proportion of its discharge as bed-flow rather than surface flow, and so develop a more seasonal pattern of surface discharge (Schumm, 1977). This pattern of channel aggradation is quite common in the Himalaya and has, no doubt, contributed to the widespread belief that rivers are "drying up." Indeed, in ecological terms, this is the net result, since the water has become less accessible. Because the higher river beds mean higher flood levels, increased flood hazard also becomes possible.

A final consequence of forest thinning is increased landslide activity. The forest trees have dense root systems which may play an important structural role in the hillsides. When the trees are removed, their roots rot away and the stability of the hillside declines. The consequence is increasing landslide frequency. Landslide debris is a major source of sediment in Himalayan stream channels and landslides are an important agency in forest destruction, whether created by forest destruction or other human activities, notably road construction (plate 20.2). Once triggered, even small landslides may demonstrate the capacity to expand progressively year by year across the Himalayan hillslopes to become huge "chronic" landslides generating large amounts of debris each monsoon season (Haigh et al., 1988b).

In sum, the environmentalist thesis is that forest degradation in the Himalaya, resulting from economic exploitation by outsiders, but exacerbated by overgrazing and development processes in the Hills, is causing a dramatic change in the hydrological regime. This is manifested by a reduction in evapotranspiration and increased surface water discharge during storms, causing increased flooding, increased soil erosion, increased landslide activity, increased sediment delivery to streams, increased channel deposition, increased flood levels, and increased channel width, causing decreased surface discharge in favor of bed-flow. At the same time, erosion and surface compaction of the soil causes reduced soil depth, soil water storage capacity, and infiltration capacity, often resulting in reduced groundwater recharge and reduced dry season flows. So deforestation in the Himalaya and on its fringes is causing erosion, landslides, floods, and the drying up of springs, and its impacts are transmitted from the headwaters down channel, perhaps as far as the Bay of Bengal (see Reiger, 1976; Haigh, 1981, 1984b; Agarwal et al., 1982; Myers, 1984, 1986).

Countering "the myths" of Himalayan environmental degradation

Recent research, coming mainly from teams working in the Middle Hills of Nepal, has begun to question the accuracy of the environmentalist model outlined above. The teams claim that both the extent and impact of the forest degradation may be much exaggerated. For them, "a haze of uncertainty" surrounds much of the information which is produced as evidence – for example, estimates of per capita fuelwood consumption in the Himalaya vary by a factor of 26 (Thompson and Warburton, 1985; Ives, 1987). Extrapolation based on such evidence is inevitably meaningless (Negi, 1983).

Their attack begins on figures for deforestation, a word which has been too loosely applied. Processes as diverse as timber harvesting and the lopping of tree branches for firewood have been counted as "deforestation," along with forest conversion to permanent agriculture, grazing, and even commercial tree crops (Hamilton and King, 1983; Hamilton, 1987). There has been a tendency to ignore the fact that replanting and regeneration is mitigating the gross impact of forest removal (Gilmour, 1988). In many locations, the balance between forest and agriculture seems to have remained fairly stable for many centuries.

The hydrological impact of forest thinning is proportional to the area of the catchment affected, and this is relatively small. There are few reliable studies from the region which can verify real change on a large scale. The notion that forests "act like sponges to soak up and store water" is argued to be nonsense. Forests remove water from a catchment by evapotranspiration and their impact is greatest during the dry season, when the water demand from the vegetation is proportionally the greatest (Hamilton and Pearce, 1987). Forest thinning, which does not allow massive soil loss, may actually improve the hydrological regime by leaving more moisture in the soil and to recharge the groundwaters (Bruijnzeel, 1990). Correlations between land area deforested and reduced streamflows are not cause and effect, though causation may be accomplished by subsequent land-use conversion (Hamilton, 1985, p. 682). Hewlett (1982b) reviewed the worldwide evidence of hydrological changes from forest cutting in headwaters and floods in the lower basin and found no cause–effect relationship. Any increase in flooding close to a cleared area quickly disappears downstream (Hamilton, 1985, p. 684).

Severe floods are not affected by deforestation as much as by other variables in the catchment. The capacity of forest soils to absorb storm rainfall is quite limited. Once the forest soils are saturated, they will behave like any other saturated surface. Further, even though soils compacted by deforestation and grazing have lower hydraulic conductivities rates (50 versus 500 mm h^{-1} in undisturbed forest), results from Nepal suggest that, when the number of rainfall events that exceed this surface infiltration capacity is taken into account, the net contribution to surface runoff in an extreme event is relatively small (about 1 mm h^{-1}) and not enough to induce major flood problems (Gilmour et al., 1987). In fact, the rise and fall of major flood peaks is more seriously affected by changes in the area available for the spread of flood waters, and variations in the interaction between the rainfall and the catchment topography, than by vegetation type (Pearce and McKerchar, 1979). Greater flood plain occupancy, channel constriction, the alteration of human structures, and the construction of roads, drains and ditches that speed the movement of waters are the major human impacts (Hamilton, 1985; Megahan, 1988). Finally, there is a huge degree of natural variability in the strength and severity of the monsoon rains. It is human nature to believe that today's floods are worse than those in fading memory of the past, but there is no factual evidence to support this notion (plate 20.3).

Although forest degradation can lead to accelerated erosion and landslide activity, it is necessary to put this increase into perspective. Mass wasting remains the dominant hillslope process and geological factors are the major determinants of slope stability. Deforestation is linked to surface erosion, gullying, and shallow landslides (less than 3 m deep), and locally this is important (Ramsay, 1987, p. 239). However, the bulk of the sediment load in most Himalayan rivers comes from huge landslides resulting from rapid tectonic uplift of the mountains, seismic activity, and the rapid incision of the rivers. This incision cuts into the footslopes of the steep Himalayan hillsides and causes landslide activity. In most locations, natural processes dwarf the tiny impacts of human activities (Carson, 1985; Pearce, 1986).

Plate 20.3 One of the most controversial environmental issues is whether Himalayan deforestation is aggravating the flood hazard in the South Asian lowlands downstream.

Stream channel sedimentation may have several further causes. Below the mountains, there has been a massive increase in tubewell irrigation. This has lowered the groundwater table on the fringes of the plains (Someshashekhara Reddy, 1989). India's groundwater pumpage exceeded 40 million hectare meters in the 1970s and is projected to reach 150 million hectare meters by 2001 (Dewan, 1990). Rivers flowing from the hills may now soak through the floor of their own channels toward the receding groundwater table. Increases in the amount of water extracted directly for irrigation and water supply further reduce the volume of surface waters, which can move less sediment, which is thus deposited on bed of the river channel. The rising beds of the rivers and increased agricultural use of riverside lands encourages local communities to construct levées. The levées constrict the floodplain, and its capacity to store flood waters, and so transmit a greater depth of floodwater downstream. The net impact of such processes is greatest on the Gangetic Plains, which implies that little of any possible hydrological change due to deforestation in the Himalaya affects lands any distance from the mountains.

Supporters of this view argue that the problems of the Himalaya are easily exaggerated and over-dramatized. Much of what is written about the environmental impact of deforestation is unproven and arguments that link the deforestation of the Himalaya to floods thousands of kilometers away are tenuous. The Himalayan system of land management remains basically sound, but it is accepted that in some areas in the mountains there are problems. However, a great deal of careful scientific study is needed to establish the true situation. It is not appropriate to talk about impending disaster or to leap into any radical program of action until the true situation is known (Ives and Messerli, 1989).

A third eye: the Himalayan viewpoint

Opponents of the environmentalist thesis of Himalayan devastation often write in support of the Himalayan villagers who are inappropriately blamed for the wanton destruction of the hills and for problems on the Gangetic Plains (Ives and Messerli, 1989). However, as a group, the foreign scientists are much less generous toward their colleagues in the Himalayan universities, whose work they tend to ignore. For example, Ives and Messerli's book includes around 325 bibliographic citations but just 37 published in India and Pakistan (Ives and Messerli, 1989). Many of the protagonists, both for and against the environmentalists' thesis, are based in Western countries. Those who live in the Himalaya have a rather different perspective on the mountains where they live and on the environmental changes which affect their homes and families. In recent years, the Himalaya has developed a large and active indigenous research community, thanks to the creation of new universities and research centers staffed by local scholars.

Much of the output of this expanding scientific community is regarded with scepticism by outsiders. In truth, Himalayan research is loaded toward commentary rather than field study and quality control tends to be poor. Further, because of the exorbitant cost of Western publications, Himalayan scientists have restricted access to work from the outside world. Outsiders criticize the work of the Himalayan scientists for imprecision, technical deficiencies, and "unreliability." In turn, Himalayan scientists dismiss much foreign research as the hit-and-run output of outsiders with insufficient experience and a superficial understanding of local conditions (see Chalise, 1986). Doth six months in the field, and a PhD thesis from a Western university, an international expert make? The results from the Himalaya's own scientists add a third, independent, and distinctive perspective on the debate as well as a wealth of technical detail.

There are at least two totally independent scientific groupings in the Himalaya. Direct experience has demonstrated that it is not possible to exchange scientific papers across the India–Pakistan border – packages are confiscated. So Indian and Pakistani teams work on their similar problems in isolation. The focus here will be the results from Uttarakhand, India (Haigh et al., 1990). However, it is instructive to compare these results with those produced across the border in Pakistan because, despite their respective isolation, the two communities are in broad agreement (Haigh, 1991).

Case study: Uttarakhand

Most of the work of the Himalayan scientists comes from the front ranges of Uttarakhand, Garhwal, and Kumaun, in the Hill Districts of Uttar Pradesh, India. This region, covering 52 000 km² and an altitudinal range of 300–8000 m, lies off the western border of Nepal (Singh, 1968; Joshi et al., 1983) (figure 20.1 and table 20.1). The inspiration for much of Uttarakhand's environmental science is the activism of a local campaigning group, the Chipko Movement (Mishra and Tripathi, 1978; Bahuguna, 1985; Haigh, 1988). Chipko is dedicated to improving the condition of rural communities and its leaders lean to a Gandhian, Sarvodaya ("green" in latterday Western parlance) model of development leading to an "economy of limited wants" (Doctor, 1965). Local politicians complain that the Gandhians have "hijacked" popular concern about escalating problems in the rural economy and about exploitation from the Plains. Chipko's political power base is the "invisible" workforce of Uttarakhand, the village women, who are responsible for most of the agricultural and domestic activities outside the cash economy (Jain, 1984; Shiva, 1986). The stresses of rural life, including fuelwood and water shortages, are felt most acutely by such village women. Their protests provide the motive power for Chipko campaigning against forest exploitation, alcohol, surface mining, and bonded labor.

Research from several parts of Uttarakhand demonstrates that the current agro-ecosystem, which depends on the forest for a large percentage of its energy, is no longer in balance. Confiscation of forest for commercial or state conservation purposes has forced many villages to over-exploit that which remains (Pant, 1922). Singh et al. (1984, p. 85) calculate that each hectare of agricultural land demands the productivity of 5–18 hectares of good forest to sustain its fertility and productivity (plates 20.4 and 20.5). The forest supplies fodder – hence manure, fuelwood, and timber for tools and housing.

Table 20.1 Forest cover in Uttarakhand, north India

Source	Area surveyed (km²)	Forest canopy over 60% (%)
Gupta (1979)	52 000	22
Sharma (1984)	27 000	24
Tiwari et al. (1986)		4.4
Tiwari and Singh (1987)	20	6

Administration: India, Hill Districts of Uttar Pradesh: Uttarkashi, Dehra Dun, Tehri Garhwal, Chammoli, Pauri, Pithoragarh, Almora and Nainital.

Plate 20.4 Well managed agricultural terraces, as here in Nepal, can help reduce runoff and sheet erosion.

Official statistics indicate that 15 percent of Uttarakhand is arable agriculture, 46 percent government-managed forest, 20 percent forest managed by local communities, and 20 percent private or community-controlled scrub, pasture or wasteland (Shah, 1982). The 40 percent of the land in community or private control is the most seriously degraded. Uttarakhand's population of 4.8 million (1981) is expanding at a rate of around 2.3 percent per annum. In its wake, the area classed as agricultural land expands at about 1.5 percent per annum and the livestock population of around 3.4 million cattle-units increases at 0.18 percent per annum (Shah, 1982; Singh et al., 1984, p. 81; Dadhwal et al., 1989). The bulk of the population and its expansion is found, not in the hills proper, but in the Terai zone of the Himalayan fringe (Joshi et al., 1983).

Deforestation

Originally, forest may have covered around 60 percent of the land area in Uttarakhand. By the 1980s, according to analyses of satellite imagery, the forest cover was between 29 and 37.5 percent (Gupta, 1979; Tiwari et al., 1986). Air photographic analyses of 27 000 km^2 found that 24 percent of the forest had a crown closure of over 60 percent, 14.5 percent had medium crown density, and 25 percent had a low crown density or was degraded to scrub or pasture (Sharma, 1984). Satellite imagery confirmed that around

Plate 20.5 Failure to maintain terraces, combined on this hillslope at Kamaun, Uttarakhand, with overgrazing, can quickly lead to slope degradation.

20–25 percent of Uttarakhand had closed canopy forest (Gupta, 1979), though Tiwari and colleagues (1986) prefer that "good forest" (crown closure over 60 percent) covers only 4.4 percent (table 20.1). Tiwari and Singh (1987) surveyed 200 000 ha in Kumaun using air photographs and field checking. Some 45 percent was not forested, 23 percent being used for agriculture and 15 percent being scrub/grassland. Only 2 percent was called wasteland, but 85 percent of the agricultural land was said to suffer from accelerated erosion. The agricultural area was increasing at a rate of 1.5 percent per annum at the expense of forest. Forest covered 55 percent of the land but only 6 percent had a crown canopy better than 60 percent and 21 percent had a canopy of less than 40 percent, which indicates forest thinning.

In the hills, change is most dramatic where there is recent road construction (Bahuguna et al., 1981/3). Road construction contributes to forest destruction through clearance during construction, triggering clearance landslides, damage due to road clearance debris, and indirectly through improving access (Haigh, 1982, 1984a; Haigh et al., 1987). Government sources have suggested that as much as 25 percent of all timber felling on the hills is illegal and despite an active program of planting, even Government forests decline in area – by 5 percent between 1965 and 1980 (Kumar, 1983). As in Nepal, the rate of forest decline increases toward the plainward fringe of the hills (HMG Nepal, 1983). Malla (1985) reports that between 1954 and 1981, 60 percent of the tropical forest in the Nepalese Terai disappeared (cf. Kashmir, see Kawosa, 1990). In Uttarakhand, more than 40 000 ha of swamp and forest in the Terai fringe were converted to agriculture after independence and the area continues to develop rapidly (Singh, 1968; Joshi et al., 1983).

Rainfall interception and surface wash

In 1981/2, Pathak et al. (1985) studied rainfall interception at six forest sites, five with a canopy closure in excess of 70 percent, one with 38 percent, near Nainital. The canopy intercepted between 8 and 25 percent of the rainfall. At Dehra Dun, in the foothills, pine forest (1156 trees ha^{-1}) intercepted 22 percent of the rainfall, and densely coppiced Sal (*Shorea robusta*) as much as 34 percent (Dabral and Subba Rao, 1968; Ghosh and Subba Rao, 1979). Bruijnzeel (1990, p. 176) suggests that 14–25 percent is typical for tropical montane forest. Pathak et al. (1985) found that stemflow accounted for less than 1 percent of rainfall (cf. 0.5–2.0 percent average: Bruijnzeel, 1990, p. 176). However, at Dehra Dun, stem flows equivalent to 4 percent of rainfall were measured in the plantation of pine (*Pinus roxburghii*) and 6–9 percent in coppiced Sal (Dabral et al., 1963; Dabral and Subba Rao, 1968; Ghosh et al., 1980).

The leaf litter layer intercepted 7–10 percent of the rainfall in the Nainital forest sites (Pathak et al., 1985). Ghosh et al. (1980) estimate litter interception as 5 percent, while Dabral et al. (1963) measured 7.6 percent beneath pine, and around 9 percent beneath Sal and teak (*Tectona grandis*). Most of the remaining rainwater is intercepted by the undergrowth or goes to infiltration (59–84 percent: Pathak et al., 1985). Ghosh et al. (1980) found infiltration rates beneath a dense, mature, plantation of Sal of 38 mm h^{-1} to be twice those on neighboring agricultural lands. Soni et al. (1985) recorded 96 mm h^{-1} beneath Sal, 36 mm h^{-1} beneath pine, and 10 mm h^{-1} beneath teak, compared to 8 mm h^{-1} on grassland and only 3 mm h^{-1} under pine where undergrowth had been burnt to improve grazing. Further into the hills, Mohan and Gupta (1983) also found infiltration rates under thin forest (12.5 mm h^{-1}) to be lower than those on terraced crop land (33.6 mm h^{-1}) and grassland (21.5 mm h^{-1}). Mohan and Gupta attribute the low infiltration capacity of their degraded forest soil to surface compaction and poor litter production. Elsewhere, clearfelled areas were shown to have less organic matter in the soil, and a lower porosity and water-holding capacity than areas where some forest was retained (Singh and Gupta, 1989).

Surface wash proved very rare on the forested slopes. It accounted for less than 0.8 percent of the rain falling on the densely forested, 25 m^2, plots of Pathak et al. (1985), and only 1.3 percent on the one thinly forested site. However, a great deal more water seemed to be moving through the upper layers of the soil (Singh et al., 1983). Near Almora, Rawat and Rawat (1993) confirm that rainfalls of intensities 2.3 mm h^{-1} or less caused no runoff erosion in pine forest in contrast to agricultural and barren lands. At Dehra Dun, an 8 percent grassed slope of silt/clay loam yielded 21 percent of rainfall as runoff, while compacted bare fallow yielded 71 percent (Dhruva Narayana, 1987).

In Uttarakhand, the carrying capacity for grazing animals is thought to be exceeded by a factor of 2.5 or so, but the pressures are greater in more densely peopled areas (Sharma, 1981; Dadhwal et al., 1989). Trampling and compaction can reduce the infiltration capacities of soil by as much as 50 percent according to one Dehra Dun case study (Dhruva Narayana and Sastry, 1985). Infiltration rates are positively correlated with soil porosity, negatively with bulk density/compaction, and they are affected by soil chemistry/ soil type (Soni et al., 1985; Grewal et al., 1990). However, other things being equal, the greater is the grazing pressure, the higher is the bulk density and the lower is the infiltration rate. Bruijnzeel and Brenner (1989, p. 90) caution that surface wash is more likely in the Shivalik and foothill areas because of their high grazing pressures.

Singh, Pandey, and Pathak (Singh et al., 1983; Pandey et al., 1983; Pathak et al., 1984) compared sediment yields from forested and deforested (25 m^2) runoff plots near Nainital. They estimate soil losses due to surface and sub-surface erosion as 32 kg ha^{-1} from oak forest and 62 kg ha^{-1} from non-forest sites. Rawat and Rawat (1993) established their runoff plots on 3° east-facing slopes over mica schist near Almora, to compare surface runoff erosion under pine forest, grassland and barren land. Results from 26 natural rainstorm events found that the agricultural lands lost over three times and the grassed barren land more than twice as much sediment as the forest.

Desiccation

The moist broadleaved forests of the Lesser Himalaya may return as much as 50 percent of the rainfall as evapotranspiration (Seth and Khan, 1960). Planting eucalypts on a denuded watershed near Dehra Dun reduced total runoff by 28 percent (Mathur et al.,

1976). Himalayan "Chir" pine has a lower transpiration rate than local broadleaves but, being evergreen, it consumes more water through the year (Raturi and Dabral, 1986). Ecological pressure is causing pine, characteristic of higher altitudes, to replace broadleaves on many lower and warmer slopes. Locally, it had been suggested that the leaf area index was the best guide to runoff variability (Misra et al., 1979). In the Shivalik foothills, it was found that runoff from 24° slope, 9 ha watershed, declined from 38 to 31 percent of rainfall as grass productivity increased from 7 to 9 quintals ha^{-1} (Agnihotri et al., 1985). Later studies stress the hydrophysical properties of the soil–grass interaction (Grewal et al., 1990). However, taken together, the Uttarakhand results demonstrate that more vegetation means less water available for runoff.

This contradicts the local belief that, throughout the region, springs are drying up as the forest disappears. Almora, the traditional capital of Kumaun from AD 1560, straddles a high ridge with steeply sloping sides. Its demand for timber has much reduced local forest cover. Enshrined in local place-names is evidence that the ridge once possessed 360 springs. Today, hardly a tenth of that number still function (Joshi et al., 1983). Surface mining of the Krol Limestone belt above Dehra Dun may have reduced the number of springs from 76 to 42 (Saxena et al., 1990, p. 258). Degradation of local oak forests was blamed for reductions in discharge reported from more than half of the 60 springs investigated by Singh and Rawat (1985). A more systematic survey of the springs in the Gaula River basin below Nainital found that 45 percent had gone dry in recent memory (Valdiya and Bartarya, 1989a; Bartarya, 1991). Local environmentalists think that more than half of Uttarakhand's springs have dried up in the past 20 years and that the number of villages suffering water scarcity has increased by 40 percent in the past 13 years (Uttarakhand Seva Nidhi, 1986). Local workers blame the problem on accelerated soil erosion, the reduced depth of soil on deforested hillsides, and consequent reduced capacity for soil water-storage (Bartarya, 1988).

Base-flow in major rivers has also been reduced, with dry season lows in the Gaula River being decreased by up to a third in the past decade (Valdiya and Bartarya, 1989a; Bartarya, 1988). Privately, water engineers at Almora suggest that the same may be true of the Kosi River. Studies in the Nana Kosi Basin indicate that streams are much reduced in deforested areas (Rawat and Bisht, 1986; Rawat, 1990).

Forests, floods, and landslides

There are no definitive results for the hydrological consequences of forest conversion in the central Himalaya, though much has been done in the Shivalik foothills and at lower elevations on the Himalayan fringe (Bruijnzeel and Brenner, 1989, p. 91; Haigh et al., 1990). At Dehra Dun, a Sal forest watershed was converted to agriculture as part of a paired catchment study. Runoff volume increased by 15 percent and the peak rate of runoff by 72 percent relative to the undisturbed catchment. The effect was eliminated when the basin was treated with soil conservation structures including bunds and check dams (Dhruva Narayana and Sastry, 1985). Another test assessed the impacts of forest thinning on paired 6 ha Sal forest catchments on a 20–30° slope. Removing one tree in five did not affect the volume of flow but flood peaks increased by more than 8 percent. The effect was greatest in the first year after thinning, when the canopy interception was reduced from 18 to 13 percent. Planting eucalypts in a degraded catchment, also near Dehra Dun, reduced discharge by 28 percent and the peak flow by 73 percent (Mathur et al., 1976; cf. Bruijnzeel, 1990, p. 85).

In 1978, the Mussoorie Bypass, built across a steep (42°) hillside, blocked four parallel, ephemeral stream channels each draining a 1 km^2 catchment. Two of these drained deforested scrubland on the urban fringe and two drained protected forest. During the monsoon, the unfinished road-bed functioned as an efficient sediment trap. By its close, some 50 m^3 of debris lay on the road-edge beneath each forest catchment compared to 230 and 370 m^3 of debris beneath the deforested grassland and scrub catchments (Haigh, 1982a). Two years sampling work in sub-catchments of the Nana Kosi near Almora convinced Rawat (1989) that agricultural land generates 17 times, deforested catchments 9 times, and pine forested areas 4.8 times as much river channel suspended sediment as mixed oak forest (table 20.2).

Hill roads cause many landslides and rate as important sediment sources, and tens of thousands of kilometers of new roads have been built in recent decades. Estimates of sediment generation from the roadways of the Central Himalaya run from 430 to 500 m^3 km^{-1} year^{-1} (Patnaik, 1981; Valdiya, 1987). In some areas, and some years, the volume of sediment produced may be much higher than this (Bansal and Mathur, 1976; Haigh, 1984a; Haigh et al., 1987, 1988a). Valdiya and Bartarya (1989b) found that a fifth of the landslides in the Gaula Basin below Nainital occurred along roads. Road construction has been called the most important development impact (Bruijnzeel and Brenner, 1989, p. 109), though the total contribution to river loads may be quite small (Arni Gadh: less than 10 percent according to Haigh, 1979b).

Table 20.2 Ratios of soil loss in Uttarakhand, north India

Source	Pine forest	Oak forest	Agriculture	Other
Runoff plots				
Pathak et al. (1985)	1	–	–	1.9
Rawat and Rawat (1993)	1	–	3.1	2.1
River sediment loads				
Haigh (1982)	–	1	–	5–7
Rawat (1989)	4.8	1	17	9

Some workers argue that 20–25 percent of all landslides may be due to other than natural causes (Laban, 1979). Landslides are not all discrete events but emerge and evolve over many years. "Chronic" landslides which cause perennial problems may evolve from quite small trigger events (Bansal and Mathur, 1976; Haigh et al., 1988b). There is widespread concern that landslide activity is increasing (Bhatt, 1980; Valdiya, 1985). Meijerink (1974), working in the Aglar valley of Garhwal, calculates that superficial debris slides generate only 5 kg ha^{-1} year^{-1} of debris, of which just a fifth reaches the river channel. He estimates the catchments's long-term geological sediment yield as 90–450 kg ha^{-1} year^{-1}. By contrast, soil losses as high as 156 000 kg ha^{-1} year^{-1} have been recorded on 8 percent slope tilled fallow at Dehra Dun (Anantharaman et al., 1984). Dhruva Narayana (1987) estimates sediment yield from the catchment of the controversial Tehri High Dam as 19.4 kg ha^{-1} year^{-1}. However, the sediment loads of Himalayan rivers can be very high. Das (1987) recorded monsoon sediment concentrations in excess of 2 kg m^{-3} in rivers above the new Dam. Subramanian (1978, 1979; Subramanian and Dalavi, 1978; Abbas and Subramanian, 1984) adds that 90 percent of sediment mass transfer in the Ganges comes from the Himalaya, where erosion rates are among the highest on earth, and that human impact may not be ruled out. However, few workers are willing to rate the overall human impact on sediment production higher than 10 percent.

It may be expected that many rivers aggrade their beds naturally as well as with human assistance by some centimeters each year – local environmentalists estimate 10 cm year^{-1} (Bahuguna, 1978; cf. 15–30 cm year^{-1} in southeast Nepal, Valdiya, 1987, p. 176; 50–135 cm year^{-1} in Puthimari River, Bhutan, Samar Singh, 1982). Some suggest that unconstrained channels are also becoming wider and less sinuous (Kollmannsperger, 1977). Sediment-choked channels are not uncommon in the hills in areas close to construction activity. The sediment is supplied by deforestation, road construction, and clearance, and by landslides due to the channel processes. A survey of a 1 km reach of the Company Gadh below Mussoorie in 1978 found nine fresh landslides with greater than 20 m^3 of debris, in total contributing more than 1000 m^3. However, the most spectacular illustration of this process is found at Rajpur on the Mussoorie–Dehra Dun road. Here, in 1919, the Kaulagarh Bridge cleared the bed of a small perennial stream by 19.5 m. Just upstream, deforestation had triggered landsliding, which dumped an enormous volume of sediment in the channel. By 1979, the bridge cleared a wide boulder run by just 1 m (cf. Nossin, 1971; Haigh, 1984b; Agarwal and Narain, 1985) (plate 20.6). In the absence of historical knowledge, geological causes might have provided the easiest explanation for this feature.

Landslide activity is the cause of some of the greatest flood disasters in the region. Landslides can create temporary impoundments of narrow valleys. When this fails there may be dramatic releases of flood waters that dwarf even the greatest rainfall flood (Dogra, 1983, p. 11; Valdiya, 1985, 1987).

Plate 20.6 The Kaulagarh bridge, Dehra Dun, in 1919 (a) and in 1979 (b). The bridge supports have been almost completely buried by river aggradation.

Discussion

The interaction between forests, hillslope processes, and hydrology is a complex affair. Some argue that too many assertions about the impacts of deforestation in the Himalaya are based on extrapolation from textbook theory, rather than measured reality (Ives and Messerli, 1989), or concern political ambition more than fact (Haigh, 1989a, b; Bruijnzeel, 1991, p. 36). The data produced for different processes and consequences of forest conversion and degradation vary substantially from one study to another. There is reason to doubt the quality of some of the data marshalled by teams on all sides of this debate, if only because research tends to select exceptional situations for study at the expense of the less spectacular. Over-generalization and the necessity to extrapolate from small case studies to the entire mountain chain bedevils all those involved.

This last point may lie at the heart of the discrepancies between results from Uttarakhand, Pakistan, and the Middle Hills of Nepal. Dramatic change in the Himalaya is associated with new development and this, in general, is greatest along the Himalayan fringe, the Terai and front ranges. In Nepal, the most dramatic losses of forest are recorded in the Terai (26–28 percent from 1964 to 1977/8; HMG Nepal, 1983, p. 21). In Swat, large-scale deforestation followed that nation's integration into Pakistan in the mid-1960s (Haigh, 1991). Chipko padyatra (footmarch) teams report that, in the Uttarakhand interior, the "mass destruction" of trees continues (Bahuguna et al., 1981/3), but away from the roads less change occurs. As for the scientific debate, the results from Uttarakhand tend to support the broad outline of the environmentalist thesis. Certainly, the more general overviews provided by Uttarakhand scientists to local periodicals lend universal support to the notion that environmental change is fast, furious, and negative (Valdiya, 1985, 1986; Moddie, 1985; Shah, 1991).

In the outside world, the most recent surveys of international research on the hydrology of tropical forest conversion and highland–lowland interactions in southern Asia adopt a more neutral stance (Bruijnzeel and Brenner, 1989; Bruijnzeel, 1990). Forests use more water than crops or grassland, so forest conversion consistently increases water yield (110–825 mm depending on local conditions) (Bruijnzeel, 1990, p. 179). Where forest conversion increases overland flow, including saturated overland flow, it usually also increases flood peaks. The impacts of deforestation are greatest on the smaller floods and least on the largest (Bruijnzeel, 1990, p. 103). Since channel response is modelled against the mean annual flood, this implies considerable impacts on river channel morphology (Schumm, 1977). Bruijnzeel (1990, p. 180) agrees that flow volumes do accumulate downstream although local increases are much moderated by lag effects and spatial patterns of rainfall, and both sedimentation and flood damage are much influenced by groundwater extraction, increased floodplain utilization, and changes in river cross-sectional morphology (cf. Government of India, 1980).

More research and more reliable data are required (Ives and Messerli, 1989; Bruijnzeel and Brenner, 1989). Three areas seem to demand urgent study. First, if deforestation is causing increased flooding and desiccation of the landscape, one reason is the reduction of the soil water-storage capacity by erosion. Runoff in major floods is controlled by soil water-storage rather than vegetation cover (Bruijnzeel, 1990, p. 180). Forest conversion tends to lead to a reduction in soil depth and storage capacity (Bruijnzeel, 1990, p. 181). In the Himalaya, there is a saying that "when the trees are cut down, the rocks grow out of the ground." Soils in deforested areas tend to be less deep than those of equivalent forested slopes; however, more measured evidence is needed. Second, at present, evidence for river channel/floodplain response to the supposed new conditions is limited to a few case studies. More systematic survey would be useful. Finally, there remains the problem of quantifying the overall human impact on water and sediment yield from the Himalaya (Government of India, 1980).

Conclusion

Research produced by the Himalaya's own research communities generally supports the environmentalists' model of rapid and destructive environmental change. There is little independent evidence for the antithetical arguments proposed by the Western scientific teams working in Nepal's Middle Hills. However, it remains plausible that the debate is basically caused by a failure to recognize the variation in environmental conditions across the Himalayan region (Ives and Messerli, 1989; Shah, 1991; Haigh and Krecek, 1991; Karan, 1991). Reports of catastrophic environmental change for the most part do come from areas in the front line of development, such as the Lesser Himalayan Front Ranges, Terai, and Shivaliks. Away from the new roads, mines, and larger settlements, in the heart of the Himalaya, the rate of change may be relatively slow. Certainly, the traditional system of land husbandry is viable as long as the forest, which provides 90 percent of its energy needs, survives (Ralhan et al., 1991). Nevertheless, there is now more than enough data from areas like Uttarakhand to prove that human impact is sufficient to effect major and destructive environmental change. The possibility remains that changes in hydrological regime and sediment load affect areas outside the hills, even as far as Bangladesh, though the significance of these impacts compared to those of other developments on the Plains is debatable. However, the fact that major environmental change is occurring in the hills cannot be ignored and nor can the possibility that this could lead to greater problems in the future.

The reasons for this debate may perhaps, sadly, lie outside the scope of environmental science in the world of politics. Research and commentary are seldom purely objective. There is usually a reason why a particular project and area are selected for study. In environmental science, detachment from the corrupting worldly influences of tradition, prejudice, and politics is very difficult to achieve (Gandhi, 1927/9). The Himalayan debate, at one level, is about resources, grants, and perhaps also stresses across international borders. It is always wise to wonder who benefits and who loses out (cf. Haigh, 1989a, b). In microcosm, the debate in the Himalaya mirrors many larger discussions concerning global environmental change. The arguments address the foundations of scientific evidence and the guidelines which determine who and what we should believe.

Further reading

Bartarya, S.K. (1991) Watershed management strategies in Central Himalaya: the Gaula River Basin, Kumaun, India. *Land Use Policy*, 8 (3), 177–84.

Bruijnzeel, L.A. (1990) *Hydrology of Moist Tropical Forests and Effects of Conversion: State of Knowledge Review*. Paris: UNESCO IHP Humid Tropics Programme.

Bruijnzeel, L.A. and Brenner, C.N. (1989) *Highland–Lowland Interactions in the Ganges-Brahmaputra River Basin: a Review of the Published Literature*. Kathmandu: ICIMOD Occasional Paper 11.

Haigh, M.J. (1984) Deforestation and disaster in northern India. *Land Use Policy*, 1 (3), 187–98.

Hamilton, L.S. (1987) What are the impacts of Himalayan deforestation on the Ganges–Brahmaputra lowlands and delta? Assumptions and facts. *Mountain Research and Development*, 7 (3), 256–63.

Ives, J.D. (1989) Deforestation in the Himalayas. *Land Use Policy*, 6 (3), 187–93.

Ives, J.D. and Messerli, B. (1989) *The Himalayan Dilemma: reconciling Development and Conservation*. London and New York: Routledge/ United Nations University.

Valdiya, K.S. (1986) Accelerated erosion and landslide prone zones in the Central Himalayan region. In J.S. Singh (ed.), *Environmental Regeneration in the Himalaya*. Nainital: CHEA and Gyanodaya Prakashan, 12–39.

Bibliography

Abbas, N. and Subramanian, V.S. (1984) Erosion and sediment transport in the Ganges River Basin (India). *Journal of Hydrology*, 69, 173–82.

Adam, J.G., Echard, M. and Lescot, M. (1972) Plantes médicinales Haussa de l'Ader (République du Niger). *Journ. d'Agron. trop. et de Bot. appl.*, 19, 259–399.

Adams, J.M., Faure, H., Faure-Denard, L., McGlade, J.M. and Woodward, F.I. (1990) Increases in terrestrial carbon storage from the Last Glacial Maximum to the present. *Nature*, 348, 711–14.

Adams, J.M. and Woodward, F.I. (1992) The past as a key to the future: the use of palaeoenvironmental understanding to predict the effects of man on the biosphere. *Advances in Ecological Research*, 22, 257–314.

Adams, V.D. and Lamarra, V.A. (eds) (1983) *Aquatic Resources Management of the Colorado River Ecosystem*. Ann Arbor, MI: Ann Arbor Science.

Adams, W.M. and Hughes, F.M.R. (1990) Irrigation development in desert environments. In A.S. Goudie (ed.), *Techniques for Desert Reclamation*. Chichester: Wiley, 251–65.

Adey, W.H. (1978) Coral reef morphogenesis: a multidimensional model, *Science*, 202, 831–7.

Adey, W.H. and Burke, R.B. (1977) Holocene biotherms of Lesser Antilles – geologic control of development. *American Association of Petroleum Geologists Studies in Geology*, 4, 67–81.

Adhémar, J.A. (1842) *Révolutions de la Mer*. Paris.

Agarwal, A., Chopra, R. and Sharma, K. (1982) *The State of India's Environment: A Citizen's Report*. New Delhi: Centre for Science and Environment.

Agarwal, A. and Narain, S. (1985) *The State of India's Environment: The Second Citizen's Report*. New Delhi: Centre for Science and Environment.

Agarwal, A. and Narain, S. (1991) *Global Warming in an Unequal World: A Case of Environmental Colonialism*. New Delhi: Centre for Science and Environment.

Agassiz, L. (1840) *Etudes sur les glaciers*. Neuchatel.

Agnew, C. (1990a) Green belt around the Sahara. *Geographical Magazine*, 62 (4), 26–30.

Agnew, C. (1990b) Spatial aspects of drought in the Sahel. *Journal of Arid Environments*, 18, 279–93.

Agnihotri, Y., Dubey, L.N. and Dayal, S.K.N. (1985) Effect of vegetation cover on runoff from a watershed in Shivalik foothills. *Indian Journal of Soil Conservation*, 13 (1), 10–13.

Aguilar, J. (1987) Soil forming processes. In M. de Boodt and R. Hartmann (eds), *Eremology*. Rijksuniversiteit Gent, 83–94.

Akeroyd, A.V. (1972) Archaeological and historical evidence for subsidence in southern Britain, *Philosophical Transactions of the Royal Society, London, A.*, 272, 151–69.

Allan, J.A. (1991) Global biogeography: who is responsible? (Guest Editorial). *Journal of Biogeography*, 18, 121–2.

Allott, T.E.H. (1991) The reversibility of lake acidification: a diatom study from the Round Loch of Glenhead, Galloway, Scotland. Unpublished PhD thesis.

Allott, T.E.H., Harriman, R. and Battarbee, R.W. (1992) Reversibility of lake acidification at the Round Loch of Glenhead, Galloway, Scotland. *Environmental Pollution*, 77, 219–25.

Almer, B., Dickson, W., Ekstrom, C., Hornstrom, E. and Miller, U. (1974) Effects of acidification on Swedish lakes. *Ambio*, 3, 30–6.

Aloni Komanda (1978) Le role des termites dans la mise en place des sols de plateau dans le Shaba méridional. *Geo-Eco-Trop*, 2, 81–93.

Alves, D. (1991) Rate of deforestation in Brazilian Amazon. Unpublished note to 1991 Global Change Institute, Snowmass, CO, July 1991.

Amoros, C. and Petts, G.E. (eds) (1993) *Fluvial Hydrosystems*. London: Chapman and Hall.

Anantheraman, M.S., Saxena, P.B. and Pandey, B.K. (1984) The role of sediments and morphogenetic processes in soil formation, its depletion and conservation: a case study from the Dehradun Valley, Garhwal Himalaya. *Current Trends in Geology*, 5, 143–54.

Andersen, A.N. and Lonsdale, W.M. (1990) Herbivory by insects in Australian tropical savanna: a review. *Journal of Biogeography*, 17, 433–44.

Anderson, N.J., Battarbee, R.W., Appleby, P.G., Stevenson, A.C., Oldfield, F., Darley, J. and Glover, G. (1986) Palaeolimnological evidence for the recent acidification of Loch Fleet, Galloway. Palaeoecology Research Paper no. 17. London: University College.

Anisimov, O.A. (1989) Changing climate and permafrost distribution in the Soviet Arctic. *Physical Geography*, 10, 285–93.

Anon. (1989) Hydro development in Venezuela. *Water Power and Dam Construction*, December, 16–24.

Appleby, P.G., Nolan, P.J., Gifford, D.W., Godfrey, M.J., Oldfield, F., Anderson, N.J. and Battarbee, R.W. (1986) ^{210}Pb dating by low background gamma counting. *Hydrobiologia*, 143, 21–7.

Archer, S. (1989) Have Southern Texas savannas been converted to woodlands in recent history? *American Naturalist*, 134, 545–61.

Arkin, P.A. and Ardanuy, P.E. (1989) Estimating climatic-scale precipitation from space: a review. *Journal of Climate*, 2, 1229–38.

Arrhenius, S. (1896) On the influence of carbonic acid in the air upon the temperature of the ground. *Philosophical Magazine*, 41, 237–76.

Arrhenius, S. (1903) *Lehrbuch der kosmischen Physik*, vol. 2. Leipzig: Hirzel.

Ashish, M. (1979) Agricultural economy of the Kumaun Hills – threat of ecological disaster, *Economic and Political Weekly*, 14 (25), 1058–64.

Asrar, G. (1990) Mission to Planet Earth. In M.G. Coulson (ed.), *Proceedings 16th Annual Conference of the Remote Sensing Society*. University College, Swansea, i–v.

Atkinson, T.C., Briffa, K.R. and Coope, D.R. (1987) Seasonal temperatures in Britain during the past 22,000 years reconstructed using beetle remains. *Nature*, 325, 587–92.

Aubréville, A. (1949) Erosion et 'bovalisation' en Afrique noire française. *L'Agronomie Tropicale*, 2, 339–57.

Avery, M.I. and Haines-Young, R.H. (1990) Population estimates for dunlin *Calidris alpina* derived from remotely sensed satellite imagery for the Flow Country of northern Scotland. *Nature*, 344, 860–2.

AWRG (1988) *Acidity in United Kingdom Freshwaters*. Second Report of the UK Acid Waters Review Group. London: HMSO.

Bahuguna, S.L. (1978) Himalayan trauma: forests, faults, floods – Chipko seeks a new policy. *National Workshop on the Integrated Development of the Ganga-Brahmaputra-Barak Basin*. New Delhi: Gandhi Peace Foundation Working Paper 29, 1–8.

Bahuguna, S.L. (1985) Peoples response to ecological crises in the hill areas. In J. Bandyopadhayay, N.D. Jayal, U. Schoetti and Chhatrapati Singh (eds), *India's Environment – Crises and Responses*. Natraj: Dehra Dun UP, 217–26.

Bahuguna, S.L., Negi, D.S. and Upmanyu, K. (1981/3) Padyatra from Kashmire to Kohima. *Himalaya Man and Nature*, 5 (3), 3–6; 5 (8), 3–9; 6 (8), 6–7.

Baker, V.R. (1978) Adjustments of fluvial systems to climate and source terrain in tropical and subtropical environments. In A.D. Miall (ed.), *Fluvial Sedimentology*. Calgary: Can. Soc. Petrol. Geol., 211–30.

Balling, R.C.Jr and Idso, S.B. (1989) Historical temperature trends in the United States and the effect of urban population growth. *Journal of Geophysical Research*, 94, 3359–63.

Bandyopdahyay, J. and Shiva, V. (1987) Chipko: rekindling India's forest culture. *The Ecologist*, 17 (1), 26–34.

Bansal, R.C. and Mathur, H.N. (1976) Landslides: the nightmare of the hill roads. *Soil Conservation Digest*, 4 (1), 36–7.

Barbira-Scazzochio, F. (ed.) (1980) *Land, People and Planning in Contemporary Amazonia*. Cambridge: Centre of Latin American Studies.

Bard, E., Hamelin, B. and Fairbanks, R.G. (1990) U-Th ages obtained by mass spectrometry in corals from Barbados: sea level during the past 130,000 years. *Nature*, 346, 456–8.

Barone, J.B., Ashbaugh, L.L., Kusko, B.H. and Cahill, T.A. (1981) The effects of Owens Dry Lake on air quality in the Owens Valley with implications for the Mono Lake area. In American Chemical Society Symposium Series, 167, *Atmospheric Aerosol: Source/Air Quality Relationships*. Los Angeles: American Chemical Society, 327–46.

Barrett, P.J. (1991) Antarctica and global climate change – a geological perspective. In C.M. Harris and B. Stonehouse (eds), *Antarctica and Global Climatic Change*. London: Belhaven.

Barrie, L.A. and Hales, J.M. (1984) The spatial distributions of precipitation acidity and major ion wet deposition in North America during 1980. *Tellus*, 36B, 333–55.

Barry, J.Cl. (1982) La frontière méridionale du Sahara entre le Adrar des Iforas et Tombouctou. *Ecol. Médit.*, 7, 99–124.

Barry, J.Cl., Celles, J.Cl. and Manière, R. (1976) Le problème des divisions bio-climatiques et floristiques au Sahara central. Note II. Le Sahara central et méridional. *Naturalia monspeliensia, (Bot)*, 26, 211–42.

Barry, J.Cl., Celles, J.C. and Musso, J. (1986) Le problème des divisions bio-climatiques et floristiques du Sahara. Note V du Sahara au Sahel. Un essay de definition de cette marche african aux alentours. *Ecol. Médit.* 12, 187–235.

Barry, R.G. and Chorley, R.J. (1987) *Atmosphere, Weather and Climate*,

5th edn. London: Methuen.

Bartarya, S.K. (1988) Geohydrological and geomorphological studies of the Gaula River Basin, District Nainital, with special reference to the problem of erosion. Kumaun University, Department of Geology, Nainital, UP, Unpublished PhD thesis.

Bartarya, S.K. (1991) Watershed management strategies in Central Himalaya: the Gaula River Basin, Kumaun, India. *Land Use Policy*, 8 (3), 177–84.

Barth, H. (1857–8) *Reisen und Entdeckungen in Nord und Central Afrika*. Gotha.

Bartlein, P.J., Prentice, I.C. and Webb, T. (1986) Climatic response surfaces from pollen data for some eastern North American taxa. *Journal of Biogeography*, 13, 35–57.

Bartlein, P.J., Webb, T. and Fleri, E. (1984) Holocene climatic change in the northern midwest: pollen-derived estimates. *Quaternary Research*, 222, 361–74.

Bartone, C.R. (1990) Water quality and urbanization in Latin America. *Water International*, 15, 3–14.

Battarbee, R.W. (1984) Diatom analysis and the acidification of lakes. *Philosophical Transactions of the Royal Society, London, B*, 305, 451–77.

Battarbee, R.W. (1990) The causes of lake acidification with special reference to the role of acid deposition. *Philosophical Transactions of the Royal Society, London, B*, 327, 339–47.

Battarbee, R.W., Anderson, N.J., Appleby, P.G., Flower, R.J., Fritz, S.C., Haworth, E.Y., Higgitt, S., Jones, V.J., Kreiser, A., Munro, M.A.R., Natkanski, J., Oldfield, F., Patrick, S.T., Richardson, N.G., Rippey, B. and Stevenson, A.C. (1988b) *Lake Acidification in the United Kingdom 1800–1986: Evidence from Analysis of Lake Sediments*. London: Ensis.

Battarbee, R.W. and Charles, D.F. (1987) The use of diatom assemblages in lake sediments as a means of assessing the timing, trends, and causes of lake acidification. *Progress in Physical Geography*, 11, 552–80.

Battarbee, R.W., Flower, R.J., Stevenson, A.C. and Rippey, B. (1985) Lake acidification in Galloway: a palaeoecological test of competing hypotheses. *Nature*, 314, 350–2.

Battarbee, R.W., Flower, R.J., Stevenson, A.C., Jones, V.J., Harriman, R. and Appleby, P.G. (1988a) Diatom and chemical evidence for reversibility of acidification of Scottish lochs. *Nature*, 332, 530–2.

Battarbee, R.W., Flower, R.J., Stevenson, A.C., Patrick, S.T. and Juggins, S. (1989a) The relationship between diatoms and surface water quality in the Hoylandet area of Nord Trondelag, Norway. Report to the Norwegian Research Council for Science and the Humanities (NAVF). London: Ensis.

Battarbee, R.W., Mason, B.J., Renberg, I. and Talling, J.F. (eds) (1990) *Palaeolimnology and Lake Acidification*. London: The Royal Society.

Battarbee, R.W., Stevenson, A.C., Rippey, B., Fletcher, C., Natkanski, J., Wik, M. and Flower, R.J. (1989b) Causes of lake acidification in Galloway, south-west Scotland: a palaeoecological evaluation of the relative roles of atmospheric contamination and catchment change for two acidified sites with non-afforested catchments. *Journal of Ecology*, 77, 651–72.

Baulig, H. (1935) *The Changing Sea Level*. London: Institute of British Geographers.

Baulin, V.V., Belopukhova, Ye.B. and Danilova, N.S. (1984) Holocene permafrost in the USSR. In A.A. Velichko, H.E. Wright and C.W. Barnosky (eds), *Late Quaternary Environments of the Soviet Union*.

Harlow: Longman, 87–91.

Baulin, V.V. and Danilova, N.S. (1984) Dynamics of Late Quaternary permafrost in Siberia. In A.A. Velichko, H.E. Wright and C.W. Barnosky (eds), *Late Quaternary Environments of the Soviet Union*. Harlow: Longman, 69–77.

Bawa, K.S., Ashton, P.S. and Mohd Nor, S. (1990) Reproductive ecology of tropical forest plants: management issues. In K.S. Bawa and M. Hadley (eds), *Reproductive Ecology of Tropical Forest Plants*. MAB, Vol. 7. Paris: UNESCO/Carnforth, UK: Parthenon, 3–13.

Bawa, K.S. and Hadley, M. (eds) (1990) *Reproductive Ecology of Tropical Forest Plants*. MAB, Vol. 7. Paris: UNESCO/Carnforth, UK: Parthenon.

Bayliss-Smith, T.P. (1988) The role of hurricanes in the development of reef islands, Ontong Java Atoll, Solomon Islands. *Geographical Journal*, 154, 377–91.

Beadle, C.L., Long, S.P., Imbamba, S.K., Hall, D.O. and Olembo, R.J. (1985) *Photosynthesis in Relation to Plant Production in Terrestrial Environments*. Oxford: Tycooly Publishing.

Beaumont, P. (1989) *Environmental Management and Development in Drylands*. London: Routledge.

Beebee, T.J.C., Flower, R.J., Stevenson, A.C., Patrick, S.T., Appleby, P.G., Fletcher, C., Marsh, C., Natkanski, J., Rippey, B. and Battarbee, R.W. (1990) Decline of the Natterjack Toad *Bufo calamita* in Britain: palaeoecological, documentary and experimental evidence for breeding site acidification. *Biological Conservation*, 53, 1–20.

Begon, M., Harper, J.L. and Townsend, C.R. (1990) *Ecology: Individuals, Populations and Communities*, 2nd edn. Boston and Oxford: Blackwell Scientific.

Bell, M.G. (1982) The effects of land-use and climate on valley sedimentation. In A. Harding (ed.), *Climatic Change in Later Prehistory*. Edinburgh: Edinburgh University Press, 127–42.

Bell, M.G. (1983) Valley sediments as evidence of prehistoric land-use on the South Downs. *Proceedings of the Prehistoric Society*, 49, 119–50.

Bell, M.G. (1986) Archaeological evidence for the date, cause and extent of soil erosion on the chalk. In C.P. Burnham and J.I. Pitman (eds), *Soil Erosion* (SEESOIL 3), 72–83.

Belsky, A.J. (1990) Tree/grass ratios in East African savannas; a comparison of existing models. *Journal of Biogeography*, 17, 483–9.

Berger, A., Gallée, H. and Mélice, J.C. (1991) The Earth's future climate at an astronomical timescale. In C.M. Goodess and J.P. Palutikof (eds), *Future Climate Change and Radioactive Waste Disposal*. NIREX Radioactive waste disposal NSS/R257. Norwich: UEA, 148–65.

Berger, A., Imbrie, J., Hays, J., Kukla, G. and Salzman, B. (eds) (1984) *Milankovitch and Climate*. Dordrecht: Reidel.

Berger, A.L. (1992) Astronomical theory of paleoclimates and the last glacial-interglacial cycle. *Quaternary Science Reviews*, 11, 571–82.

Berglund, B.E. (ed.) (1986) *Handbook of Holocene Palaeoecology and Palaeohydrology*. Chichester: Wiley.

Bernus, E. (1981) Touaregs Nigériens. Unité culturelle et diversité régionale d'un peuple pasteur. *Mém. ORSTOM*, 94, 508.

Bernus, E. (1983) Désertification dans la région d'Eghazer et Azawak, Niger. In J.A. Mabutt and C. Floret (eds), *Etudes de cas sur la désertification*. Paris: UNESCO, 118–51.

Bhatt, C.P. (1980) *Ecosystem of the Central Himalayas and Chipko Movement*. Gopeshwar, U.P.: Dashauli Gram Swarajya Sangh.

Billings, W.D., Luken, J.O., Martensen, D.A. and Peterson, K.M. (1982) Arctic Tundra: a source or sink for atmospheric carbon dioxide in a changing environment? *Oecologia*, 53, 7–11.

Billings, W.D., Luken, J.O., Martensen, D.A. and Peterson, K.M. (1983) Increasing atmospheric carbon dioxide: possible effects on arctic tundra. *Oecologia*, 58, 286–9.

Birks, H.J.B., Juggins, S. and Line, J.M. (1990b) Lake surface-water chemistry reconstructions from palaeolimnological data. In B.J. Mason (ed.), *The Surface Waters Acidification Programme*. Cambridge: Cambridge University Press, 301–13.

Birks, H.J.B., Line, J.M., Juggins, S., Stevenson, A.C. and ter Braak, C.J.F. (1990a) Diatoms and pH reconstruction. *Philosophical Transactions of the Royal Society, London, B*, 327, 263–78.

Biswas, A.K. (1990) Conservation and management of water resources. In A.S. Goudie (ed.), *Techniques for Desert Reclamation*. Chichester: Wiley, 199–218.

Biswas, M.R. and Biswas, A.K. (1980) *Desertification*. Oxford: Pergamon Press.

Blackmore, A.C., Mentis, M.T., and Scholes, R.J. (1990) The origin and extent of nutrient-enriched patches within a nutrient-poor savanna in South Africa. *Journal of Biogeography*, 17, 463–70.

Blaikie, P. (1985) *The Political Economy of Soil Erosion in Developing Countries*. London: Methuen.

Blandford, D.C. (1981) Rangelands and soil erosion research: a question of scale. In R.P.C. Morgan (ed.), *Soil Conservation: Problems and Prospects*. Chichester: Wiley, 105–21.

Bloom, A.L. (1974) Geomorphology of reef complexes. In L.F. Laporte (ed.), *Reefs in Time and Space*. Tulsa, OK: Society of Economic Paleontologists and Mineralogists, Special Publication no. 18.

Blume, H.P., Petermann, Th. and Vahrson, W.G. (1985) Klimabezogene Deutung rezenter und releiktischer Eigenschaften von Wüstenböden. *Geomethodica*, 10, 91–121.

Boardman, J. (1987) A land farmed into the ground. *The Guardian*, 18 December.

Boardman, J. (1990) Soil erosion on the South Downs: a review. In J. Boardman, I.D.L. Foster and J.A. Dearing (eds) *Soil Erosion on Agricultural Land*. Chichester: Wiley, 87–105.

Boer, M.M., Koster, E.A. and Lundberg, H. (1990) Greenhouse impact in Fennoscandia – preliminary findings of a European Workshop on the effects of climatic change. *Ambio*, 19, 2–10.

Boerboom, J.H.A. and Jonkers, W.B.J. (1987) Land use in tropical rainforest areas. In C.F. van Beusekom, C.P. van Goor and P. Schmidt (eds), *Wise Utilisation of Tropical Rain Forest Lands*. Tropenbos Science Series 1. Ede, Netherlands: Tropenbos, 30–45.

Boerboom, J.H.A. and Wiersum, K.F. (1983) Human impact on tropical moist forest. In W. Holzner, M.J.A. Werger and I. Ikumsuma (eds), *Man's Impact on Vegetation*. The Hague: W. Junk, 83–106.

Bonnefille, R., Roeland, J.C. and Guiot, J. (1990) Temperature and rainfall estimates for the past 40,000 years in equatorial Africa. *Nature*, 346, 347–9.

Booth, W. (1991) Sahara is shrinking. *Guardian Weekly*, 28 July, 17.

Bormann, F.H. and Likens, G.E. (1970) The nutrient cycles of an ecosystem. *Scientific American*, 223, 92–101.

Bornkamm, R. and Kehl, H. (1985) Pflanzensoziologische Zonen in der Marmarika. *Flora*, 176, 141–51.

Borzenkova, I.I. and Zubakov, V.A. (1984) Climatic optimum of the Holocene as a model of the global climate at the beginning of the 21st century. *Meteorol, i. Gidrolog.*, N8, 69–77.

Bottomley, M., Folland, C.K., Hsiung, J., Newell, R.E. and Parker, D.E. (1990) *Global Ocean Surface Temperature Atlas (GOSTA)*. Joint UK Meteorological Office/Massachusetts Institute of Technology Project. London: HMSO.

Boulos, L. (1987) *Medical Plants of North Africa*. Algenoc, MI: Reference Publ.

Bourne, R. (1978) *Assault on the Amazon*. London: Gollancz.

Boutrais, J. (1980) Arbre et le boeuf en zone soudano-guinéenne. *Cah. ORSTOM, sér Sci. Hum*, 17, 235–46.

Bouwman, A.F. (ed.) (1990) *Soils and the Greenhouse Effect*. Chichester: Wiley.

Bradley, R. (1984) *Quaternary Paleoclimatology*. London: Allen and Unwin.

Bradley, R.S., Diaz, H.F., Eisheid, J.K., Jones, P.D., Kelly, P.M. and Goodess, C.M. (1987) Precipitation fluctuations over Northern Hemisphere land areas since the mid-19th century. *Science*, 237, 171–5.

Bradley, R.S. and Groisman, P.Ya. (1989) Continental scale precipitation variations in the 20th century. In B. Sevruk (ed.), *Precipitation Measurement*. Proceedings of the WMO/IASH/ETH International Workshop on Precipitation Measurements, St Moritz, Switzerland, December 3–7, 1989.

Bradley, R.S. and Jones, P.D. (1985) Data bases for isolating the effects of increasing carbon dioxide concentration. In M.C. MacCracken and F.M. Luther (eds), *Detecting the Climatic Effects of Increasing Carbon Dioxide*. Washington, DC: US DoE, DoE/ER-0235, 29–54.

Bradley, R.S. and Jones, P.D. (eds) (1992) *Climate since A.D. 1500*. London: Harper and Collins.

Brand, J. (1988) The transformation of rainfall energy by a tropical rain forest canopy in relation to soil erosion. *Journal of Biogeography*, 15, 41–8.

Bremer, H. (1981) Reliefformen und reliefbildende prozesse in Sri Lanka. *Relief, Boden, Paläoklima (Zur Morphogenese in den feuchten Tropen)*, 1, 7–183.

Briffa, K.R., Bartholin, T.S., Eckstein, D., Jones, P.D., Karlen, W., Schweingruber, F.H. and Zetterberg, P. (1990) A 1,400-year tree-ring record of summer temperatures in Fennoscandia. *Nature*, 346, 434–39.

Broecker, W.S. (1987) Unpleasant surprises in the greenhouse. *Nature*, 328, 123–6.

Broecker, W.S. and Denton, G.H. (1990) The role of ocean-atmosphere reorganisations in glacial cycles. *Quaternary Science Reviews*, 9, 305–42.

Broecker, W.S., Peteet, D.M. and Rind, D. (1985) Does the ocean-atmosphere system have more than one stable mode of operation? *Nature*, 315, 21–6.

Brown, A.G. (1987) Long-term sediment storage in the Severn and Wye catchments. In K.J. Gregory, J. Lewin and J.B. Thornes (eds), *Palaeohydrology in Practice*. Chichester: Wiley, 307–32.

Brown, A.G. and Barber, K.E. (1985) Late Holocene palaeoecology and sedimentary history of a small lowland catchment in central England. *Quaternary Research*, 24, 87–102.

Brown, B.E. and Suharsono (1990) Damage and recovery of coral reefs affected by El Niño related seawater warming in the Thousand Islands, Indonesia. *Coral Reefs*, 8, 163–70.

Brown, C.S., Meier, M.F. and Post, A. (1982) Calving speed of Alaskan tidewater glaciers with applications to the Columbia Glacier, Alaska. US Geological Survey Prof. Paper, 1258–C.

Brown, D.J.A. and Turnpenny, A.W.H. (1988) Effects on fish. In M.R. Ashmore, J.N.B. Bell and C. Garretty (eds), *Acid Rain and Britain's Natural Ecosystems*. London: Centre for Environmental Technology, Imperial College, 87–96.

Brown, H.C.P. and Thomas, V.G. (1990) Ecological considerations for the future of food security in Africa. In C.A. Edwards, R. Lal, P. Madden, R.H. Miller and Gar House (eds), *Sustainable Agricultural Systems*. Ankeny: Soil and Water Conservation Society, 353–77.

Brown, M.J.F. (1974) A development consequence – disposal of mining waste on Bougainville, Papua New Guinea. *Geoforum*, 18, 19–27.

Brown, R.L. (ed.) (1985) *State of the World*, World Watch Institute. New York: W.W. Norton.

Bruce, P.A. (1986) *Economic History of Virginia in the Seventeenth Century, vols 1 and 2*. London: Macmillan.

Brugger, R.J. (1988) *Maryland, a Middle Temperament*. Baltimore: Johns Hopkins University Press.

Bruijnzeel, L.A. (1990) *Hydrology of Moist Tropical Forests and Effects of Conversion: State of Knowledge Review*, Paris: UNESCO IHP Humid Tropics Programme.

Bruijnzeel, L.A. (1991) Hydrological impacts of tropical forest conversion. *Nature and Resources (UNESCO)*, 27 (2), 36–46.

Bruijnzeel, L.A. and Brenner, C.N. (1989) *Highland–Lowland Interactions in the Ganges-Brahmaputra River Basin: a Review of the Published Literature*. Kathmandu: ICIMOD Occasional Paper 11.

Bruneau de Mire, Ph. and Gillet, G. (1956) Contribution à l'étude de la flore de l'Air. *Journ. d'Agron. trop. et de Bot. appl.*, 3, 221–47; 7–8, 422–42; 11, 701–60; 12, 857–86.

Brush, G.S. (1984a) Patterns of recent sediment accumulation in Chesapeake Bay (Virginia–Maryland, USA) tributaries. *Chemical Geology*, 44, 227–42.

Brush, G.S. (1984b) Stratigraphic evidence of eutrophication in an estuary. *Water Resources Research*, 20, 531–41.

Brush, G.S. (1986) Geology and paleoecology of Chesapeake Bay: a long-term monitoring tool for management. *Journal of the Washington Academy of Sciences*, 76, 146–60.

Brush, G.S. (1989) Rates and patterns of estuarine sediment accumulation. *Limnology and Oceanography*, 34, 1235–46.

Bryant, E. (1987) CO_2-warming, rising sea-level and retreating coasts: review and critique. *Australian Geographer*, 18, 101–13.

Bryson, R.A. and Barreis, D.A. (1967) Possibility of major climatic modifications and their implications: northwest India, a case for study. *Bulletin of the American Meteorological Society*, 48, 136–42.

Bryson, R.A., Wendland, W.M., Ives, J.D. and Andrews, J.T. (1969) Radiocarbon isochrones on the disintegration of the Laurentide ice sheet. *Arctic and Alpine Research*, 1, 1–13.

Budd, W.F. and Smith, I.N. (1981) The growth and retreat of ice sheets in response to orbital radiation changes. In *Sea Level, Ice and Climatic Change*. International Association of Hydrological Sciences publication number 131, 369–410.

Buddemeier, R.W. and Kinzie, R.A. (1976) Coral growth. *Annual Reviews in Oceanography and Marine Biology*, 14, 183–225.

Buddemeier, R.W. and Smith, S.V. (1988) Coral reef growth in an era of rapidly rising sea level: predictions and suggestions for long-term research. *Coral Reefs*, 7, 51–6.

Büdel, J. (1965) *Die Relieftypen dar Fluchenspul-Zone Sud-Indiens am Ostabfall Dekan gegen Madras*. Bonn: Colloquium Geograficum 8.

Büdel, J. (1981) *Klima Geomorphologie, two vols*, Berlin: Borntraeger.

Budyko, M. and Izrael, Y.A. (eds) (1987) *Anthropogenic Climatic Changes*. Leningrad: L. Gidrometeoizdat.

Budyko, M.I. (1969) The effect of solar radiation variations on the climate of the Earth. *Tellus*, 21, 611–19.

Burkis, R.H. and Black, C.C. (eds) (1976) CO_2 *Metabolism and Plant Productivity*. Baltimore: Johns Hopkins University Press.

Burley, J. (1990) The conservation and use of plant resources in drylands. In A.S. Goudie (ed.), *Techniques for Desert Reclamation*. Chichester: Wiley, 199–218.

Burns, J.C., Coy, J.S., Tervet, D.J., Harriman, R., Morrison, B.R.S. and Quine, C.P. (1984) The Loch Dee project: a study of the ecological effects of acid precipitation and forest management on an upland catchment in south-west Scotland. *Fish Management*, 15, 145–68.

Burrows, W.H., Carter, J.O., Scanlan, J.C. and Anderson, E.R. (1990) Management of savannas for livestock production in north-east Australia: contrasts across the tree-grass continuum. *Journal of Biogeography*, 17, 503–12.

Busson, F. (1965) *Plantes alimentaires de l'Ouest Africain*. Thèse, Univ. d'Aix-Marseille.

Butzer, K.W. (1980) Adaptation to global environmental change. *Professional Geographer*, 32, 269–78.

Callendar, G.S. (1938) The artificial production of CO_2 and its influence on temperature. *Quarterly Journal of the Royal Meteorological Society*, 64, 223–37.

Cambray, J.A. and Jubb, R.A. (1977) Dispersal of fishes via the Orange-Fish tunnel, South Africa. *Journal of the Limnological Society of South Africa*, 3 (1), 33–5.

Cameron, N.G. (1990) Representation of diatom communities by fossil assemblages in Loch Fleet, Galloway, Scotland. Unpublished PhD Thesis, University of London.

Cameron, W.M. and Pritchard, D.W. (1963) Estuaries. In M.N. Hill (ed.), *The Sea, vol. 2*. New York: Wiley, 306–424.

Carothers, S.W. and Dolan, R. (1982) Dam changes on the Colorado River, *Nat. Hist.*, 91 (1), 74–83.

Carothers, S.W. and Johnson, R.R. (1983) Status of the Colorado River Ecosystem in Grand Canyon National Park and Glen Canyon National Recreation Area. In V.D. Adams and V.A. Lamarra (eds), *Aquatic resources management of the Colorado River Ecosystem*. Ann Arbor, MI: Ann Arbor Scientific Publications.

Carson, B. (1985) *Erosion and Sedimentation Processes in the Nepalese Himalaya*. Kathmandu: ICIMOD Occasional Paper 1.

Cermak, V. (1971) Underground temperature and inferred climatic temperature of the past millennium. *Palaeogeography, Palaeoclimatology, Palaeoecology*, 10, 1–19.

Cess, R.D. and Potter, G.L. (1988) A methodology for understanding and intercomparing atmospheric climate feedback processes within general circulation models. *Journal of Geophysical Research*, 93, 8305–14.

Chalise, R.S. (1986) Constraints on resources and development in the mountainous regions of South Asia. In S.C. Joshi, M.J. Haigh, Y.P.S. Pangtey, D.R. Joshi and D.D. Dani (eds), *Nepal Himalaya: Geoecological Perspectives*. Nainital, India: H.R. Publishers, 12–26.

Chamberlin, T.C. (1895) Glacial phenomena of North America. In J. Geikie (ed.), *The Great Ice Age*. New York: Appleton, 724–55.

Chamberlin, T.C. (1897) A group of hypotheses bearing on climatic changes. *Journal of Geology*, 5, 653–83.

Chamberlin, T.C. (1898) The influence of great epochs of limestone formation upon the constitution of the atmosphere. *Journal of Geology*, 6, 609–21.

Chamberlin, T.C. (1899) An attempt to frame a working hypothesis of the cause of glacial periods on an atmospheric basis. *Journal of Geology*, 7, 454–584, 667–85, 751–87.

Chappell, J. and Polach, H. (1991) Postglacial sealevel rise from a coral record at Huon Peninsula, Papua New Guinea. *Nature*, 349, 147–9.

Charles, D.F., Battarbee, R.W., Renberg, I., van Dam, H. and Smol, J.P. (1989) Palaeoecological analysis of lake acidification trends in North America and Europe using diatoms and chrysophytes. In S.A. Norton, S.E. Lindberg and A.L. Page (eds), *Acid Precipitation Vol. 4. Soils, Aquatic Processes and Lake Acidification*. New York: Springer-Verlag, 207–76.

Charney, J., Quirk, W.J., Crow, S.H. and Kornfield, J. (1977) A comparative study of the effects of albedo change on drought in the semi-arid regions. *Journal of Atmospheric Science*, 34, 1366–85.

Cheng Guodong (1983) Vertical and horizontal zonation of high-altitude permafrost. In *Proceedings Fourth International Conference on Permafrost, Fairbanks, Alaska*. Washington DC: National Academy Press, 136–41.

Chyurlia, J.P. (1984) *Water Resources Report*. Ottawa: Kenting Earth Sciences.

Clark, J.A. and Primus, J.A. (1987) Sea level changes resulting from future retreat of ice sheets: an effect of CO_2 warming of the climate. In M.J. Tooley and I. Shennan (eds), *Sea Level Changes*. Oxford: Basil Blackwell, 356–70.

Clark, M.J. (ed.) (1988) *Advances in Periglacial Geomorphology*. Chichester: Wiley.

Clayton, K. (1991) Scaling environmental problems. *Geography*, 76, 2–15.

CLIMAP project members (1976) The surface of ice age earth. *Science*, 191, 1131–7.

CLIMAP project members (1984) The last interglacial ocean. *Quaternary Research*, 21, 123–224.

Cloudsley-Thompson, J.L. (1977) Reclamation of the Sahara. *Environmental Conservation*, 5, 115–19.

Cloudsley-Thompson, J.L. (1978) Human activities and desert expansion. *Geographical Journal*, 144, 416–23.

Cochrane, T.T. and Sanchez, P.A. (1982) Land resources, soils and their management in the Amazon region: a state of knowledge report. In S.B. Hecht (ed.), *Amazonia: Agriculture and Land Use Research*. Cali, Colombia: CIAT, 137–209.

COHMAP members (1988) Climatic changes of the last 18,000 years: observations and model simulations. *Science*, 241, 1043–52.

Coker, A.M., Thompson, P.M., Smith, D.I. and Penning-Rowsell, E.C. (1989) The impact of climate change on coastal zone management in Britain: a preliminary analysis. *The Publications of the Academy of Finland 9/89, Conference on Climate and Water 2*, 148–60.

Coles, S.L. and Jokiel, P.L. (1977) Effects of temperature on photosynthesis and respiration in hermatypic corals. *Marine Biology*, 43, 209–16.

Coles, S.L. and Jokiel, P.L. (1978) Synergistic effects of temperature, salinity and light on the hermatypic coral *Montipora verrucosa*. *Marine Biology*, 49, 187–95.

Colman, S.M., Halka, J.P., Hobbs III, C.H., Mixon, R.B. and Foster, D.S. (1990) Ancient channels of the Susquehanna River beneath Chesapeake Bay and the Delmarva Peninsula. *Geological Society of America Bulle-*

tin, 102, 1268–79.

Colwell, R.N. (1985) *Manual of Remote Sensing*, 2nd edn. Falls Church, VA: American Society of Photogrammetry.

Cooke, R.U. and Reeves, R.W. (1976) *Arroyos and Environmental Change in the American South West*. Oxford: Oxford University Press.

Cooper, S.R. and Brush, G.S. (1991) Long-term history of Chesapeake Bay anoxia. *Science*, 206, 298–306.

Cosby, B.J., Hornberger, J.N., Galloway, J.N. and Wright, R.F. (1985) Freshwater acidification from atmospheric deposition of sulphuric acid: a quantitative model. *Environmental Science and Technology*, 19, 1144–9.

Cosgrove, D. and Petts, G. (eds) (1990) *Water, Engineering and Landscape*. London: Belhaven Press.

Crane, A.J. and Cocks, A.T. (1987) The transport, transformation and deposition of airborne emissions from power stations. *CEGB Research*, no. 20, 3–15.

Croll, J. (1867) *Climate and Time*. New York: Appleton and Co.

Cronin, T.M. (1983) Rapid sea level and climate change: evidence from continental and island margins. *Quaternary Science Reviews*, 1, 177–214.

Croome, R.L., Tyler, P.A., Walker, K.F. and Williams, W.D. (1976) A limnological survey of the River Murray in the Albury-Wodonga area. *Search*, 7 (1), 14–17.

Cross, A. (1990) *Tropical Deforestation and Remote Sensing: the Use of NOAA/AVHRR Data over Amazonia*. Geneva: UNEP/GRID, Final Report to the EEC (DG VIII) Article B946/88.

Crowley, T.J. (1988) Paleoclimate modelling. In M.E. Schlesinger (ed.), *Physically-based Modelling and Simulation of Climate and Climate Change, part II*. Dordrecht: Kluwer, 883–949.

CSIRO (1990) *Regional Impact of the Enhanced Greenhouse Effect on the Northern Territory*. Melbourne: CSIRO.

Dabral, B.G., Prem Nath and Swarup, R. (1963) Some preliminary investigations on rainfall interception by leaf litter. *Indian Forester*, 89, 112–16.

Dabral, B.G. and Subba Rao, B.K. (1968) Interception studies in Chir and teak plantations – New Forest. *Indian Forester*, 94, 540–51.

Dadhwal, K.S., Pratrap Narain and Dhyani, S.K. (1989) Agroforestry systems in the Garhwal Himalayas of India. *Agroforestry Systems*, 7, 213–25.

Dahl, A.L. (1985) Status and conservation of South Pacific coral reefs. *Procedings, 5th International Coral Reef Congress, Tahiti*, 6, 509–13.

Dahl-Jensen, D. and Johnsen, S.J. (1986) Palaeotemperatures still exist in the Greenland Ice Sheet. *Nature*, 320, 250–2.

Dai, Y.S. (1987) The engineering geological characteristics of the loess and soil erosion in the Middle Reaches of the Huanghe River. In T.S. Liu (ed.), *Aspects of Loess Research*. Beijing: China Ocean Press, 432–6.

Daly, R.A. (1934) *The Changing World of the Ice Age*. New Haven, CT: Yale University Press.

Dansereau, P. (1947) Zonation et succession sur la restinga de Rio de Janeiro. I. Halosere. *Révue Canadiènne de Biologie*, 6, 448–77.

Dansgaard, W., Johnsen, S.J., Clausen, H.B. and Langway, C.C. Jr (1971) Climatic record revealed by the Camp Century Ice Core. In K.K. Turekiam (ed.), *The Late Cenozoic Glacial Ages*. London and New Haven, CT: Yale University Press, 37–56.

Dansgaard, W., White, J.W.C. and Johnsen, S.J. (1989) The abrupt termination of the Younger Dryas climate event. *Nature*, 339, 532–3.

Dargie, T.C.D. and Furley, P.A. (1993) Land use and land cover changes in northern Roraima. In P.A. Furley (ed.), *The Forest Frontier: Settlement and Change in Brazilian Roraima*. London: Routledge.

Darkoh, M.B.K. (1991) Land degradation and resource management in Kenya. *Desertification Control Bulletin*, 19, 61–72.

Das, S.M. (1987) Tehri Dam and Upper Ganga silt pollution. *Himalaya: Man and Nature*, 10 (9), 49–51, 54.

Davies, P.J., Marshall, J.F. and Hopley, D. (1985) Relationships between reef growth and sea level in the Great Barrier Reef. *Proceedings, 5th International Coral Reef Congress, Tahiti*, 3, 95–103.

Davies, P.J. and Montaggioni, L.F. (1985) Reef growth and sea level change: the environmental signature. *Proceedings, 5th International Coral Reef Congress, Tahiti*, 3, 477–515.

Davis, M.B. (1976) Erosion rates and land use history in Southern Michigan. *Environmental Conservation*, 3, 139–48.

Davis, M.B. (1986) Lags in the response of forest vegatation to climatic change. In C. Rosenzweig and R. Dickinson (eds), *Climate-Vegetation Interactions*. Greenbelt, MD: NASA Conf. Publ. 2440.

Dearing, J.A. (1991) Erosion and land use. In B.E. Berglund (ed.), The Cultural Landscape during 6000 Years in Southern Sweden. *Ecological Bulletins*, 41, 283–92.

Dearing, J.A. (1992) Sediment yields and sources in a Welsh upland lake-catchment during the past 800 years. *Earth Surface Processes and Landforms*, 17, 1–22.

Dearing, J.A., Alstrom, K., Bergman, A., Regnell, J. and Sandgren, P. (1990) Past and present erosion in southern Sweden. In J. Boardman, I.D.L. Foster and J.A. Dearing (eds), *Soil Erosion on Agricultural Land*. Chichester: Wiley, 173–91.

Dearing, J.A., Elner, J.K. and Happey-Wood, C.M. (1981) Recent sediment flux and erosional processes in a Welsh upland lake catchment based on magnetic susceptibility measurements. *Quaternary Research*, 16, 356–72.

Dearing, J.A. and Foster, I.D.L. (1986) Limnic sediments used to reconstruct sediment yields and sources in the English Midlands since 1765. In V. Gardiner (ed.), *International Geomorphology, vol. 1*. Chichester: Wiley, 853–68.

Dearing, J.A., Hakansson, H., Lieddberg-Jonsson, B., Persson, A., Skansjo, S., Widholm, D. and El-Daoushy, F. (1987) Lake sediments used to quantify the erosional response to land use change in southern Sweden. *Oikos*, 50, 60–78.

Décamps, H., Capblanq, J., Casanova, H. and Tourenq, J.M. (1979) Hydrobiology of some regulated rivers in south-west France. In J.V. Ward and J.A. Stanford (eds), *The Ecology of Regulated Streams*. New York: Plenum Press, 273–88.

de Deckker, P., Colin, J-P. and Peypouquet, J-P. (eds) (1988) *Ostracoda in the Earth Sciences*, Amsterdam: Elsevier.

Deevey, E.S. (1969) Coaxing history to conduct experiments. *BioScience*, 19, 40–3.

Deevey, E.S., Rice, D.S., Rice, P.M., Vaughan, H.H., Brenner, M. and Flannery, M.S. (1979) Mayan urbanism: impact on a tropical karst environment. *Science*, 206, 298–306.

Degens, E.T., Kempe, S. and Richey, J.E. (eds) (1991) *Biogeochemistry of Major World Rivers*, SCOPE 42. Chichester: Wiley.

Delibrias, C. and Laborel, J. (1971) Recent variations of the sea level along the Brazilan coast. *Quaternaria*, 14, 459.

Denham, D. (1822–5) *Mission to the Niger. The Bornu Mission*. Reprint,

Cambridge, 1966.

Denton, G.H., Bockheim, J.G., Wilson, S.C. and Stuiver, M. (1989) Late Wisconsin and early Holocene glacial history, inner Ross Embayment, Antarctica. *Quaternary Research*, 31, 151–82.

Denton, G.H. and Hughes, T.J. (eds) (1981) *The Last Great Ice Sheets*. New York: Wiley.

Denton, G.H. and Hughes, T.J. (1983) Milankovitch theory of ice ages; hypothesis of ice sheet linkage between regional insolation and global climate. *Quaternary Research*, 20, 124–44.

Denton, G.H., Hughes, T.J. and Karlen, W. (1986) Global ice sheet system interlocked by sea level. *Quaternary Research*, 26, 3–26.

Denton, G.H., Prentice, M.L., Kellogg, D.E. and Kellogg, T.B. (1984) Late Tertiary history of the Antarctic ice sheet: evidence from the Dry Valleys. *Geology*, 12, 263–7.

Derbyshire, E. and Mellors, T.W. (1986) Loess. In P.G. Fookes and P.R. Vaughan (eds), *A Handbook of Engineering Geomorphology*. Glasgow: Surrey University Press and Blackie and Sons Ltd, 258–69.

Derbyshire, E. and Mellors, T.W. (1988) Geological and geotechnical characteristics of some loess and loessic soils from China and Britain: a comparison. *Engineering Geology*, 25, 135–75.

Derbyshire, E., Wang, J.T., Jin, Z.X., Billard, A., Muxart, T., Egels, Y., Kasser, M., Jones, D.K.C. and Owen, L. (1991) Landslides in the Gansu loess of China. *Catena*, supplement 20, 119–45.

Deshmukh, I. (1986) *Ecology and Tropical Biology*. Palo Alto, CA and Oxford: Blackwell Scientific.

Detwiler, R.P. and Hall, C.A.S. (1988) Tropical forests and the global carbon cycle. *Science*, 239, 42–7.

Devisse, J. (1990) Impact des industries métallurgiques sur l'environnements de l'Homme en Afrique. In Symp. impact anthrop. et changement climat. abrupt, Sfax, Tunisie.

Devoy, R.J.N. (1979) Flandrian sea level changes and vegetational history of the lower Thames estuary. *Philosophical Transactions of the Royal Society, London, B*, 285, 355–407.

Dewan, M.L. (1990) Indian monsoons and Himalayan water conservation. *Himalaya: Man and Nature*, 14 (5), 17–21; 14 (6), 4–8.

Dhruva Narayana, V.V. (1987) Downstream impacts of soil conservation in the Himalayan region. *Mountain Research and Development*, 7 (3) 256–63.

Dhruva Narayana, V.V. and Abrol, I.P. (1981) Effect of reclamation of alkali soils on water balance. In R. Lal and E.W. Russell (eds), *Tropical Agricultural Hydrology*. Chichester: Wiley, 283–98.

Dhruva Narayana, V.V. and Sastry, G. (1985) Soil conservation in India. In S.A. El-Swaify et al. (eds), *Soil Erosion and Conservation*. Ankeny, IA: Soil Conservation Society of America, 3–9.

Diamond, J.M. (1975) The island dilemma: lessons of modern biogeography study for the design of natural reserves. *Biological Conservation*, 7, 129–46.

Diamond, J.M. (1986) The design of a nature reserve system for Indonesian New Guinea. In M. Soulé (ed.), *Conservation Biology: the Science of Scarcity and Diversity*. Sunderland, MA: Sinauer Associates, 485–503.

Diaz, H.F., Bradley, R.S. and Eischeid, J.K. (1989) Precipitation fluctuations over global land areas since the late 1800s. *Journal of Geophysical Research*, 94, 1195–210.

Dickinson, R.E. and Henderson-Sellers, A. (1988) Modelling tropical deforestation: a study of GCM land surface parametrisations. *Quarterly*

Journal of the Royal Meteorological Society, 114, 439–62.

Digerfeldt, G. (1972) The post-glacial development of Lake Trummen. *Folia Limnologica Scandinavica*, 16.

Din, S.H.S. (1977) Effects of the Aswan High Dam on the Nile flood on the estuarine and coastal circulation pattern along the Mediterranean Egyptian coast. *Limnology and Oceanography*, 22, 194–207.

Doake, C.S.M. and Vaughan, D.G. (1991) Rapid disintegration of the Wordie Ice shelf in response to atmospheric warming. *Nature*, 350, 328–30.

Doctor, A.H. (1965) *Sarvodaya: a Political and Economic Study*. Bombay: Asia.

Dogra, B. (1983) *Forests and People: a Report on the Himalaya*. New Delhi: Dogra.

Dollard, C.J., Unsworth, M.H. and Harvey, M.J. (1983) Pollutant transfer in upland regions by occult precipitation. *Nature*, 302, 241–3.

Douglas, I. (1967) Erosion of granite terrains under tropical rainforest in Australia, Malaysia and Singapore. *Publications de l'Association Internationale d'Hydrologie Scientifique*, 75, 31–9.

Douglas, I. (1968) Erosion in the Sungai Bombak catchment, Selangor, Malaysia. *Journal of Tropical Geography*, 26, 1–16.

Douglas, I. (1981) Soil conservation measures in river basin planning. In S.K. Saha and C.J. Barrow (eds), *River Basin Planning: Theory and Practice*. Chichester: Wiley, 49–73.

Douglas, I. (1989) Land degradation, soil conservation and the sediment load of the Yellow River, China: review and assessment. *Land Degradation and Rehabilitation*, 1, 141–51.

Douglas, I. and Spencer, T. (eds) (1985) *Environmental Change and Tropical Geomorphology*. London: Allen and Unwin.

Dregne, H.E. (1986) Desertification of arid lands. In F. El-Baz and M.H.A. Hassan (eds), *Physics of Desertification*. Dordrecht: Nijhoff, 4–34.

Dregne, H.E. and Tucker, C.J. (1988) Desert encroachment. *Desertification Control Bulletin*, 16, 16–19.

Dubief, F. (1959–63) *Le climat du Sahara, 2 vols*. Algiers: Inst. de Rech. Sahara.

Duggin, M.J. and Robinove, C.J. (1990) Assumptions implicit in remote sensing data acquisition and analysis. *International Journal of Remote Sensing*, 11, 1669–94.

Dumont, H.J. (1986) The Nile River System. In B.R. Davies and K.F. Walker (eds), *The Ecology of River Systems*. Dordrecht: Dr W. Junk Publishers.

Dupire, M. (1962) Les nomades et leur betail. *L'Homme*, 1, 22–39.

Dustan, P. (1975) Growth and form of the reef-building coral *Montastrea annularis*. *Marine Biology*, 3, 101–7.

Eastman, J.R. and McKendry, J.E. (1991) *Change and Time Series Analysis*. UNITAR, Explorations in Geographic Information Systems Technology, Volume 1. United Nations Institute for Training and Research, Switzerland.

Easton, W.H. and Olson, E.A. (1976) Radiocarbon profile of Hanauma reef, Oahu, Hawaii. *Geological Society of America Bulletin*, 87, 711–19.

Eddison, J. and Green, C. (1988) *Romney Marsh: Evolution, Occupation, Reclamation*. Oxford: Oxford University Committee for Archaeology, Monograph 24.

Eddy, J.A. (1976) The Maunder Minimum. *Science*, 192, 1189–202.

Eddy, J.A., Meier, M.F. and Roots, E.F. (eds) (1988) *Arctic Interactions, Recommendations for an Arctic Component in the International Geosphere-Biosphere Programme*. Boulder, CO: Office for Interdisciplinary Earth Studies.

Ehrich, C. (1980) The next flood knocking at the door: a foreigner's view on the ecological crisis in India. In T.V. Singh and J. Kaur (eds), *Studies in Himalayan Ecology*. New Delhi: Himalayan Books, 196–200.

El-Ashry, M.T. and Gibbons, D.C. (eds) (1988) *Water and Arid Lands of the Western United States*. Cambridge: Cambridge University Press.

El-Ashry, M.T., Schilfgaarde, J. van and Schiffman, S. (1985) Salinity pollution from irrigated agriculture. *Journal of Soil and Water Conservation*, 40, 48–52.

Elkington, J. (1987) *The Shrinking Planet*. US Information Technology and Sustainable Development. Washington, DC: World Resources Institute.

Emanuel, K.A. (1986) An air-sea interaction theory for tropical cyclones. Part I: Steady-state maintenance. *Journal of Atmospheric Science*, 43, 585–604.

Emanuel, K.A. (1987) The dependence of hurricane intensity on climate. *Nature*, 326, 483–5.

EMEP (1991) Calculated budgets for airborne acidifying components in Europe, 1985, 1987, 1988, 1989 and 1990. *EMEP MCS-W Report*, 1/91.

Emiliani, C. (1955) Pleistocene temperatures. *Journal of Geology*, 19, 538–78.

Englefield, G.J.H., Tooley, M.J. and Zong, Y. (1990) *An Assessment of the Clwyd Coastal Lowlands after the Floods of February 1990*. Durham: Environmental Research Centre.

Engstrom, D.R. and Hansen, B.C.S. (1985) Postglacial vegetational change and soil development in southeastern Labrador as inferred from pollen and chemical stratigraphy. *Canadian Journal of Botany*, 63, 543–61.

Erwin, T.L. (1988) The tropical forest canopy: the heart of biological diversity. In E.O. Wilson (ed.), *Biodiversity*. Washington, DC: National Academy Press, 123–9.

Esch, D.C. and Osterkamp, T.E. (1990) Cold regions engineering: climatic warming concerns for Alaska. *Journal of Cold Regions Engineering*, 4, 6–14.

Evans, R. (1990) Soil erosion: its impact on the English and Welsh landscape since woodland clearance. In J. Boardman, I.D.L. Foster and J.A. Dearing (eds), *Soil Erosion on Agricultural Land*. Chichester: Wiley, 231–54.

Evans, T. (1990) History of Nile flows. In P.P. Howell and J.A. Allan (eds), *The Nile: Resource Evaluation, Resource Management, Hydropolitics and Legal Issues*. London: SOAS/RGS, 5–39.

Fairbanks, R.G. (1989) A 17,000-year glacio-eustatic sea level record: influence of glacial melting rates on the Younger Dryas event and deep-ocean circulation. *Nature*, 342, 637–42.

Fairbridge, R.W. (1961) Eustatic changes in sea level. *Physics and Chemistry of the Earth*, 4, 99–185.

Fairbridge, R.W. (1983) Isostasy and eustasy. In D.E. Smith and A.G. Dawson (eds), *Shorelines and Isostasy*. London: Academic Press, 3–25.

Fairbridge, R.W. (1987) The spectra of sealevel in a Holocene time frame. In M.R. Rampino, J.E. Sanders, W.S. Newman and L.K. Königsson (eds), *Climate: History, Periodicity and Predictability*. New York: Van Nostrand Company, 127–42.

Fairbridge, R.W. (1989) Crescendo events in sea level change. *Journal of Coastal Research*, 5, 2–6.

Fairbridge, R.W. and Jelgersma, S. (1990) Sea level. In R. Paepe, R.W. Fairbridge, S. Jelgersma and M.A. Pool (eds), *Greenhouse Effect, Sea Level and Drought*. Dordrecht: Kluwer Academic Publishers, 117–43.

Falkenmark, M. (1989) The massive water scarcity now threatening Africa – why isn't it being addressed? *Ambio*, 18, 112–18.

FAO (1982) *Tropical Forest Resources*. Rome: FAO Forestry Paper 30.

FAO (1986) Tropical forest management – a status report. *Unasylva*, 39, 156.

FAO (1989) *Studies on the Volume and Yield of Tropical Forest Stands, 1. Dry Forest Formations*. Rome: FAO Forestry Paper 51/1.

Farmer, G., Wigley, T.M.L., Jones, P.D. and Salmon, M. (1989) *Documenting and Explaining Recent Global-mean Temperature Changes*. Final Report to NERC Contract GR3/6565, Climatic Research Unit, Norwich.

Faure, H. and Faure-Desnard, L. (1990) Future forcing on the global hydrological cycle. In F. Blacas, N. Petit-Maire and J. Riser (eds), *IGCP 252 Deserts, Past and Future Evolution*, Scientific reports, CNRS Lab. de Géologie du Quaternaire, Marseille, 25–27.

Fearnside, P.M. (1985) *The Carrying Capacity of the Brazilian Rainforest*. New York: Columbia University Press.

Fearnside, P.M. (1987) Deforestation and international economic development projects in Brazilian Amazonia. *Conservation Biology*, 1, 3.

Fearnside, P.M. (1988) Jari at age 19: lessons for Brazil's silvicultural plans at Carajas. *Interciencia*, 13 (1), 12–24.

Fearnside, P.M. (1990) Deforestation in Brazilian Amazonia. In G.M. Woodwell (ed.), *The Earth in Transition: Patterns and Processes of Biotic Impoverishment*. Cambridge: Cambridge University Press, 211–38.

Fearnside, P.M., Tardin, A.T. and Meira, L.G. (1990) *Deforestation in the Brazilian Amazon*. Brasilia: National Secretariat of Science and Technology.

Felix-Katz, J. de (1980) Analyse eco-énergetique d'un élevage nomade (Touareg) au Niger, dans la région de Azawak. *Ann. Géogr.*, 491, 57–72.

Fells, E. and Keller, R. (1973) World register on man-made lakes. In W.C. Ackermann, G.F. White and E.B. Worthington (eds), *Man-made Lakes: Their Problems and Environmental Effects*. Washington, DC: American Geophysical Union, 43–9.

Ferrians, O.J., Kachadoorian, R. and Greene, G.W. (1969) Permafrost and related engineering problems in Alaska. *Geological Survey Professional Paper*, 678.

Fifield, R. (1984) The shape of Earth from space. *New Scientist*, 1430, 46–50.

Firth, C.R. and Haggart, B.A. (1989) Loch Lomond stadial and Flandrian shorelines in the inner Moray Firth area, Scotland. *Journal of Quaternary Science*, 4, 37–50.

Fisk, D.A. and Done, T.J. (1985) Taxonomic and bathymetric patterns of bleaching in corals, Myrmidion Reef (Queensland). *Proceedings, 5th International Coral Reef Congress, Tahiti*, 6, 149–54.

Flaherty, D.C. (1991) *The Disappearing Tropical Forests*. Paris: MAB/IHP – UNESCO.

Fleagle, R.G. and Businger, J.A. (1963) *An Introduction to Atmospheric Physics*. New York: Academic Press.

Flint, E.P. and Richards, J.F. (1991) Historical analysis of changes in land use and carbon stock of vegetation in south and south-east Asia. *Canadian Journal of Forest Research*, 21, 91–110.

Flohn, H. (1974) Background of a geophysical model of the initiation of the next glaciation. *Quaternary Research*, 4, 385–404.

Flohn, H. (1977) Climate and energy: a scenario to a 21st century problem.

Climatic Change, 1, 5–20.

Flower, R.J. and Battarbee, R.W. (1983) Diatom evidence for recent acidification of two Scottish lochs. *Nature*, 20, 130–3.

Flower, R.J., Battarbee, R.W. and Appleby, P.G. (1987) The recent palaeolimnology of acid lakes in Galloway, south-west Scotland: diatom analysis, pH trends and the role of afforestation. *Journal of Ecology*, 75, 797–824.

Flower, R.J., Battarbee, R.W., Natkanski, J., Rippey, B. and Appleby, P.G. (1988) The recent acidification of a large Scottish loch located partly within a National Nature Reserve and Site of Special Scientific Interest. *Journal of Applied Ecology*, 25, 715–24.

Flower, R.J., Nawas, R. and Dearing, J.A. (1984) Sediment supply and accumulation in a small Moroccan lake: an historical perspective. *Hydrobiologia*, 112, 81–92.

Folland, C.K. (1988) Numerical models of the raingauge exposure problem, field experiments and an improved collector design. *Quarterly Journal of the Royal Meteorological Society*, 114, 1485–516.

Folland, C.K., Karl, T. and Vinnikov, K.Ya. (1990) Observed climate variations and change. In J.T. Houghton, G.J. Jenkins and J.J. Ephraums (eds), *Climate Change: the IPCC Scientific Assessment*. Cambridge: Cambridge University Press, 195–238.

Folland, C.K. and Parker, D.E. (1990) Observed variations of sea surface temperature. In M.E. Schlesinger (ed.), *Climate–Ocean Interaction*. Dordrecht: Kluwer, 21–52.

Folland, C.K., Parker, D.E. and Kates, F.E. (1984) Worldwide marine temperature fluctuations 1856–1981. *Nature*, 310, 670–3.

Fontferrier, Cpt. (1923) Etude historique sur le mouvement caravanier dans le cercle d'Agadez. *Bull. Comit. d'étud. hist. et scient. de l'AOF*, 302–14.

Fortuin, J.P.F. and Oerlemans, J. (1990) Paramaterization of the annual surface temperature and mass balance of Antarctica. *Annals of Glaciology*, 14, 78–84.

Foster, I.D.L., Dearing, J.A. and Grew, R. (1988) Lake-catchments: an evaluation of their contribution to studies of sediment yield and delivery processes. *Sediment Budgets (Proc. Porto Alegre Symp.) IAHS Publn*, 174, 413–24.

Foster, I.D.L., Dearing, J.A., Grew, R. and Orend, K. (1990) The sedimentary data base: an appraisal of lake and reservoir based studies of sediment yield. *Erosion Transport and Deposition Processes (Proc. Jerusalem Workshop) IAHS Publn*, 189, 119–43.

Frank, N.L. and Hussain, S.A. (1971) The deadliest tropical cyclone in history? *Bulletin of the American Meteorological Society*, 52, 438–44.

Frankel, O. and Soule, M.E. (1981) *Conservation and Evolution*. Cambridge: Cambridge University Press.

Frederiksen, P., Langaas, S. and Mbaye, M. (1990) NOAA-AVHRR and GIS based monitoring of fire activity in Senegal – a provisional methodology and potential applications. In J.G. Goldammer (ed.), *Fire in the Tropical Biotas*. Berlin: Springer, 400–17.

Freeland, W.J. (1990) Large herbivorous mammals: exotic species in Northern Australia. *Journal of Biogeography*, 17, 445–9.

French, H.M. (1976) *The Periglacial Environment*. London and New York: Longman.

French, H.M. (1987) Permafrost and ground ice. In K.J. Gregory and D.E. Walling (eds), *Human Activity and Environmental Processes*. Chichester: Wiley, 237–69.

Friedli, H., Loetscher, H., Oeschger, H. Siegenthalet, U. and Stauffer, B.

(1986) Ice core record of the $^{13}C/^{12}C$ record of atmospheric CO_2 in the past two centuries. *Nature*, 324, 237–8.

Fritz, S.C., Juggins, S., Battarbee, R.W. and Engstrom, D.R. (1991) Reconstruction of past changes in salinity and climate using a diatom-based transfer function. *Nature*. 352, 706–8.

Frost, P., Medina, E., Menaut, J.-C., Solbrig, O., Swift, M. and Walker, B.H. (1986) Responses of savannas to stress and disturbance: a proposal for a collaborative programme of research. *Biology International (Special Issue)*, 10. Paris: IUBS.

Fu, B.J. (1983) Problems on ecological balance in the Loess Plateau. *Ecological Science*, 2 (1), 44–8 (in Chinese).

Fuller, R.M., Parsell, R.J., Oliver, M. and Wyatt, G. (1989) Visual and computer classifications of remotely-sensed images. A case study in Cambridgeshire. *International Journal of Remote Sensing*, 10, 193–210.

Funder, S. (1989) Quaternary geology of the ice-free areas and adjacent shelves of Greenland. In R.J. Fulton (ed.), *Quaternary Geology of Canada and Greenland*. Ottawa: Geological Survey of Canada, 1, 743–92.

Furley, P.A. (1986) Radar surveys for resource evaluation in Brazil: an illustration from Rondonia. In M.J. Eden and J.T. Parry (eds), *Remote Sensing and Tropical Land Management*. Chichester: Wiley, 79–99.

Furley, P.A. (1990) The nature and sustainability of Brazilian Amazon soils. In D. Goodman and A. Hall (eds), *The Future of Amazonia*. London: Macmillan, 309–59.

Furley, P.A. and Newey, W.W. (1983) *Geography of the Biosphere*. London: Butterworths.

Furley, P.A., Proctor, J. and Ratter, J.A. (eds) (1992) *Nature and Dynamics of Forest-Savanna Boundaries*. London: Chapman and Hall.

Galloway, J.M., Likens, G.E., Keene, W.C. and Miller, J.M. (1982) The composition of precipitation in remote areas of the world. *Journal of Geophysical Research*, 87, 8771–86.

Gandhi, M.K. (1927/9) *The Story of My Experiments with Truth*. Ahmedabad: Navajivan.

Garner, H.F. (1974) *The Origin of Landscapes*. Oxford/New York: Oxford University Press.

Gasse, F., Talling, J.F. and Kilham, P. (1983) Diatom assemblages in East Africa: classification, distribution and ecology. *Révue d'Hydrobiologie Tropicale*, 16, 3–34.

Gasse, F., Téhet, R., Durand, A., Gibert, E. and Fontes, J.-C. (1990) The arid-humid transition in the Sahara and the Sahel during the last deglaciation. *Nature*, 346, 141–6.

Gast, M. (1968) *Alimentation des populations de l'Ahaggar*. Paris: Mém. du CRAPE.

Gates, W.L., Rowntree, P.R. and Zeng, Q-C. (1990) Validation of climate models. In J.T. Houghton, G.J. Jenkins and J.J Ephraums (eds), *Climate Change. The IPCC Scientific Assessment*. Cambridge: Cambridge University Press, 93–130.

Gauthier-Pilters, H. and Dagg, A.I. (1981) *The Camel*. Chicago: University of Chicago Press.

Geertz, C. (1969) Two types of ecosystem. In A.P. Vayda (ed.), *Environment and Cultural Behavior: Ecological Studies in Cultural Anthropology*. Garden City, NJ: The Natural History Press, 3–28.

Gentry, A.H. (1982) Patterns of Neotropical plant species diversity. *Evolutionary Biology*, 15, 1–85.

Gentry, A.H. (1986) Endemism in tropical v. temperate plant communities. In M. Soulé (ed.), *Conservation Biology: the Science of Scarcity and*

Diversity. Sunderland, MA: Sinauer Associates, 153–81.

Gerasimov, I.P. and Gindin, A.M. (1977) The problem of transferring runoff from the northern and Siberian rivers to the arid regions of the European USSR, Soviet Central Asia and Kazakhstan. In G.F. White (ed.), *Environmental Effects of Complex River Development*. Boulder, CO: Westview, 59–70.

Ghosh, R.C. and Subba Rao, B.K. (1979) Forest and floods. *Indian Forester*, 105 (4), 249–59.

Ghosh, R.C., Subba Rao, B.K. and Ramola, B.C. (1980) Interception studies in Sal (*Shorea robusta*) coppice forest. *Indian Forester*, 106 (8), 513–14.

Gilles, P.J. (1977) Environmental planning at Bougainville Copper. In J.H. Winslow (ed.), *The Melanesian Environment*. Canberra: ANU Press, 358–64.

Gilmour, D.A. (1988) Not seeing the trees for the forest: a reappraisal of the deforestation crisis in two Hill districts of Nepal. *Mountain Research and Development*, 8, 343–50.

Gilmour, D.A., Bonell, M. and Cassells, D.S. (1987) The effects of forestation on soil hydraulic properties in the Middle Hills of Nepal. *Mountain Research and Development*, 7, 239–49.

Gliessman, S.R. (1990) The ecology and management of traditional farming systems. In M.A. Altieri and S.B. Hecht (eds), *Agroecology and Small Farm Development*. Boca Raton, FL: CRC Press, 13–18.

Glynn, P.W. (1984) Widespread coral mortality and the 1982/83 El Niño warming event. *Environmental Conservation*, 11, 133–46.

Glynn, P.W. and D'Croz, L. (1990) Experimental evidence for high temperature stress as the cause of El Niño-coincident coral mortality. *Coral Reefs*, 8, 181–91.

Glynn, P.W., Cortes, J., Gusman, H.M. and Richmond, R.J. (1988) El Niño (1982–83) associated coral mortality and relationship to sea surface temperature deviations in the tropical eastern Pacific. *Proceedings, 6th International Coral Reef Symposium, Townsville*, 3, 237–43.

Godwin, H. (1940) Studies of the postglacial history of British vegetation, III. Fenland pollen diagrams, IV. Postglacial changes of relative land and sealevel in the English Fenland. *Philosophical Transactions of the Royal Society, London, B*, 230, 239–303.

Gold, L.W. and Lachenbruch, A.H. (1973) Thermal conditions in permafrost – a review of North American literature. In *North American contribution, Permafrost 2nd International Conference* (Yakutsk, USSR, 13–28 July 1973), Washington, DC: National Academy of Science, 2–23.

Goldammer, J.G. (ed.) (1990) *Fire in the Tropical Biota: Ecosystem Processes and Global Challenges*. Ecological Studies 84. Berlin: Springer-Verlag.

Goldemberg, J., Johansson, T., Reddy, A. and Williams, R. (1988) *Energy for a Sustainable World*. New Delhi: Wiley Eastern.

Golden, F. and Amfitheotrof, E. (1982) Making rivers run backward. *Time*, 14 June, 80.

Gomez-Pompa, A., Whitmore, T.C. and Hadley, M. (eds) (1991) *Rain Forest Regeneration and Management*. MAB Series 6. Paris: UNESCO/ Carnforth, UK: Parthenon.

Gong, S.Y. and Jiang, D.Q. (1979) Soil erosion and its control in small watersheds of the Loess Plateau. *Scientia Sinica*, 22, 1302–13.

Goodland, R.J.A. (1980) Environmental ranking of Amazonian development projects. In F. Barbira-Scazzochio (ed.), *Land, People and Planning in Contemporary Amazonia*. Cambridge: Centre of Latin American

Studies.

Goodland, R.J.A. (1982) *Tribal Peoples and Economic Development.* Washington, DC: World Bank International Bank for Reconstruction and Development.

Goodland, R.J.A. and Irwin, H.S. (1975) *Amazon Jungle: Green Hell to Red Desert.* Amsterdam: Elsevier.

Goodrich, L.E. (1982) The influence of snow cover on the ground thermal regime. *Canadian Geotechnical Journal*, 19, 421–32.

Goodwin, C.W., Brown, J. and Outcalt, S.A. (1984) Potential responses of permafrost to climatic warming. In J.H. McBeath (ed.), *The Potential Effects of Carbon-dioxide Induced Climatic Changes in Alaska.* Proceedings of a Conference, University of Fairbanks-Alaska, Miscellaneous Publications, 83 (1), 92–105.

Gore, J. and Petts, G.E. (1989) *Alternatives in Regulated River Management.* Boca Raton, FL: CRC Press.

Goreau, T.F. and Goreau, N.I. (1959) The physiology of skeleton formation in corals II. Calcium deposition by hermatypic corals under various conditions in the reef. *Biological Bulletin*, 117, 239–50.

Goreau, T.J. (1990) Coral bleaching in Jamaica. *Nature*, 343, 417–19.

Goreau, T.J. and Macfarlane, A.H. (1990) Reduced growth rate of *Montastrea annularis* following the 1987–1988 coral-bleaching event. *Coral Reefs*, 8, 211–16.

Gorham, E. (1958) The influence and importance of daily weather conditions in the supply of chloride, sulphate and other ions to fresh-waters from atmospheric precipitation. *Philosophical Transactions of the Royal Society, London, B*, 241, 147–78.

Gottschalk, L.D. (1945) Effects of soil erosion on navigation in upper Chesapeake Bay. *Geographic Review*, 35, 319–38.

Gottschalk, M.K.E. (1971) *Stormvloeden en rivieroverstromingen in Nederland: Deel I, De periode voor 1400.* Assen: van Gorcum.

Gottschalk, M.K.E. (1975) *Stormvloeden en rivieroverstromingen in Nederland: Deel II, De periode 1400–1600.* Assen: van Gorcum.

Gottschalk, M.K.E. (1977) *Stormvloeden en rivieroverstromingen in Nederland: Deel III, De periode 1600–1700.* Assen: van Gorcum.

Goudie, A. (1990) *The Human Impact on the Natural Environment*, 3rd edn. Oxford: Blackwell.

Goudie, A.S. (1981) Desertification. In R.J. Johnston (ed.), *The Dictionary of Human Geography*, Oxford: Blackwell, 77.

Goudie, A.S. (ed.) (1990) *Techniques for Desert Reclamation.* Chichester: Wiley.

Gould, S.J. (1990) *Wonderful Life. The Burgess Shale and the Nature of History.* London: Hutchinson Radius.

Government of India (1980) *Report of the National Commission on Floods.* New Delhi: Department of Irrigation.

Gowlett, J.A.J., Hairns, J.W.K., Walton, D.A., Wood, B.A. (1981) Early archaeological sites, hominid remains and traces of fire from Chesowanja, Kenya. *Nature*, 294, 125–9.

Graf, W.L. (1987) An American Stream. *Geographical Magazine*, October, 504–9.

Grainger, A. (1990) *The Threatening Desert: Controlling Desertification.* London: Earthscan.

Granger, O. (1978) Increased variability in California precipitation. *Annals Association of American Geographers*, 69, 533–43.

Grant, P.J. (1981) Recently increased tropical cyclone activity and inferences concerning coastal erosion and inland hydrological regimes in New Zealand and eastern Australia. *Climatic Change*, 3, 317–22.

Graus, R.R., Macintyre, I. and Herchenroder, B.E. (1985) Computer simulation of the Holocene facies history of a Caribbean fringing reef (Galeta Point, Panama). *Proceedings, 5th International Coral Reef Congress, Tahiti*, 3, 317–22.

Grebenard, D. (1985) *Le néolithique final et les débuts de la métallurgie*. Etudes Nigériennes, 49. Paris: Niamey.

Grewal, S.S., Juneja, M.L., Mittal, S.P., Agnihotri, Y., and Bansal, R.C. (1990) Effect of soil hydrophysical properties on erosion, seepage, and grass yield from a Shivalik watershed selected for water resources development. *Indian Journal of Soil Conservation*, 18 (1/2): 46–55.

Grigg, R.W. (1988) Palaeoceanography of coral reefs in the Hawaiian-Emperor chain. *Science*, 240, 1737–43.

Grigg, R.W. and Epp, D. (1989) Critical depth for the survival of coral islands: effects on the Hawaiian archipelago. *Science*, 243, 638–41.

Grove, J.M. (1988) *The Little Ice Age*. London and New York: Methuen.

Guetter, P.J. and Kutzbach, J.E. (1990) A modified Köppen classification applied to model simulations of glacial and interglacial climates. *Climatic Change*, 16, 193–215.

Guiot, J., Pons, A., de Beaulieu, J.L. and Reille, M. (1989) A 140,000-year continental climate reconstruction from two European pollen records. *Nature*, 338, 309–13.

Gupta, P.N. (1979) *Afforestation, Integrated Watershed Management, Torrent Control and Land Use Development for U.P. Himalaya and Siwaliks*. Lucknow: Uttar Pradesh Forest Department.

Haggart, B.A. (1987) Relative sea level changes in the Moray Firth area, Scotland. In M.J. Tooley and I. Shennan (eds), *Sea Level Changes*. Oxford: Basil Blackwell, 67–108.

Haigh, M.J. (1979a) Landslide sediment accumulations on the Mussoorie-Tehri road, Garhwal Lesser Himalaya. *Indian Journal of Soil Conservation*, 7 (1), 1–4.

Haigh, M.J. (1979b) Environmental geomorphology of the Landour-Mussoorie area, U.P. *Himalayan Geology*, 9, 657–88.

Haigh, M.J. (1981) Floods and erosion in North India. *Himalaya: Man and Nature*, 5 (1), 21–5; 5 (2), 4–9.

Haigh, M.J. (1982a) A comparison of sediment accumulations beneath forested and deforested microcatchments, Garhwal Himalaya. *Himalayan Research and Development*, 1, 118–20.

Haigh, M.J. (1982b) Road development and rural stress in the Indian Himalaya. *Nordia*, 16, 135–40.

Haigh, M.J. (1984a) Landslide prediction and highway maintenance in the Lesser Himalaya, India. *Zeitschrift für Geomorphologie, Suppl. Bd*, 51, 17–38.

Haigh, M.J. (1984b) Deforestation and disaster in northern India. *Land Use Policy*, 1 (3), 187–98.

Haigh, M.J. (1988) Understanding Chipko: the Himalayan people's movement for forest conservation. *International Journal of Environmental Studies*, 31, 99–110.

Haigh, M.J. (1989a) Water erosion and its control: case studies from South Asia. In K. Ivanov and D. Pechinov (eds), *Water Erosion*. Paris: UNESCO Technical Documents in Hydrology IHP-III, Project 2.6, 1–38.

Haigh, M.J. (1989b) Problems of headwater management in the Central Himalaya. In J. Krecek, H. Grip, M.J. Haigh and A. Hocevar (eds), *Headwater Control*. 2. Plsen: WASWC/IUFRO/CSVTS, 280–94.

Haigh, M.J. (1991) Reclaiming forest lands in the Himalaya: notes from Pakistan. In G.S. Rajwar (ed.), *Advances in Himalayan Ecology*. New

Delhi/Houston: Today and Tomorrow/Scholarly (Recent Researches in Ecology, Environment and Pollution 6), 199–210.

Haigh, M.J. and Krecek, J. (1991) Headwater management: problems and policies. *Land Use Policy*, 8 (3), 171–6.

Haigh, M.J., Rawat, J.S. and Bartarya, S.K. (1987) Impact of hill roads on downslope forest cover. *Himalaya: Man and Nature*, 11 (4), 2–3.

Haigh, M.J., Rawat, J.S. and Bartarya, S.K. (1988a) Environmental correlations of landslide frequency along new highways in the Himalaya: preliminary results. *Catena*, 15, 539–53.

Haigh, M.J., Rawat, J.S. and Bartarya, S.K. (1988b) Entropy minimising landslide systems. *Current Science*, 57, 1000–2.

Haigh, M.J., Rawat, J.S. and Bisht, H.S. (1990) Hydrological impact of deforestation in the Central Himalaya. International Association of Hydrological Sciences, Publication 190 (Hydrology of Mountainous Areas), 419–33.

Hallock, P. and Schlager, W. (1986) Nutrient excess and the demise of coral reefs and carbonate platforms. *Palaios*, 1, 389–98.

Halpern, D., Hayes, S.P., Lectmaa, A., Hansen, D.V. and Philander, S.G.H. (1983) Oceanographic observations of the 1982 warming of the tropical eastern Pacific. *Science*, 221, 1173–5.

Hamilton, L.S. (1985) Overcoming myths about soil and water impacts on tropical forest land uses. In S.A. El-Swaify et al. (eds), *Soil Erosion and Conservation*. Ankeny, IA: Soil Conservation Society of America, 680–90.

Hamilton, L.S. (1987) What are the impacts of Himalayan deforestation on the Ganges-Brahmaputra lowlands and delta? Assumptions and facts. *Mountain Research and Development*, 7, 256–63.

Hamilton, L.S. and King, P.N. (1983) *Tropical Forested Watersheds: Hydrologic and Soils Responses to Major Uses and Conversion*. Boulder, CO: Westview Press.

Hamilton, L.S. and Pearce, A.J. (1986) Biophysical aspects in watershed management. In K.W. Easter, J.A. Dixon and M.M. Hufschmidt (eds), *Watershed Resource Management – an Integrated Framework with Case Studies from Asia and the Pacific*. Boulder, CO: Westview Studies in Water Policy and Management 10, 33–52.

Hamilton, L.S. and Pearce, A.J. (1987) Soil and water impacts of deforestation. In J. Ives and D.C. Pitt (eds), *Deforestation: Social Dynamics in Watersheds and Mountain Ecosystems*. London: Routledge.

Hansen, J. and Lacis, A.A. (1990) Sun and dust versus greenhouse gases: an assessment of their relative roles in global climate change. *Nature*, 346, 713–19.

Hansen, J. and Lebedeff, S. (1987) Global trends of measured surface air temperature. *Journal of Geophysical Research*, 92, 13345–72.

Hansen, J. and Lebedeff, S. (1988) Global surface air temperatures: update through 1987. *Geophysical Research Letters*, 15, 323–6.

Hansen, J., Fung, I., Lacis, A., Rind, D., Lebedeff, S., Ruedy, R. and Russell, G. (1988) Global climate changes as forecast by Goddard Institute for Space Studies three-dimensional model. *Journal of Geophysical Research*, 93, D8, 9341–64.

Hansen, J., Johnson, D., Lacis, A., Lebedeff, S., Lee, P., Rind, D. and Russell, G. (1981) Climate impact of increasing carbon dioxide. *Science*, 213, 957–66.

Hansen, J., Russell, G., Rind, D., Stone, P., Lacis, A., Lebedeff, S., Ruedy, R. and Travis, L. (1983) Efficient three-dimensional global models for climate studies: Models I and II. *Monthly Weather Review*, 111, 609–22.

Hare, F.K. (1961) *The Restless Atmosphere*. Hutchinson University Press.

Harlan, J.R. (1989a) Wild grass seed harvesting in the Sahara and Sub Sahara of Africa. In D.R. Harris and G.C. Hillman (eds), *Foraging and Farming*. Cambridge: Cambridge University Press, 79–98.

Harlan, J.R. (1989b) The tropical African cereals. In D.R. Harris and G.C. Hillman (eds), *Foraging and Farming*. Cambridge: Cambridge University Press, 335–43.

Harmelin-Vivien, M.L. and Laboute, P. (1986) Catastrophic impact of hurricanes on atoll outer reef slopes in the Tuamotu (French Polynesia). *Coral Reefs*, 5, 55–62.

Harriman, R. and Morrison, B.R.S. (1982) The ecology of streams draining forested and non-forested catchments in an area of central Scotland subject to acid precipitation. *Hydrobiologia*, 88, 251–63.

Harriman, R., Morrison, B.R.S., Caines, L.A., Collen, P. and Watt, A.W. (1987) Long term changes in fish populations of acid streams and lochs in Galloway, south-west Scotland. *Water, Air and Soil Pollution*, 32, 89–112.

Harriott, V.J. (1985) Mortality rates of scleractinian corals before and during a mass bleaching event. *Marine Ecology Progress Series*, 21, 81–8.

Harris, D.R. (1966) Recent plant invasions in the arid and semi-arid southwest of the United States. *Annals of the Association of American Geographers*, 56, 408–22.

Harris, D.R. (1980) Tropical savanna environments: definition, distribution, diversity and development. In D.R. Harris (ed.), *Human Ecology in Savanna Environments*. London: Academic Press, 3–27.

Harris, R. (1987) *Satellite Remote Sensing*. London: Routledge and Kegan Paul.

Harris, S.A. (1986) *The Permafrost Environment*. London and Sydney: Croom Helm.

Harris, S.A., French, H.M., Heginbottom, J.A., Johnston, G.H., Ladanyi, B., Sego, D.C. and Van Everdingen, R.O. (1988) *Glossary of Permafrost and Related Ground-ice Terms*. Technical Memorandum No. 142, Ottawa: National Research Council of Canada, Permafrost Subcommittee.

Harrison, W., Malloy, R.J., Rusnack, G.A. and Terasmae, J. (1965) Possible late Pleistocene uplift at Chesapeake Bay entrance. *Journal of Geology*, 73, 201–29.

Harriss, R.C., Gorham, E., Sebacher, D.I., Bartlett, K.B. and Flebbe, P.A. (1985) Methane flux from northern peatlands. *Nature*, 315, 652–4.

Hassan, F.A. (1981) Historical Nile floods and their implications for climatic change. *Science*, 212, 1142–5.

Hastenrath, S., Ming-Chin, W. and Pao-Shin, C. (1984) Toward the monitoring and prediction of north-east Brazil droughts. *Quarterly Journal of the Royal Meteorological Society*, 118, 411–25.

Hauhs, M. (1990) Lange Bramke: an ecosystem study of a forested watershed. In D.C. Adriano and A.M. Havas (eds), *Acidic Precipitation Vol. 1 Case Studies*. New York: Springer-Verlag, 275–305.

Hay, J. (1973) Salt cedar and salinity on the Upper Rio Grande. In M.T. Farvar and J.P. Milton (eds), *The Careless Technology*. London: Tom Stacey,.288–300.

Hays, J.D., Imbrie, J. and Shackleton, N.J. (1976) Variations in the earth's orbit: pacemaker of the ice ages. *Science*, 235, 1156–67.

Heaps, N.S. (1983) Storm surges, 1967–1982. *Geophysical Journal of the Royal Astronomical Society*, 74, 331–76.

Hecht, S. and Cockburn, A. (1990) *The Fate of the Forest*. Harmondsworth:

Penguin Books.

Hecht, S.B. (1981) Deforestation in the Amazon Basin: magnitude, dynamics and soil resource effects. *Studies in Third World Societies*, 13, 6–108.

Hecht, S.B., Norgaard, R. and Possio, G. (1988) The economics of cattle ranching in eastern Amazon. *Interciencia*, 13, 233–40.

Helldén, U. (1984) Land degradation and land productivity monitoring – needs for an integrated approach. In A. Hjort (ed.), *Land Management and Survival*. Uppsala: Scandinavian Institute of African Studies.

Helldén, U. (1987) An assessment of woody biomass, community forests, land use and soil erosion in Ethiopia. *Lund Studies in Geography*, 14, 75.

Heller, F. and Liu, T.S. (1982) Magnetostratigraphical dating of loess deposits in China. *Nature*, 300, 431–3.

Heller, F. and Liu, T.S. (1984) Magnetism of Chinese loess deposits. *Journal of Geophysical Research*, 77, 125–41.

Hemming, J. (ed.) (1985) *Change in the Amazon Basin*, vols 1 and 2. Manchester: Manchester University Press.

Henderson-Sellers, A. (1991) Policy advice on greenhouse-induced climatic change: the scientist's dilemma. *Progress in Physical Geography*, 15, 53–70.

Henderson-Sellers, A. and McGuffie, K. (1987) *A Climate Modelling Primer*. Chichester: Wiley.

Henriksen, A. (1979) A simple approach for identifying and measuring acidification of fresh water. *Nature*, 278, 542–5.

Henriksen, A., Lien, L. and Traaen, T.S. (1990) Critical loads for surface waters – chemical criteria for inputs of strong acids. Norwegian Institute for Water Research, Acid Rain Research Report 22/1990.

Henriksen, A., Lien, L., Traaen, T.S., Sevaldrud, I.S., and Brakke, D.F. (1988) Lake acidification in Norway – present and predicted status. *Ambio*, 17, 259–66.

Hernandez-Avila, M.L., Roberts, H.H. and Rouse, L.J. (1977) Hurricane-generated waves and coastal boulder rampart formation. *Proceedings, 3rd International Coral Reef Symposium, Miami*, 2, 71–8.

Heuse, G. (1987) Planetary engineering and desert reclamation. In M. de Boodt and R. Hartmann (eds), *Eremology*. Gent, 415–24.

Hewlett, J.D. (1982a) *Principles of Forest Hydrology*. Athens: University of Georgia Press.

Hewlett, J.D. (1982b) Forests and floods in the light of recent investigation. In *Canadian Hydrology Symposium, Proceedings*. Ottawa: National Research Council, 545–60.

Hilbert, D.W., Prudhomme, T.I. and Oechel, W.C. (1987) Response of tussock tundra to elevated carbon dioxide regimes: analysis of ecosystem CO_2 flux through non-linear modelling. *Oecologia*, 72, 466–72.

Hill, R.D. (1990) Environments, production and resources. In J.I. Furtado, W.B. Morgan, J.R. Pfafflin and K. Ruddles (eds), *Tropical Resources: Ecology and Development*. Chur, Switzerland: Harwood Publ., 1–38.

Hinnov, L.A., Brush, G.S. and Brush, L.M. (1991) Time series analysis of a sedimentary core from the upper Chesapeake Bay. *Eos*, 72, 151–2.

HMG Nepal (1983) *The Forests of Nepal – a Study of Historical Trends and Projections to 2000*. Kathmandu: His Majesty's Government, Ministry of Water Resources, Water and Energy Commission, Report 4/2/200783/1/1.

HMSO (1991) *Climatological and Environmental Effects of Rainforest Destruction*. London: Environment Committee, House of Commons.

Hobbs, J.E. (1988) Recent climatic change in Australia. In S. Gregory (ed.),

Recent Climatic Change. London: Belhaven Press, 285–97.

Hoegh-Guldberg, O. and Smith, G.J. (1989) Light, salinity, and temperature and the population density, metabolism and export of zooxanthellae from *Stylophora pistillata* and *Seriatopora hystrix. Journal of Experimental Marine Biology and Ecology,* 129, 279–303.

Hoffman, J.S., (1984) Estimates of future sea level rise. In M.C. Barth and J.G. Titus (eds), *Greenhouse Effect and Sea-level Rise: a Challenge for This Generation.* New York: Van Nostrand Reinhold, 79–103.

Hoffmann, J.S. Keyes, D. and Titus, J.G. (1983) *Projecting Future Sea Level Rise.* Report PM-221. Washington, DC: US Environmental Protection Agency.

Hoffman, J.S., Wells, J.B. and Titus, J.G. (1986) Future global warming and sea level rise. In G. Sigbjarnason (ed.), *Iceland Coastal and River Symposium.* Reykjavik: National Energy Authority, 245–66.

Holland, G.J., McBride, J.L. and Nicholls, N. (1988) Australian region tropical cyclones and the greenhouse effect. In G.I. Pearson (ed.), *Greenhouse: Planning for Climatic Change.* Melbourne: CSIRO, 438–55.

Hollis, G.E. (1978) The falling levels of the Caspian and Aral seas. *The Geographical Journal,* 144, 62–80.

Holm-Nielsen, L.B., Nielsen, I.C. and Alsey, H.B. (1989) *Tropical Forests, Botanical Dynamics, Speciation and Diversity.* London: Academic Press.

Holt, T., Kelly, P.M. and Cherry, B.S.G. (1984) Cryospheric impacts of Soviet River Diversion Schemes. *Annals of Glaciology,* 5, 61–8.

Hopley, D. (1986) Corals and reefs as indicators of paleosea levels, with special reference to the Great Barrier Reef. In O. van de Plassche (ed.), *Sea-level Research.* Norwich: Geo Books, 195–228.

Hopley, D. (1987) Holocene sealevel changes in Australia and the Southern Pacific. In R.J.N. Devoy (ed.), *Sea Surface Studies.* London: Croom Helm, 375–405.

Hopley, D. and Kinsey, D.W. (1988) The effects of a rapid short-term sea-level rise on the Great Barrier Reef. In G.I. Pearson (ed.), *Greenhouse: Planning for Climatic Change.* Melbourne: CSIRO, 189–201.

Houghton, J.T., Jenkins, G.J. and Ephraums, J.J. (eds) (1990) *Climate Change, the IPCC Scientific Assessment.* Cambridge: Cambridge University Press.

Houghton, R.A. (1990a) The future role of tropical forests in affecting carbon dioxide concentration of the atmosphere. *Ambio,* 19, 204–9.

Houghton, R.A. (1990b) The global effects of deforestation. *Environmental Science and Technology,* 24, 414–36.

Houghton, R.A. (1991) Release of carbon to the atmosphere from degradation of forests in tropical Asia. *Canadian Journal of Forest Research,* 21, 132–42.

Houghton, R.A., Boone, R.D., Fruei, J.E., Hobbie, J.M., Palm, C.A., Peterson, G.R., Shaver, G.M., Woodwell, G.M., Moore, B., Skole, D.L. and Myers, N. (1987) The flux of carbon from terrestrial ecosystems to the atmosphere in 1980 due to changes in land use: geographic distribution of the total flux. *Tellus,* 39B, 122–39.

Houghton, R.A., Lefkowitz, D.S. and Skole, D.L. (1991) Changes in the landscape of Latin America between 1850 and 1985: I Progressive loss of forests. *Forest Ecology and Management,* 38, 143–72.

Howells, G.D. and Brown, D.J.A. (1987) The Loch Fleet Project, SW Scotland. *Transactions of the Royal Society of Edinburgh: Earth Sciences,* 78, 241–8.

Hubbard, D.K. (1988) Controls of modern and fossil reef development: Common ground for biological and geological research. *Proceedings,*

6th International Coral Reef Symposium, Townsville, 1, 243–52.

Hudson, J.H. (1981) Growth rates in *Montastrea annularis*: a record of environmental change in Key Largo coral reef marine sanctuary, Florida. *Bulletin of Marine Science*, 31, 444–59.

Hughes, T. (1981) The weak underbelly of the West Antarctic ice sheet. *J. Glaciol.*, 27, 518–25.

Hughes, T., Denton, G.H. and Grosswald, M.G. (1977) Was there a late-Wurm Arctic Ice Sheet? *Nature*, 266, 596–602.

Hulme, M. (1992) A land-based 1951–80 global precipitation climatology for model evaluation. *Climate Dynamics*, 7, 57–72.

Hulme, M., Marsh, R. and Jones, P.D. (1992) Global changes in a humidity index between 1931–60 and 1961–90. *Climate Research*, 2, 1–22.

Hulme, M., Wigley, T.M.L. and Jones, P.D. (1990) Limitations of regional climate scenarios for impact analysis. In M.M. Boer and R.S. De Groot (eds), *Landscape-ecological Impact of Climatic Change*. Proceedings of a European Conference, Lunteren, The Netherlands, 3–7 December 1989. Amsterdam: IOS Press, 111–29.

Huntley, B. (1990a) Studying global change: the contribution of Quaternary palynology. *Palaeogeography, Palaeoclimatology, Palaeoecology (Global and Planetary Change)*, 82, 53–61.

Huntley, B. (1990b) Lessons from climates of the past. In J. Leggett (ed.) *Global Warming. The Greenpeace Report*. Oxford: Oxford University Press, 133–48.

Huntley, B. and Prentice, I.C. (1988) July temperatures in Europe from pollen data, 6000 years before present. *Science*, 241, 687–90.

Huntley, B.J. (1982) Southern African savannas. In B.J. Huntley and B.H. Walker (eds), *Ecology of Tropical Savanna*. Berlin: Springer, 101–19.

Hutchinson, G.E., Bonatti, E., Cowgill, U.M. and Goulden, C.E. (1970) Ianula: an account of the history and development of the Lago di Monterosi, Latium, Italy. *Transactions of the American Philosophical Society (NS)*, 60, 1–78.

ICIHI (Independent Commission on International Humanitarian Issues) (1986) *The Vanishing Forest*. London: Zed Books.

Idso, S.B. (1989a) *Carbon Dioxide and Global Change: Earth in Transition*. Tempe, AZ: IBR Press.

Idso, S.B. (1989b) Greenhouse warrming or Little Ice Age demise: a critical problem for climatology. *Theor. and Appl. Clim.*, 39, 54–6.

IGBP (1990) *Global Change. Report No. 12, The International Geosphere Programme: a Study of Global Change. The Initial Core Projects*. Stockholm: IGBP Secretariat, Royal Swedish Academy of Sciences.

Imbrie, J. and Imbrie, K.P. (1979) *Ice Ages. Solving the Mystery*. London: Macmillan.

Imbrie, J., McIntyre, A. and Mix, A. (1989) Oceanic response to orbital forcing in the Late Quaternary: observational and experimental strategies. In A. Berger, S. Schneider and J.-Cl. Duplessy (eds), *Climate and Geo-Sciences*. Dordrecht: Kluwer, 121–64.

Ingram, J. (1990) *The Role of Trees in Maintaining and Improving Soil Productivity: a Review*. London: Commonwealth Science Council.

Ireland, S. (1987) The Holocene sedimentary history of the coastal lagoons of Rio de Janeiro state, Brazil. In M.J. Tooley and I. Shennan (eds), *Sea-level Changes*. Oxford: Basil Blackwell, 25–66.

IUCN (1980) *World Conservation Strategy*. Gland, Switzerland: International Union for the Conservation of Nature, United Nations Development Programme, and World Wildlife Fund.

IUCN (1991) *Caring for the Earth: A Strategy for Sustainable Living*.

Gland, Switzerland: International Union for the Conservation of Nature, United Nations Development Programme, and World Wide Fund for Nature.

Ives, J.D. (1987) The Himalaya–Ganges problem – the theory of Himalayan degradation: its validity and application challenged by recent research. *Mountain Research and Development*, 7, 181–3, 189–99.

Ives, J.D. (1989) Deforestation in the Himalayas. *Land Use Policy*, 6 (3), 187–93.

Ives, J.D., Andrews, J.T. and Barry, R.G. (1975) Growth and decay of the Laurentide ice sheet and comparisons with Fenno-Scandia. *Naturwissenschaften*, 62, 118–25.

Ives, J.D. and Messerli, B. (1989) *The Himalayan Dilemma: Reconciling Development and Conservation*. London and New York: Routledge/ United Nations University.

Jaap, W. (1985) An epidemic zooxanthellae explosion at Middle Sambo Reef, Florida Keys. *Bulletin of Marine Science*, 29, 414–22.

Jackson, M.G. (1986) A strategy for improving the production of livestock in the hills of Uttar Pradesh. In J.S. Singh (ed.), *Environmental Regeneration in the Himalaya*. Nainital: CHEA and Gyanodaya Prakashan.

Jacobsen, T. and Adams, R.M. (1958) Salt and silt in ancient Mesopotamian agriculture. *Science*, 128, 1251–8.

Jaeger, F. (1956) Zur Gliederung und Benennung des tropischen Graslandgürtels. *Geogr. Berichte*, 1, 36–44.

Jain, S. (1984) Women and peoples ecological movement: case study of women's role in the Chipko Movement of Uttar Pradesh. *Economic and Political Weekly*, 29 (41), 1788–94.

Janzen, D.H. (1988) Tropical dry forests. In E.O. Wilson (ed.), *Biodiversity*. Washington, DC: National Academy Press, 130–6.

Jelgersma, S. (1961) Holocene sea level changes in the Netherlands. *Mededelingen van de Geologische Stichting Serie C. VI. 7*, 1–100.

Jelgersma, S., Oele, E. and Wiggers, A.J. (1979) Depositional history and coastal development in the Netherlands and adjacent North Sea since the Eemian. In E. Oele, R.T.W. Schuttenhelm and A.J. Wiggers (eds), *The Quaternary History of the North Sea*. Uppsala: Acta Universitatis Upsaliensis, Symposia Universitatis Upsaliensis Annum Quisgentesimum Celebrantis 2, 115–42.

Jenkins, A., Whitehead, P.G., Cosby, B.J. and Birks, H.J.B. (1990) Modelling long-term acidification: a comparison with diatom reconstructions and the implications for reversibility. *Philosophical Transactions of the Royal Society, London, B*, 327, 209–14.

Jenkinson, D.S., Adams, D.E. and Wild, A. (1991) Model estimates of CO_2 emissions from soil in response to global warming. *Nature*, 351, 304–6.

Jenne, R. (1975) Data sets for meteorological research, *NCAR Technical Note TN/1A-III*. Boulder, CO: National Center for Atmospheric Research.

Jensen, K.W. and Snekvik, E. (1972) Low pH levels wipe out salmon and trout populations in southern Norway. *Ambio*, 1, 223–5.

Johnson, C., Knowles, R. and Colchester, M. (1989) *Rainforests: Land Use Options for Amazonia*. London: OUP/WWF.

Jokiel, P.L. and Coles, S.L. (1990) Response of Hawaiian and other Indo-Pacific reef corals to elevated temperature. *Coral Reefs*, 8, 155–62.

Jones, P.D. (1988a) Hemispheric surface air temperature variations: recent trends and an update to 1987. *Journal of Climate*, 1, 654–60.

Jones, P.D. (1988b) The influence of ENSO on global temperatures. *Climate Monitor*, 17, 80–9.

Jones, P.D. and Bradley, R.S. (1992) Climate since the commencement of instrumental records. In R.S. Bradley and P.D. Jones (eds), *Climate since AD 1500*. London: Routledge, 246–68.

Jones, P.D., Groisman, P.Ya., Coughlan, M., Plummer, N., Wang, W-C. and Karl, T.R. (1990) Assessment of urbanization effects in time series of surface air temperature over land. *Nature*, 347, 169–72.

Jones, P.D., Raper, S.C.B., Santer, B.D., Cherry, B.S.G., Goodess, C.M., Kelly, P.M., Wigley, T.M.L., Bradley, R.S. and Diaz, H.F. (1985) A gridpoint surface air temperature data set for the Northern Hemisphere. US Department of Energy, Report No. TR022, Washington, DC.

Jones, P.D. and Wigley, T.M.L. (1990) Global warming trends. *Scientific American*, August, 66–73.

Jones, P.D. and Wigley, T.M.L. (1991) The global temperature record for 1990. *DoE Research Summary*, No. 10, CDIAC, Oak Ridge.

Jones, P.D., Wigley, T.M.L. and Wright, P.D. (1986) Global temperature variations, 1861–1984. *Nature*, 322, 430–4.

Jones, V.J., Stevenson, A.C. and Battarbee, R.W. (1989) The acidification of lakes in Galloway, south-west Scotland: a diatom and pollen study of the post-glacial history of the Round Loch of Glenhead. *Journal of Ecology*, 77, 1–23.

Jong, K. (ed.) (1979) *Biological Aspects of Plant Genetic Resource Conservation in South East Asia*. University of Aberdeen: Institute of South East Asian Ecology.

Jordan, C.F. (1985) *Nutrient Cycling in Tropical Forest Ecosystems*. New York: Wiley.

Jordan, C.F. (1987) *Amazonian Forests*. Ecological Studies, 60. New York: Springer Verlag.

Joshi, S.C., Joshi, D.R. and Dani, D.D. (1983) *Kumaun Himalaya – A Geographical Perspective on Resource Development*. Nainital: Gyanodaya Prakashan.

Joussiffe, A. (1990) Quadafi's Herculean effort. *Geographical Magazine*, 62, 5, 38–42.

Junk, W.J., Bayley, P.B. and Sparks, R.E. (1989) The flood-pulse concept in river-floodplain systems. *Canadian Special Publication on Fisheries and Aquatic Sciences*, 106, 110–27.

Justice, C.O., Townshend, J.R.G., Holben, B.N. and Tucker, C.J. (1985) Analysis of the phenology of global vegetation using meterological satellite data. *International Journal of Remote Sensing*, 6, 1271–318.

Kadomura, H. (ed.) (1989) *Savannization Processes in Africa*, I. Tokyo: Faculty of Science, Metropolitan University.

Kahrl, W.L. (1982) *Water and Power. The Conflict over Los Angeles' Water Supply in the Owens Valley*. Berkeley: University of California Press.

Karan, P.P. (1991) Biology, environmental conservation and development in Bhutan Himalaya. In G.S. Rajwar (ed.), *Advances in Himalayan Ecology*, New Delhi/Houston: Today and Tomorrow/Scholarly (Recent Researches in Ecology, Environment and Pollution 6), 167–97.

Karl, T.R. and Jones, P.D. (1989) Urban bias in area-averaged surface air temperature trends. *Bulletin of the American Meteorological Society*, 70, 265–70.

Kaufman, Y. (1989) The atmospheric effect on remote sensing and its correction. In G. Asrar (ed.), *Theory and Applications of Optical Remote Sensing*. New York: Wiley, 336–428.

Kaufman, Y.J., Setzer, A., Justice, C., Tucker, C.J., Pereira, M.C., and Fung, I. (1990) Remote sensing of biomass burning in the tropics. In J.G. Goldammer (ed.), *Fire in the Tropical Biota*. Berlin: Springer, 371–99.

Kawosa, M.A. (1990) State of environment in Jammu and Kashmir.

Himalaya: Man and Nature, 14 (7), 12–16.

Keeling, C.D., Bacastow, R.B., Carter, A.F., Piper, S.C., Whorf, T.P., Heimann, M., Mook, W.G. and Roeloffzen, H. (1989) A three dimensional model of atmospheric CO_2 transport based on observed winds: 1. Analysis of observational data. In D.H. Peterson (ed.), *Aspects of Climate Variability in the Pacific and Western Americas*. Geophysical Monograph 55, Washington, DC: American Geophysical Union, 165–236.

Kellman, M.C. (1969) Some environmental components of shifting cultivation in upland Mindanao. *Journal of Tropical Geography*, 29, 40–56.

Kellogg, W.W. (1978) Global influences of mankind on climate. In J. Gribbin (ed.), *Climatic Change*. Cambridge: Cambridge University Press, 205–27.

Kelly, M. (1980) *The Status of the Neoglacial in Western Greenland*. Grønlands geologiske undersøgelse, 96.

Kelly, P.M. and Sear, C. (1984) Climatic impact of explosive volcanic eruptions. *Nature*, 311, 740–3.

Kennett, J.P. (1977) Cenozoic evolution of Antarctic glaciation, the circum-Antarctic ocean and their impact on global palaeoceanography. *Journal of Geophysical Research*, 82, 3843–60.

Kesel, R.W. (1985) Tropical fluvial geomorphology. In A.F. Pitty (ed.), *Themes in Geomorphology*. London: Croom Helm, 102–21.

Khondaker, M.S. (1989) Perception of and response to floods in Bangladesh. Unpublished PhD thesis, University of Manchester.

Kidson, C. (1982) Sea-level changes in the Holocene. *Quaternary Science Reviews*, 1, 121–51.

Kidson, C. and Heyworth, A. (1973) The Flandrian sea-level rise in the Bristol Channel. *Proceedings of the Ussher Society*, 2 (6), 565–84.

Kidson, C. and Heyworth, A. (1979) Sea level. In K. Suguio, T.R. Fairchild, L. Martin and J.-M. Flexor (eds), *Proceedings of the 1978 International Symposium on Coastal Evolution in the Quaternary*. São Paulo, Universidade de São Paulo, 1–28.

King, L. (1984) Permafrost in Skandinavien Untersuchungsergebnisse aus Lappland, Jotunheimen und Dovre/Rondane. *Heidelberger Geographische Arbeiten*, 76.

King, L. (1986) Zonation and ecology of high mountain permafrost in Scandinavia. *Geografiska Annaler*, 68A, 131–9.

King-Hele, D.G., Brookes, C.J. and Cook, G.E. (1980) The pear shaped section of the Earth. *Nature*, 286, 377–8.

Kinsey, D.W. (1983) Standards of performance in primary production and carbon turnover. In D.J. Barnes (ed.), *Perspectives on Coral Reefs*. Manuka: B. Clouston/AIMS, 209–20.

Klute, G. (1990) Arbeit bei den Kel Adagh, PhD thesis, Bayreuth.

Kneip, L.M. and Pallestrini, L. (1982) Estudo de artefatos Liticos e osseos das populacoes pre-historicas de Itaipu-Niteroi, R.J. In K. Suguio, M.R. Mousinho de Meis and M.G. Tessler (eds), *Atas IV Simposio do Quaternrio ho Brasil*. Rio de Janeiro: Sociedade Brasileira de Geologia, 431–42.

Knowlton, N., Lang, J.C. and Keller, B.D. (1990) Case study of natural population collapse: post-hurricane predation on Jamaican staghorn corals. *Smithsonian Contributions to the Marine Sciences*, 31, 1–25.

Koerner, R.M. (1989) Ice core evidence for extensive melting of the Greenland ice sheet in the last Interglacial. *Science*, 244, 964–8.

Kollmannsperger, F. (1977) *Long Range Landscape Changes under the Influence of Man*. Eschborn, Germany: GTZ.

Koster, E.A. and Boer, M.M. (eds) (1989) Landscape ecological impact of climatic change on Boreal/(Sub)Arctic regions, with emphasis on Fenno-scandia, Report prepared for the European Conference on Landscape Ecological Impact of Climatic Change, Lunteren, The Netherlands, 3–7 December 1989.

Kotlyakov, V.M. (1991) The Aral Sea basin: a critical environmental zone. *Environment*, 33, 4–9 and 36–8.

Kraft, J.C. (1971) Sedimentary facies patterns and geologic history of a Holocene marine transgression. *Geological Society of America Bulletin*, 82, 2131–58.

Küchler, A.W. (1964) *The Potential Natural Vegetation of the Coterminous United States*. New York: American Geographical Society.

Kuenan, P.H. (1950) *Marine Geology*. New York: Wiley.

Kuhlmann, D.H.H. (1985) The protection role of coastal forests on coral reefs. *Proceedings, 5th International Coral Reef Congress*, 6, 503–8.

Kumar, V. (1983) Trends and economic analysis of U.P. Hill forests. *Journal of Rural Development*, 2 (1), 134–7.

Kutzbach, J.E. (1980) Estimates of past climate at Paleolake Chad, North Africa, based on a hydrological and energy balance model. *Quaternary Research*, 14, 210–23.

Kutzbach, J.E. and Guetter, P.J. (1986) The influence of changing orbital parameters and surface boundary conditions on climate simulations for the last 18 000 years. *Journal of Atmosphere Science*, 39, 1177–88.

Kutzbach, J.E. and Street-Perrott, F.A. (1985) Milankovitch forcing of fluctuations in the level of tropical lakes from 18 to 0 kyr BP. *Nature*, 317, 130–4.

Laban, P. (1979) *Landslide Occurrence in Nepal*. Kathmandu: Integrated Watershed Management Project, Phewa Tal Report SP/13.

Labeyrie, J., Lalou, C. and Delibrias, G. (1969) Etude des transgressions marines sur l'atoll de Mururoa par la datation des differents niveaux de corail. *Cahier Pacifique*, 3, 59–68.

Laborel, J. (1986) Vermetid gastropods as sealevel indicators. In O. van de Plassche (ed.), *Sea-level Research*. Norwich: Geo Books, 281–310.

Lachenbruch, A.H. and Marshall, B.V. (1986) Changing climate: geothermal evidence from permafrost in the Alaskan Arctic. *Science*, 234, 689–96.

Lachenbruch, A.H., Cladhouhos, T.T. and Saltus, R.W. (1988) Permafrost temperature and changing climate. In K. Senneset (ed.), *Permafrost Proceedings, Vol. 3, Fifth International Conference, August 2–5 1988*, Trondheim, Norway: Tapir Publishers, 9–17.

Lachenbruch, A.H., Sass, J.J., Marshall, B.V. and Moses, T.H. (1982) Permafrost, heat flow, and the geothermal regime at Prudhoe Bay, Alaska. *Journal of Geophysical Research*, 87–B11, 9301–16.

Lai, F.S. (1991) Sediment and solute yields from logged steep upland catchments in Peninsula Malaysia. Unpublished PhD Thesis, School of Geography, University of Manchester.

Lal, R. (1990) Water erosion and conservation. In A.S. Goudie (ed.), *Techniques for Desert Reclamation*. Chichester: Wiley, 161–98.

Lal, R., Sanchez, P.A. and Cummings, R.W. (eds) (1986) *Land Clearing and Development in the Tropics*. Rotterdam: A.A. Balkema.

LaMarche, V.C. (1973) Holocene climatic variations inferred from tree line fluctuations in the White Mountains, California. *Quaternary Research*, 3, 632–60.

Lamb, D. (1990) *Exploiting the Tropical Rain Forest: an Account of Pulpwood Logging in Papua, New Guinea*. MAB, Vol. 3. Paris: UNESCO/Carnforth, Lancashire: Parthenon.

Lamb, H.H. (1966) Climate in the 1960s. *Geographical Journal*, 132,

183–212.

Lamb, H.H. (1977) *Climate: Past, Present and Future. Vol. II: Climate History and the Future*. London and New York: Methuen.

Lamb, H.H. (1980) Climatic fluctuations in historical times and their connexion with transgressions of the sea, storm floods and other coastal changes. In A. Verhulst and M.K.E. Gottschalk (eds), *Transgressies en occupatiegeschiedenis in de Kustgebieden van Nederland en Belgie*. Ghent: Belgisches Centrum voor Landelijke Geschiedenis, 66, 251–84.

Lamb, H.H. (1982) *Climate, History and the Modern World*. London: Methuen.

Lamb, H.H. (1984) Climate in the last thousand years: natural climatic fluctuations and change. In H. Flohn and R. Fantechi (eds), *The Climate of Europe Past, Present and Future*. Dordrecht: D. Reidel.

Landsberg, H.E. (1981) *The Urban Climate*. Orlando: Academic Press.

Langley, S.P. (1884) Researches on solar heat. *Professional Papers of the Signal Service*, no. 15.

Lanly, J-P. (1982) *Tropical Forest Resources*, FAO Forestry Paper 30. Rome: FAO.

Lanly, J-P. (1985) Defining and measuring shifting cultivation. *Unasylva*, 37, 17–21.

Lasker, H.R., Peters, E.C. and Coffroth, M.A. (1987) Bleaching of reef coelenterates in the San Blas Islands, Panama. *Coral Reefs*, 3, 183–90.

Lean, J. and Warrilow, D.A. (1989) Simulation of the regional impact of Amazon deforestation. *Nature*, 342, 411–13.

Legates, D.R. and Willmott, C.J. (1990) Mean seasonal and spatial variability in gauge-corrected, global precipitation. *International Journal of Climatology*, 10, 111–28.

Legg, T.P. (1989) Removal of urbanisation effects from the Central England temperature series, *Long-range Forecasting and Climate Research Memorandum, No. 33*. Bracknell: UK Meteorological Office.

Leggett, J. (ed.) (1990) *Global Warming: the Greenpeace Report*. Oxford: Oxford University Press.

Le Houerou, H.N. (ed.) (1980) *Les fourrageres ligneux en Afrique*. Addis Ababa: Centre pour l'élevage en Afrique.

Le Houerou, H.N. (1989) *The Grazing Land Ecosystems of the African Sahel*. Ecological Studies, 75. Berlin: Springer-Verlag.

Leivestad, H., Hendrey, G., Muniz, I.P. and Snekvik, E. (1976) Effects of acid precipitation on freshwater organisms. In F.H. Braekke (ed.), *Impact of Acid Precipitation on Forest and Freshwater Ecosystems in Norway*. Oslo: SNSF Research Report, 86–111.

Lelong, F., Roose, E., Aubert, G., Fauck, R. and Pedro, G. (1984) Géodynamique actuelle de differents sols à vegetation naturelle ou cultivés d'Afrique de l'ouest. *Catena*, 11, 343–76.

le Tacon, F. and Harley, J.L. (1990) Deforestation in the tropics and proposals to arrest it. *Ambio*, 19, 372–8.

Lézine, A-M. (1989) Late Quaternary vegetation and climate of the Sahel. *Quaternary Research*, 32, 317–34.

Lhote, H. (1959) *The Search for the Tassili Frescoes*. London, tr. A. Houghton Brodick.

Li, K.M. (1985) Problems on harnessing the Loess Plateau. *Natural Resource*, 2, 21–6 (in Chinese).

Lichty, R.G., Macintyre, I.G. and Stuckenrath, R. (1982) *Acropora palmata* reef framework: a reliable indicator of sea level in the western Atlantic for the past 10,000 years. *Coral Reefs*, 1, 125–30.

Likens, G.E. and Davis, M.B. (1975) Post-glacial history of Mirror Lake and its watershed in New Hampshire, U.S.A.: an initial report. *Int. Ver.*

Theor. Angew. Limnol. Verh., 19, 982–93.

Lin, Z.G. and Liang, W.M. (1982) Engineering properties and zoning of loess and loess-like soils in China. *Canadian Geotechnical Journal*, 19, 76–91.

Linell, K.A. (1973) Long-term effects of vegetative cover on permafrost stability in an area of discontinuous permafrost. In *North American Contribution, Permafrost Second International Conference, Yakutsk, USSR, 1973*, Washington, DC: National Academy of Sciences, 688–93.

Lisitzin, E. (1974) *Sea Level Changes*. Amsterdam: Elsevier.

Liu, T.S. (1964) *Loess on the Middle Reaches of the Yellow River*. Beijing: Science Press (in Chinese).

Liu, T.S. (1985) *Loess and the Environment*. Beijing: China Ocean Press.

Liu, T.S., An, Z.S., Yuan, B.Y. and Han, J.M. (1985) The loess-palaeosol sequence in China and climatic history. *Episodes*, 8, 21–8.

Liu, T.S. and Chang, T.H. (1964) The huangtu (loess) of China. *Report of the Sixth INQUA Congress*, 4, 503–24.

Loffler, E. (1977) *Geomorphology of Papua New Guinea*. Canberra: ANU Press.

Lohnes, R.A. and Handy, R.L. (1968) Slope angles in friable loess. *Journal of Geology*, 76, 247–58.

Long, D., Dawson, A.G. and Smith, D.E. (1989) A Holocene tsunami deposit in eastern Scotland. *Journal of Quaternary Science*, 4, 61–6.

Lorius, C., Jouzel, J., Raynaud, D., Hansen, J. and Le Treut, H. (1990) The ice-core record: climate sensitivity and future greenhouse warming. *Nature*, 347, 139–45.

Louwe Koojimans, L.P. (1974) *The Rhine/Meuse Delta: Four Studies on Its Prehistoric Occupation and Holocene Geology*. Leiden: E.J. Brill.

Love, G. (1988) Cyclone storm surges: post greenhouse. In G.I. Pearson (ed.), *Greenhouse: Planning for Climatic Change*. Melbourne: CSIRO, 202–215.

Lovejoy, T.E., Rylands, A.B., Malcolm, J.R., Quintela, C.E., Harper, L.H., Brown, K.S., Powell, A.H., Powell, G.V.N., Schubart, H.O.R. and Hays, M.B. (1986) Edge and other effects of isolation on Amazon forest fragments. In M.E. Soulé (ed.), *Conservation Biology: the Science of Scarcity and Diversity*. Sunderland, MA: Sinauer Associates, 257–85.

Lovelock, J.E. (1979) *Gaia, a New Look at Life on Earth*. Oxford: Oxford University Press.

Lovelock, J. (1988) *The Ages of Gaia*. Oxford: Oxford University Press.

Lowe-McConnell, R.H. (1986) The fish of the Amazon system. In B.R. Davies and K.F. Walker (eds), *The Ecology of River Systems*. Dordrecht: Dr W. Junk, 339–52.

Ludwig, G., Müller, H. and Streif, H. (1981) New dates on Holocene sea-level changes in the German Bight. In S.D. Nio, R.T.E. Schuttenhelm and Tj. C.E. van Weering (eds), *Holocene Marine Sedimentation in the North Sea*. Special publication of the International Association of Sedimentologists, 5. Oxford: Blackwell Scientific, 211–19.

Ludwig, J.A. (1988) *Shrubkill User's Guide*. Deniliquin, New South Wales: CSIRO, Division of Wildlife and Ecology.

Ludwig, J.A. (1990) Shrubkill: a decision support system for management burns in Australian Savannas. *Journal of Biogeography*, 17, 547–50.

Lugo, A.E. (1988) Estimating reductions in the diversity of tropical forest species. In E.O. Wilson (ed.), *Biodiversity*. Washington, DC: National Academy Press, 6, 58–70.

Lugo, A.E., Ewel, J.J., Hecht, S.B., Murphy, P.C., Padoch, C., Schmink, M.C. and Stone, D. (1987) *People and the Tropical Forest*. A Research Report from MAB Program. Superintendent of Documents, US Government Printing Service, Washington.

Luning, H.A. (1987) The need for tropical rain forests and their products. In C.F. van Beusekom, C.P. van Goor and P. Schmidt (eds), *Wise Utilisation of Tropical Rain Forests*. Ede, Netherlands: Tropenbos, 15–29.

Luthin, J.N. and Guymon, G.L. (1974) Soil moisture-vegetation-temperature relationships in central Alaska. *Journal of Hydrology*, 23, 233–46.

Mabbutt, J.A. (1985) Desertification in the world's rangelands. *Desertification Control Bulletin*, 12, 1–11.

MacArthur, R.H. and Wilson, E.O. (1967) *The theory of island biogeography*. Princeton, NJ: Princeton University Press.

McBeath, J.H. (ed.) (1984) *The Potential Effects of Carbon Dioxide-induced Climate Changes in Alaska*. Proceedings of a Conference, University of Alaska-Fairbanks, Miscellaneous Publications, 83–1.

MacCracken, M.C. and Ghan, S.J. (1988) Design and use of zonally-averaged climate models. In M.E. Schlesinger (ed.), *Physically-Based Modelling and Simulation of Climate and Climatic Change*. Boston: Kluwer, 755–812.

MacDicken, K.G. and Vergara, N.T. (eds) (1990) *Agroforestry: Classification and Management*. New York: Wiley.

McGregor, D.F.M. (1980) An investigation of soil erosion in the Colombian rainforest zone. *Catena*, 7, 265–73.

Macintyre, I., Burke, R.B. and Stuckenrath, R. (1979) Thickest recorded Holocene section, Isla Perez core hole, Alacran Reef, Mexico. *Geology*, 5, 749–54.

Macintyre, I., Burke, R.B. and Stuckenrath, R. (1981) Core holes in the outer fore reef off Carrie Bow Cay, Belize: A key to the Holocene history of the Belizean barrier reef complex. *Proceedings, 4th International Coral Reef Symposium, Manila*, 1, 567–74.

Mackay, J.R. (1975) The stability of permafrost and recent climatic change in the Mackenzie Valley, NWT. *Geological Survey of Canada Paper*, 75-1B, 173–6.

McKeon, G.M., Day, K.A., Howden, S.M., Mott, J.J., Orr, D.M., Scattini, W.J. and Weston, E.J. (1990) Northern Australian savannas: management for pastoral production. *Journal of Biogeography*, 17, 355–72.

Mackinnon, J. and K., Child, G. and Thorsell, J. (1986) *Managing Protected Areas in the Tropics*. Gland, Switzerland: International Union for the Conservation of Nature (IUCN).

McNeely, J.H. (1988) *Economics and Biological Diversity*. Gland, Switzerland: IUCN.

Mackereth, F.J.H. (1966) Some chemical observations on post-glacial lake sediments. *Philosophical Transactions of the Royal Society, London, B*, 250, 165–213.

Maddox, J. (1990) Clouds and global warming. *Nature*, 247, 329.

Mahaffy, M.W. (1976) A three-dimensional numerical model of ice sheets: test on the Barnes Ice Cap, Northwest Territories. *Journal of Geophysical Research*, 81, 1059–66.

Mahar, D.J. (1989) *Government Policies and Deforestation in Brazil's Amazon Region*. Washington, DC: World Bank.

Maitland, P.S., Lyle, A.A. and Campbell, R.N.B. (1988) *Acidification and Fish in Scottish Lochs*. Peterborough: Institute of Terrestrial Ecology, Natural Environment Research Council.

Maley, J. (1977) Paleoclimates of the Central Sahara during the early Holocene. *Nature*, 269, 573–77.

Malingreau, J.P., Stephens, G. and Fellows, L. (1985) Remote sensing and forest fires: Kalimantan and Borneo. *Ambio*, 14, 314–21.

Malingreau, J.P., Tucker, C.J. (1988) Large scale deforestation in the south eastern Amazon Basin. *Ambio*, 17, 49–55.

Malingreau, J.P., Tucker, C.J. and Laporte, N. (1989) AVHRR for monitoring global tropical deforestation. *International Journal of Remote Sensing*, 10, 855–68.

Malla, K.B. (1985) *Watershed Degradation in Nepal*. Kathmandu: National Remote Sensing Centre.

Malo, A.R. and Nicholson, E. (1990) A study of rainfall and vegetation dynamics in the African Sahel using normalized difference vegetation index. *Journal of Arid Environments*, 19, 1–21.

Manabe, S. (1971) Estimates of future changes of climate due to increase of CO_2 in the air. In W.H. Matthews et al. (eds), *Man's Impact on Climate*. Cambridge, MA: MIT Press, 249–64.

Manabe, S. and Broccoli, A.J. (1985) The influence of continental ice sheets on the climate of an ice age. *Journal of Geophysical Research*, 90, 2167–90.

Manabe, S. and Bryan, K. (1985) CO_2-induced change in a coupled ocean-atmosphere model and its paleoclimatic implications. *Journal of Geophysical Research*, 90, 11689–707.

Manabe, S. and Hahn, D.G. (1977) Simulation of the tropical climate of an ice age. *Journal of Geophysical Research*, 82, 3889–911.

Manabe, S. and Wetherald, R.T. (1967) Thermal equilibrium of the atmosphere with a given distribution of relative humidity. *Journal of Atmospheric Science*, 21, 361–85.

Manabe, S. and Wetherald, R.T. (1975) The effects of doubling the CO_2 concentration on the climate of a GCM. *Journal of Atmospheric Science*, 32, 3–15.

Manley, G. (1953) The mean temperature of Central England, 1698 to 1952. *Quarterly Journal of the Royal Meteorological Society*, 79, 242–61.

Manley, G. (1974) Central England temperatures: monthly means 1659 to 1973. *Quarterly Journal of the Royal Meteorological Society*, 100, 389–405.

Maragos, J.E. (1978) Coral growth: geometrical relationships. In D.R. Stoddart and R.E. Johannes (eds), *Coral Reefs: Research Methods*. Paris: UNESCO, 469–84.

Maragos, J.E., Evans, C. and Holthus, P. (1985) Reef corals in Kaneohe Bay six years before and after termination of sewage discharges (Oahu, Hawaiian Archipelago). *Proceedings, 5th International Coral Reef Congress, Tahiti*, 4, 189–94.

Marchant, D. and Denton, G.H. (1990) The climatic significance and isotopic age of an early Pliocene polar desert pavement, Dry Valleys, Antarctica. *Antarctic Journal US*.

Marsh, G.P. (1864) *Man and Nature*. New York: Scribner.

Marsh, J.G. and Martin, T.V. (1982) The SEASAT altimeter mean surface model. *Journal of Geophysical Research*, 87, 3269–80.

Marshall, J.F. and Davies, P.J. (1982) Internal structure and Holocene evolution of One Tree Reef, Southern Great Barrier Reef. *Coral Reefs*, 1, 21–8.

Marshall, J.F. and Jacobsen, G. (1985) Holocene growth of a mid-Pacific atoll: Tarawa, Kiribati. *Coral Reefs*, 4, 11–17.

Marten, G.G. (1990) Small scale agriculture in South East Asia. In M.A. Altieri and S.B. Hecht (eds), *Agroecology and Small Farm Development*. Boca Raton, FL: CRC Press, 183–202.

Martin, J.H. and Fitzwater, S.E. (1988) Iron deficiency limits phytoplankton growth in the north-east Pacific subarctic. *Nature*, 331, 341–3.

Mather, P.M. (1987) *Computer Processing of Remotely-Sensed Images*. Chichester: Wiley.

Mather, P.M. (1992) Remote sensing and the detection of change. In M.C. Whitby (ed.), *Land Use Change: the Causes and Consequences*. NERC Institute of Terrestrial Ecology Symposium no. 27. London: HMSO, 60–8.

Mathur, H.N., Ram Babu, Joshie, P. and Singh, B. (1976) Effect of clear felling and reforestation on runoff and peak flow in small watersheds. *Indian Forester*, 102, 219–26.

Mathur, H.N. and Sajwan, S.S. (1978) Vegetation characteristics and their effect on runoff and peak flow in small watersheds. *Indian Forester*, 104, 398–406.

Matthes, F.E. (1939) Report of the Committee on Glaciers, April 1939. *Transactions of the American Geophysical Union*, 20, 518–23.

Maude, A. (1970) Shifting cultivation and population growth in Tonga. *Journal of Tropical Geography*, 31, 57–64.

Megahan, W.F. (1988) Effects of forest roads on watershed function in mountainous areas. In A.S. Balasubramanian, S. Chandra, D.T. Bergado and Prinya Nutalaya (eds), *Environmental Geotechnics and Problematic Soils and Rocks*. Rotterdam: A.A. Balkema, 335–48.

Meier, M.F. (1984) Contribution of small glaciers to global sea level. *Science*, 226, 1418–21.

Meier, M.F. (1990) Reduced rise in sea level. *Nature*, 343, 115–16.

Meijerink, A.M.J. (1974) *Photo-hydrological Reconnaissance Surveys*. Amstelveen: Vrije Universiteit te Amsterdam, Academische Proefschrift.

Mensching, H. (1981) Is the desert spreading? *Applied Geography and Development*, 27, 7–18.

Mercer, J.H. (1978) West Antarctic ice sheet and CO_2 greenhouse effect: a threat of disaster. *Nature*, 271, 321–5.

Mermel, T.W. (1990) The world's major dams and power plants. *Water Power and Dam Construction*, May, 52–60.

Merrick, L.C. (1990) Crop genetic diversity and its conservation in traditional ecosystems. In M.A. Altieri and S.B. Hecht (eds), *Agroecology and Small Farm Development*. Boca Raton, FL: CRC Press, 3–12.

Meybeck, M. and Helmer, R. (1989) The quality of rivers: from pristine stage to global pollution. *Palaeogeography, Palaeoclimatology, Palaeoecology (Global and Planetary Change)*, 75, 283–309.

Micklin, P.P. (1986) The Status of the Soviet Union's North-South Water transfer projects before their abandonment in 1985–86. *Soviet Geography*, 27, 287–329.

Micklin, P.P. (1988) Desiccation of the Aral Sea. *Science*, 241, 1170–6.

Middleton, A. (1990) Wind erosion and dust storm control. In A.S. Goudie (ed.), *Techniques for Desert Reclamation*. Chichester: Wiley, 87–108.

Middleton, N.J. (1985) Effect of drought on dust production in the Sahel. *Nature*, 316, 431–4.

Miehe, S. (1988) Vegetation ecology of the Jebel Marra massif in the semi-arid Sudan. *Diss. Bot.*, 113.

Mikolajewicz, U., Santer, B.D. and Maier-Reimer, E. (1990) Ocean response to greenhouse warming. *Nature*, 345, 589–93.

Milankovitch, M. (1930) Mathematische Klimalehre und astronomische Theorie der Klimaschwankungen. In W. Köppen and R. Geiger (eds), *Handbuch der Klimatologie*. 1(A). Berlin.

Miller, R.R. (1958) Origin and affinities of the freshwater fish fauna of western North America. *Zoogeography*, 51, 187–222.

Milliman, J.D., Broadus, J.M. and Gable, F. (1989) Environmental and economic implications of rising sea level and subsiding deltas: the Nile and Bengal examples. *Ambio*, 18, 340–5.

Milner, N.J. and Varallo, P.V. (1990) Effects of acidification on fish and fisheries in Wales. In R.W. Edwards, A.S. Gee and J.H. Stoner (eds), *Acid Waters in Wales*. Dordrecht: Kluwer, 121–44.

Minckley, W.L. and Deacon, J.E. (1968) Southwestern fishes and the enigma of "endangered species". *Science*, 159, 1424–32.

Mishra, A. and Tripathi, S. (1978) *Chipko Movement: Uttarakhand Women's Bid to Save Forest Wealth*. New Delhi: Gandhi Peace Foundation.

Misra, P.R., Sud, A.D., Madan Lal, and Singh, K. (1979) *Central Soil and Water Conservation*. Research and Training Institute, Dehra Dun, Annual Report 99.

Mitchell, J.F.B. (1989) The "greenhouse" effect and climate change. *Review of Geophysics*, 27, 115–39.

Mitchell, J.F.B. (1990) Greenhouse warming: is the Holocene a good analogue? *Journal of Climate*, 3, 1177–92.

Mitchell, J.F.B., Manabe, S., Meleshko, V. and Tolioka, T. (1990) Equilibrium climate change – and its implications for the future. In J.T. Houghton, G.J. Jenkins and J.J. Ephraums (eds), *Climate Change. The IPCC Scientific Assessment*. Cambridge: Cambridge University Press, 131–64.

Mitchell, J.F.B., Senior, C.A. and Ingram, W.J. (1989) CO_2 and climate: a missing feedback. *Nature*, 341, 132–4.

Mitchell, J.F.B. and Zeng Qingcun (1991) Climate change prediction. In J. Jäger and H.I. Ferguson (eds), *Climate Change: Science, Impacts and Policy*, Proceedings of the Second World Climate Conference. Cambridge: Cambridge University Press, 59–70.

Mix, A.C. (1989) Influence of productivity variations on long-term atmospheric CO_2. *Nature*, 337, 541–4.

Moddie, A.D. (1985) Development with desertification: U.P. Hills Bhimtal's water system: facts, policies, lacuna, and responses, etc. *Central Himalayan Environment Association, Bulletin*, 2, 1–36.

Moffatt, I. (1991) Possible ecological impacts of the greenhouse effect on the Northern Territory, Australia. *Global Ecology and Biogeography Letters*, 1, 102–7.

Mohan, S.C., and Gupta, R.K. (1983) Infiltration rates in various land uses from a Himalayan watershed in Tehri Garhwal. *Indian Journal of Soil Conservation*, 11 (2/3), 1–4.

Möller, F. (1963) On the influence of changes in the CO_2 concentration in air on the radiation balance of the earth's surface and on the climate. *Journal of Geophysical Research*, 68, 3877–86.

Monod, Th. (1954) Modes "contracte" et "diffus" de végétation saharienne. In J.L. Cloudsley-Thompson (ed.), *Biology of Deserts*. London, 35–44.

Monod, Th. (1986) The Sahel zone north of the equator. In M. Evenari and J. Noy Meir (eds), *Hot Desert and Arid Shrubland*. Oxford, 203–43.

Montaggioni, L.F. (1988) Holocene reef growth history in mid-plate high volcanic islands. *Proceedings, 5th International Coral Reef Congress, Tahiti*, 3, 455–60.

Moran, E. (1990) Private and public colonisation schemes in the Amazon. In D. Goodman and A. Hall (eds), *The Future of Amazonia*. London: Macmillan, 70–89.

Mörner, N.A. (1969) The late Quaternary history of the Kattegatt Sea and the Swedish West Coast. *Sveriges Geologiska Undersokning*, C. 640, 1–487.

Mörner, N.A. (1980) The Fennoscandian Uplift: geological data and their geodynamical implication. In N.A. Mörner (ed.), *Earth Rheology, Isostasy and Eustasy*. Chichester: John Wiley, 251–84.

Mörner, N.A. (1987) Models of global sealevel changes. In M.J. Tooley and I. Shennan (eds), *Sea-level Changes*. Oxford: Basil Blackwell, 332–55.

Morrison, B.R.S. and Battarbee, R.W. (1988) Effects on freshwater flora and fauna (excluding fish). In M.H. Ashmore, N.J.B. Bell and C. Garretty (eds), *Acid Rain and Britain's Natural Ecosystems*. London: Centre for Environmental Technology, Imperial College, 71–86.

Mortimore, M. (1987) Shifting sands and human sorrow: social response to drought and desertification. *Desertification Control Bulletin*, 14, 1–14.

Mortimore, M. (1989) *Adapting to Drought. Farmers, Famines and Desertification in West Africa*. Cambridge: Cambridge University Press.

Mortlock, R.A., Charles, C.D., Froelich, P.N., Zibello, M.A., Saltzman, J., Hays, J.D. and Burckle, L.H. (1991) Evidence for lower productivity in the Antarctic Ocean during the last glaciation. *Nature*, 351, 220–3.

Mott, J.J., Williams, J., Andrew, M.H. and Gillison, A.N. (1985) Australian savanna ecosystems. In J.C. Tothill and J.J. Mott (eds), *Ecology and Management of the World's Savannas*. Canberra: Australian Academy of Science, 56–82.

Muehe, D. (1979) Sedimentology and topography of a high energy coastal environment between Rio de Janeiro and Cabo Frio Brazil. *Anais de Academia brasileira de ciencias*, 51, 473–81.

Muniz, I.P. (1991) Freshwater acidification: its effects on species and communities of freshwater microbes, plants and animals. *Proceedings of the Royal Society of Edinburgh*, 97B, 227–54.

Muzik, K. (1985) Dying coral reefs of the Ryukyu archipelago. *Proceedings, 5th International Coral Reef Congress, Tahiti*, 6, 483–9.

Myers, N. (1979) *The Sinking Ark*. Oxford: Pergamon.

Myers, N. (1980) *Conversion of Tropical Moist Forests*. Report to National Academy Science, Washington, DC: National Research Council.

Myers, N. (1984) The Himalayas: an influence on 500 million people. *Earthwatch*, 3 (7), 3–6.

Myers, N. (1985) *The Primary Sources*. London: W.W. Norton.

Myers, N. (1986) Environmental repercussions of deforestation in the Himalayas. *Journal of World Forestry Resource Management*, 2, 63–72.

Myers, N. (1988a) Tropical deforestation and mega-extinction spasms. In M.E. Soulé (ed.), *Conservation Biology: The Science of Scarcity and Diversity*. Sunderland, MA: Sinauer Associates, 394–409.

Myers, N. (1988b) Tropical deforestation and climatic change. *Environmental Conservation*, 15, 293–8.

Myers, N. (1989) *Deforestation Rates in Tropical Forests and Their Climatic Implications*. London: Friends of the Earth.

Myers, N. (1990) Tropical forests. In J. Leggett (ed.), *Global Warming: the Greenpeace Report*. Oxford: Oxford University Press, 372–99.

Myers, N. and Tucker, R. (1987) Deforestation in Central America: Spanish legacy and North American consumers. *Environment Review*, 11, 55–71.

Nachtigal, G. (1879) *Sahara und Sudan*, 2 vols. Berlin.

Nair, P.K.R. (1990) *The Prospects for Agroforestry in the Tropics*. Washington, DC: World Bank Technical Paper 131.

Nakada, M. and Lambeck, K. (1987) Glacial rebound and relative sea-level variations: a new appraisal. *Geophysical Journal of the Royal*

Astronomical Society, 90, 171–224.

Napier, T.L. (1990) The evolution of US soil-conservation policy: from voluntary adoption to coercion. In J. Boardman, I.D.L. Foster and J.A. Dearing (eds), *Soil Erosion on Agricultural Land*. Chichester: Wiley, 627–44.

NASA (1988) *Earth System Science*. Report of the Earth System Sciences Committee, NASA Advisory Council, Washington, DC.

NASA (undated) *From Pattern to Process: The Strategy of Earth Observing System*. Science Steering Committee Report, Washington, DC.

National Research Council (U.S.) (1987) *Responding to Changes in Sea Level: Engineering Implications*. Washington, DC: National Academy Press.

Nautiyal, J.C. and Babor, P.S. (1985) Forestry in the Himalayas – how to avert an environmental disaster. *Interdisciplinary Science Reviews*, 10, 27–41.

Neftel, A., Moor, E., Oeschger, H. and Stauffer, B. (1985) Evidence from polar ice cores for the increase in atmospheric CO_2 in the past two centuries. *Nature*, 315, 45–7.

Negi, S.S. (1983) Need for a conservation based rural energy strategy in the Himalayas. *Journal of Rural Development*, 2, 134–43.

Nekrasov, I.A. (1984) Dynamics of the cryolithozone in the Northern Hemisphere during the Pleistocene. In *Permafrost Fourth International Conference Proceedings, Fairbanks-Alaska*. Washington, DC: National Academy Press, 903–6.

Nelson, R. and Holben, B. (1986) Identifying deforestation in Brazil using multiresolution satellite data. *International Journal of Remote Sensing*, 7, 429–48.

Neumann, A.C. and Macintyre, I. (1985) Reef response to sea level rise: keep-up, catch-up or give-up. *Proceedings, 5th International Coral Reef Congress, Tahiti*, 3, 105–10.

Newson, M. (1992) Twenty years of systematic physical geography: issues for a "New Environmental Age." *Progress in Physical Geography*, 16, 209–21.

Ngeh, P., Furley, P.A. and Grace, J. (submitted) Effects of land clearing methods on soil physical and chemical properties in a lowland tropical rain forest at Mbalmayo, Cameroon.

Nicholaides, J.J., Bandy, D.E., Sanchez, P.A. and Villachica, J.H. (1984) From migrating to continuous agriculture in the Amazonian Basin. In *Improved Productions System as an Alternative to Shifting Agriculture*. Rome: FAO Soils Bulletin 53, 141–68.

Nicolaisen, J. (1963) *Ecology and Culture of the Pastoral Tuareg*. Copenhagen: National Museum.

Nicholas, F.J. and Glasspole, J. (1931) *British Rainfall (1931)*. UK Meteorological Office.

Nichols, H. (1975) Palynological and paleoclimatic study of the Late Quaternary displacement of the Boreal forest-tundra ecotone in Keewatin and Mackenzie, N.W.T., Canada. Institute of Arctic and Alpine Research, University of Colorado, Occasional Paper, 15.

Nicholson, F.H. (1978) Permafrost modification by changing the natural energy budget. In *Proceedings Vol. 1 Third International Permafrost Conference, Edmonton, Canada*. Canada: National Research Council, 61–7.

Nicholson, F.H. and Granberg, H.B. (1973) Permafrost and snow cover relationships near Schefferville. In *Proceedings Second International Permafrost Conference, Yakutsk, North American Contribution*. Washington, DC: National Academy of Sciences, 151–8.

Nicholson, S.E. (1978) Climatic variations in the Sahel and other African regions during the past five centuries. *Journal of Arid Environments*, 1, 3–24.

Nieuwenhuijzen, M.E. and Koster, E.A. (1989) Assessing the effects of climatic change on permafrost – a short literature survey. In E.A. Koster and M.M. Boer (eds), *Landscape Ecological Impact of Climatic Change on Boreal/(Sub)Arctic Regions, with Emphasis on Fennoscandia*. Luntern, The Netherlands: Discussion Report LICC Conference, 3–7 December, 79–93.

Nilsson, J. and Grennfelt, P. (eds) (1988) *Critical Loads for Sulphur and Nitrogen*. Report from a workshop held at Skokloster, Sweden 19–24 March 1988. UN/ECE and Nordic Council of Ministers, 1988. Nord 1988, 15.

Nisbet, E.G. (1989) Some northern sources of atmospheric methane: production, history and future implications. *Canadian Journal of Earth Sciences*, 26, 1603–11.

NOAA (1990) *Climate Assessment for 1989: Selected Indicators of Global Climate*. Asheville: CAC/NCDC.

Nortcliff, S. and Dias, A.C. (1988) The change in soil physical conditions resulting from forest clearances in the humid tropics. *Journal of Biogeography*, 15, 61–6.

Nossin, J.J. (1971) Outline of the geomorphology of the Doon Valley, northern U.P., India. *Zeitschrift für Geomorphologie N.F*, 12, 18–50.

Nullet, D. (1989) Recent climate history of Hawaii. *Pacific Science*, 43, 96–101.

Nunn, P.D. (1990a) Warming of the South Pacific region since 1880; evidence, causes and implications. *Journal of Pacific Studies*, 15, 35–50.

Nunn, P.D. (1990b) Recent environmental changes on Pacific islands. *Geographical Journal*, 156, 127–40.

Nye, P.H. and Greenland, D.J. (1960) *The Soil under Shifting Cultivation*. London: Commonwealth Bureau of Soils Technical Communication 51.

Obeid, M. and Seif ed Din (1970) Ecological studies on the vegetation of the Sudan. I. *Acacia senegal* (L). Willd. and its natural regeneration. *Journal of Applied Ecology*, 7, 507–18.

Odén, S. (1968) *The Acidification of Air Precipitation and Its Consequences in the Natural Environment*. Stockholm: Energy Committee Bulletin, 1. Swedish Natural Sciences Research Council.

Oechel, W.C., Riechers, G.H. and Grulke, N. (1987) *The Effect of Elevated Atmospheric CO_2 and Temperature on an Arctic Ecosystem*. Washington, DC: US Department of Energy, Carbon Dioxide Research Division.

Oerlemans, J. (1988) Simulation of historic glacier variations with a simple climate-glacier model. *Journal of Glaciology*, 34, 333–41.

Oerlemans, J. (1989) A projection of future sea level. *Climatic Change*, 15, 151–74.

Oerlemans, J. (1991) The mass balance of the Greenland ice sheet: sensitivity to climate change as revealed by energy balance modelling. *The Holocene*, 1, 40–9.

Oke, T.R. (1978) *Boundary Layer Climates*. London: Methuen.

O'Keefe, J.H. and De Moor, F.C. (1988) Changes in the physico-chemistry and benthic invertebrates of the Great Fish River, South Africa, following an inter-basin transfer of water. *Regulated Rivers*, 2, 39–55.

Oldfield, F. (1977) Lakes and their drainage basins as units of sediment-based ecological study. *Progress in Physical Geography*, 1, 460–504.

Oldfield, F. and Maher, B. (1984) A mineral magnetic approach to erosion studies. In Report of Conference on Drainage Basin Erosion and

Sedimentation, Newcastle, NSW, Australia.

Oldfield, F., Worsley, A.T. and Appleby, P.G. (1985) Evidence from lake sediments for recent erosion rates in the highlands of Papua New Guinea. In I. Douglas and T. Spencer (eds), *Environmental Change and Tropical Geomorphology*. London: Allen and Unwin, 185–96.

Oldfield, M.L. and Alcorn, J.B. (eds) (in press) *Biological Resources under Traditional Management*. Boulder, CO: Westview Press.

Ormerod, S.J., Allinson, N., Hudson, D. and Tyler, S.J. (1986) The distribution of breeding dippers (*Cinclus cinclus* (L.) Aves) in relation to stream acidity in upland Wales. *Freshwater Biology*, 16, 501–7.

Orth, R.J. and Moore, K.A. (1984) Distribution and abundance of submerged aquatic vegetation in Chesapeake Bay: an historical perspective. *Estuaries*, 7, 531–40.

Osterkamp, T.E. (1984a) Response of Alaskan permafrost to climate. In *Fourth International Conference on Permafrost, Fairbanks-Alaska, 1983. Final Proceedings*. Washington, DC: National Academy Press, 145–52.

Osterkamp, T.E. (1984b) Potential Impact of a warmer climate on permafrost in Alaska. In J.H. McBeath (ed.), *The Potential Effects of Carbon Dioxide-induced Climate Changes in Alaska*. Proceedings of a conference, University of Alaska-Fairbanks, Miscellaneous Publications, 83–1, 106–13.

Owens, P.N. (1990) Valley sedimentation at Slapton, South Devon, and its implications for the estimation of lake sediment-based erosion rates. In J. Boardman, I.D.L. Foster and J.A. Dearing (eds), *Soil Erosion on Agricultural Land*. Chichester: Wiley, 193–200.

Pandey, A.N., Pathak, P.C. and Singh, J.S. (1983) Water sediment and nutrient movement in forested and non-forested catchments in Kumaun Himalayas. *Forest Ecology and Management*, 71, 19–29.

Pant, G.B. (1922) *Forest Problem in Kumaun*. Nainital: Gyanodaya Prakashan (1986 edition).

Pant, G.B. and Hingane, L.S. (1988) Climatic change in and around the Rajasthan Desert during the twentieth century. *Journal of Climatology*, 8, 391–401.

Parry, M. (1990) *Climate Change and World Agriculture*, London: Earthscan.

Passmore, J. (1974) *Man's Responsibility for Nature*. Duckworth: London.

Pathak, P.C., Pandey, A.N. and Singh, J.S. (1984) Overland flow, sediment output and nutrient loss from certain forest sites in central Himalaya, India. *Journal of Hydrology*, 71, 239–51.

Pathak, P.C., Pandey, A.N. and Singh, J.S. (1985) Apportionment of rainfall in Central Himalayan forests, India. *Journal of Hydrology*, 76, 319–51.

Pathak, P.C. and Singh, J.S. (1984) Nutrients in precipitation components for pine and oak forests in Kumaun Himalaya. *Tellus*, 36B, 44–9.

Patnaik, N. (1981) Role of soil conservation and afforestation for flood moderation. *International Conference on Flood Disasters, Preprints*. New Delhi: Indian National Science Academy, 1–17.

Patrick, S.T., Waters, D., Juggins, S. and Jenkins, A. (eds) (1991) *The United Kingdom Acid Waters Monitoring Network: Site Descriptions and Methodology Report*. London: Ensis Ltd.

Pauli, H.W. (1978) Soil conservation in Queensland – problems and progress. *Journal of Soil Conservation Service of NSW*, 34, 106–16.

Payne, A.J., Sugden, D.E. and Clapperton, C.M. (1989) Modeling the growth and decay of the Antarctic Peninsula ice sheet. *Quaternary Research*, 31, 119–34.

Pearce, A.J. (1986) *Erosion and Sedimentation*. Hawaii: East-West Center, Environment and Policy Institute Working Paper.

Pearce, A.J. and Mckerchar, A.I. (1979) Upstream generation of storm runoff. In D.L. Murray and P.J. Ackroyd (eds), *Physical Hydrology – the New Zealand Experience*. Wellington: NZ Hydrological Society, 165–92.

Pearce, D., Barbier, E. and Markandya, A. (1990) *Sustainable Development*. Aldershot: E. Elgar.

Pearman, G.I. and Fraser, P.J. (1988) Sources of increased methane. *Nature*, 332, 489–90.

Peck, A.J. (1978) Salinisation of non-irrigated soils and associated streams: a review. *Australian Journal of Soil Research*, 16, 157–68.

Peel, D.A. and Mulvaney, R. (1988) Air temperature and snow accumulation in the Antarctic Peninsula during the past 50 years. *Annals of Glaciology*, 11, 206–7.

Peltier, W.R. and Tushingham, A.M. (1989) Global sea level rise and the greenhouse effect: might they be connected? *Science*, 244, 806–10.

Penck, A. and Brückner, E. (1909) *Die Alpen im Eiszeitalter*. Leipzig.

Pennington, W. (Mrs. T.G. Tutin) (1981) Records of a lake's life in time: the sediments. *Hydrobiologia*, 79, 197–219.

Pennington, W., Haworth, E.Y., Bonny, A.P. and Lishman, J.P. (1972) Lake sediments in northern Scotland. *Philosophical Transactions of the Royal Society, London, B*, 264, 191–294.

Pereira, H.C. (1989) *Policy and Practice in the Management of Tropical Watersheds*. London and Boulder, CO: Belhaven/Westview Press.

Persson, R. (1974) *World Forest Resources. Review of the World's Forest Resources in the Early 1970s*. Stockholm: Royal College of Forestry.

Peters, C.M., Gentry, A.H. and Mendelsohn, R.O. (1989) Valuation of an Amazonian rainforest. *Nature*, 339, 655–6.

Peters, R.L. (1991) Consequences of global warming for biological diversity. In R.L. Wyman (ed.), *Global Climate Change and Life on Earth*. London and New York: Routledge, 99–118.

Petersen, K.S. (1981) The Holocene marine transgression and its molluscan fauna in the Skagerrak Limfjord region, Denmark. In S.D. Nio, R.T.E. Schuttenhelm and Tj. C.E. van Weering (eds), *Holocene Marine Sedimentation in the North Sea*. Special publications of the International Association of Sedimentologists, 5. Oxford: Blackwell Scientific, 497–503.

Peterson, K.M., Billings, W.D. and Reynolds, D.R. (1984) Influence of water table and atmospheric CO_2 concentration on the carbon balance of arctic tundra. *Arctic and Alpine Research*, 16, 331–5.

Petit-Maire, N., Fontugne, M. and Rouland, C. (1991) Atmospheric methane ratio and environmental changes in the Sahara and Sahel during the last 130 Kyrs. *Palaeogeography, Palaeoclimatology, Palaeoecology*, 86, 197–204.

Petit-Maire, N. and Riser, J. (eds) (1983) *Sahara ou Sahel? Quaternaire récent du bassin de Taoudenni (Mali)*. Paris: Librairie du Muséum.

Petit-Maire, N. et al. (1988) *Le Sahara à l'Holocène: Mali 1:1,000,000*. Marseille: CNRS Lab. de Géologie du Quaternaire.

Petrere M. (1989) River fisheries in Brazil: a review. *Regulated Rivers*, 4, 1–16.

Petts, G.E. (1990) Regulation of Large Rivers: problems and possibilities for environmentally-sound river development in South America. *Interciencia*, 15, 388–95.

Petts, G.E., Moller, H. and Roux, A.L. (eds) (1989a) *Historical Changes of Large Alluvial Rivers: Western Europe*. Chichester: Wiley.

Petts, G.E., Imhof, J.G., Manny, B.A., Maher, J.F.B. and Weisberg, S.B. (1989b) Management of fish populations in large rivers: a review of tools and aproaches. *Canadian Special Publication of Fisheries and Aquatic Sciences*, 106, 578–88.

Péwé, T.L. (1982) Geologic hazards of the Fairbanks area, Alaska. *Division of Geological and Geophysical Surveys Special Report*, 15, Collete, Alaska.

Péwé, T.L. (1983) Alpine permafrost in the contiguous United States: a review. *Arctic and Alpine Research*, 15, 145–56.

Pfister, C. (1990) Meteorological anomalies in the last 700 years amplitude and frequency. In W. Jakobi (Chairman) *Changing Weather Patterns*. Cologne: GerlingKonzern Globale, 21–33.

Pickup, G. (1984) Geomorphology of tropical rivers. Landforms, hydrology and sedimentation in the Fly and Lower Purori, Papua New Guinea. *Catena Supplement*, 5, 1–17.

Pillsbury, A.F. (1981) The salinity of rivers. *Scientific American*, 15, 54–65.

Pimental, D., Terhune, E.C., Dyson-Hudson, R., Rochereau, S., Samis, R., Smith, E., Denman, D., Reifschneider, D. and Shephard, M. (1976) Land degradation: effects on food and energy resources. *Science*, 194, 149–55.

Pirazzoli, P.A. (1989) Present and near-future global sea-level changes. *Palaeogeography, Palaeoclimatology, Palaeoecology*, 75, 241–58.

Pirazzoli, P.A. and Montaggioni, L.F. (1986) Late Holocene sea-level changes in the Northwest Tuamotu Islands, French Polynesia. *Quaternary Research*, 25, 350–68.

Pissart, A. (1990) Advances in periglacial geomorphology. *Zeitschrift für Geomorphologie*, Supplement Band 79, 119–31.

Pollard, W.H. and French, H.M. (1980) A first approximation of the volume of ground ice, Richards Islands, Pleistocene Mackenzie Delta, Northwest Territories, Canada. *Canadian Geotechnical Journal*, 17, 509–16.

Poore, D. (ed.) (1989) *No Timber without Trees*. London: Earthscan.

Popper, W. (1951) *The Cairo Nilometer*. Berkeley: University of California Press.

Posey, D.A. and Balee, W. (eds) (1989) Resource management in Amazonia: indigenous and folk strategies. *Advances in Economic Botany*, 7, 1–240.

Post, W.M., Emanuel, W.R., Zinke, P.J. and Stangenberger, A.G. (1982) Soil Carbon Pools and world life zones. *Nature*, 298, 156–9.

Potter, G.L., Ellsaesser, H.W., MacCracken, M.C. and Luther, F.M. (1979) Performance of the Lawrence Livermore Laboratory zonal atmospheric model. In *Proceedings GARP JOC Study Conference on Climate Models, GARP Publication Series No. 22 (Vol. 2)*. Geneva: WMO, 995–1001.

Potter, L. (1987) Degradation, innovation and social welfare in the Riam Kiwi Valley, Kalimantan, Indonesia. In P. Blaikie and H. Brookfield (eds), *Land Degradation and Society*. London: Longman, 164–76.

Prell, W.L. (1982) Oxygen and carbon isotope stratigraphy for the Quaternary of Hole 502B: evidence for two modes of isotopic variability. In W.L. Prell et al., *Initial Reports DSDP 68*. Washington, DC: US Government Printing Office, 455–64.

Preuss, H. (1979) Progress in computer evaluation of sea level data within I.G.C.P. Project No. 61. In K. Suguio, T.R. Fairchild, L. Martin and J.-M. Flexor (eds), *Proceedings of the 1978 International Symposium on Coastal Evolution in the Quaternary*. São Paulo, Universidade de São Paulo, 104–34.

Pritchard, D.W. (1967) Observations of circulation in coastal plain estuaries. In G.H. Lauff (ed.), *Estuaries*. Washington, DC: American Association for the Advancement of Science, 37–44.

Probert-Jones, J.R. (1984) On the homogeneity of the annual temperature of Central England since 1659. *Journal of Climatology*, 3, 241–54.

Prudhomme, T.I., Oechel, W.C., Hastings, S.J. and Lawrence, W.T. (1984) Net ecosystems gas exchange at ambient and elevated carbon dioxide concentrations in tussock tundra at Toolik Lake, Alaska: an evaluation of methods and initial results. In J.H. McBeath (ed.), *The Potential Effects of Carbon Dioxide-induced Climate Changes in Alaska*. Proceedings of a conference, University of Alaska-Fairbanks, Miscellaneous Publications, 83–1, 155–63.

Pugh, D. (1990) Sea-level: change and challenge. *Nature and Resources*, 26, 36–46.

Qian, N. (1990) Fluvial processes in the lower Yellow River after levée breaching at Tongwaxiang in 1855. *International Journal of Sediment Research*, 5, 1–13.

Quezel, P. (1958) *Mission botanique au Tibesti*. Inst. de Réch. Sahar. Mém. 4, Algiers.

Quezel, P. (1965) La végétation du Sahara du Tchad à Mauretanie. *Geotanica Selecta* 2.

Rai, S.C., Ahmad, A. and Rawat, J.S. (1989) *Environmental Degradation Due to Population Pressure, Migration and Settlement: a Case Study of Central Himalaya*. Kosi, Almora: G.B. Pant Institute of Himalayan Environment and Development.

Ralhan, P.K., Negi, G.C.S. and Singh, S.P. (1991) Structure and function of the agroforestry system in the Pithoragarh district of Central Himalaya: an ecological viewpoint. *Agriculture, Ecosystems and Environment*, 35, 283–96.

Ramanathan, V. (1988a) The greenhouse theory of climate change: a test by an inadvertent global experiment. *Science*, 240, 293–9.

Ramanathan, V. (1988b) The radiative and climatic consequences of the changing atmospheric composition of trace gases. In F.S. Rowland and I.S.A. Isaksen (eds), *The Changing Atmosphere*. Chichester: Wiley, 159–86.

Ramanathan, V., Cess, R.D., Harrison, E.F., Minnis, P., Barkstrom, B.R., Ahmad, E., and Hartman, D. (1989) Cloud radiative forcing and climate: results from the Earth Radiation Budget Experiment. *Science*, 243, 57–63.

Ramanathan, V., Pitcher, E.J., Malone, R.C. and Blackmon, M.L. (1983) The response of a spectral GCM to refinements in radiative processes. *Journal of Atmospheric Science*, 40, 605–30.

Ramsay, W.J.H. (1987) Deforestation and erosion in the Himalaya – is the link myth or reality? *International Association of Hydrological Sciences, Publication*, 167, 239–50.

Randell, L. (1990) Land transformation in an African savanna. In M.G. Coulson (ed.), *Remote Sensing and Global Change*. Proceedings 16th Annual Conference of the Remote Sensing Society, University College, Swansea, 140–9.

Rankin, J.M. (1985) Forestry in the Brazilian Amazon. In G.T. Prance and T.E. Lovejoy (eds), *Key Environments: Amazonia*. Oxford: Pergamon Press, 369–92.

Rapp, A. (1987) Reflections on desertification 1977–1987: problems and prospects. *Desertification Control Bulletin*, 15, 27–33.

Rasmusson, E.M. (1987) Global climate change and variability: effects on drought and desertification in Africa. In M.H. Glantz (ed.), *Drought and*

Hunger in Africa. Cambridge: Cambridge University Press, 3–22.

Rasool, S.I. (1987) Potential of remote sensing for the study of global change. *Advances in Space Research*, 7, 1–97.

Rasool, S.I. and Schneider, S.H. (1971) Atmospheric CO_2 and aerosols: effects of large increases on global climate. *Science*, 173, 138–41.

Rathuri, A.S. and Dabral, B.G. (1986) Water consumption by Chir pine (*Pinus roxburghii*); Banj Oak (*Quercus incana*); Sal (*Shorea robusta*); and Ipil-ipil (*Leucaena leucocephala*) in juvenile stage. *Indian Forester*, 112, 711–33.

Rawat, A.S. (1983) *Garhwal Himalaya – a Historical Survey 1815–1947.* Delhi: Eastern Book Linkers, 105–36.

Rawat, J.S. (1989) *Channel Network Capacity and Geochemical Properties of Water of the Nana Kosi Basin, Kumaun Lesser Himalaya.* Almora: Kumaun University, Department of Geography, DST Project Final Report.

Rawat, J.S. and Bisht, H.S. (1986) *Channel Network Capacity and Geochemical Properties of Water of the Nana Kosi Basin, Kumaun Lesser Himalaya, India.* Almora: Kumaun University, Department of Geography.

Rawat, M.S. (1990) Physical landscapes and their degradational processes: a study in environmental geomorphology. Unpublished PhD thesis, Kumaun University, Department of Geography, Almora.

Rawat, M.S. and Rawat, J.S. (1993) Anthropogenic transformations of sheetwash erosion; experimental study in Kumaun Himalaya. *Geomorphology*, 7.

Redclift, M. (1990) Developing sustainably: designating ecological zones. *Land Use Policy*, 8, 202–16.

Reeh, I.V. (1989) Dynamic and climatic history of the Greenland ice sheet. In R.J. Fulton (ed.), *Quaternary Geology of Canada and Greenland.* Ottawa: Geological Surveys of Canada, 1, 795–822.

Reichelt, R. (1989) L'hydraulique pastorale et la désertification au Sahel des nomades en Afrique de l'ouest, réalités et perspectives. *Geol. Jahrb.* C., 52, 3–32.

Ren, M.E. and Shi, Y.L. (1986) Sediment discharge of the Yellow river (China) and its effect on the sedimentation of the Bahai and Yellow Sea. *Continental Shelf Research*, 6, 785–810.

Renberg, I. and Battarbee, R.W. (1990) The SWAP Palaeolimnology Programme: a synthesis. In B.J. Mason (ed.), *The Surface Waters Acidification Programme.* Cambridge: Cambridge University Press, 281–300.

Renberg, I. and Segerström, U. (1981) Applications of varved lake sediments in palaeoenvironmental studies. *Wahlenbergia*, 7, 125–33.

Repetto, R. (1990) Deforestation in the Tropics. *Scientific American*, 262, 18–24.

Repetto, R. and Gillis, M. (eds) (1988) *Public Policies and the Misuse of Forest Resources.* World Resources Institute, Cambridge University Press.

Reynolds, S.G. (1990) The influence of forest-clearance methods, tillage, and slope runoff on soil chemical properties and banana plant yields in the South Pacific. In J. Boardman, I.D.L. Foster and J.A. Dearing (eds), *Soil Erosion on Agricultural Land.* Chichester: Wiley, 339–50.

Rhoades, J.D. (1990) Soil salinity: causes and controls. In A.S. Goudie (ed.), *Techniques for Desert Reclamation.* Chichester: Wiley, 109–34.

Rieger, H.C. (1976) Floods and droughts – the Himalaya and the Ganges Plain as an Ecological System. In *Mountain Environment and Development.* Kathmandu: SATA, 13–29.

Rind, D., Goldberg, R., Hansen, J., Rosenzweig, C. and Ruedy, R. (1990)
Potential evapotranspiration and the likelihood of future drought.
Journal of Geophysical Research, 95 (D7), 9983–10004.

Riseborough, D.W. and Burn, C.R. (1988) Influence of an organic mat on
the active layer. In *Permafrost, Vol. 1, Fifth International Conference on
Permafrost in Trondheim, Norway, 1988*, Rapir Publishers, 633–8.

Ritchie, J.C. (1987) *Postglacial Vegetation of Canada*. Cambridge:
Cambridge University Press.

Ritchie, J.C., Eyles, C.H. and Haynes, C.V. (1985) Sediment and pollen
evidence for an early to mid Holocene humid period in the eastern
Sahara. *Nature*, 314, 352–5.

Ritchie, J.C. and Haynes, C.V. (1987) Holocene vegetation zonation in the
eastern Sahara. *Nature*, 330, 645–7.

Roberts, N. (1989) *The Holocene: an Environmental History*. Oxford:
Blackwell.

Roberts, N. (1990) Ups and downs of African lakes. *Nature*, 346, 107.

Roberts, N. and Barker, P. (1993) Landscape stability and biogeomorphic
response to past and future climatic shifts in inter-tropical Africa. In
D.S.G. Thomas and R.J. Allison (eds), *Environmental Sensitivity*.
Chichester: Wiley, 65–82.

Roberts, N. and Wright, H.E. (1993) Vegetational, lake-level and climatic
history of the Near East and Southwest Asia. In H.E. Wright, J.E.
Kutzbach, T. Webb, W.F. Ruddiman, F.A. Street-Perrott and P.J.
Bartlein (eds), *Global Climates since the Last Glacial Maximum*.
Minneapolis: University of Minnesota Press.

Robin, G. de Q. (1988) The Antarctic Ice Sheet, its history and response to
sea level and climatic changes over the past 100 million years. *Palaeogeography, Palaeoclimatology and Palaeoecology*, 67, 31–50.

Robinson, J.M. (1989) On uncertainty in the computation of global
emissions from biomass burning. *Climatic Change*, 14, 243–61.

Robinson, T.W. (1965) Introduction, spread and areal extent of salt cedar
(*Tamarix*) in the western States. United States Geological Survey, 491-A,
Washington, DC.

Rockwell, R.C. (1991) Culture and global change as driving forces in
global land-use cover change. Unpublished paper presented at 1991
Global Change Institute, Snowmass, CO, July 1991.

Roe, G.M. (1991) Report on the breach to the sea wall in Towyn, North
Wales on 26 February, 1990. In *Coastal Protection*. London: Planning
and Transport Research and Computation (International) Co. Ltd,
Education and Research Services, 1–17.

Rogers, C.S. (1985) Degradation of Caribbean and western Atlantic coral
reefs and decline of associated fisheries, *Proceedings, 5th International
Coral Reef Congress, Tahiti*, 6, 491–6.

Rohlfs, G. (1874–1875) *Quer durch Afrika. Reise von Mittelmeer nach
dem Tchad See und zum Golf von Guinea*, 2 vols. Leipzig.

Rosen, B.R. (1982) Darwin, coral reefs and global geology. *Bioscience*, 32,
519–25.

Rosen, B.R. (1984) Reef coral biogeography and climate through the late
Cainozoic: just islands in the sun or a critical pattern of islands? In P.J.
Brenchley (ed.), *Fossils and Climate*. Chichester: Wiley-Interscience,
201–62.

Rothlisberger, F. (1986) *10,000 Jahre Gletschergeschichte der Erde*.
Aarau: Verlag Sauerlander.

Rowntree, P.R. (1990) predicted climate changes under "greenhouse-gas"
warming. In M. Jackson, B.V. Ford-Lloyd and M.L. Parry (eds),
Climatic Change and Plant Genetic Resources. London and New York:

Belhaven, 18–33.

Rowntree, P.R. (1991) Climatic effects of deforestation. In HMSO, *Climatological and Environmental Effects of Rain Forest Destruction*. House of Commons Select Committee on the Environment, London, Appendix 2, 158–9.

Ruddiman, W.F. and McIntyre, A. (1981) The mode and mechanism of the last deglaciation: oceanic evidence. *Quaternary Research*, 16, 125–34.

Rudolf, B., Hauschild, H., Reiss, M. and Schneider, U. (1991) Operational global analysis of monthly precipitation totals planned by the GPCC. *Dynamics of Atmosphere and Oceans*, 16, 17–32.

Sabadell, J.E., Risley, E.M., Jorgensen, H.T. and Thornton, B.S. (1982) *Desertification in the United States: Status and Issues*. Washington, DC: Bureau of Land Management.

Sader, S.A., Powell, G.V. and Rappole, J.H. (1991) Migratory bird habitat monitoring through remote sensing. *International Journal of Remote Sensing*, 12, 363–72.

St John, T.V. (1985) Mycorrhizae. In G.T. Prance and T.E. Lovejoy (eds), *Key Environments; Amazonia*. Oxford: Pergamon, 277–83.

Salati, E. (1987) The forest and the hydrological cycle. In R.E. Dickinson (ed.), *The Geophysiology of Amazonia*. New York: Wiley, 273–96.

Salati, E. and Vose, P.B. (1989) Amazon Basin: A system in equilibrium. *Science*, 225, 129–37.

Salati, E., Vose, P.B. and Lovejoy, T.E. (1986) Amazon rainfall, potential effects of deforestation and plans for future research. In G.T. Prance (ed.), *Tropical Rainforest and the World Atmosphere*. Boulder, CO: Westview Press, 61–74.

Saltzman, B. (1985) Paleoclimatic modelling. In A.D. Hecht (ed.), *Paleoclimate Analysis and Modelling*. New York: Wiley, 341–93.

Samar Singh (1982) *Soil Conservation Problem: Approach and Progress in India*. New Delhi: Government of India, Ministry of Agriculture.

Saxena, P.B., Ghildyal, S., and Ghildyal, S. (1990) A study of water resource systems – their uses and conservation for agricultural development in Dehra Dun Valley, Central Himalaya. In Tej Vir Singh (ed.), *Geography of the Mountains*. New Delhi: Heritage Publishers (Contributions to Indian Geography 12), 250–65.

Sayer, J.A. and Whitmore, T.C. (1991) Tropical moist forest destruction and species extinction. *Biological Conservation*, 55, 199–213.

Schaake, P. (1982) Ein Beitrag zurn Preubenjahr 1981 en Berlin. *Beilage zur Berliner Wetterkarte*, des Instituts für Meteorologie der Frien Universitat Berlin, 32/82, SO4/82.

Schelske, C.L. (1975) Silica and nitrate depletion as related to rate of eutrophication in Lakes Michigan, Huron, and Superior. In A.D. Hasler (ed.), *Coupling of Land and Water Systems*. Berlin: Springer-Verlag, Ecological Studies, 10, 277–98.

Schelske, C.L., Conley, D.J., Stoermer, E.F., Newberry, T.L. and Campbell, C.D. (1986) Biogenic silica and phosphorus accumulation in sediments as indices of eutrophication in the Laurentian Great Lakes. *Hydrobiologia*, 143, 79–86.

Schiffers, H. (1951) Begriff, Grenze und Gliederung der Sahara. *Petermanns Geogr. Mittlg.*, 95, 239–46.

Schlesinger, M.E. (1989) Model projections of the climatic changes induced by increased atmospheric CO_2. In A. Berger, S. Schneider, and J.-Cl. Duplessy (eds), *Climate and Geo-Sciences*. Dordrecht: Kluwer, 275–415.

Schlesinger, M.E., Gates, W.L. and Han, Y-J. (1985) The role of the ocean in CO_2-induced climate warming: preliminary results from the OSU coupled atmosphere-ocean general circulation model. In J.C.J. Mihoul (ed.), *Coupled Ocean-Atmosphere Models*. Amsterdam: Elsevier Oceanography Series 40, 447–78.

Schlesinger, M.E. and Xu (1989) CO_2-induced climate change simulated by an improved atmospheric GCM/mixed-layer ocean model: results for the $1 \times CO_2$ climate, Appendix, NSF proposal.

Schlesinger, M.E. and Zhao, Z.-C. (1989) Seasonal climatic changes induced by doubled CO_2 as simulated by the OSU atmospheric GCM/mixed-layer ocean model. *Journal of Climate*, 2, 459–95.

Schlesinger, W.H., Reynolds, J.F., Cunningham, G.L., Huenneke, L.F., Jarrell, W.M., Virginia, R.A. and Whitford, W.G. (1990) Biological feedbacks in global desertification. *Science*, 247, 1043–8.

Schmidt, R. (1987) Tropical rain forest management. *Unasylva*, 39 (156), 2–17.

Schneider, S.H. (1989) *Global Warming: Are We Entering the Greenhouse Century?* San Francisco: Sierra Club Books.

Schulz, E. (1987) Die holozäne vegetation der zentralen Sahara (N-Mali, N-Niger, SW-Libyen). *Palaeoecology of Africa*, 18, 143–61.

Schulz, E. (1988) Der Südrand der Sahara. *Würzb. Geogr. Arb.*, 69, 167–210.

Schulz, E. (1990) Holocene environment of the central and southern Sahara. *Hydrobiologia*, 214, 359–66.

Schulz, E. and Adamou, A. (1988) Die Vegetation des Air Gebirge in Nord-Niger und ihre traditionelle Nutzung. *Giessener Beitr. zur Entwicklungsforschung* I, 17, 75–86.

Schulz, E., Joseph, A., Baumhauer, R., Schultze, E. and Sponholz, B. (1990) Upper Pleistocene and Holocene history of the Bilma region (Kawar, NE-Niger), CIFEG Occ. Publ. Orleans, 281–4.

Schulz, E. and Pomel, S. (1993) *La Bordure sud du Sahara. Du desert à la savane*. Bordeaux: CEGET.

Schumm, S.A. (1977) *The Fluvial System*. New York: Wiley.

Schumm, S.A., Harvey, M.D. and Watson, C.C. (1984) *Incised Channels, Morphology, Dynamics and Control*. Littleton, CO: Water Resources Publications.

Scoffin, T.P. et al. (1980) Calcium carbonate budget of a fringing reef on the west coast of Barbados. Part II. Erosion, sediments and internal structure. *Bulletin of Marine Science*, 30, 475–508.

Scoffin, T.P. and Dixon, J.E. (1983) The distribution and structure of coral reefs: one hundred years since Darwin. *Biological Journal of the Linnean Society*, 20, 11–38.

Scott Barr, E. (1961) The infrared pioneers, I. Sir William Herschel. *Infrared Physics*, 1, 1–41.

Sear, C.B., Kelly, P.M., Jones, P.D. and Goodess, C.M. (1987) Global surface temperature responses to major volcanic eruptions. *Nature*, 330, 365–7.

Sebacher, D.J., Harris, R.C., Bartlett, K.B., Sebacher, S.M. and Grice, S.S. (1986) Atmospheric methane sources: Alaskan tundra bogs, and alpine fen, and a subarctic boreal marsh. *Tellus*, 38B, 1–10.

Secrett, C. (1986) The environmental impact of Transmigration. *The Ecologist*, 16, 2/3.

Seignobos, Chr. (1982) Matières grasses, parcs et civilisations agraires (Tchad et Nord-Cameroun.). *Cahiers ORSTOM*, 35, 229–69.

Sellers, W.D. (1969) A global climate model based on the energy balance of the Earth-atmosphere system. *Journal of Applied Meteorology*, 8, 392–400.

Semtner, A.J. (1984) The climatic response of the Arctic Ocean to Soviet river diversions. *Climatic Change*, 6, 109–30.

Serrïo, E.A. and Toledo, J.M. (1988) Sustaining pasture-based production systems in the tropics, paper presented to the MAB on "Conversion of tropical forest to pasture in Latin America." Mexico, Oaxaca (October 1988).

Seth, S.K. and Khan, M. (1960) An analysis of the soil moisture regime in sal (*Shorea robusta*) forest of Dehra Dun with reference to natural regeneration. *Indian Forester*, 86, 398–406.

Setzer, A.W., Pereira, M.C., Pereira Jnr, A.C. and Almeida, S.A.O. (1988) Relatorio de Atividades do Projeto IBDF-INPE, SEQE 1987. Instituto de Pesquisas Espacciais, Publ. No. 4534-RPE/565, São Paulo, São João dos Campos.

Seubert, C.E., Sanchez, P.A. and Velverde, C. (1977) Effects of land clearing methods and soil properties of an Ultisol and crop performance in the Amazon jungle of Peru. *Tropical Agriculture*, 54 (4), 307–21.

Sevaldrud, I.H., Muniz, I.P. and Kalvenes, S. (1980) Loss of fish populations in southern Norway. Dynamics and magnitude of the problem. In D. Drabløs and A. Tollan (eds), *Ecological Impact of Acid Precipitation*. Norway: SNSF-project, 350–1.

Sevruk, B. (1982) Methods of correcting for systematic error in point precipitation measurement for operational use. Operational Hydrology Report No. 21, WMO No. 589.

Shackleton, N.J., Berger, A. and Peltier, W.R. (1990) An alternative astronomical calibration of the lower Pleistocene time scale based on ODP site 677. *Transactions of the Royal Society, Edinburgh, Earth Science*, 81, 251–61.

Shackleton, N.J., Imbrie, J. and Pisias, N.G. (1988) The evolution of oceanic oxygen-isotope variability in the North Atlantic over the past three million years. *Philosophical Transactions of the Royal Society, London, B*, 318, 679–88.

Shackleton, N.J. and Opdyke, N.D. (1973) Oxygen isotope and palaeo-magnetic stratigraphy of equatorial Pacific core V28-238. *Quaternary Research*, 3, 39–55.

Shah, S.L. (1982) *Socioeconomic, Technological, Organisational and Institutional Constraints in the Afforestation of Civil, Soyam, Usar, and Waste Lands for Resolving the Fuel Wood Crisis in the Hill Districts of Uttar Pradesh*. Almora: Vivekananda Laboratory for Hill Agriculture ICAR.

Shah, S.L. (1991) Ecodevelopment in Uttarakhand: concepts and strategies, emerging lessons from action research in Khulgad micro watershed in Almora District. *Central Himalayan Environment Association, Bulletin*, 4, 1–28.

Sharma, K.C. (1981) Economic analysis of grassland development in Naurar watershed, Ramganga catchment in Hills of Uttar Pradesh. Pantnagar, U.P. G.B. Pant University of Agriculture and Technology, Department of Agricultural Economics, unpublished MSc thesis.

Sharma, M.K. (1984) Ecological status of the Himalaya. *Deccan Geographer*, 22, 499–504.

Shaw, J. (1989) Drumlins, subglacial meltwater floods, and ocean responses. *Geology*, 17, 853–6.

Shennan, I. (1982) Interpretation of Flandrian sea level data from the Fenland, England. *Proceedings of the Geologists Association*, 93, 53–63.

Shennan, I. (1986a) Flandrian sealevel changes in the Fenland, I the geographical setting and evidence of relative sealevel changes. *Journal of Quaternary Science*, 1, 119–54.

Shennan, I. (1986b) Flandrian sealevel changes in the Fenland II, tendencies of sea-level movement, altitudinal changes and local and regional factors. *Journal of Quaternary Science*, 1, 155–79.

Shennan, I. (1989) Holocene crustal movements and sealevel changes in Great Britain. *Journal of Quaternary Science*, 4, 77–89.

Shennan, I., Tooley, M.J., Davis, M.J. and Haggart, B.A. (1983) Analysis and interpretation of Holocene sealevel data. *Nature*, 302, 404–6.

Sheridan, D. (1981) *Desertification of the United States*. Washington, DC: Council on Environmental Quality.

Shine, K.P. and Henderson-Sellers, A. (1983) Modelling climate and the nature of climate models: a review. *Journal of Climatology*, 3, 81–94.

Shiva, V. (1986) Chipko: trees, water and women. *Sanity (CND)*, April, 30–3.

Shmida, A. (1985) Biogeography of the desert flora. In M. Evenari and J. Noy Meir (eds), *Hot Desert and Arid Shrubland*. Oxford, 23–77.

Simberloff, D.S. (1984) Mass extinctions and the destruction of moist tropical forest. *Zh. Obshch. Biol.*, 45, 767–78.

Singh, A. (1989) Digital change detection techniques suing remotely-sensed data. *International Journal of Remote Sensing*, 10, 989–1005.

Singh, A.K. and Rawat, D.S. (1985) Depletion of oak forests threatening springs: an explanatory survey. *National Geographical Journal of India*, 31, 44–8.

Singh, D., Joshi, R.D., Chopra, S.K. and Singh, A.B. (1974) Quaternary history of vegetation and climate of the Rajasthan desert, India. *Philosophical Transactions of the Royal Society, London, B*, 267, 467–501.

Singh, J.S., Pandey, A.N. and Pathak, P.C. (1983) A hypothesis to account for the major pathway of soil loss from Himalaya. *Environmental Conservation*, 10, 343–5.

Singh, J.S., Pandey, U. and Tiwari A.K. (1984) Man and forest: a Central Himalayan case study. *Ambio*, 132, 80–7.

Singh, L.R. (1968) Tarai region of Uttar Pradesh. In R.L. Singh (ed.), *India – Regional Studies*. Calcutta: Indian National Committee on Geography (21st IGU Congress), 106–36.

Singh, R.P. and Gupta, M.K. (1989) Infiltration studies under different forestry management systems in western Himalaya. *Indian Forester*, 115, 706–13.

Sinun, W. (1991) Hillslope hydrology, hydrogeomorphology and hydrochemistry of an equatorial lowland rainforest, Danum Valley, Malaysia. Unpublished MSc thesis, University of Manchester.

Skillings, R.F. and Techeyan, N.O. (1979) *Economic Development Prospects of the Amazon Region of Brazil*. Baltimore: Center of Brazilian Studies, Johns Hopkins University.

Skowronek, A. (1987) *Böden als Indikator klimagesteuerter Landformung in der zentralen Sahara*. Relief Boden Paläoklima. Stuttgart.

Slater, P.N. (1979) A re-examination of Landsat MSS. *Photogrammetric Engineering*, 11, 1479–85.

Smil, V. (1984) *The Bad Earth: Environmental Degradation in China*. New York: Sharpe.

Smith, G.I. and Street-Perrott, F.A. (1983) Pluvial lakes of the western United States. In S.C. Porter (ed.), *Late Quaternary Environments of the United States, vol. 1, The Late Pleistocene*. Harlow: Longman, 190–212.

Smith, M.W. (1975) Microclimatic influences on ground temperatures and permafrost distribution, Mackenzie Delta, Northwest Territories. *Canadian Journal of Earth Sciences*, 12, 1421–38.

Smith, M.W. (1988) The significance of climatic change for the permafrost environment. In K. Senneset (ed.), *Proceedings Vol. 3, Fifth*

International Conference on Permafrost in Trondheim, Norway, 1988. Tapire Publishers, 18–23.

Smith, M.W. (1990) Potential responses of permafrost to climatic change. *Journal of Cold Regions Engineering*, 4, 29–37.

Smith, M.W. and Riseborough, D.W. (1984) Permafrost sensitivity to climatic change. In *Permafrost Fourth International Conference Proceedings, Fairbanks-Alaska.* Washington, DC: National Academy Press, 1178–83.

Smith, N. (1971) *A History of Dams.* London: Peter Davies.

Smith, R.A. (1852) On the air and rain of Manchester. *Memoirs and Proceedings of the Manchester Literary and Philosophical Society*, 2, 207–17.

Smith, S. (1980) The environmental adaptation of nomads in the West African Sahel. A key to understanding prehistoric pastoralists. In M.J. Williams and H. Faure (eds), *The Sahara and the Nile.* Rotterdam: Balkema, 476–88.

Smith, S.V. (1978) Coral reef area and the contribution of reefs to processes and resources of the world's oceans. *Nature*, 273, 225–6.

Smith, S.V. and Kinsey, D.W. (1978) Calcification and organic carbon metabolism as indicated by carbon dioxide. In D.R. Stoddart and R.E. Johannes (eds), *Coral Reefs: Research Methods.* Paris: UNESCO, 469–84.

Smyth, C. and Jennings, S. (1990) Late Bronze Age-Iron Age valley sedimentation in East Sussex, southern England. In J. Boardman, I.D.L. Foster and J.A. Dearing (eds), *Soil Erosion on Agricultural Land.* Chichester: Wiley, 273–84.

Snowball, I. and Thompson, R. (1990) A mineral magnetic study of Holocene sedimentation in Lough Catherine, Northern Ireland. *Boreas*, 19, 127–46.

Solbrig, O. (1990) Savannas and global change. *Biology International*, 21, 15–19.

Solbrig, O. (1991) Savanna modelling for global change. *Biology International* (Special Issue), 24, 1–47.

Someshashekara Reddy, S.T. (1989) Declining groundwater levels in India. *Water Resources Development*, 5, 185–90.

Sommer, A. (1976) Attempt at an assessment of the world's tropical forests. *Unasylva*, 28 (112/113), 5–25.

Soni, P., Sanjay Naithani, and Mathur, H.N. (1985) Infiltration studies under different vegetation cover. *Indian Journal of Forestry*, 8, 170–3.

Soulé, M. (ed.) (1986) *Conservation Biology: the Science of Scarcity and Diversity.* Sunderland, MA: Sinauer Associates.

South Commission (1990) *The Challenge to the South: the Report of the South Commission.* Oxford: Oxford University Press.

Space Science Board (1988) *Space Science in the 21st Century: Mission to Planet Earth.* Washington, DC: National Research Council, National Academy Press.

Spencer, R.W. and Christy, J.R. (1990) Precise monitoring of global temperature trends from satellites. *Science*, 247, 1558–62.

Spencer, T. and Douglas, I. (1985) The significance of environmental change: diversity, disturbance and tropical ecosystems. In I. Douglas and T. Spencer (eds), *Environmental Change and Tropical Geomorphology.* Chichester: Wiley, 13–33.

Spittler, G. (1989) Wüste, Wildnis und Zivilisation, die Sicht der Kel Ewey. *Paideuma*, Frobenius Institut, Frankfurt, 35, 237–87.

Squire, G.R. (1990) Effects of changes in climate and physiology around the dry limits of agriculture in the tropics. In M. Jackson, B.V.

Ford-Lloyd and M.L. Parry (eds), *Climatic Change and Plant Genetic Resources*. London and New York: Belhaven, 116–47.

Stanford, J.A. and Ward, J.V. (1986a) The Colorado River System. In B.R. Davies and K.F. Walker (eds), *The Ecology of River Systems*. Dordrecht: Dr W Junk Publishers, 353–74.

Stanford, J.A. and Ward, J.V. (1986b) Fish of the Colorado system. In B.R. Davies and K.F. Walker (eds), *The Ecology of River Systems*. Dordrecht: Dr W Junk Publishers, 385–402.

Stark, N.M. and Jordan, C.F. (1978) Nutrient retention by the root mat of an Amazonian rainforest. *Ecology*, 59, 434–7.

Stearn, C.W. and Scoffin, T.P. (1977) Carbonate budget of a fringing reef, Barbados. *Proceedings, Third International Coral Reef Symposium, Miami*, 2, 471–6.

Stearn, C.W., Scoffin, T.P. and Martindale, W. (1977) Calcium carbonate budget of a fringing reef on the west coast of Barbados. Part I. Zonation and productivity. *Bulletin of Marine Science*, 27, 479–510.

Steers, J.A. (1953) The east coast floods, January 31–February 1, 1953. *Geographical Journal*, 119, 280–95.

Sternberg, H.O. (1975) *Amazon River of Brazil*. New York: Springer Verlag.

Stetson, C.W. (1956) *Washington and His Neighbors*. New York: Garrett and Massie.

Stevenson, A.C., Juggins, S., Birks, H.J.B., Anderson, D.S., Anderson, N.J., Battarbee, R.W., Berge, F., Davis, R.B., Flower, R.J., Haworth, E.Y., Jones, V.J., Kingston, J.C., Kreiser, A.M., Line, J.M., Munro, M.A.R. and Renberg, I. (1991) *The Surface Waters Acidification Project Palaeolimnology Programme: Modern Diatom/Lake-Water Chemistry Dataset*. London: Ensis.

Stewart, R.W., Kjerfve, B., Milliman, J. and Dwivedi, S.N. (1990) Relative sea-level change: a critical evaluation. *UNESCO Reports in Marine Science*, 54, 1–21.

Stiles, D. (1984) Desertification: a question of linkage. *Desertification Control Bulletin*, 11, 1–6.

Stocker, O. (1962) Steppe-Wüste-Savanne. *Veroff. Geobot. Inst. ETH Zürich (Rübel)*, 37, 234–43.

Stoddart, D.R. (1973) Coral reefs: the last two million years. *Geography*, 58, 313–23.

Stoddart, D.R. (1985) Symposium no. 3 – Hurricane effects on coral reefs: conclusion. *Proceedings, 5th International Coral Reef Congress*, 3, 349–50.

Stoddart, D.R. (1990) Coral reefs and islands and predicted sea-level rise. *Progress in Physical Geography*, 14, 521–36.

Stolz, W. (1988) Analysis of large format camera photographs of the Po Delta, Italy, for topographic and thematic mapping. *International Journal of Remote Sensing*, 9, 1705–14.

Storey, A., Folland, C.K. and Parker, D.E. (1986) A homogenous archive of daily Central England temperature from 1772 to 1985, and new monthly average values from 1974 to 1985. *Synoptic Climatology Branch Memorandum No. 107*. Bracknell: UK Meteorological Office.

Stott, P. (1988a) The forest as Pheonix: towards a biogeography of fire in mainland South East Asia. *Geographical Journal*, 154, 337–50.

Stott, P. (1988b) Savanna forest and seasonal fire in South East Asia. *Plants Today*, 1, 196–200.

Stott, P. (1989) Lessons from an ancient land: Kakadu National Park, Northern Territory, Australia. *Plants Today*, 2, 121–5.

Stott, P. (1990) Stability and stress in the savanna forests of mainland

South-East Asia. *Journal of Biogeography*, 17, 373–83.

Stott, P. (1991) Recent trends in the ecology and management of the world's savanna formations. *Progress in Physical Geography*, 15, 18–28.

Stouffer, R.J., Manabe, S. and Bryan, K. (1989) Interhemispheric asymmetry in climate response to a gradual increase in atmospheric carbon dioxide. *Nature*, 342, 660–2.

Street, F.A. (1980) The relative importance of climate and local hydrogeological factors in influencing lake level fluctuations. *Palaeoecology of Africa*, 12, 137–58.

Street-Perrott, F.A. (1991) General circulation (GCM) modelling of palaeoclimates: a critique. *The Holocene*, 1, 74–80.

Street-Perrott, F.A. (1992) Tropical wetland sources. *Nature*, 355, 23–4.

Street-Perrott, F.A. and Harrison, S.P. (1985) Lake levels and climate reconstruction. In A.D. Hecht (ed.), *Paleoclimate Analysis and Modeling*. Chichester: Wiley, 291–340.

Street-Perrott, F.A., Mitchell, J.F.B., Marchand, D.S. and Brunner, J.S. (1991) Milankovitch and albedo forcing of the tropical monsoons: a comparison of geological evidence and numerical simulations for 9,000 yr BP. *Transactions of the Royal Society, Edinburgh, Earth Science*, 81, 407–27.

Street-Perrott, F.A. and Perrott, R.A. (1990) Abrupt climate fluctuations in the tropics: the influence of Atlantic Ocean circulation. *Nature*, 343, 607–12.

Street-Perrott, F.A. and Roberts, N. (1983) Fluctuations in closed basin lakes as an indicator of past atmospheric circulation patterns. In A. Street-Perrott, M. Beran and R. Ratcliffe (eds), *Variations in the Global Water Budget*. Dordrecht: Reidel, 331–45.

Street-Perrott, F.A., Roberts, N. and Metcalfe, S. (1985) Geomorphic implications of late Quaternary hydrological and climatic changes in the Northern Hemisphere tropics. In I. Douglas and T. Spencer (eds), *Environmental Change and Tropical Geomorphology*. London: George Allen and Unwin, 165–83.

Streif, H. (1990) Quaternary sealevel changes in the North Sea, an analysis of amplitudes and velocities. In P. Brosche and J. Sunderman (eds), *Earth's Rotation from Eons to Days*. Berlin: Springer Verlag, 201–14.

Study of Man's Impact on Climate (SMIC) (1971) *Inadvertent Climate Modification*. Cambridge, MA: MIT Press.

Stuiver, M. and Quay, P.D. (1980) Changes in atmospheric carbon-14 attributed to a variable sun. *Science*, 207, 11–19.

Stuth, J.W., Conner, J.R., Hamilton, W.T., Riegel, D.A., Lyons, B.G., Myrick, B.R. and Couch, M.J. (1990) RSPM: a resource systems planning model for integrated resource management. *Journal of Biogeography*, 17, 531–40.

Styles, K.A. and Hansen, A. (1989) Territory of Hong Kong (Geotechnical Area Studies Programme Report XII), Geotechnical Control Office, Hong Kong.

Subramanian, V.S. (1978) Input by Indian rivers into the world oceans. *Indian Academy of Science, Proceedings*, 87A, 77–88.

Subramanian, V.S. (1979) Chemical and suspended sediment characteristics of rivers of India. *Journal of Hydrology*, 44, 37–55.

Subramanian, V.S. and Dalavi, R.A. (1978) Some aspects of stream erosion in the Himalaya. *Himalayan Geology*, 8, 822–34.

Suess, E. (1888) *Das Antlitz der Erde*. Translated by H.B.C. Solas, 1906. Oxford: Clarendon Press.

Sugden, D.E. (1987) The polar and glacial world. In M.J. Clark, K.J.

Gregory and A.M. Gurnell (eds), *Horizons in Physical Geography*. London: Macmillan, 214–31.

Sugden, D. and Payne, A.J. (1990) Topography and ice sheet growth. *Earth Surface Processes and Landforms*, 15, 625–39.

Sun, J.Z. (1988) Environmental geology in loess areas of China. *Environmental Geology and Water Science*, 12, 49–61.

Sutcliffe, D.W. (1983) Acid precipitation and its effects on aquatic systems in the English Lake District (Cumbria). *Report Freshwater Biological Association*, 51, 30–62.

Sutcliffe, D.W., Carrick, T.R., Heron, J., Rigg, E., Talling, J.F., Woof, C. and Lund, J.W.G. (1982) Long-term and seasonal changes in the chemical composition of precipitation and surface waters of lakes and tarns in the English Lake District. *Freshwater Biology*, 12, 451–506.

Swain, A.M., Kutzbach, J.E. and Hastenrath, S. (1983) Estimates of Holocene precipitation for Rajasthan, India, based on pollen and lake-level data. *Quaternary Research*, 19, 1–17.

Swift, B.L. (1984) Status of riparian ecosystems in the United States. *Water Resources Bulletin*, 20, 223–9.

Swift, J. (1973) Disaster and a Sahelian nomad economy. In D. Dalby and R.J. Harrison Church (eds), *Drought in Africa*. London: Centre for African Studies, 71–8.

Tardin, A.T. and da Cunha, R.P. (1990) *Evaluation of Deforestation in Legal Amazonia Using LANDSAT TM Images*. Instituto de Pesquisas Espaciais (INPE) Publ. No. 5015-RPE/609, São Paulo, São Jose dos Campos.

Teller, J.T. (1900) Meltwater and precipitation runoff to the North Atlantic, Arctic and Gulf of Mexico from the Laurentide ice sheet and adjacent regions during the Younger Dryas. *Paleoceanography*, 5, 897–905.

Tempany, H.A. (1949) *The Practice of Soil Conservation in the British Colonial Empire*. London: Commonwealth Bureau of Soil Science Technical Communication, 41.

Thie, J. (1974) Distribution and thawing of permafrost in the southern part of the discontinuous permafrost zone in Manitoba. *Arctic*, 27, 189–200.

Thom, B.G. and Chappell, J. (1975) Holocene sea levels relative to Australia. *Search*, 6, 90–3.

Thom, B.G. (1973) The dilemma of high interstadial sea levels during the last glaciation. *Progress in Physical Geography*, 6, 233–59.

Thompson, M. and Warburton, M. (1985) Uncertainty on a Himalayan scale. *Mountain Research and Development*, 5, 115–35.

Thompson, L.G., Mosley-Thompsom, E., Dansgaard, W. and Grootes, P.M. (1986) The Little Ice Age as recorded in the stratigraphy of the tropical Quelccaya ice cap. *Science*, 234, 361.

Thompson, R. and Oldfield, F. (1986) The Rhode River, Chesapeake Bay, an integrated catchment study. In R. Thompson and F. Oldfield (eds), *Environmental Magnetism*. London: Allen and Unwin, 185–97.

Thornes, J. (1990) The interaction of erosional and vegetational dynamics in land degradation: spatial outcomes. In J.B. Thornes (ed.), *Vegetation and Erosion*. Chichester: John Wiley, 41–53.

Tiwari, A.K., and Singh, J.S. (1987) Analysis of forest land use and vegetation in a part of the Central Himalaya using aerial photographs. *Environmental Conservation*, 14, 233–44.

Tiwari, A.K., Saxena, A.K. and Singh, J.S. (1986) Inventory of forest biomass for Indian Central Himalaya. In J.S. Singh (ed.), *Environmental Regeneration in the Himalaya*. Nainital: CHEA and Gyanodaya Prakashan, 236–47.

Toky, O.P. and Ramakrishman, P.S. (1983) Secondary succession following slash and burn agriculture in north-eastern India. I. Biomass, litterfall
and productivity. *Journal of Ecology*, 71, 735–45.

Tolba, M.K. (1987) *Sustainable Development: Constraints and Opportunities*. London: Butterworth.

Tooley, M.J. (1974) Sealevel changes during the last 9000 years in
northwest England. *Geographical Journal*, 140, 18–42.

Tooley, M.J. (1978) *Sea-level Changes*. Oxford: Clarendon Press.

Tooley, M.J. (1979) Sea level changes during the Flandrian Stage and the
implications for coastal development. In K. Suguio, T.R. Fairchild, L.
Martin and J.M. Fleor (eds), *Proceedings of the 1978 International
Symposium on Coastal Evolution in the Quaternary*. São Paulo: Universidade de São Paulo.

Tooley, M.J. (1981) Methods of reconstruction. In I.G. Simmons and M.J.
Tooley (eds), *The Environment in British Prehistory*. London: Duckworth, 1–48.

Tooley, M.J. (1982) Sea-level changes in northern England. *Proceedings of
the Geologists Association*, 93, 43–51.

Tooley, M.J. (1985) Climate, sea-level and coastal changes. In M.J. Tooley
and G.M. Sheail (eds), *The Climatic Scene*. London: George Allen and
Unwin, 206–34.

Tooley, M.J. (1987) Sea level studies. In M.J. Tooley and I. Shennan (eds),
Sea-level Changes. Oxford: Basil Blackwell, 1–24.

Tooley, M.J. (1989) Global sea levels: floodwaters mark sudden rise.
Nature, 342, 20–1.

Tooley, M.J. and Jelgersma, S. (eds) (1992) *Impacts of Sea Level Rise on
European Coastal Lowlands*. Oxford: Basil Blackwell.

Toupet, Ch. (1963) Evolution de la nomadisation en Mauretanie sa-
heliènne. In Bataillon (ed.), *Nomades et nomadismes au Sahara*. Paris:
UNESCO, 67–79.

Tousson, O. (1925) L'Histoire du Nil. *Memoires de Institute Egypte*, 9,
366–404.

Townshend, J.R.G. (1981) The spatial resolving power of earth resource
satellites. *Progress in Physical Geography*, 5, 681–98.

Townshend, J.R.G. and Justice, C.O. (1986) Analysis of the dynamics of
African vegetation using the normalised difference vegetation index.
International Journal of Remote Sensing, 7, 1435–53.

Tracey, J.I.Jr and Ladd, H.S. (1974) Quaternary history of Eniwetok and
Bikini atolls, Marshall Islands. *Proceedings, 2nd International Coral
Reef Symposium, Brisbane*, 2, 537–50.

Trefethen, P. (1972) Man's impact on the Columbia River. In R.T.
Oglesby, C.A. Carlson and J.A. McCann (eds), *River Ecology and Man*.
New York: Academic Press, 77–98.

Troels-Smith, J. (1955) Characterisation of unconsolidated sediments.
Danmarks Geologiske Undersøgelse IV, 3, 10, 1–73.

Troll, C. (1952) Das Pflanzenkleid der Tropen in seiner Abhängigkeit von
Klima Boden und Mensch. *Deutsch. Geographentag*, Frankfurt, 35–66.

Trollope, W.S.W. (1982) Ecological effects in South African Savannas. In
B.J. Huntley and B.H. Walker (eds), *Ecology of Tropical Savannas*.
Ecological Studies 42. Berlin: Springer-Verlag, 292–306.

Turner, R.E. and Rabalais, N.N. (1991) Changes in Mississippi River
water quality this century. *Bioscience*, 41, 140–7.

Turner, R.M. and Karpiscak, M.M. (1980) *Recent Vegetation Changes
Along the Colorado River Between Glen Canyon Dam and Lake Mead,
Arizona*. US Geological Survey Professional Paper 1132.

Tyler, S.J. and Ormerod, S.J. (1988) Effects on birds. In M.H. Ashmore,

N.J.B. Bell and C. Garretty (eds), *Acid Rain and Britain's Natural Ecosystems.* London: Centre for Environmental Technology, Imperial College, 97–109.

Tyson, P.D. (1986) *Climatic Change and Variability in Southern Africa.* Cape Town: Oxford University Press.

UK EODC (1991) *UK ERS-1 Reference Manual.* Farnborough: RAE.

Ullmann, I. (1985) Tagesgänge von Transpiration und stomatärer Leitfähigkeit sahelisch und saharischer Akazien in der Trockenzeit. *Flora,* 176, 383–409.

UN (1987) *The Prospects of World Urbanization.* New York: United Nations.

Underwood, J., Donald, A.P. and Stoner, J.H. (1987) Investigations into the use of limestone to combat acidification in two lakes in west Wales. *Journal of Environmental Management,* 24, 29–40.

UNEP (1990) *The State of the Marine Environment.* Reports and studies No. 39. UNEP Regional Seas and Studies No. 115.

Uttarakhand Seva Nidhi (1986) *Workshop on 7th Five Year Plan for Kumaun Hills.* Nainital: Consul.

Valdiya, K.S. (1985) Himalayan tragedy – big dams, seismicity, erosion and drying up of springs in the Himalayan region. *Central Himalayan Environment Association Bulletin,* 1, 1–24.

Valdiya, K.S. (1986) Accelerated erosion and landslide prone zones in the Central Himalayan region. In J.S. Singh (ed.), *Environmental Regeneration in the Himalaya.* Nainital: CHEA and Gyanodaya Prakashan, 12–39.

Valdiya, K.S. (1987) *Environmental Geology: the Indian Context.* New Delhi: Tata McGraw Hill.

Valdiya, K.S. and Bartarya, S.K. (1989a) Diminishing discharges of mountain springs in a part of Kumaun Himalaya. *Current Science,* 58, 416–26.

Valdiya, K.S. and Bartarya, S.K. (1989b) Problem of mass-movements in a part of Kumaun Himalaya. *Current Science,* 58, 486–91.

Valentin, G. (1981) Organisation pelliculaires superficielles de quelques sols de région subdésértique (Agadez, Niger). Dynamisme et conséquence sur économie den eau. Thèse Univ. Paris VII, 259.

van Andel, T.H., Zangger, E. and Demitrack, A. (1990) Land use and soil erosion in prehistoric and historical Greece. *Journal of Field Archaeology,* 17, 379–96.

van de Plassche, O. (ed.) (1986) *Sea Level Research: a Manual for the Collection and Evaluation of Data.* Norwich: Geo Books.

van der Veen, C.J. (ed.) (1987) *Dynamics of the West Antarctic Ice Sheet.* Dordrecht: Kluwer.

van Ijssel, W.J. and Sombroek, W.G. (1987) Spatial variability of tropical rain forest and forest lands. In C.F. van Beusekome, C.P. van Goor and P. Schmidt (eds), *Wise Utilisation of Tropical Rain Forest Lands.* Ede, Netherlands: Tropenbos, 68–80.

van Maydell, E. (1983) Arbres et arbustes du Sahel. *Schriftenreihe GTZ,* 147.

Vannote, R.L., Minshall, G.W., Cummins, K.W., Sedell, J.R. and Cushing, C.E. (1980) The river continuum concept. *Canadian Journal of Fisheries and Aquatic Sciences,* 37, 130–7.

van Straaten, L.M.J.U. (1954) Radiocarbon datings and changes of sea level at Velzen (Netherlands). *Geologie en Mijnbouw,* 16, 247–53.

Velichko, A.A. and Nechayev, V.P. (1984) Late Pleistocene permafrost in European USSR. In A.A. Velichko, H.E. Wright and C.W. Barnosky (eds), *Late Quaternary Environments of the Soviet Union.* Harlow:

Longman, 79–86.

Verstraete, M.H. (1986) Defining desertification: a review. *Climatic Change*, 9, 5–18.

Viereck, L.A. and Van Cleve, K. (1984) Some aspects of vegetation and temperature relationships in the Alaskan Taiga. In J.H. McBeath (ed.), *The Potential Effects of Carbon Dioxide-induced Climatic Changes in Alaska*. Proceedings of a Conference. University of Alaska-Fairbanks, Miscellaneous Publications, 83–1, 129–42.

Villachia, H., Silva, J.E., Peres, J.R. and da Rocha C.M.C. (1990) Sustainable agricultural systems of the humid tropics of South America. In C.A. Edwards, R. Lal, P. Madden, R.H. Miller and G. House (eds), *Sustainable Agricultural Systems*. Ankeny, IA: Soil and Water Conservation Society, 391–437.

Vine, H. (1966) Tropical soils. In C.C. Webster and P.N. Wilson (eds), *Agriculture in the Tropics*. London: Longman, 28–67.

Vinnikov, K. Ya., Groisman, P.Ya. and Lugina, K.M. (1990) The empirical data on modern global climate changes (temperature and precipitation). *Journal of Climate*, 3, 662–77.

Visser, W.A. (ed.) (1980) *Geological Nomenclature*. Utrecht: Bohn, Scheltema and Holkema.

Vogg, R. (1987) Die Böden des saharo-sahelischen Nordens der Republik Mali. *Stuttgarter Geograph. Studien*, 106, 225–48.

Vogt, J. (1990) Aspects of historical erosion in Western Europe. In P. Brimblecombe and C. Pfister (eds), *The Silent Countdown*. Berlin: Springer-Verlag.

Volk, O.H., and Geiger, E. (1970) Schaumböden als Ursache der Vegetationslosiskeit in ariden Gebieten. *Zeitschrift für Geomorphologie N.F.*, 14, 79–95.

Völkel, J. (1989) Geomorphologische und pedologische Untersuchungen zum quartären Klimawandel in den Dünengebieten SE-Nigers, Südsahara und Sahel. *Bonner Geogr. Abhandl.*, 79, 285.

von Ufford, H.A.Q. (1953) The disastrous storm surge of 1 February. *Weather*, 8, 116–20.

Voronov, A.G., Malkhazova, S.M. and Komarova, L.V. (1983) Evaluation of the medical-geographic setting in the Midland region of the USSR and possible changes resulting from the proposed interbasin transfer of water from Siberia and central Asia. *Soviet Geography*, 24, 503–15.

Voropaev, G.V. and Velikanov, A.L. (1985) Partial Southward Diversion of Northern and Siberian Rivers. In G.N. Golubev and A.K. Biswas (eds), *Large-Scale Water Transfers: Emerging Environmental and Social Experiences*. Oxford: Tycooly, 67–83.

Vrstraete, M.M., Beelward, A.S. and Kennedy, P.J. (1990) The Institute for Remote Sensing Applications contribution to the global change research programme. In M.G. Coulson (ed.), *Remote Sensing and Global Change*. Proc. 16th Annual Conference of the Remote Sensing Society, University College, Swansea, iv–xii.

Walker, B.H. (1987) A general model of savanna structure and function. In B.H. Walker (ed.), *Determinants of Tropical Savannas*. Oxford: IUBS Monograph Series, 3, 1–12.

Walker, B.H. and Menaut, J.-C. (eds) (1988) *Research Procedure and Experimental Design for Savanna Ecology and Management*. RSSD Australia Publication 1. Melbourne: CSIRO for IUBS/Unesco MAB.

Walker, B.H. and Noy-Meir, I. (1982) Aspects of the stability and resilience of savanna ecosystems. In B.J. Huntley and B.H. Walker (eds), *Ecology of Tropical Savannas*. Berlin: Springer, 556–90.

Walker, D.A. and Edwards, G. (1983) C3, C4: *Mechanisms, and Cellular and Environmental Regulation of Photosynthesis.* Oxford: Blackwell Scientific.

Walker, J. and Rowntree, P.R. (1977) The effect of soil moisture on circulation and rainfall in a tropical model. *Quarterly Journal of the Royal Meteorological Society,* 103, 29–46.

Walsh, R.P.D., Hulme, M. and Campbell, M.D. (1988) Recent rainfall changes and the impact on hydrology and water supply in the semi-arid zone of the Sudan. *Geographical Journal,* 154, 181–98.

Walter, H. (1964) *Die Vegetation der Erde in ökophysiologischer Betrachtung* I, Jena: Gustav Fischer Verlag.

Warren, H. and Agnew, C. (1988) An assessment of desertification and land degradation in arid and semi-arid areas. London: ZIED paper, no. 2.

Warren, A. and Maizels, J.K. (1976) *Ecological Change and Desertification.* London: University College.

Warrick, R. and Farmer, G. (1990) The greenhouse effect, climatic change and rising sea levels: implications for development, *Transactions, Institute of British Geographers,* n.s., 15, 5–20.

Warrick, R. and Oerlemans, J. (1990) Sea level rise. In J.T. Houghton, G.J. Jenkins and J.J. Ephraums (eds), *Climate Change. IPCC Scientific Assessment.* Cambridge: Cambridge University Press, 257–81.

Washburn, A.L. (1979) *Geocryology. A Survey of Periglacial Processes and Environments.* London: E. Arnold.

Washington, W.M. and Meehl, G.A. (1984) Seasonal cycle experiment on the climate sensitivity due to a doubling of CO_2 with an atmospheric general circulation model coupled to a simple mixed-layer ocean model. *Journal of Geophysical Research,* 89, 9475–503.

Washington, W.M. and Meehl, G.A. (1989) Climate sensitivity due to increased CO_2 experiments with a coupled atmosphere and ocean general circulation model. *Climate Dynamics,* 4, 1–38.

Watson, A. (1990) The control of blowing sand and mobile desert dunes. In A.S. Goudie (ed.), *Techniques for Desert Reclamation.* Chichester: Wiley, 35–85.

Watt, W.D., Scott, D. and Ray, S. (1979) Acidification and other chemical changes in Halifax County lakes after 21 years. *Limnology and Oceanography,* 24, 1154–61.

Webb, P.N. and Harwood, D.M. (1987) The Sirius formation of the Beardmore Glacier region. *Ant. J. US,* 22, 8–12.

Webb, P.N., Harwood, D.M., McKelvey, B.C., Mercer, J.H. and Stott, L.D. (1984) Cenozoic marine sedimentation and ice volume variation on the East Antarctic craton. *Geology,* 12, 287–91.

Webb, T. and Wigley, T.M.L. (1985) What past climates can indicate about a warmer world. In M.C. MacCracken and F.M. Luther (eds), *Projecting the Climatic Effects of Increasing Carbon Dioxide.* Washington, DC: United States Department of Energy, 239–57.

Webb, T., Cushing, E.J. and Wright, H.E. (1984) Holocene changes in the vegetation of the Midwest. In H.E. Wright (ed.), *Late Quaternary Environments of the United States, vol. 2, The Holocene.* Harlow: Longman, 142–65.

Wehmeier, E. (1980) Desertification processes and groundwater utilization in the northern Nefzaoua, Tunisia. *Stuttgarter Geographische Studien,* 95, 125–43.

Wei Min Hao, Mei-Huey Liu and Crutzen, P.J. (1990) Estimates of annual and regional releases of CO_2 and other trace gases to the atmosphere from fires in the tropics, based on the FAO statistics for the period 1975–1980. In J.G. Goldammer (ed.), *Fire in the Tropical Biota.* Berlin:

Springer, 440–62.

Weidick, A. and Ten Brink, N.W. (1974) Greenland ice sheet history since the last glaciation. *Quaternary Research*, 4, 429–40.

Weidick, J.T. (1984) Review of glacier changes in West Greenland. Zeit Gletscherk. *Glaziogeol*, 21, 301–9.

Welcomme, R. (1979) *Fisheries Ecology of Floodplain Rivers*. Harlow: Longman.

Wells, D.E., Gee, A.S. and Battarbee, R.W. (1986) Sensitive surface waters – a UK perspective. *Water Air and Soil Pollution*, 31, 631–68.

Wendland, W.M. (1977) Tropical storm frequencies related to sea surface temperature. *Journal of Applied Meteorology*, 16, 477–81.

Werner, P.A. (ed.) (1988) *Greenhouse 88: Planning for Climate Change Issues for the Top End*. Darwin: Conservation Commission of the Northern Territory.

Werner, P.A. (ed.) (1991) *Savanna Ecology and Management: Australian Perspectives and Intercontinental Comparisons*. RSSD Publication 5. Oxford: Blackwell Scientific.

Whitley, J.R. and Campbell, R.S. (1974) Some aspects of water-quality and biology of the Missouri River. *Transactions of the Missouri Academy of Sciences*, 8, 60–72.

Whitmore, T.C. (1990) *An Introduction to Tropical Rain Forests*. Oxford: Clarendon Press.

Whitmore, T.C. and Prance, G.T. (eds) (1987) *Biogeography and Quaternary History in Tropical America*. Oxford: Clarendon Press.

Whittaker, R.H. (1975) *Communities and Ecosystems*, 2nd edn. New York: Macmillan.

Wickens, G.E. (1976) Speculations on long distance dispersal and the flora of Jebel Marra, Sudan republic. *Kew Bulletin*, 31, 105–50.

Wickland, D.E. (1989) Future directions for remote sensing in terrestrial ecological research. In G. Asrar (ed.), *Theory and Applications of Optical Remote Sensing*. New York: Wiley, 691–724.

Wigley, T.M.L. (1988) When will equilibrium CO_2 results be relevant? *Climate Monitor*, 17, 99–106.

Wigley, T.M.L. (1989) Possible climatic change due to SO_2 derived cloud condensation nuclei. *Nature*, 339, 365–7.

Wigley, T.M.L. and Barnett, T.P. (1990) Detection of the greenhouse effect. In J.T. Houghton, G.J. Jenkins and J.J. Ephraums (eds), *Climate Change. IPCC Scientific Assessment*. Cambridge: Cambridge University Press, 243–5.

Wigley, T.M.L. and Kelly, P.M. (1990) Holocene climatic change, ^{14}C wiggles and variations in solar irradiance. *Philosophical Transactions of the Royal Society, London, A*, 330, 547–60.

Wigley, T.M.L., Lough, J.M. and Jones, P.D. (1984) Spatial patterns of precipitation in England and Wales and a revised homogenous England and Wales precipitation series. *Journal of Climatology*, 4, 1–26.

Wigley, T.M.L. and Raper, S.C.B. (1987) Thermal expansion of sea water associated with global warming. *Nature*, 330, 127–31.

Wigley, T.M.L. and Raper, S.C.B. (1990) Natural variability of the climate system and detection of the greenhouse effect. *Nature*, 344, 324–7.

Williams, E.H.Jr and Bunkley-Williams, L. (1988) Bleaching of Caribbean coral reef symbionts in 1987–88. *Proceedings, 6th International Coral Reef Symposium*, 3, 313–18.

Williams, E.H.Jr and Bunkley-Williams, L. (1990) The world-wide coral reef bleaching cycle and related sources of coral mortality. *Atoll Research Bulletin*, 335, 1–71.

Williams, E.H.Jr, Goenaga, C. and Vicete, V. (1987) Mass bleachings on

Atlantic coral reefs. *Science*, 237, 877–8.

Williams, L.D. and Wigley, T.M.L. (1983) A comparison of evidence for late Holocene summer temperature variations in the northern hemisphere. *Quaternary Research*, 20, 286–307.

Williams, P.J. (1979) *Pipelines and Permafrost: Physical Geography and Development in the Circumpolar North*. Harlow and New York: Longman.

Williams, P.J. and Smith, M.W. (1989) *The Frozen Earth. Fundamentals of Geocryology*. Cambridge: Cambridge University Press, Studies in Polar Research.

Williams, W.D. and Aladin, N.V. (1991) The Aral Sea: recent limnological changes and their conservation significance. *Aquatic Conservation*, 1, 3–23.

Willoughby, H.E., Masters, J.M. and Landsca, C.W. (1989) A record minimum sea level pressure observed in Hurricane Gilbert. *Monthly Weather Review*, 117, 2824–8.

Wilshire, H. (1980) Human causes of accelerated wind erosion in California's deserts. In D.R. Coates and J.D. Vitek (eds), *Geomorphic Thresholds*. New York: Dowden, Hutchinson and Ross, 415–33.

Wilson, C.A. and Mitchell, J.F.B. (1987) A doubled CO_2 climate sensitivity experiment with a global climate model including a simple ocean. *Journal of Geophysical Research*, 92, 13315–43.

Wilson, E.O. (ed.) (1988) *Biodiversity*. Washington, DC: National Academy Press.

Winstanley, D. (1973) Rainfall patterns and general atmospheric circulation. *Nature*, 245, 190–4.

Winter, W.H. (1987) Using fire and supplements to improve cattle production from monsoon tallgrass pastures. *Tropical Grasslands*, 21, 71–81.

Wolman, M.G. and Schick, A.P. (1967) Effects of construction on fluvial sediment, urban and suburban areas of Maryland. *Water Resources Research*, 3, 451–64.

Woodley, J.D., Chornesky, E.A., Clifford, P.A., Jackson, J.B.C., Kaufman, L.S., Knowlton, N., Lang, J.C., Pearson, M.P., Porter, J.W., Rooney, M.C., Rylaarsdam, K.W., Tunnicliffe, V.J., Wahle, C.M., Wulff, J.L., Curtis, A.S.G., Dallmeyer, M.D., Jupp, B.P., Koehl, M.A.R., Neigel, J. and Sides, E.M. (1981) Hurricane Allen's impact on Jamaican coral reefs. *Science*, 214, 749–55.

Woodroffe, C.D. (1990) The impact of sea-level rise on mangrove shorelines. *Progress in Physical Geography*, 14, 483–520.

Woodroffe, C.D., McLean, R., Polach, H. and Wallensky, E. (1990) Sea level and coral atolls: Late Holocene emergence in the Indian Ocean. *Geology*, 18, 62–6.

Woodruff, S.D., Slutz, R.J., Jenne, R.J. and Steurer, P.M. (1987) A comprehensive ocean-atmosphere data set. *Bulletin of the American Meteorological Society*, 68, 1239–50.

Woodward, F.I. and Rochefort, L. (1991) Sensitivity analysis of vegetation diversity to environmental change. *Global Ecology and Biogeography Letters*, 1, 7–23.

Woodwell, G.M. (ed.) (1984) *The Role of Terrestrial Vegetation in the Global Carbon Cycle: Measurement by Remote Sensing*. New York: Wiley.

Woodwell, G.M. (ed.) (1990) *The Earth in Transition: Patterns and Processes of Biotic Impoverishment*. Cambridge: Cambridge University Press.

Woolhouse, H.W. (1990) Aspects of photosynthetic biochemistry and climate change. In M. Jackson, B.V. Ford-Lloyd and M.L. Parry (eds), *Climatic Change and Plant Genetic Resources*. London and New York: Belhaven, 34–9.

World Resources Institute (WRI) (1985) *Tropical Forests: a Call to Action*. Washington, DC: WRI.

World Resources Institute (WRI) (1990) *World Resources Report 1990–91*. Washington, DC: WRI.

Worster, D. (1979) *Dust Bowl. The Southern Plains in the 1930s*. New York: Oxford University Press.

Wray, D.A. and Cope, F.W. (1948) *Geology of Southport and Formby*. Memoirs of the Geological Survey of the United Kingdom. London: HMSO.

Wright, H.E., Kutzbach, J.E., Webb, T., Ruddiman, W.F., Street-Perrott, F.A. and Bartlein, P.J. (eds) (1993) *Global Climates for 9000 and 6000 Years Ago*. Minneapolis: University of Minnesota Press.

Wright, R.F. and Hauhs, M. (1991) Reversibility of acidification: soils and surface waters. *Proceedings of the Royal Society of Edinburgh, B*, 97, 169–91.

Wright, R.F. and Snekvik, E. (1978) Acid precipitation: chemistry and fish populations in 700 lakes in southernmost Norway. *Verh. Internat. Verein. Limnol.*, 20, 765–77.

Yadava, P.S. (1990) Savannas of north-east India. *Journal of Biogeography*, 17, 385–94.

Yonge, C.M. and Nicholls, A.G. (1931) Studies on the physiology of corals IV. The structure, distribution and physiology of the zooxanthellae. *Scientific Reports of the Great Barrier Reef Expedition, British Museum (Natural History)*, 1, 135–76.

Young, A. (1989) *Agroforestry for Soil Conservation*. Wallingford, UK and Nairobi, Kenya: ICRAF, CAB International.

Zagwijn, W.H. (1983) Sea level changes in The Netherlands during the Eemian. *Geologie en Mijnbouw*, 62, 437–50.

Zhadin, V.I. and Gerd, S.V. (1963) *Fauna and Flora of the Rivers, Lakes and Reservoirs of the USSR*. Jerusalem: Israel Program for Scientific Translations.

Zhang, R. and Xie, S.N. (1990) Prognosis of aggradation in the lower Yellow River by historic analysis of the morphology of its abandoned ancient channel. *International Journal of Sediment Research*, 5, 15–29.

Zimmerman, B.L. and Bierregaard, R.O. (1986) Relevance of the equilibrium theory to island biogeography and species/area relations to conservation with a case from Amazonia. *Journal of Biogeography*, 13, 133–43.

Zwally, H.J. (1989) Growth of Greenland ice sheet: interpretation. *Science*, 246, 1589–91.

Index